Design of Deep Braced Excavation and Earth Retaining Systems Under Complex Built Environment

Wengang Zhang · Hanlong Liu

Design of Deep Braced Excavation and Earth Retaining Systems Under Complex Built Environment

Theories and Case Studies

Science Press
Beijing

Springer

Wengang Zhang
Chongqing University
Chongqing, China

Hanlong Liu
Chongqing University
Chongqing, China

ISBN 978-981-16-5319-3 ISBN 978-981-16-5320-9 (eBook)
https://doi.org/10.1007/978-981-16-5320-9

Jointly published with Science Press, Beijing, China
The print edition is not for sale in China Mainland. Customers from China Mainland please order the print book from: Science Press.

This Springer imprint is published by the registered company Springer Nature Singapore Pte Ltd.
The registered company address is: 152 Beach Road, #21-01/04 Gateway East, Singapore 189721, Singapore

Foreword

Deep excavation is a crucial phase in the construction of superstructures and underground transportation system. In the congested urban area, the construction of braced excavation needs to ensure not only the stability of excavation itself, but also the integrity and serviceability of surrounding buildings. It has become the mainstream of deep braced excavation to design and construct with inspection of the induced deformation.

Incautious deep braced excavation can induce excessive soil displacement and consequent subsidence or uneven settlement of ground facilities behind the wall. Such deformation problems and control measures have been widely reported. The discussions of abundant completed projects can provide engineers guidance to raise more reliable designs and appropriate schemes for deep excavations. For deep excavations in soft soils, the matters of poor design and the consequent large deformation problem still exist due to the incomprehensible understanding of the soils' natural mechanical properties. Empirical observations or simulations using numerical methods such as finite-element or finite-difference procedures have been widely adopted to present excavation-induced displacement prediction. However, the design of deep excavation as well as the earth retaining system should comprise of the use of different methods such as the empirical methods, the observational approaches, the numerical simulation, the probabilistic and inverse analysis, the case report, and failure case analysis.

This book covers the scopes and methods mentioned above, including the basic flow chart for excavation design and construction; the basal heave stability check; estimation of excavation responses including the retaining wall deflection, ground surface settlement, as well as the strut forces; the reliability analysis and back analysis; effect of excavation on adjacent infrastructures and some protection measures; failure case reports; and causal analysis. It tries to cope with design practice, implemented via the recent publications from the authors in reputable journals.

This book comprises eleven chapters. Chapter 1 overviews the relevant topics. Chapter 2 provides the overall stability analysis, in which the strength reduction technique was adopted for the calculation of factor of safety against basal heave. Chapter 3 presents the determination of the lateral earth pressure as well as the strut forces. Chapters 4 and 5 focus on the estimation of the retaining wall deflection and the ground surface settlements, respectively. Chapter 6 gives probabilistic analysis on excavation responses while Chap. 7 performs the analysis about one-strut failure, which directly triggers the Nicoll Highway collapse in Singapore and Xianghu Metro Station collapse in Hangzhou. Chapter 8 focuses on the jet grouting slabs and jet grouting piles as ground improvements for the control of ground deformation. Chapter 9 shows the protection of adjacent infrastructures. Chapter 10 provides the case reports and back analysis while Chapter 11 summarizes some excavation failures worldwide.

I am pleased to witness the publication of this book and hope that the authors would bring us more findings in deep excavation and earth retaining system design when they reprint it.

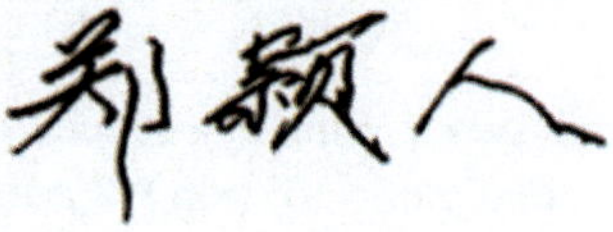

April 2021

Yingren Zheng
Academician of Chinese Academy of Engineering
Chongqing, China

Preface

Excavation is an essential segment for foundation engineering, i.e., in the construction of the basements of high-rise buildings, shopping malls, subways or mass rapid transit systems, etc. Nowadays, deep excavation under a complex built environment is more common than years ago due to the rapid urbanization and the ever-increasing speed of urban infrastructure development. Though books on the design of earth retaining systems introduce the basic solutions, numerical or instrumental analysis, and design procedures, they are generally too simple to cope with current analysis and design in engineering practice. With economic development and near further urbanization, excavation goes deeper, longer, and larger in scale; sometimes it is conducted in difficult soils or complicated built environments. These conditions require advanced analysis and design methods as well as modern construction techniques.

Aimed at both theoretical soundness and practical application, this book covers a greater scope, including the basic design flow chart, basal heave stability check, the apparent earth pressure as well as the strut force determination, the retaining wall deflection and the ground settlement induced by braced excavation under complicated built environment, the groundwater drawdown effect, the probabilistic reliability analysis, the investigation on the effect of deep braced excavation on adjacent infrastructures, the protection measures such as jet grouting slabs or piles, the case reports, and the back analysis method, as well as case failure and causal analysis. From basic to advanced, it tries to attain theoretical rigorousness and consistency. On the other hand, this book also tries to cope with design practice, implemented by the recent publications from the authors. Therefore, it can be readily used for analysis and design in actual excavations.

In the process of writing this book, the authors received enthusiastic help and invaluable assistance from many people, which is deeply appreciated. The first author would like to express his special thanks to Mr. Wang Wei, Miss Li Yongqin, Mr. Hong Li, Dr. Zhang Runhong, Miss Li Hongrui, and Mr. Gu Xin, for their help in editing the book draft and professional page proofreading. The first author is grateful to his Ph.D. supervisor, Prof. Goh Anthony Teck Chee of Nanyang Technological University, for years of supervision and help, as well as the communication, even though he left Singapore for almost five years. Professor Goh leads him toward early career development.

The support from the National Fund for Academic Publication in Science and Technology, National Key R&D Program of China (Project No. 2019YFC1509605); the National Natural Science Foundation of China (Project No. 52078086); Program of Distinguished Young Scholars, Natural Science Foundation of Chongqing, China (Project No. cstc2020jcyj-jq0087); and Chongqing Construction Science and Technology Plan Project (Project No. 2019-0045) is greatly acknowledged.

Finally, the first author would like to thank his family for their love, understanding, and support.

Being deeply loved by someone gives you strength, while loving someone deeply gives you courage.

by Miyazaki Hayao

Chongqing, China Wengang Zhang

Contents

Symbols and Abbreviations[1]

Symbols

c	Cohesion
φ	Friction angle
φ'_m	Effective friction angle after reduction
c'_m	Effective cohesion after reduction
s_u	Undrained shear strength
$s_{u,m}$	Undrained shear strength after reduction
FS	Factor of safety against basel heave
K_a	Coefficient of active earth pressure
K_p	Coefficient of passive earth pressure
σ_a	Total active earth pressure (horizontal) acting on the retaining wall
σ_p	Total passive earth pressure (horizontal) acting on the retaining wall
σ'_a	Effective active earth pressure acting on the retaining wall
σ'_p	Effective passive earth pressure acting on the retaining wall
K_0	Coefficient of at-rest earth pressure
K_0^{nc}	K_0-value for normal consolidation (default $K_0^{nc} = 1 - sin\varphi$)
σ_y	Vertical stress at a point in the ground
σ_x	Horizontal stress at a point in the ground
σ_t	Tension cut-off and tensile strength
φ_{cv}	Critical state friction angle
γ	Bulk density
$\gamma\prime$	Effective unit weight of soil
γ_m	Moisture unit weight of soil
γ_{unsat}	Unit weight above phreatic level
γ_{sat}	Unit weight below phreatic level
γ_w	Unit weight of water

[1]The following list is provided for ease of reference for those symbols and abbreviations that are most frequently used in this textbook. It is, therefore, not exhaustive. Nonetheless, all symbols and abbreviations used in this book are defined at their first mention in the main text.

γ_{ave}	Average unit weight of the soil
H_e	Excavation depth
c_u	Undrained shear strength of soil
E	Elastic Young's modulus
E_{50}	Triaxial loading stiffness
E_{ur}	Triaxial unloading stiffness
E_{oed}	Oedometer loading stiffness
E^{ref}	Reference Young's modulus
p^{ref}	Reference pressure
S_H	Horizontal strut spacing
S_V	Vertical strut spacing
h_{avg}	Average vertical strut spacing
α	Wall stiffness coefficient
S	System stiffness
ν	Poisson's ratio
ν_{ur}	Poisson's ratio for unloading–reloading
G	Shear modulus
G^*	Rate of increase in soil shear modulus with depth
G_0^{ref}	Reference shear modulus at very small strains
N_{pre}	Load on the strut before strut failure
N_{post}	Load on the strut after strut failure
N_{fail}	Load on the failed strut before failure
R_{inter}	Interface reduction factor
σ_a	Allowable stress of the steel
λ	Short-term magnified factor of the allowable stress
ρ_s	Dimensionless system stiffness
δ_v	Ground surface settlement
δ_{vm}	Maximum ground surface settlement
δ_{hm}	Wall deflection
δ_{hm}	Maximum wall deflection
N_b	Stability number of soil, defined as $\gamma H_e / s_u$
N_{cb}	Critical stability number against basal heave
$\gamma_{0.7}$	Threshold shear strain at which $G_s = 0.722G_0$
β	Reliability index

Abbreviations

ANN	Artificial neural network
ANOVA	Analysis of variance
APD	Apparent pressure diagram
BF	Basic function
COV	Coefficient of variation
DPL	Distributed prop load

FEL	Final excavation level
FORM	First-order reliability method
JGP	Jet grouting pile
LR	Logarithmic regression
MARS	Multivariate adaptive regression splines
MCS	Monte Carlo simulation
OSF	One-strut failure
PDF	Probability density function
RFI	Relative fit improvement
SLS	Serviceability limit state
ULS	Ultimate limit state

Chapter 1
Overview

Analysis of deep excavations is usually required before going into design and construction. Deep excavation analysis is a typical soil–structure interaction topic. Soil is a nonlinear, inelastic, and anisotropic material. Its behavior is normally affected by water contents, stress history, etc. Some types of soils have the characteristics of consolidation and creep, for which the performances are greatly influenced by time and space. Theoretically, analysis of deep excavation involves simulations of elastoplastic behavior of soil, interface behavior between soil and retaining walls, and the excavation process. Some simulation theories are not fully developed, and some are too complicated to be used in engineering practice. Thus, to analyze exclusively based on contemporary theories is not practical at the present stage. To sum up, reasonable excavation analysis, at the current stage, should make use of conventional soil mechanics and simple structural mechanics, combined with numerical simulation methods, along with appropriate modifications according to the field observation.

A complete deep excavation design includes a retaining system, a strutting system, a dewatering system, an excavation procedure, a monitoring system, building protection measures, etc. Figure 1.1 illustrates the general flow charts of deep excavation design, which can be paraphrased as follows.

1.1 Geological Investigation and Soil Tests

Deep excavation projects include the construction of building basements, subway stations, etc., of which the depths may range from several meters to tens of meters. The process of excavation may encounter different kinds of the soil underneath the same excavation site from soft clay to hard stratum. Special geographical location often encounters complex geological conditions, e.g., the closer the construction site to a slope, the more complicated the geological condition. In short, the geological

W. Zhang and H. Liu, *Design of Deep Braced Excavation and Earth Retaining Systems Under Complex Built Environment*,
https://doi.org/10.1007/978-981-16-5320-9_1

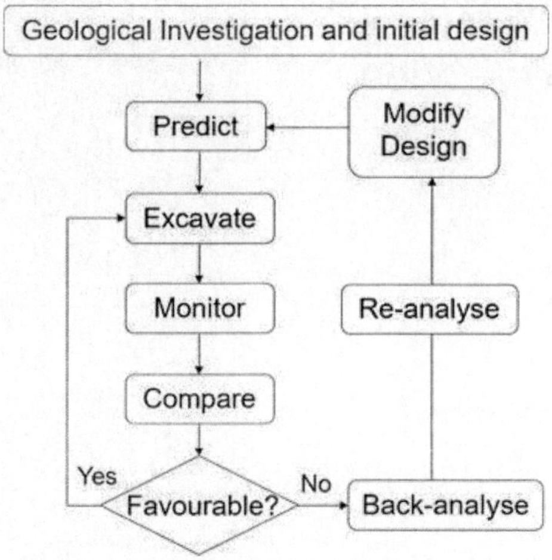

Fig. 1.1 Flow chart for analysis and design of deep excavation

condition determines the type and construction of retaining walls and greatly influences the excavation behavior as well. In addition to the geological condition, the distribution of groundwater also contributes to the excavation behavior. For example, it may fall below the hydrostatic water pressure in an urban area because of the long-term overuse of groundwater. On hillsides, there might exist an artesian aquifer, which is of a rather high pore water pressure. In seaside areas, seawater may permeate into the soils and tides will make the water level fluctuate daily. In summary, it is inevitable to encounter all kinds of complex geological and hydrological conditions under which deep excavation activities are conducted.

In addition to the geological structure and hydrogeological conditions of the site, the properties of the site soil also play an important role in influencing the site soil behavior, therefore, affecting the safety of the excavation itself and responses of the surrounding environment. Site soil properties include physical and mechanical properties. Due to the volatile and lengthy geological processes, i.e., deposition, sedimentation, metamorphism, weathering, and biological effect, and so on, the physical and mechanical properties of soil and rock mass show complex characteristics, such as spatial variability and anisotropy. The spatial variability of soil indicates the soil behavior, and properties are significantly different from one location to another in the three-dimensional space even within the same formation. The proposal of the spatial variability concept greatly helps in evaluating and realizing spatial uncertainty of soil and rock mass in a wide range of engineering sites. As for anisotropy, undisturbed soil shows characteristics of inherent anisotropy due to the directional arrangement of irregular soil particles during the sedimentation process. Besides, the mechanical properties of clay are significantly dependent on the principal stress direction, which is further dominated by the loading/unloading direction. Under different stress states, the stress–strain relationship for each principal stress direction shows significant anisotropy, which is defined as the stress-induced anisotropy. The resulted anisotropy embodies multiplex characteristics of soil structure. For more efficient utilization of underground space, complicated iterations of loading–unloading–reloading is unavoidably encountered during the geoengineering process. A reasonable and proper description of the soil inherent and stress-induced anisotropy is essential in simulating complex geological conditions. While in most of the existing analyses, the soil behavior has been idealized, neglecting the soil's complex characteristics to reduce computational efforts. However, comprehensive consideration of these features can reflect the realistic physical environment that the engineering activities rely on, which is of great significance in assessing accurate construction responses. Therefore, in order to obtain the complex physical and mechanical properties of rock and soil mass, a variety of tests need to be conducted on site and also in the laboratory.

To sum up, the construction of deep excavation is greatly manipulated by the complicated site conditions and soil properties. Therefore, it is necessary to carry out detailed geological investigations and soil tests before the design stage to fully understand the general situation of the site. There are various soil tests for deep excavations. These include tests of basic soil behavior, such as unit weight, specific weight, water content, and the Atterberg limit, and tests of mechanical behavior, such as consolidation coefficient and strength. According to the information from the soil tests, engineers can judge whether the soil is a drained or an undrained material. Since the strength differs significantly between drained and undrained materials, the choice of analysis methods and retaining walls varies accordingly. The more precise the results of soil tests, the more reasonable the analysis results would be and the more economical the retaining and excavation systems would be. However, the merits of soil tests cannot be seen directly and, therefore, are generally neglected.

The field/in situ soil testings include the following: Standard Penetration Tests (SPTs), Permeability Tests, and the Pressuremeter Tests while the geotechnical laboratory tests carried out on soil samples recovered from the site include the following:

(i) Classification tests

- Bulk density tests
- Dry density tests
- Specific gravity tests
- Moisture content determination
- The Atterberg Limits determination (liquid and plastic limits, and plastic index)
- Particle size distribution tests

(ii) Soil strength determination

- Unconsolidated undrained triaxial compression tests
- Consolidated undrained triaxial compression tests with pore water pressure measurement

(iii) Consolidation tests

- 1D Oedometer tests.

The geological laboratory tests carried out on rock samples recovered from the site include the following:

(i) Unconfined compressive strength test
(ii) Point load test
(iii) Cerchar test, Paddle test, and Norwegian test
(iv) Slake durability test
(v) Petrographic test.

1.2 Conditions of Ground and the Adjacent Properties

From the perspective of mechanics, deep excavation necessarily gives rise to the movement of the soils near the excavation site. However, if the movement or settlement is too large, it will damage neighboring buildings or public facilities, underground or aboveground. Some buildings or facilities are especially sensitive to settlement, a little of which may bring about cracks in beams or columns, while others can stand more settlement. The allowable settlement of a building or a facility is highly correlated with its foundation type, construction material, structural type, and age. Therefore, investigation of the condition of adjacent properties and public facilities before designing an excavation project is required to determine the allowable settlement, which in turn determines the type of retaining and strutting systems and the selection of auxiliary methods.

According to the shape, area, and elevation of the excavation site, along with geological conditions, the distribution of groundwater, and conditions of neighboring properties, a provisional retaining method and an excavation method can be determined. Therefore, it is necessary to have a thorough understanding of the conditions of the excavation site, including the site location, the regional geology, the hydrogeological setting, and the topography as well as the anticipated ground conditions.

1.3 Design Criteria

Whether an excavation is successful is significant to the lives and properties of many people. Thus, an appropriate design criterion must be selected before design.

A deep excavation design criterion should include, at least, the method of stability analysis, the methods of simplified and advanced deformation and stress analyses, a dewatering scheme, the design of structural components, property protection, etc. Chapters 2–4 discuss the calculation of basal heave stability, deformation, and structural force control involved in common design specifications. Chapter 5 introduces the related issues of pre-excavation dewatering and induced ground settlement with reference to requirements of different design criteria. Chapter 6 presents the probabilistic analysis on excavation serviceability limit state respecting serviceability criterion. This book explores theories, as well as their applications in deep excavation, from both theoretical and practical perspectives. Getting familiar with the contents should not only help the reader to understand excavation behavior, design, and analyses, but also help develop a suitable excavation design criterion.

1.4 Types of Excavation Support Systems

Excavations for urban building development and civil engineering works are generally carried out up to the property boundary and usually require batters that are not stable without structural support. Some of the structural systems which have been commonly used to provide lateral support to the ground include the following:

(a) sheet pile wall,
(b) soldier pile wall,
(c) diaphragm wall,
(d) caisson wall, and
(e) grouted pile wall.

The retaining wall types mentioned above can be further divided into the following three categories, based on the form of support provided:

(a) cantilevered or unbraced wall,
(b) strutted or braced wall, and
(c) tied-back or anchored wall.

The cantilevered wall depends upon its own strength and embedment to carry the soil and water loadings, while strutted and tied-back walls transfer much of these loads to additional support elements. Cantilevered walls are limited to relatively shallow excavations unless designed to be massive and embedded in stiff soils. However, under such circumstances, the strutted or tied-back wall is likely to be more economic than the cantilevered wall. In a strutted excavation, there is an economic incentive to the contractor to excavate as far as possible below the supports. The reasonable strutting layout needs to balance economy and safety. In Chap. 7 of this book, the load transfer mechanism and stability check of single strut failure in the deep excavation are discussed. For tie-back systems, the excavation floor at any given time is a necessary and convenient platform for the installation of the anchors, which is usually carried out at the correct levels due to difficulties in providing access later. In this respect, tie-back systems have an advantage over strutted systems in that they automatically avoid the overexcavation. Although an open excavation area is created by the tie-back systems, their adoption is not always possible due to encroachment into adjacent property.

1.5 Auxiliary Methods and Protection of Neighboring Properties

A deep excavation may have difficulty meeting design criteria, due to poor geological conditions or deteriorated adjacent buildings. Even if it reaches the criterion, it may be rather expensive. Auxiliary methods can help solve the dilemma. These include soil improvement methods, counterfort methods, cross wall methods, micropiles, underpinning methods, etc. Chapter 8 is, therefore, provided for the design of auxiliary methods.

Due to the unbalanced earth pressures on the two sides of a retaining wall, excavation produces a displacement of retaining walls and settlement of the ground. The buildings and public facilities within the influence zones of the settlement may have a differential settlement. When settlement goes beyond the allowable amount, the buildings or public facilities nearby may be subject to cracks or be damaged. The damages may turn out to be structural or non-structural. However, hardly any proprietors will accept them, either way. To avoid such damages, the prediction of the settlement is necessary to decide whether and how to take protecting measures. There are many ways to protect neighboring properties. Some of them can be deduced from theoretical analyses. Others rely on engineering experience and empirical data. Chapter 9 introduces some useful measures to protect adjacent properties as well as the relevant analyses.

1.6 Instrumentations and Monitoring System

Despite thorough geological investigations, soil testings, rigorous analyses, and design, current excavation theories, based on different hypotheses, can hardly cope with the uncertainties involved in the geological conditions and from other sources. The actual excavation responses are generally different from the predicted responses estimated in the design stage. Therefore, excavation has to be carried out along with monitoring instruments, which tell, immediately, changes of stress and displacement generated during the whole excavation process. The engineers in charge can thereby check the safety of excavation at any time and hereby conduct back analysis to correct or improve the understanding of site conditions. Therefore, the excavation and support scheme can be adjusted in time accordingly. For large-scale excavations, the geological uncertainty increases, and a monitoring system is urgently required. Concerning the items and design of monitoring systems for deep excavations, please refer to Chap. 10.

1.7 Organization of this Book

This book is divided into 11 chapters, and the contents are summarized as follows:

This chapter introduces the flow charts of analysis and design of excavations.

Chapter 2 covers the basal heave stability as well as the calculations.

Chapter 3 introduces the lateral earth pressure acting on the retaining wall and the determination of the strut forces.

Chapter 4 focuses on the wall deflections induced by deep braced excavation and the main influential factors.

Chapter 5 analyzed the ground settlements induced by deep braced excavation and groundwater drawdown.

Chapter 6 performed the probabilistic analysis on excavation responses in view of the uncertainties of soil parameters.

Chapter 7 examined the strut responses due to one-strut failure or structural behaviors resulted from the one-column failure.

Chapter 8 investigated the effect of the jet grouting slabs and jet grouting piles in controlling ground deformations.

Chapter 9 examined the influence of deep braced excavation on adjacent upper slope, piles of buildings, and tunnels, as well as the protection measures.

Chapter 10 summarized the case reports and back analysis method.

Chapter 11 provided the summary of excavation failure cases and causal analysis.

Chapter 2
Basal Heave Stability

2.1 Introduction

Failures or collapses of excavations are disastrous at excavation sites. At worst, they endanger the safety of workers and the adjacent properties. Their influential zones are usually so large that significant ground settlement may be introduced and adjacent properties within the influential zones of settlement may be damaged at different extents. Considering this, to avoid failures or collapses is of vital importance and stability analyses are, therefore, firstly required.

Failure of an excavation may arise from the stress on the support system exceeding the strength of its materials, for example, when the strut loading exceeds the buckling resistance of struts or the bending moment of the retaining wall exceeds the threshold value, etc. It can also arise from the shear stress in soil exceeding its shear strength.

When the shear stress at a point in the soil exceeds or equalizes the shear strength of soil at the point, the point is in the failure or the impending failure state. When many failure points connect and form a slip line, the failure surface is, thus, produced. Once the failure surface forms, the excavation failure or collapse will occur. This is the overall shear failure.

The basal heave is the main overall failure mode, which arises from the weight of soil outside the excavation zone exceeding the bearing capacity of soil below the excavation bottom, causing the soil to move and the excavation bottom to heave so much that the whole excavation collapses. At present, the analysis methods for the stability of basal heave generally include limit equilibrium method, limit analysis method, reliability analysis, and deformation control analysis programmed into various numerical simulation algorithms.

W. Zhang and H. Liu, *Design of Deep Braced Excavation and Earth Retaining Systems Under Complex Built Environment*,
https://doi.org/10.1007/978-981-16-5320-9_2

2.2 Types of Factors of Safety

There are basically three methods to determine the factor of safety for stability analysis: the strength factor method, the load factor method, and the dimension factor method, which are given as follows:

(1) *Strength factor method*: The method considers the soil strength involving uncertainty and has the strength reduced by a factor of safety. If the factor of safety for the strength factor method is represented as FS_s, the parameters for the effective stress analysis are as follows:

$$\tan\varphi'_m = \frac{\tan\varphi'}{FS_s} \quad (2.1)$$

$$c'_m = \frac{c'}{FS_s} \quad (2.2)$$

The parameter for the undrained analysis is

$$s_{u,m} = \frac{s_u}{FS_s} \quad (2.3)$$

After conducting a force equilibrium or a moment equilibrium analysis with the reduced parameters c'_m, φ'_m, or $s_{u,m}$ derived as above, we can design the penetration depth.

The method locates the factor of safety at the source where the greatest uncertainty arises and is, therefore, quite a reasonable method. Since the after-reduction parameters will lead to a smaller K_p and a larger K_a, the distribution of earth pressures on the retaining wall will be skewed. As a result, the method is applicable only to stability analysis and cannot be applied to deformation analysis or stress analysis.

(2) *Load factor method*: The factor of safety for the load method, FS_l, can be defined as follows:

$$FS_l = \frac{R}{D} \quad (2.4)$$

where R represents the resistance and D is the driving force. R and D can also be the resistant moment and the driving moment, or the bearing capacity and the external force, respectively. FS_l considers uncertainty arising from the soil strength, the analysis method, and external forces synthetically. The factors of safety adopted in this chapter are mainly derived from the load factor method.

(3) *Dimension factor method*: Suppose that the retaining walls are in the limiting state and the soil strengths are fully mobilized. With the force equilibrium (the horizontal force equilibrium, the moment equilibrium, or other type of force

equilibrium), the penetration depth of retaining walls in the limiting state can be determined. The penetration depth for the design is

$$H_{p,d} = FS_d H_{p,cal} \tag{2.5}$$

where FS_d= factor of safety for the dimension factor method and $H_{p,cal}$= penetration depth computed from the limit equilibrium.

The factor of safety is usually defined as a ratio of the resistant force to driving force or as a factor to reduce the strength. Equation 2.4 is too much empirically oriented and cannot properly express the meaning of the factor of safety, leading to unreasonable results and is not recommended. Unavoidably, cross checking via other methods is essential.

2.3 Review of the Limit Equilibrium Method

The commonly used methods to assess the basal heave stability are based on the LE methods proposed by Terzaghi [1], Bjerrum and Eide [2], and Eide et al. [3]. Terzaghi's [1] method is valid for wide excavations in which the width of the excavation B is equal to or larger than the depth of the excavation H (B $\geq$ H), while the Bjerrum and Eide [2] and Eide et al. [3] methods apply for deep excavations (H > B). Figure 2.1 shows the basal heave failure mechanism assumed by Terzaghi (1943). The Terzaghi [1] basal heave factor of safety FS_T can be expressed as.

$$FS_T = \frac{5.7c_{ub}B_1}{(\gamma II + q_s)B_1 - c_{uh}H} \tag{2.6}$$

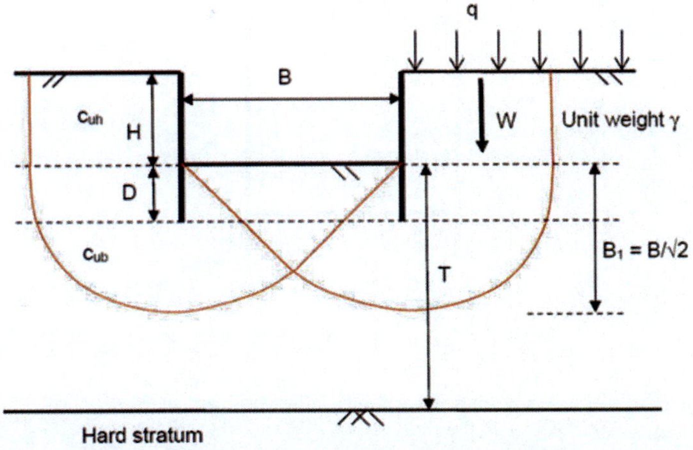

Fig. 2.1 Terzaghi's basal heave stability failure mechanism

where γ is the unit weight of the soil, c_{uh} is the undrained shear strength of the clay from the ground surface to the final depth of excavation H, c_{ub} is the undrained shear strength of the clay below the final excavation level, q is the surcharge, and B_1 is the smaller of $(B/\sqrt{2})$ or the depth to the hard stratum T. The value of 5.7 in the numerator of Eq. (2.6) corresponds to the bearing capacity factor N_c for a rough footing. As illustrated in Fig. 2.1, the failure surface extends from the ground surface to a depth of $B/\sqrt{2}$ ($\approx$0.7B) below the formation level or to the top of the underlying hard stratum whichever is shallower. Wall embedment below the excavation level D is ignored. In the Eide et al. [3] method, the failure is assumed to occur below the base of the wall and is resisted by the weight of the interior block of soil and the adhesion along the wall block–soil interface. The effects of wall stiffness are also not considered in the above-mentioned methods. Extensive numerical studies [4] have shown that the basal heave factor of safety can in some cases increase with an increasing penetration depth of the wall D, increasing wall stiffness EI, and decreasing proximity of the bedrock.

Bjerrum and Eide [2] assumes that the unloading behavior caused by excavation is analogous to the building foundation subject to an upward loading and that the shape of the failure surface is similar to the failure mode of the deep foundation. It is also assumed that the failure surface where the radius of the circular arc is equal to $B/\sqrt{2}$ is the critical failure surface and the corresponding factor of safety is the one against basal heave (Fig. 2.2).

The factor of safety can be expressed as follows:

$$F_B = \frac{c_u N_c}{\gamma \mathrm{H} + q_s} \tag{2.7}$$

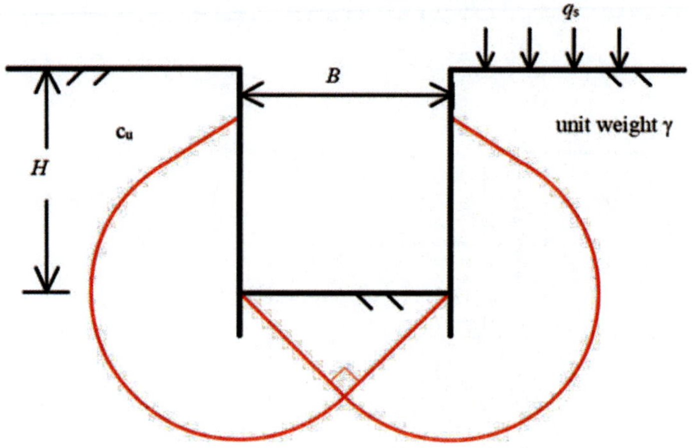

Fig. 2.2 Bjerrum' and Eide's basal heave stability failure mechanism

where q_s is the surcharge on the ground surface and N_c is Skempton's bearing capacity factor as shown in Fig. 2.3.

Another method that is commonly adopted in Japan and Chinese Taiwan is the slip circle method which involves taking moments about the lowest strut level [5]. In the slip circle method, the basal heave factor of safety FS_{slip} is assessed by taking moments about the lowest strut level (Fig. 2.4).

$$FS_{silp} = \frac{M_r}{M_d} = \frac{R\int_0^{(\pi/2)+\delta}(c_u d\theta)}{W(R/2) + q\left(R^2/2\right)} \tag{2.8}$$

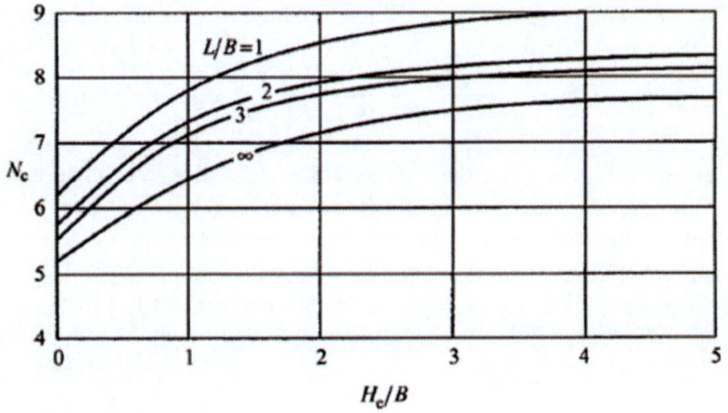

Fig. 2.3 Skempton's bearing capacity factor (Skempton, 1951)

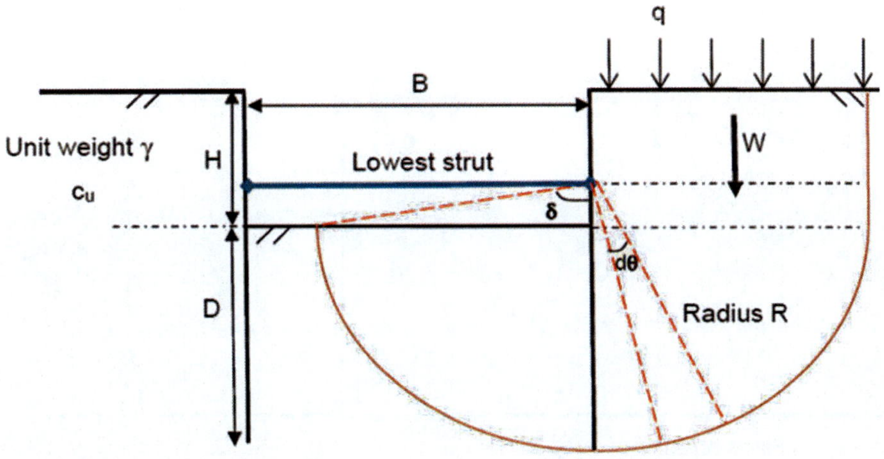

Fig. 2.4 Slip circle method for determining basal heave factor of safety

where M_r = resisting moment; M_d = driving moment; R = radius of the failure slip circle; W = weight of the soil mass behind the wall and above the excavation level; q_s = surcharge pressure; H = excavation depth; D = wall penetration depth; B = excavation width; c_u = clay undrained shear strength; and δ = angle shown in Fig. 2.4.

An alternative limit equilibrium method called the modified Terzaghi method [6] is also used to assess the basal heave factor of safety for excavations with stiff diaphragm walls. In this method, following the assumption adopted by Eide et al. [3], the failure is assumed to occur beneath the base of the wall and is partially resisted by the interior wall adhesion $f_s = \alpha c_u$, as illustrated in Fig. 2.5, in which α is the clay-wall adhesion factor. The α value typically varies from 0.5 for stiff clays to 1.0 for soft clays [7]. The factor of safety FS_{DW} can be expressed as follows:

$$FS_{DW} = \frac{N_c c_{ub} B_1 + c_{uh} H + c_{ud} D + f_s D}{(\gamma H + q_s) B_1} \tag{2.9}$$

As illustrated in Fig. 2.5, γ is the unit weight of the soil, c_{uh} is the undrained shear strength of the clay from the ground surface to the final depth of excavation H, c_{ud} is the undrained shear strength of the clay from depth H to the tip of the wall, and c_{ub} is the undrained shear strength of the clay below the final excavation level. As in Eq. (2.6), the bearing capacity factor N_c in the numerator of Eq. (2.9) is assumed as 5.7.

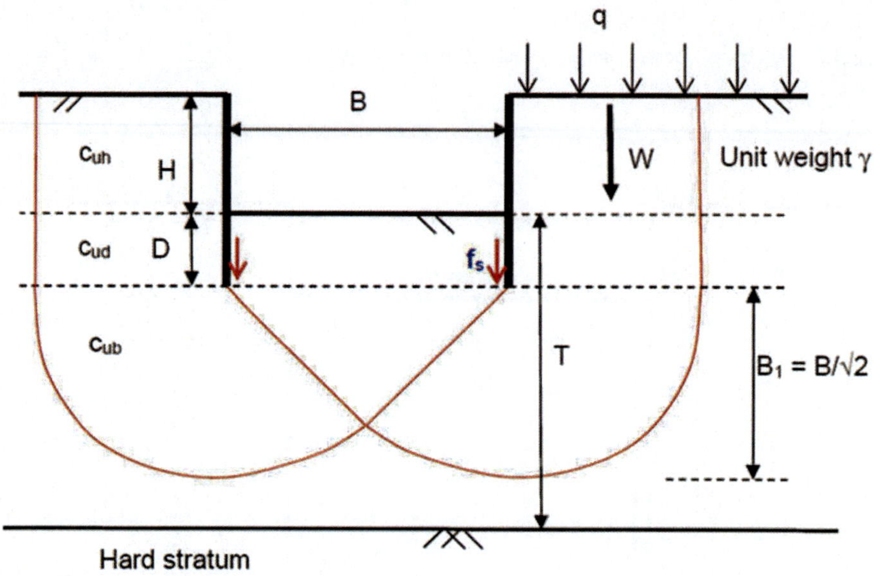

Fig. 2.5 Modified Terzaghi's method for basal heave stability for diaphragm walls

2.4 Review of Upper Bound Method

2.4.1 Basic Theory

The upper bound method is a kind of limit analysis proposed by Drucker et al. [8]. It has been firstly applied to structural engineering. Based on the characteristics of geomaterials, Chen [9] applied the upper bound method to bearing capacity analysis of foundations and stability analysis of slopes [9]. From descriptions illustrated in Drucker et al. [8], the upper bound method can be expressed as follows:

$$\int_S T_i v_i dS + \int_V X_i v_i dV \leq \int_V \sigma_{ij}\dot{\varepsilon}_{ij} dV i, j = 1, 2, 3 \tag{2.10}$$

where $\dot{\varepsilon}_{ij}$ is the plastic strain rate in the admissible velocity field; v_i is the velocity field (compatible velocity field) which is geometrically compatible with $\dot{\varepsilon}_{ij}$; X_i and T_i are the surface force on boundary S and the volume force in space V, respectively; and σ_{ij} is the stress which is satisfied with the associated flow rule.

From Eq. (2.10), energy dissipation power is not less than external power for all kinematically admissible velocity fields. Similar to the shear strength reduction method, the safety factor can be defined as the ratio of energy dissipation power to the external power.

In fact, the upper bound method is a special kind of limit equilibrium method. The difference is that the upper bound method is based on a power equation, while the limit equilibrium method is based on the mechanical equation. Similarly, there are also the Terzaghi and Prandtl failure models.

2.4.2 Upper Bound Method for the Terzaghi and Prandtl Failure Model

The upper bound method for braced excavation was proposed by Chang et al. [10]. However, in their work, the depth of support structures was not taken into account. Based on that, Ying et al. proposed a modified upper bound method for braced excavation [11]. The corresponding kinematically admissible failure mechanism and speed velocity field are shown in Fig. 2.6.

For example, the basal heave factor of safety for the Prandtl failure model can be expressed as follows:

$$F_s = \frac{dE}{dW} \tag{2.11}$$

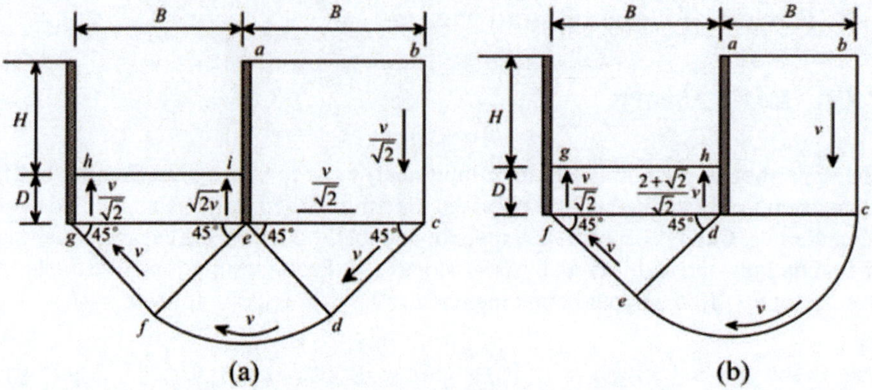

Fig. 2.6 **a** The Prandtl and Terzaghi **b** failure model for upper bound method (Adapted from [12])

where dE is energy dissipation power and dW is external power. dW can be obtained by Eq. (2.12). dE includes eight proportions: the dissipation power at surface *bc* (E_1), *cd* (E_2), *ce* (E_3), *fd* (E_5), *fg* (E_6), *gh* (E_7), *ei* (E_8), and shear region *efh* (E_4).

$$dW = \gamma HB\frac{v}{\sqrt{2}} + qB\frac{v}{\sqrt{2}} \tag{2.12}$$

$$dE_1 = S_u(H + D)\frac{v}{\sqrt{2}} \tag{2.13}$$

$$dE_2 = dE_3 = dE_6 = S_uB\frac{v}{\sqrt{2}} \tag{2.14}$$

$$dE_4 = S_u\frac{B}{\sqrt{2}}\frac{\pi}{2}v \tag{2.15}$$

$$dE_5 = \int_{\frac{\pi}{4}}^{\frac{3\pi}{4}} \int_{H+D}^{H+D+R\sin\theta} S_u v\frac{B}{\sin\theta}dzd\theta = S_u\frac{B}{\sqrt{2}}\frac{\pi}{2}v \tag{2.16}$$

$$dE_7 = S_uD\frac{v}{\sqrt{2}} \tag{2.17}$$

$$dE_8 = S_uD\sqrt{2}v \tag{2.18}$$

$$dE = \sum_{i=1}^{8} dE_i = \frac{3B + \pi B + H + 4D}{(\gamma H + q)B}S_u\frac{v}{\sqrt{2}} \tag{2.19}$$

$$F_s = \frac{H + 4D + 3B + \pi B}{(\gamma H + q)B} \tag{2.20}$$

Via Eqs. (2.11)–(2.20), the basal heave factor of safety for the Prandtl failure model can be obtained. The basal heave factor of safety for the Terzaghi failure model can be obtained by similar power equations.

2.4.3 Multi-block Upper Bound Method

On one hand, similar to the upper bound method in Sect. 2.4.1, the multi-block method is also based on kinematically admissible failure mechanism and power equations. On the other hand, for the multi-block method, the failure area is divided into more blocks (Fig. 2.7). For example, in Sect. 2.4.1, the Terzaghi failure model needs 4 blocks, and the Prandtl failure model needs 5 blocks. Generally, the multi-block method needs twenty to thirty blocks.

In Fig. 2.7, the support system was assumed to be rigid so that the failure surfaces would not cross the diaphragm wall. Similarly, it is assumed that the failure surfaces would be above the hard soil layer. From Fig. 2.7, the failure surfaces consisted of $n+1$ points and n blocks (included two quadrilateral blocks and n-2 triangle blocks). In order to guarantee the accuracy of calculation, sufficient numbers of triangle blocks were needed. Other notations were showed in Table 2.1.

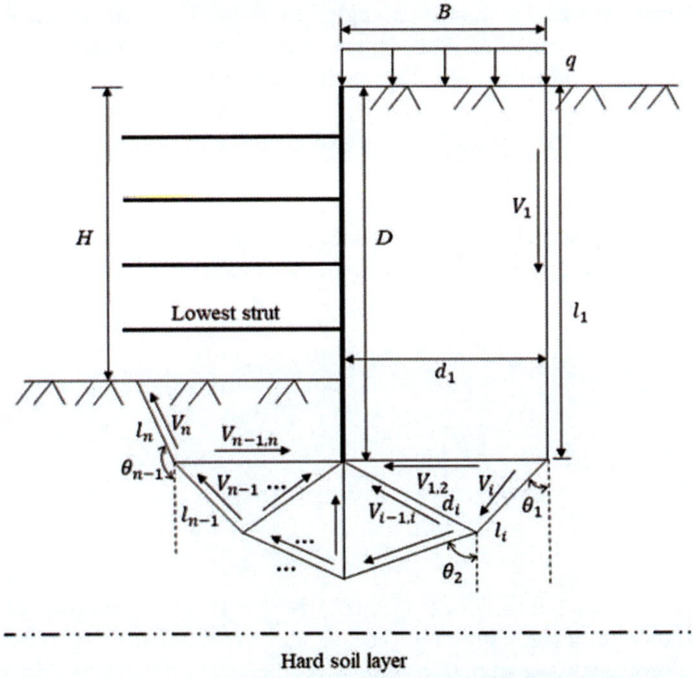

Fig. 2.7 Kinematically admissible failure mechanism of multi-block method

Table 2.1 Explanations of symbols

Symbol	Description
B/m	Width of excavation
H/m	Depth of excavation
D/m	Length of the support wall
T/m	Depth of hard soil or rock layer
d_i/m	Length of the interface between two adjacent blocks
l_i/m	Length of failure surface of each block
v_i	velocity vector in the surface of l_i
$v_{i,i+1}$	velocity vector in the surface of d_i
θ_i	Angle between v_{i+1} and negative vertical direction (assuming down is negative)
q	Upper load
w_i	Weight of ith block

Based on compatible conditions of the kinematical field, $\boldsymbol{v}_{i+1}$ and $\boldsymbol{v}_{i,i+1}$ can be calculated by Eq. (2.21):

$$\boldsymbol{v}_{i+1} = \boldsymbol{v}_i + \boldsymbol{v}_{i,i+1} i = 1, 2, \cdots, n-1 \tag{2.21}$$

In adjacent blocks, a vector triangle consists of three velocity vectors $(\boldsymbol{v}_{i+1}, \boldsymbol{v}_i, \text{ and } \boldsymbol{v}_{i,i+1})$, and direction of three velocity vectors is known and the value of one vector ($\boldsymbol{v}_i$) is known. Via Eq. (2.21), the values of $\boldsymbol{v}_{i+1}$ and $\boldsymbol{v}_{i,i+1}$ can be determined. In this chapter, all directions of vectors are obtained from failure surfaces, and it is assumed that $\boldsymbol{v}_1$ equals 1. The detailed derivation is provided in reference [13].

The dissipation power includes power at failure surfaces and interfaces between adjacent blocks; the external power includes the power of gravity and load. Safety factor can be obtained by Eqs. (2.22)–(2.24).

$$dW = \sum_{i=1}^{n} w_i v_i cos\theta_i + q v_1 d_1 \tag{2.22}$$

$$dE = \sum_{i=1}^{n} S_u l_i v_i + \sum_{i=1}^{n-1} S_u d_i v_{i,i+1} + \omega S_u v_1 l_1 \tag{2.23}$$

$$F_s = dE/dW \tag{2.24}$$

Moreover, for the multi-block method, the basal heave stability analysis is regarded as an optimization problem. The objective function is the minimum safety factor. The progressive search algorithm was adopted to determine critical failure surfaces [14]. Via multiple iterations, critical failure surfaces can be determined.

2.5 Deterministic and Reliability Analysis of Basal Heave Stability for Excavation in Spatial Variable Soils

2.5.1 Shear Strength Reduction Technique

Apart from limit equilibrium methods introduced in Sect. 2.3, the stability of the excavation can also be determined using the shear strength reduction (SR) technique. This technique has been used by various authors [15–19]. The method is now available in many commercial FE and finite-difference (FD) programs.

The factor of safety is defined here as the number by which the original shear strength parameters must be divided in order to bring the excavation to the point of failure, as given in Eqs. (2.25a) and (2.25b).

$$\tan \varphi_m' = \frac{\tan \varphi\prime}{F_\mathrm{b}} \tag{2.25a}$$

$$c_m' = \frac{c'}{F_\mathrm{b}} \tag{2.25b}$$

This definition of the factor of safety is exactly the same as that used in limit equilibrium methods, namely the ratio of restoring to driving moments.

SR technique allows the use of different factors of safety to the c' and ϕ' terms. In common practice, however, the same factor is applied to both terms. To find the true F_b, it is necessary to initiate a systematic search for the value of F_b that will just cause the excavation to fail. This is achieved by the program solving the problem repeatedly using a predefined sequence of F_b values.

2.5.2 Numerical Schemes

In this section, extensive plane strain FE analyses were carried out to assess the basal heave factor of safety for wide excavations in clay. The parametric study was performed using the FE software Plaxis [20]. A typical very refined mesh comprising of 4784 elements and 39,001 nodes is shown in Fig. 2.8. For symmetry, only half the cross-section was considered. The soil was modeled by 15-noded triangular elements. The wall structural elements were assumed to be linear elastic and were modeled by 5-noded beam elements. The interface elements that model the slippage between the soil and the wall consist of 10-noded joint elements. The shear response of the interface element is controlled by an elastic–perfectly plastic Coulomb shear strength criterion. The struts were modeled by linear elastic 3-noded bar elements. The nodes along the side boundaries of the mesh were constrained from displacing horizontally while the nodes along the bottom boundary were constrained from moving horizontally and

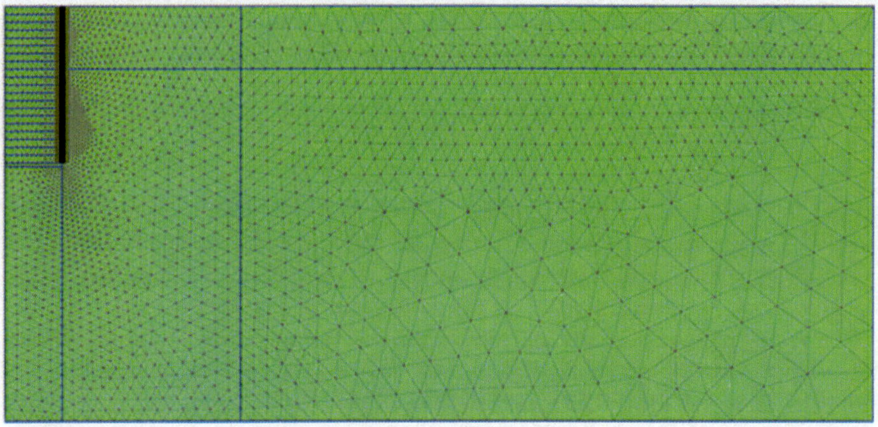

Fig. 2.8 Typical finite-element mesh (struts not shown)

vertically. The right vertical boundary extends far from the excavation (approximately 5B) to minimize the effects of the boundary restraints.

More advanced soil constitutive models such as the hardening soil model (HSM) and the hardening soil model with small strain (HSS) are increasingly used to assess the wall deflections and ground settlements [21]. However, since the focus of this part is on the basal heave failure, the Mohr–Coulomb soil model was used to model the undrained behavior of the clay. This is because even when advanced soil constitutive models are used in the Plaxis FE analysis to simulate the stage-by-stage excavation phases, during the shear strength reduction phase to determine the factor of safety, the Plaxis software adopts the Mohr–Coulomb model, without taking into consideration any stress-dependent stiffness behavior and the hardening effects. The input parameters for the elastic–plastic Mohr–Coulomb constitutive relationship include the undrained elastic modulus E_u, the undrained Poisson's ratio υ_u, the undrained friction angle ϕ_u, the dilation angle ψ, and the undrained cohesion c_u. For this study, cases with a homogeneous clay layer with constant undrained shear strength c_u and also cases with c_u increasing linearly with depth were considered. The soil is assumed to be subjected to undrained shearing during excavation. Only clays with isotropic c_u are considered. A total of 104 different cases were analyzed. The parameters varying include the undrained shear strength c_u, the width of the excavation B, the depth of the excavation H, the wall penetration depth D, the depth to the hard stratum T, and the wall stiffness EI. The range of soil, wall, and geometrical properties of the excavation that were analyzed are shown in Table 2.2.

The construction sequence is comprised of the following steps: (1) the wall is installed (wished into place) without any disturbance in the surrounding soil; and (2) the soil is excavated uniformly in 2 m intervals, and struts are installed until the final depth H is reached.

Table 2.2 Summary of soil, structural, and geometrical properties

Parameter	Range of values
Geometrical, wall, and strut properties	
Excavation width B (m)	20–100
Excavation depth H (m)	16–24
Depth of wall penetration D (m)	4–24
Depth to hard stratum T (m)	10–120
Wall stiffness EI (kNm^2/m)	8.364×10^3–4.032×10^6
Strut axial stiffness EA (kN/m)	2×10^6
Soil properties	
Soil unit weight γ (kN/m^3)	16
Soil undrained shear strength c_u (kPa)[a]	35–100
c_{u0} (kPa)[b]	5–10
m (kPa/m)[b]	1-0.1.5
Soil elastic modulus ratio Eu/cu	300
Poisson's ratio υ_u	0.495
Friction angle ϕ_u (degrees)	0
Dilation angle ψ (degrees)	0
Clay-wall adhesion factor α	0.5–1.0

[a]Constant c_u
[b]c_u linearly increasing with depth from the ground surface $c_u = c_{u0} + mz$ (kPa).

2.5.3 *Simulation Results*

The general trends of the FE analyses are highlighted in this section. Analyses with different wall thicknesses from 0.6 to 1.2 m showed minimal differences in FS_{FE}. For brevity, these results have been omitted. In all the plots in this section, only the results with 1.2 m thick diaphragm walls ($EI = 4.032 \times 10^6$ kNm^2/m) are shown. Figure 2.9a shows a typical plot of the plastic points at the end of excavation to the final excavation level for $H = 16$ m with $B = 20$ m and $D = 16$ m. Figure 2.9b shows the corresponding deformed mesh (at an exaggerated scale) after the shear strength reduction phase. The plots show a "block" type vertical movement of the soil beneath the excavation and the contribution of the shear resistance between the wall and the soil to resist basal heave instability for diaphragm walls.

The influence of the width of the excavation B and the wall penetration depth D on FS_{FE} is shown in Fig. 2.10 and Table 2.3. For comparison, the results for a flexible Frodingham sheet pile wall (EI = 8364 kNm^2/ m) are also shown. For both the sheet pile wall and diaphragm wall, the basal heave factor of safety FS_{FE} decreases with the increase of the excavation width. For the sheet pile wall, the increase of the wall penetration depth D from 4 to 8 m results in a slight increase in FS_{FE}. However,

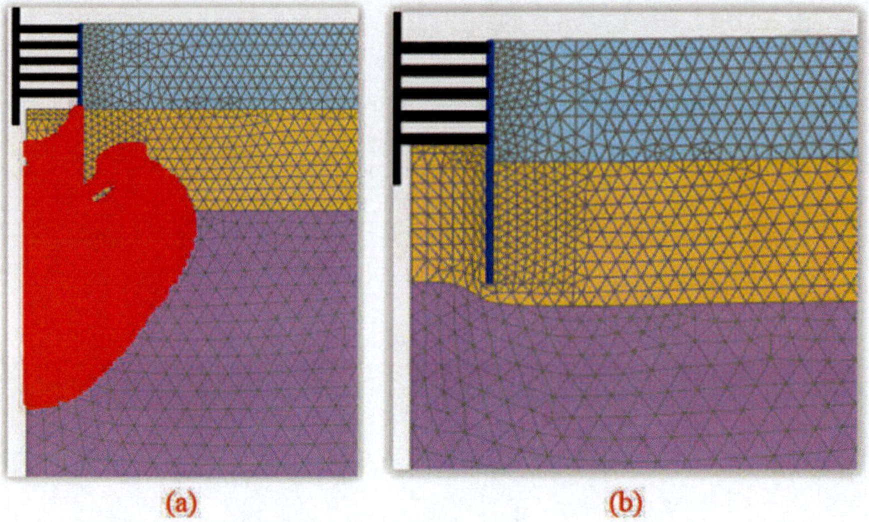

Fig. 2.9 a Plastic points after excavation to H = 16 m. b Deformed mesh after shear strength reduction analyses for H = 16 m

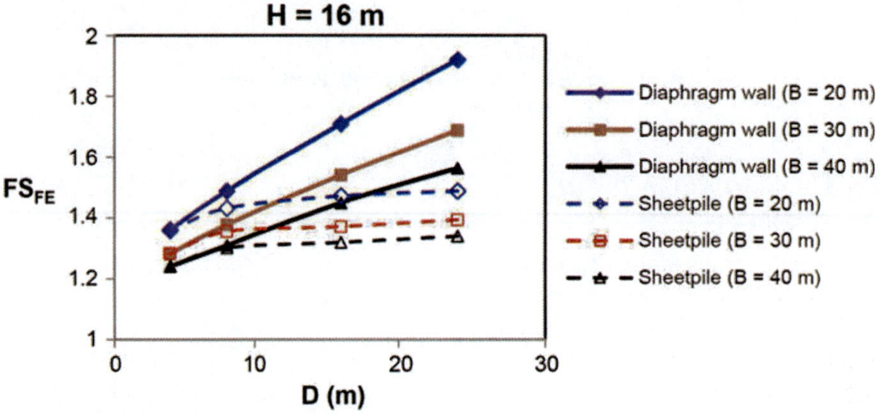

Fig. 2.10 Effects of excavation width B and wall penetration D on FS_{FE} (H = 16 m, T = 100 m, and c_u = 50 kPa)

subsequent increases in *D* result in minimal increases in FS_{FE}. As can be seen, the influence of *D* on FS_{FE} is more significant for the diaphragm wall. For example, for the case with $B = 30$ m, the FS_{FE} increased by 31% when *D* was increased from 4 to 24 m. The same trend is observed for the case of 100 m width and 24 m depth excavation as shown in Fig. 2.11.

Figure 2.12 shows that the influence of the wall interface adhesion α on FS_{FE} is approximately linear. For example, for the case with $D = 16$ m, there is a reduction

Table 2.3 Effects of excavation width B and wall penetration D on FS (T = 100 m)

γ (kN/m³)	$c_{ub} = c_{uh} = c_{ud}$ (kPa)	B (m)	H (m)	D (m)	FS_T	FS_{silp}	FS_{FE} (sheet pile wall)	FS_{FE} (diaphgram wall)
16	50	20	16	4	1.433	1.094	1.359	1.362
16	50	20	16	8	1.433	1.149	1.432	1.488
16	50	20	16	16	1.433	1.184	1.472	1.710
16	50	20	16	24	1.308	1.197	1.488	1.920
16	50	30	16	4	1.308	1.094	1.282	1.284
16	50	30	16	8	1.308	1.149	1.355	1.377
16	50	30	16	16	1.308	1.184	1.371	1.54
16	50	30	16	24	1.308	1.197	1.393	1.688
16	50	40	16	4	1.253	1.094	1.243	1.241
16	50	40	16	8	1.253	1.149	1.302	1.310
16	50	40	16	16	1.253	1.184	1.320	1.449
16	50	40	16	24	1.253	1.197	1.340	1.563

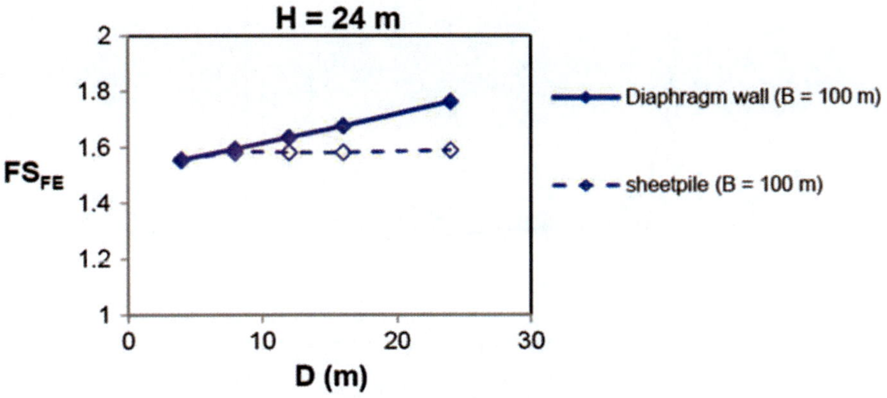

Fig. 2.11 Effects of wall penetration D on FS_{FE} (B = 100 m, H = 24 m, T = 120 m, and c_u = 100 kPa)

of FS_{FE} of approximately 10% when the wall interface adhesion α is reduced from 1.0 to 0.5. Figure 2.13 shows the influence of the depth from the wall tip to the hard stratum (T–D) on the factor of safety. As expected, the presence of the hard stratum close to the excavation significantly increases the FS_{FE}. The factor of safety increases significantly for (T– D)/B < 0.7 as the hard stratum prevents the full development of the failure surface of the soil, just as in the case of the ultimate bearing capacity of a footing with a rigid base at shallow depth. For example, for the case with D = 8 m, the FS_{FE} = 1.31 for (T–D)/B = 1.8 compared with FS_{FE} = 1.57 for (T–D)/B = 0.3. The results also highlight that for the same (T–D)/B ratio, increasing D will have an

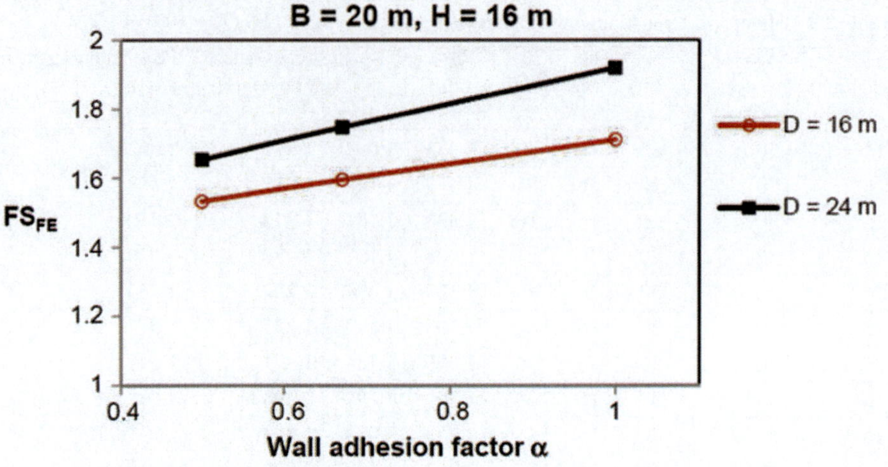

Fig. 2.12 Effects of wall interface adhesion factor on FS_{FE} (B = 20 m, H = 16 m, and cu = 50 kPa)

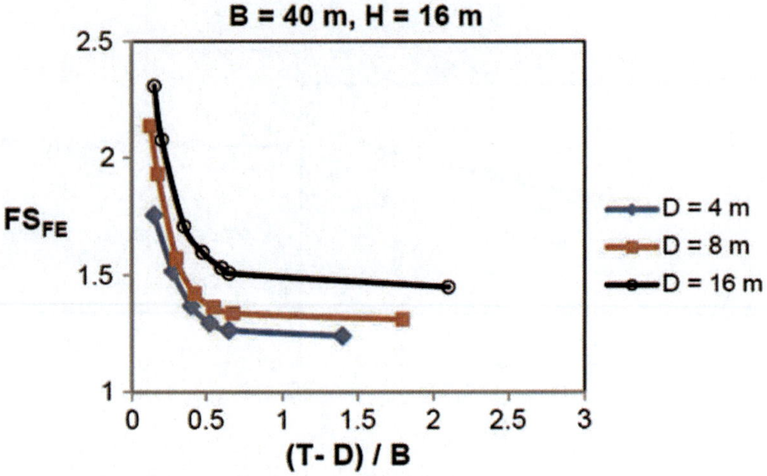

Fig. 2.13 Effects of (T–D)/B on FS_{FE} (B = 40 m, H = 16 m, and cu = 50 kPa)

influence on the factor of safety. It can be concluded that the thickness of the clay beneath the wall tip (T–D) is a key component of the resistance of the excavation system to basal heave failure.

Comparisons were carried out using the conventional LE methods described in Sect. 2.3. As indicated in Table 2.2, the predictions using the slip circle method are independent of the excavation width and increase only marginally with increasing penetration depth D. Table 2.2 shows that the FS_{Slip} predictions are lower than both FS_T and FS_{FE}. Comparison in Tables 2.2 and 2.3 indicates that the Terzaghi method gives reasonable estimates of the factor of safety compared with the finite-element predictions when the wall embedment depth is small (D = 4 m). However, generally, the Terzaghi method would give more conservative estimates of the factor of safety compared with the FE predictions for greater wall embedment depths (D > 8 m). Comparison of all the cases as shown in Figs. 2.14 and 2.15 indicates that the predictions based on the Terzaghi method (FS_T) and slip circle methods (FS_{Slip}) are generally conservative compared with FS_{FE}. The data is listed in Table 2.4.

In order to account for the influence of the hard stratum at a shallow depth, based on regression analysis of the FE results, the bearing capacity factor N_c in Eq. (2.9) has been modified as shown in Fig. 2.16 and Eq. (2.26). The proposed N_c values are in good agreement with the upper bound N_c values proposed by Chen [9] for calculating the bearing capacity of a soft clay overlying a strong clay layer as shown

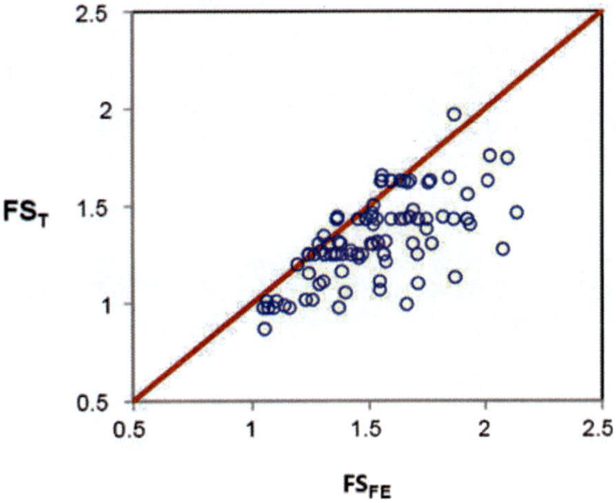

Fig. 2.14 Comparison of FS_{FE} versus FS_T

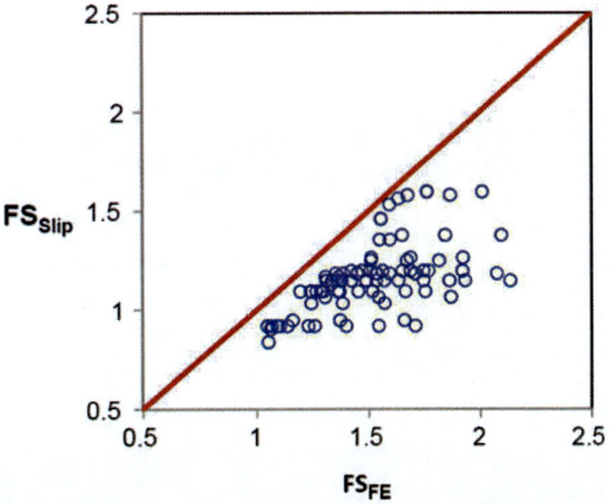

Fig. 2.15 Comparison of FS_{FE} versus FS_{Slip}

Table 2.4 Effects of (T–D)/B on FS_{FE} (B = 40 m, H = 16 m, and c_u = 50 kPa)

γ (kN/m^3)	$c_{ub} = c_{uh} = c_{ud}$ (kPa)	B (m)	H (m)	D (m)	T (m)	(T–D)/B	FS_T	FS_{FE}
16	50	40	16	4	60	1.4	1.253	1.241
16	50	40	16	4	30	0.65	1.253	1.265
16	50	40	16	4	25	0.525	1.272	1.296
16	50	40	16	4	20	0.4	1.319	1.369
16	50	40	16	4	15	0.275	1.406	1.517
16	50	40	16	4	10	0.15	1.619	1.755
16	50	40	16	8	80	1.8	1.253	1.310
16	50	40	16	8	35	0.675	1.253	1.336
16	50	40	16	8	30	0.55	1.253	1.363
16	50	40	16	8	25	0.425	1.272	1.421
16	50	40	16	8	20	0.3	1.319	1.571
16	50	40	16	8	15	0.175	1.406	1.932
16	50	40	16	8	13	0.125	1.619	2.137
16	50	40	16	16	100	2.1	1.253	1.449
16	50	40	16	16	42	0.65	1.253	1.509
16	50	40	16	16	40	0.6	1.253	1.532
16	50	40	16	16	35	0.475	1.272	1.598
16	50	40	16	16	30	0.35	1.319	1.708
16	50	40	16	16	24	0.2	1.406	2.077
16	50	40	16	16	22	0.15	1.466	2.309

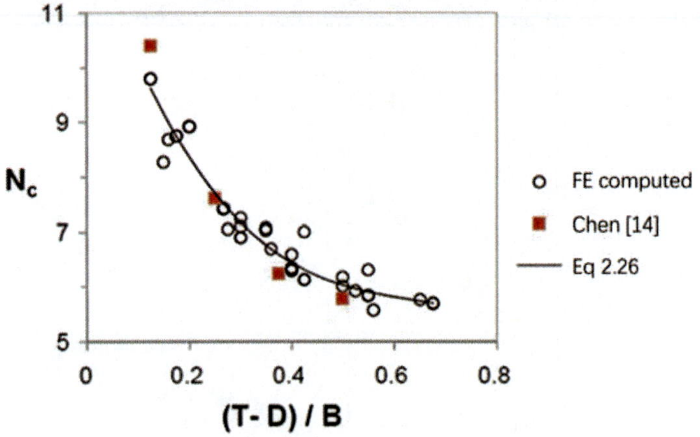

Fig. 2.16 N_c values for (T–D)/B $\leq$ 0.7

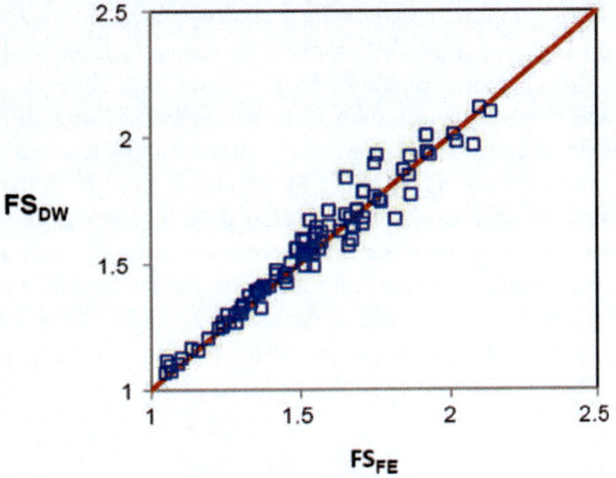

Fig. 2.17 Comparison of FS_{DW} versus FS_{FE}

in Fig. 2.16. Figure 2.17 shows the plot of FS_{DW} obtained using Eq. (2.26) versus the FS_{FE} values for all the cases which is reasonably accurate. In this study, clays with constant c_u as well as clays with c_u increasing linearly with depth were also considered. For clays with c_u increasing linearly with depth, the average c_u values were used. For example, the c_{uh} value is taken as the average c_u from the ground surface to the depth H, and the c_{ub} value is taken as the average c_u from depth H to depth $(H + B_1)$ in which $B_1 = B/\sqrt{2}$.

$$\begin{aligned} N_c &= -21.6[(T-D)/B]^3 + 42.2[(T-D)/B]^2 - 28.8[(T-D)/B] + 12.6 \quad for\ (T-D)/B \leq 0.7 \\ N_c &= 5.7 \quad for\ (T-D)/B > 0.7 \end{aligned} \tag{2.26}$$

2.5.4 *Reliability Analysis*

Since the deterministic factor of the safety method considered in the previous sections does not explicitly reflect the inherent and spatial uncertainties in the soil properties and excavation geometries, a reliability-based approach is considered in this section to assess the probability of basal heave failure using the modified Terzaghi factor of safety method and the concept of spatial factors.

2.5.4.1 Spatial Averaging Effect and the Variance Reduction Factor

Spatial averaging is a concept in which the spatial variability of the soil property such as the undrained shear strength is averaged to approximate a random variable that represents a soil parameter. The variance of the spatially averaged soil property over some domain is less than the variance at the point. As the size of the domain increases, the variance of the soil property decreases. Vanmarcke [22] proposed a dimensionless variance reduction function Γ^2 which is a function of the scale of fluctuation θ and characteristic length L, to determine the reduction in the point variance under local averaging. Vanmarcke [23] adopted the variance reduction function to consider spatial averaging in reliability analysis, via which the variances of soil parameters may be reduced by multiplying a factor known as the variance reduction factor. Successful applications of the variance reduction technique have been reported using, for example, the constant model, triangular model, and exponential model [24]. The exponential model, which is more commonly employed in the random field modeling in geotechnical engineering, is adopted in this study, which is as follows:

$$\Gamma^2 = \frac{1}{2}\left(\frac{\theta}{L}\right)\left[\frac{2L}{\theta} - 1 + \exp\left(-\frac{2L}{\theta}\right)\right] \tag{2.27}$$

in which θ = scale of fluctuation and L = characteristic length. Given the variance reduction factor Γ^2, the reduced variance σ2 Γ can be obtained with the following equation:

$$\sigma_{\Gamma}^2 = \Gamma^2\sigma^2 \tag{2.28}$$

in which σ = standard deviation of the soil parameter of concern (c_u in this study). The positive square root of the variance reduction factor is referred to herein as the standard deviation reduction factor, or simply the reduction factor (Γ), to differentiate it from the variance reduction factor Γ^2.

The scale of fluctuation θ = 2, 5, 20, 50, and 100 m and the characteristic length L = 9, 18, 27, and 36 m are considered, respectively, in this study. It should be noted that the θ values considered are commonly adopted for spatial variability analysis while the characteristic length L is closed related to the excavation geometries. The accurate determination of L involves the comparison of reduction factors that are most consistent with those back-calculated using the Monte Carlo Simulation-based Random Field Modeling approach. The expression of Eq. (2.27) indicates that the increased scale of fluctuation θ results in the significant increase of the reduction factor Γ and for the same θ value, a smaller characteristic length L would result in a larger reduction factor. In brief, a smaller scale of fluctuation, as well as a larger characteristic length, would give a lower variance reduction factor.

2.5.4.2 The Developed Spreadsheet for Spatial Variability Analysis

Figure 2.18 shows the spreadsheet setup for evaluating the spatial variability reliability index β for the modified Terzaghi expression with FS replaced by the term αc_{ud}. This spreadsheet is modified from the original version by Goh et al. [6] by introducing the spatial factors as shown in Cells B21:C24. The eight lognormally

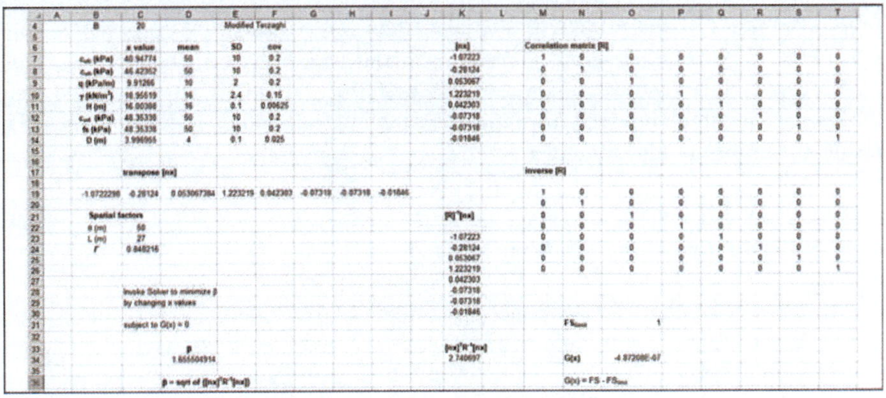

Fig. 2.18 Spreadsheet setup for evaluating the reliability index considering spatial variability

distributed and uncorrelated variables are c_{ub}, c_{uh}, q, γ, *H*, c_{ud}, α, and *D*. The width of the excavation B = 20 m is assumed to be constant in the analysis. The uncertainties in the parameters *H* and *D* could arise from construction deviations in the actual excavation depth and depth of wall penetration, respectively. For the detailed use of the developed spreadsheet, Goh et al. [6] could be referred to. For brevity, these procedures are omitted in this study.

Figure 2.19a shows the effects of different combinations of the spatial variability of the cu of the soil layers for L = 9 and 18 m, respectively. Also shown is the case without considering the spatial variability (i.e., uniform soil conditions). It is obvious that as the scale of fluctuation increases, which indicates an increasing uniform ground, the reliability index *β* values considering spatial variability converge to the constant value of 1.961. The plot shows that when only the spatial variability of c_{ub} is considered, *β* is slightly lower than when combinations of c_{ub} and c_{uh}, as well as c_{ub}, c_{uh} and c_{ud}, are considered. This again suggests that the design can be

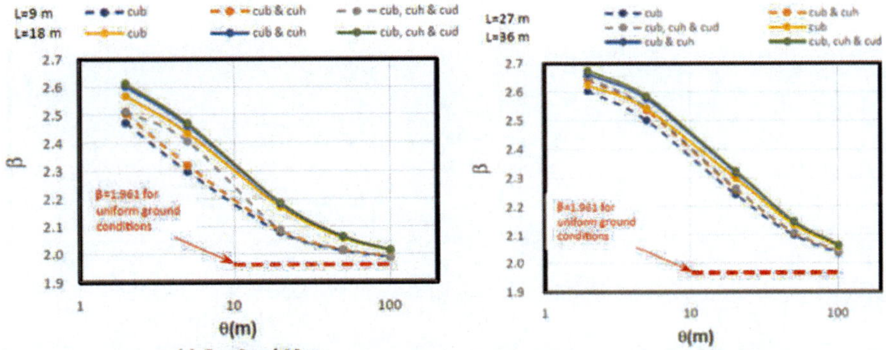

(a) L = 9 and 18 m (b) L = 27 and 36 m

Fig. 2.19 Effects of soil layers considering spatial variability too conservative if the effects of spatial variability of the soil parameters are not considered. Figure 2.19b which shows the spatial variability effects for L = 27 and 36 m, respectively, display similar trends as with Fig. 2.19a with slightly larger β for the same θ. It can be inferred from Fig. 2.19 that a larger characteristic length L can result in a higher reliability index.

2.5.4.3 Influence of Penetration Depth D

Figure 2.20 shows the effect of penetration depth on the calculated reliability index and P_f values, considering only the variation of c_{ub} since the derived β is slightly lower, as indicated previously in Sect. 2.5.4.2. It is obvious that the penetration depth has a significant influence on β and P_f values. A decrease of D from 8 to 4 m results in a reduction of up to 0.5 for β and an obvious increase of the failure probabilities.

Fig. 2.20 Influence of penetration depth on: **a** reliability index β and **b** P_f

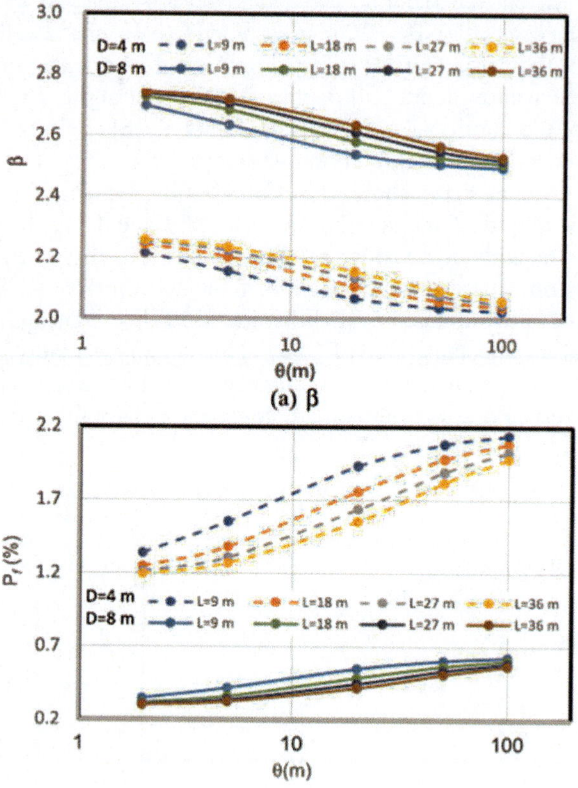

(b) P_f

Fig. 2.21 Influence of COV of cub and characteristic length L

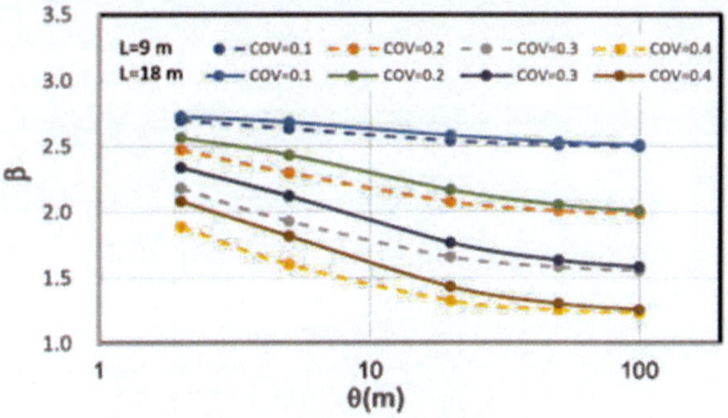

(a) L = 9 and 18 m

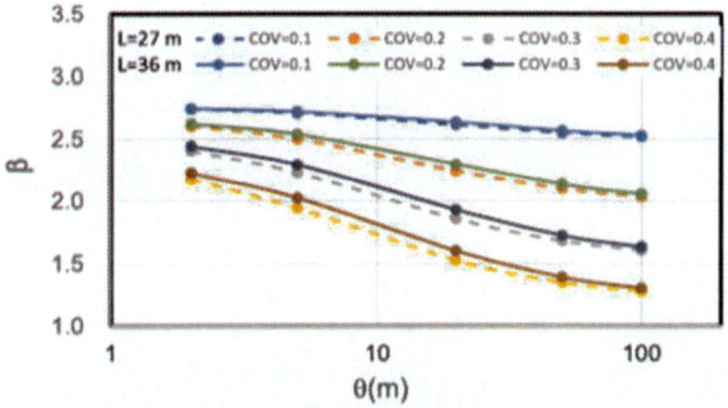

(b) L =27 and 36 m

2.5.4.4 Influence of *COV* of c_{ub}

Figure 2.21a and b shows the effects of COV of c_{ub} and characteristic length L on the reliability index β. It is obvious that for the same θ, β decreases as the COV of c_{ub} increases, and greater characteristic length results in slightly larger β values.

2.5.4.5 Influence of f_s, q, and γ

Figure 2.22a shows the influence of fs denoted by αcud. A smaller $\alpha = 0.6$ results in a reduced reliability index, compared with the default α of 1.0, since it contributes to the resistance, acting as the numerator of Eq. (2.9). Similarly, the increase of surcharge load q and the unit weight γ would result in a significant reduction of the reliability index, as shown in Fig. 2.22b and c, respectively.

Fig. 2.22 Influence of: **a** f_s, **b** q, and **c** γ on reliability index

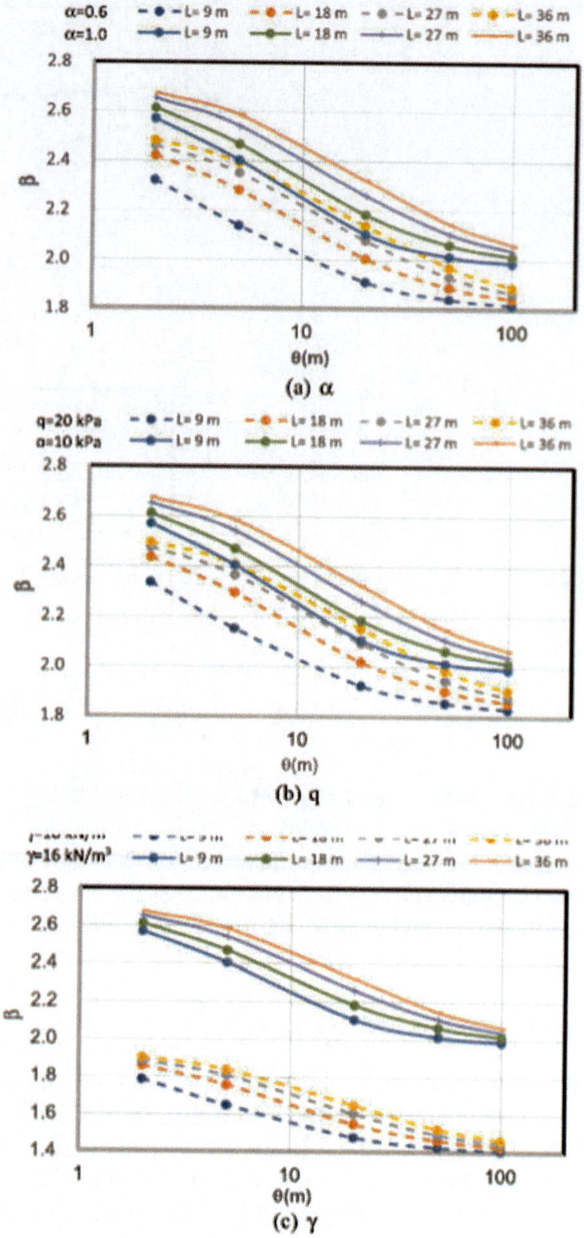

2.5.5 *Summary*

FE analyses were carried out to assess the basal heave factor of safety for excavations in soft clays supported by diaphragm walls. A total of 104 different cases with different geometrical properties of the excavation and wall and soil properties were considered. Comparing the factor of safety for sheet pile and diaphragm walls, the increase in FS_{FE} with increasing D is more significant for diaphragm walls. The basal heave factor of safety FSFE was found to decrease with the increase of the excavation width. The FE analyses also highlight that the depth to the hard stratum from the tip of the wall is a key component of the resistance of the excavation system to basal heave failure.

Comparison of the FS_{FE} values with the predictions based on the Terzaghi method and the slip circle method was also performed. Generally, the FS_T and FS_{Slip} predictions are more conservative compared with the finite-element predictions. The validity of the modified Terzaghi method in Eqs. (2.6) and (2.9) that considers the wall penetration depth and depth to the hard stratum is confirmed to give predictions that are consistent with the basal heave factor of safety obtained from FE analyses.

A reliability-based framework that considers the spatial averaging effect is proposed to assess the probability of basal heave failure using the modified Terzaghi factor of the safety method. It is found that consideration of the spatial variability would generally result in a higher reliability index (i.e., a decrease in the probability of failure). The conventional analysis assuming uniform ground conditions generally results in a more conservative design. Parametric analyses show the reliability index is significantly influenced by the spatial variability of the different soil layers, the scale of fluctuation, the characteristic length, and the penetration depth, as well as the COV of c_{ub}, f_s, q, and γ.

2.6 Evaluation of Basal Heave Stability of Braced Excavations in Anisotropic Clay

Conventional methods for estimating the basal stability of braced excavations generally do not consider the anisotropy of the soft clay, which may lead to the incorrect assessment of excavation stability [25]. This study presents the results of extensive finite-element analysis to analyze the influence of clay anisotropy on basal heave stability.

2.6.1 *Soil Anisotropic Constitutive Model*

The NGI-ADP model [26] is an anisotropic shear strength model for clay in which the undrained shear strength s_u profiles for active (A), direct simple shear (DSS), and passive (P) loading (stress paths) are given as input data. The model interpolates anisotropic undrained shear strength between s_u^C, s_u^{DSS}, and s_u^E along the failure surface according to the orientation of principal stresses as defined in Fig. 2.23.

The main soil parameters of the NGI-ADP model are G_{ur}/s_u^A (ratio of the unloading/reloading shear modulus to the plane strain active shear strength), γ_f^C (shear strain at failure under triaxial compression), γ_f^E (shear strain at failure under triaxial extension), and γ_f^{DSS} (shear strain at failure under direct simple shear). The soil strength includes $s_{u,ref}^A$ (reference plane strain active shear strength), $s_u^{C,TX}/s_u^A$ (ratio of the triaxial compressive shear strength to the plane strain active shear strength), y_{ref} (reference depth), $s_{u,inc}$ (increase in the shear strength with increasing depth), s_u^P/s_u^A (ratio of the plane strain passive shear strength to the plane strain active shear strength), τ_0/s_u^A (initial mobilization), s_u^{DSS}/s_u^A (ratio of the direct simple shear strength to the plane strain active shear strength), and Poisson's ratio ν. $s_{u,ref}^A$ represents the shear strength obtained in (plane strain) undrained active stress paths for the reference depth y_{ref}. Additional details are referred to in Brinkgreve et al. [20].

The yield criterion for the NGI-ADP model in plane strain can be expressed by

$$f=\sqrt{\left(\frac{\sigma_{yy}-\sigma_{xx}}{2}-(1-k)\tau_0-k\frac{s_u^A-s_u^P}{2}\right)^2+\left(\tau_{xy}\frac{s_u^A+s_u^P}{2s_u^{DSS}}\right)^2}-k\frac{s_u^A+s_u^P}{2}=0 \tag{2.29}$$

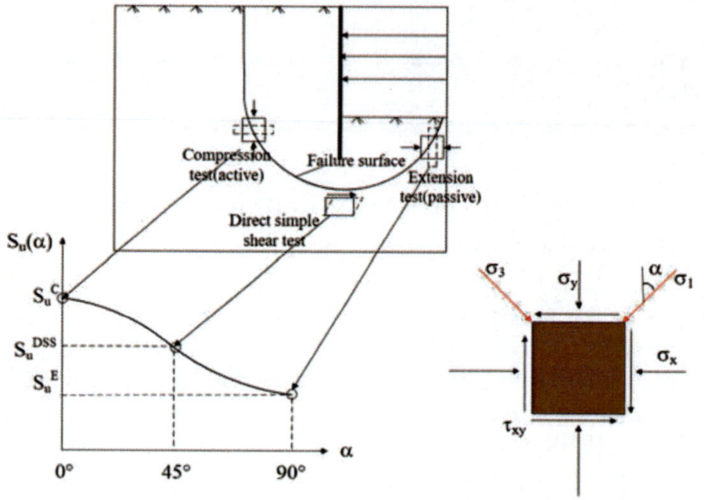

Fig. 2.23 Anisotropic undrained shear strength according to the orientation of principal stresses [26]

where $k = 2\frac{\sqrt{\gamma^p/\gamma_f^p}}{1+\gamma^p/\gamma_f^p}$ when $\gamma^p < \gamma_f^p$, in which $\gamma^p and \gamma_f^p$ are the plasxtic shear strain and the failure plastic shear strain; otherwise $k = 1$;

$$\tau_0 = 0.5(\sigma'_{v0} - \sigma'_{h0}) = 0.5\sigma'_{v0}(1 - K_0) \tag{2.30}$$

The NGI-ADP model uses elliptical interpolation between failure strain in a passive stress state, direct simple shear, and active stress state.

2.6.2 Numerical Modeling

This section introduces a metro station excavation case in Hangzhou, China, consisting of anisotropic clays. The width and length of the excavation are approximately 20 and 291 m, and the excavation depth is 17.0 m. Figure 2.24 shows the plan view of the excavation area. In this section, the results of the plane strain finite-element analysis for the Hangzhou excavation are presented to validate the use of the NGI-ADP model embedded in the finite-element software.

The excavation system consisted of 34.8 m long diaphragm walls and five levels of struts, installed at −2 m, 6.1 m, 9.9 m, 11.4 m, and 12.9 m, respectively. The groundwater table is at a depth of −2.5 m. Excavation time is short so undrained total stress analysis was considered. The soil was modeled by 15-noded triangular elements, the diaphragm wall was assumed to be linearly elastic and modeled by 5-noded beam elements, the struts were modeled by 3-noded bar elements, and the interface between the wall and the soil was assumed as rigid ($R_{inter} = 1.0$). The two side boundaries were fixed to constrain the horizontal displacement, and the bottom boundary was constrained both horizontally and vertically. The right vertical boundary has been extended far enough from the excavation area to eliminate the

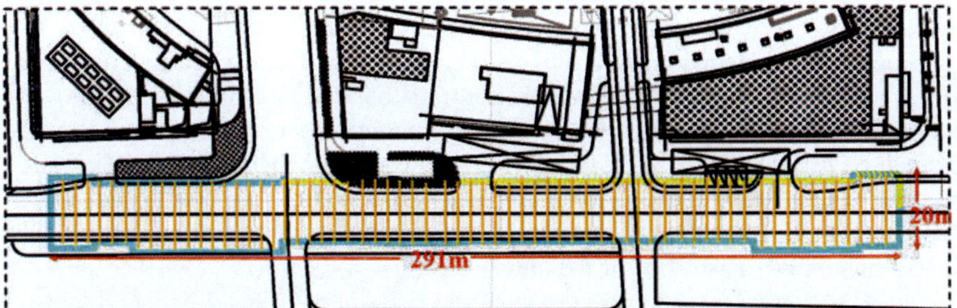

Fig. 2.24 Plan view of the excavation area

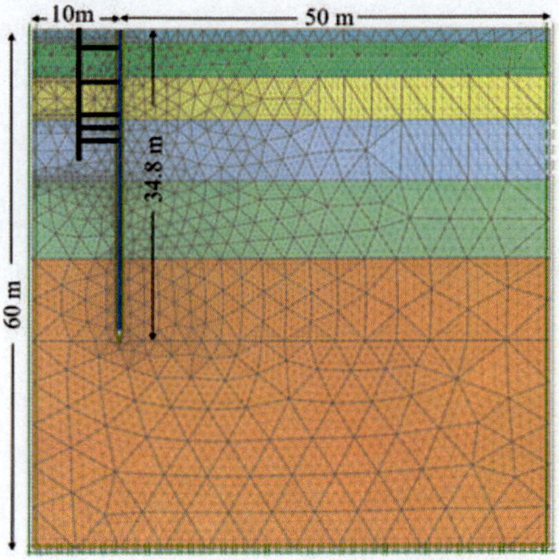

Fig. 2.25 The typical mesh of the FE model

boundary effects. A typical finite-element mesh is shown in Fig. 2.25. The soil parameters of the site are listed in Table 2.5.

As shown in Fig. 2.26, the FE calculated wall deflection is consistent with the measured inclinometer data both in terms of the shape and the maximum value.

This section describes the simplified soil profile that was used for the total stress plane strain parametric study to analyze the braced excavation performance in anisotropic clay. The FE model comprises of supporting structures including a retaining wall and four levels of struts, and 40 m thick soft clay layer overlying a 20 m thick stiff clay layer. Figure 2.27 shows the schematic cross-section of the excavation system. The struts were located at depths of 1 m, 3 m, 5 m, and 7 m below the original ground surface. The NGI-ADP constitutive model was used for the soft clay, while the Mohr–Coulomb (Undrained C) model was used for modeling the stiff clay. The excavation width B, the wall system stiffness ln(S), the final excavation depth H_e, and the penetration depth of the wall below the formation level D are shown schematically in Fig. 2.25. The strut stiffness per meter EA is 6.1×10^5 kN/m. For simplicity, the elastic modulus of the wall E_{wall} is assumed to be constant and equal to 2.8×10^7 kPa, and the rigidity (stiffness) of the wall is studied by varying the wall width b. The ranges of three critical soil parameters considered in this study including s_u^A (kPa), $G_{ur}/s_u^A(-)$, and $s_u^P/s_u^A(-)$ are also shown in Fig. 2.27. The range of s_u^P/s_u^A considered in this study is 0.4, 0.5, 0.6, 0.8, and 1.0, with $s_u^P/s_u^A = 1.0$ denoting that the clay is isotropic, and a smaller value of s_u^P/s_u^A indicating a higher degree of strength anisotropy of the clay.

Table 2.5 Soil parameters

Soil type	Depth (m)	γ (kN/m^3)	E (kPa)	e (−)	K_0 (−)	G_{ur}/s_u^A (−)	$s_{u,ref}^A$ (kPa)	$s_{u,inc}^A$ (kPa)	s_u^P/s_u^A (−)	s_u^{DSS}/s_u^A (−)
Fill	−1.5	19.0	8000	0.6	0.6	300	20	0	1	1
Clay	−5.5	19.8	7470	0.767	0.45	300	2.72	$2.48z$[a]	0.45	0.59
Silty clay1	10.5	19.2	6020	0.851	0.55	300	16.35	$2.61z$	0.41	0.56
Silty clay2	17.5	19.0	5420	0.898	0.5	600	43.75	$3.64z$	0.41	0.56
Silty clay3	26.5	18.8	4980	0.923	0.6	900	107.46	$2.64z$	0.41	0.56
Silty clay4	36.0	19.2	12,460	0.822	0.4	900	177.33	$3.17z$	0.36	0.52

[a] z is the depth of the soil below the ground surface

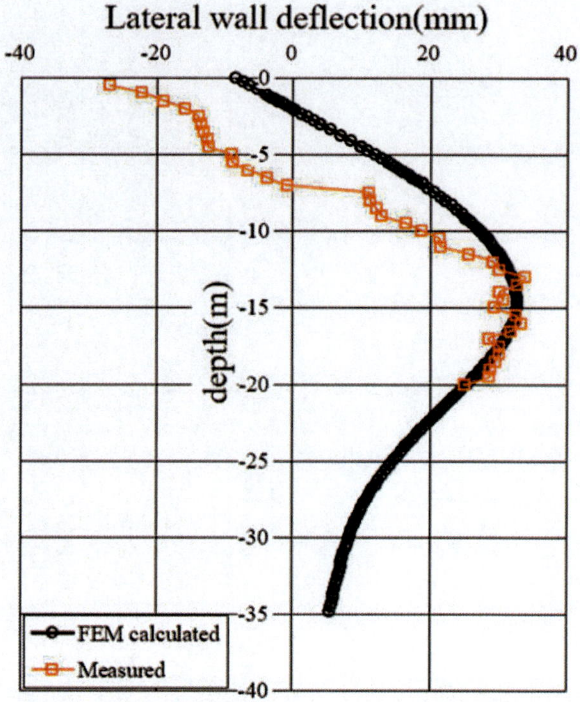

Fig. 2.26 Comparison between FEM calculated and measured wall deflection

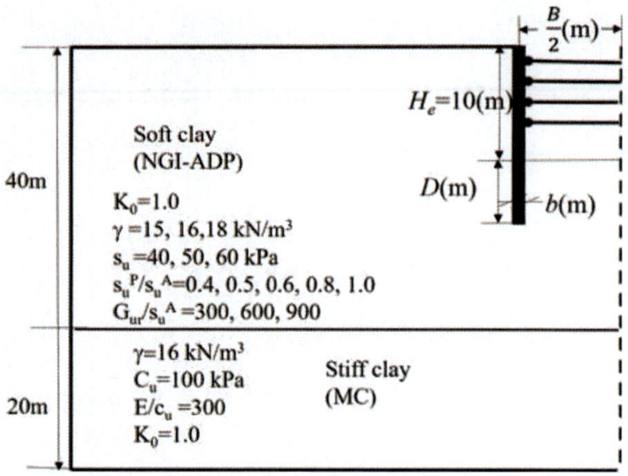

Fig. 2.27 The cross-sectional profile of the simplified FE model

From symmetry, only half of the cross-section is considered as shown in Fig. 2.25. The right vertical boundary extends 60 m from the excavation edge to minimize the effects of the boundary restraints. The nodes along the left and right boundaries were constrained horizontally and the bottom boundary was constrained both vertically and horizontally. The soil was modeled using 15-noded triangular elements, the structural elements of the wall were assumed to be linearly elastic and modeled by 5-noded beam elements, and the struts were represented by 3-noded bar elements.

The parametric study was carried out with an emphasis on the basal heave factor of safety for braced excavation in anisotropic clay, using the NGI-ADP model for the soft clay layer. The stability of the excavation was then determined using the shear strength reduction technique as detailed in Brinkgreve et al. [20]. The properties of the soft and stiff clay are listed in Table 2.6. In this study, based on the recommendations of Brinkgreve et al. [20], the $s_u^{C,TX}/s_u^A$ and τ_0/s_u^A ratios are set to their default values of 0.99 and 0.7, respectively. The structural properties of the support system are shown in Table 2.7. A total of 2523 different finite-element analyses were carried out in this study.

2.6.3 Computed Results and Analyses

Some typical soil displacement contours for the anisotropic and isotropic cases are shown in Fig. 2.28. For isotropic cases, the maximum soil displacement occurs at the center of the formation level. The influence of the s_u^P/s_u^A on the soil displacement is significant, with the maximum soil displacement for $s_u^P/s_u^A = 0.5$ approximately 175% larger than the maximum soil displacement for $s_u^P/s_u^A = 1.0$.

Figure 2.29 shows the comparison of the basal heave factor of safety FS for some typical anisotropic and isotropic cases. The results indicate that the s_u^P/s_u^A ratio has a significant influence on the FS, with a lower FS for the anisotropic soil. It also can be concluded that the wall system stiffness ln(S) has considerable influence on the FS only when the magnitude of ln(S) is small, i.e., flexible wall; while for the stiff walls, the ln(S) shows a marginal influence on the FS. In addition, ln(S) shows a larger effect on the FS when the soil is anisotropic compared with the isotropic case.

Figure 2.30 shows the influence of s_u^A and s_u^P/s_u^A on the FS. The FS increases almost linearly as s_u^P/s_u^A increases. As expected, the FS increases as the s_u^A is increased. The plot also indicates that the relationship between the FS and s_u^A is also nearly linear as inferred by the approximately equal intervals between the lines. Figures 2.31 and 2.32 show the influence of the D/H_e and G_{ur}/s_u^A on the FS, respectively. The D/H_e and G_{ur}/s_u^A both show marginal influence on the FS. Figure 2.33 which shows the effect of the soil unit weight γ indicates that the FS decreases almost linearly with the increase in γ.

Table 2.6 Soil model properties

Soft clay (NGI-ADP)	Parameter	$s_u^{C,TX}/s_u^A(-)$	${\gamma_f}^C$ (%)	${\gamma_f}^{DSS}$ (%)	${\gamma_f}^E$ (%)	υ_u (−)	s_u^{DSS}/s_u^A (−)	
	Value	0.99	0.75	1.735	3.5	0.495	$((1+s_u^P/s_u^A)/2$	
	Parameter	$\tau_0/s_u^A(-)$	R_{inter} (−)	s_u^A (kPa)	γ(kN/m^3)	G_{ur}/s_u^A (-)	s_u^P/s_u^A (-)	
	Value	0.7	1	40, 50, 60	15, 16, 18	300, 600, 900	0.4, 0.5, 0.6, 0.8, 1.0	
Stiff clay (MC)	Parameter	γ (kN/m^3)	c_u (kPa)	K_o = 1-sinϕ(-)	ϕ_u (°)	E/c_u (−)	υ_u (−)	R_{inter} (−)
	Value	16	100	1	0	300	0.495	1

Table 2.7 Parameters of the excavation system

Parameter and units	Ranges
Horizontal strut spacing $L_{spacing}$ (m)	4
Average vertical strut spacing h_{avg} (m)	2
Strut stiffness per meter EA (kN/m)	6.1×10^5
Elastic modulus of wall E_{wall} (kPa)	2.8×10^7
System stiffness ratio ln(S)[a] (−)	3.90, 4.76, 5.43, 8.06, 8.92, 10.13
Excavation width B (m)	20, 30
Final excavation depth H_e (m)	10, 16
Wall penetration depth D (m)	3, 5, 10

[a]In which $S = (EI)_{wall}/\gamma_w h_{avg}^4$, where $(EI)_{wall}$ is the wall stiffness, γ_w is the unit weight of water, and h_{avg} is the average vertical strut spacing

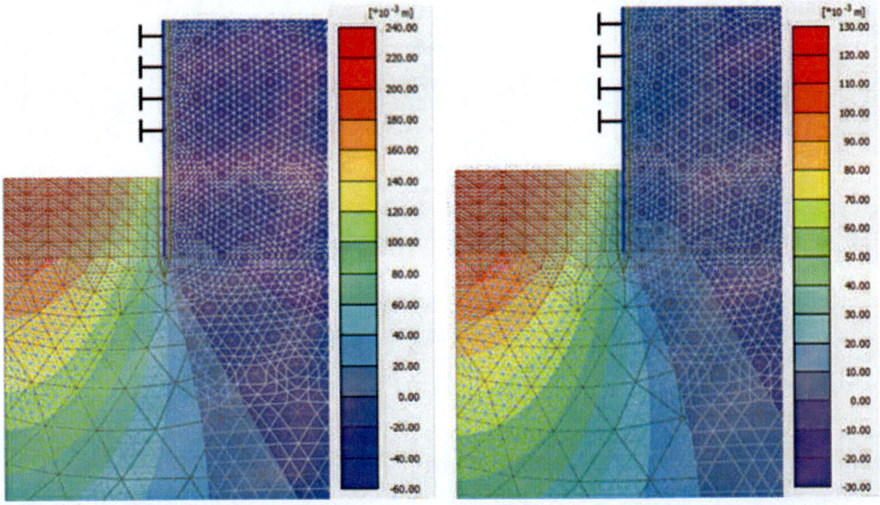

(a) s_u^P/s_u^A =0.5, $\delta_{v,max}$=221.0mm (b) s_u^P/s_u^A =1.0, $\delta_{v,max}$=126.4mm

Fig. 2.28 Comparison of typical soil displacement contours for anisotropic and isotropic cases ($B/H_e = 2$, $D/H_e = 0.5$, $\gamma = 16$ kN/m^3, $s_u^A = 40$ kPa, $G_{ur}/s_u^A = 600$, and $\ln(S) = 8.92$)

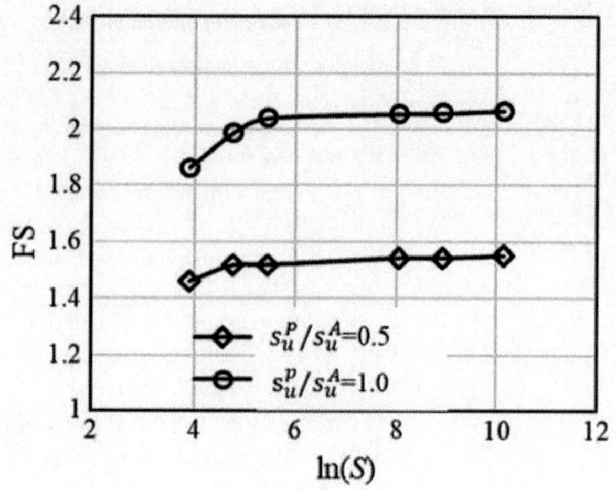

Fig. 2.29 Comparision of FS for anisotropic and isotropic cases ($B/H_e = 2$, $D/H_e = 0.5$, $\gamma = 15$ kN/m^3, $s_u^A = 50$ kPa, and $G_{ur}/s_u^A = 600$)

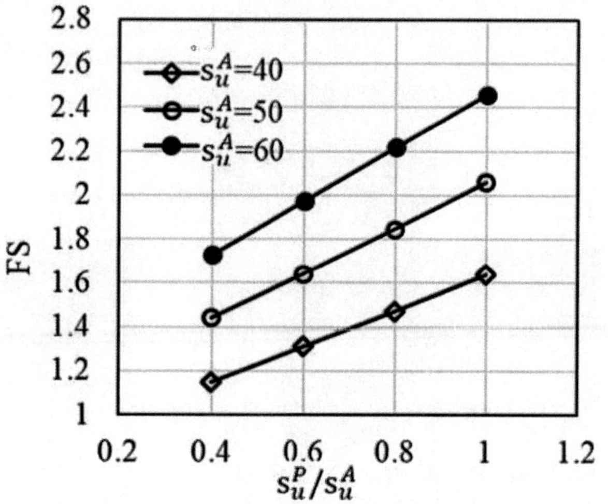

Fig. 2.30 Influence of s_u^A and s_u^P/s_u^A on the FS ($B/H_e = 2$, $D/H_e = 0.5$, $\gamma = 16$ kN/m^3, ln(S) = 8.92, and $G_{ur}/s_u^A = 600$)

2.6.4 Estimation of FS

Based on the numerical results of a total of 2523 hypothetical cases, the logarithmic regression (LR) model is used to develop a simple predictive model to determine the FS against basal heave for braced excavations in anisotropic clay, with the coefficient of determination of $R^2 = 0.8751$. The logarithmic regression FS equation is as follows:

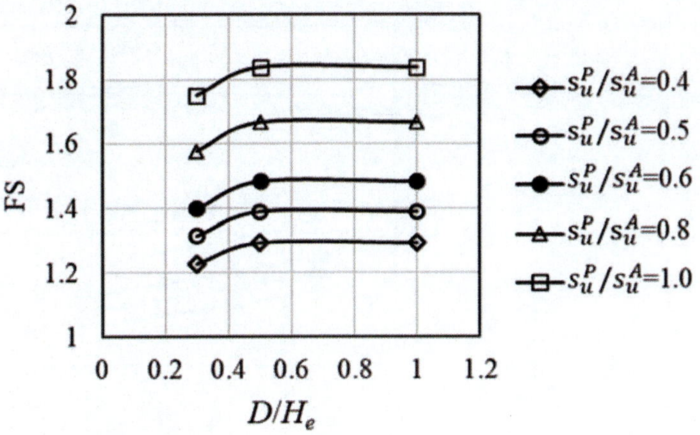

Fig. 2.31 Influence of D and s_u^P/s_u^A on the FS ($B/H_e = 2$, $\gamma = 16$ kN/m^3, ln(S) = 8.06, $s_u^A = 50$ kPa, and $G_{ur}/s_u^A = 600$)

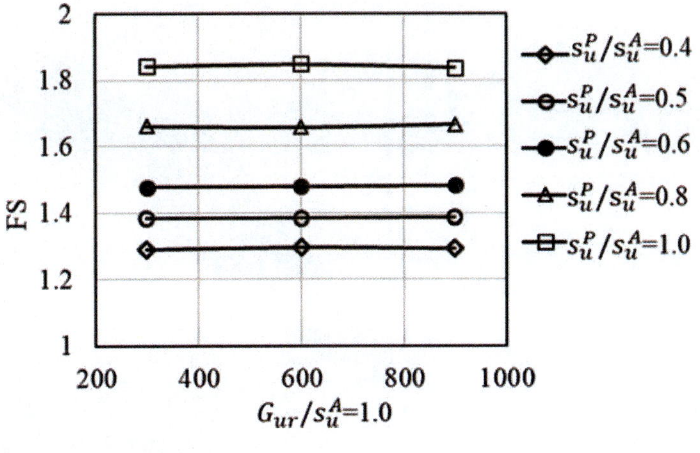

Fig. 2.32 Influence of G_{ur}/s_u^A and s_u^P/s_u^A on the FS ($B/H_e = 2$, $D/H_e = 0.5$, $\gamma = 18$ kN/m^3, ln(S) = 8.06, and $s_u^A = 50$ kPa)

$$FS_{LR} = 0.506(D/H_e)^{0.0601}(B/H_e)^{0.2157}\ln(S)^{0.0374}(s_u^A)^{0.9289}\gamma^{(-0.8652)} (G_{ur}/s_u^A)^{(-0.0197)}(s_u^P/s_u^A)^{0.3596} \quad (2.31)$$

The comparison of the factor of safety computed by the finite-element analysis $FS_{\_FEM}$ and $FS_{\_LR}$ is shown in Fig. 2.34.

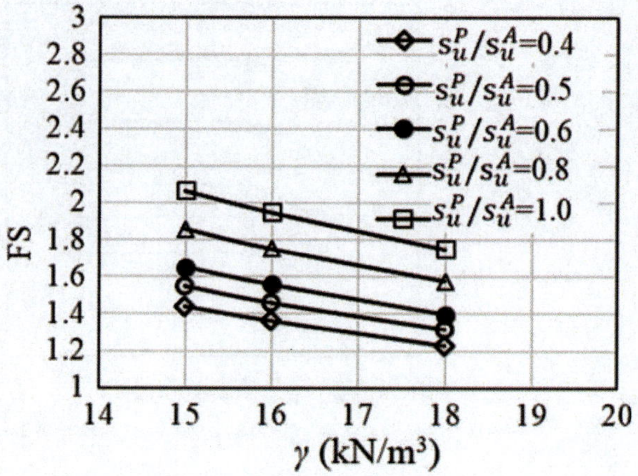

Fig. 2.33 Influence of γ and s_u^P/s_u^A on the FS ($B/H_e = 2$, $D/H_e = 0.5$, $\ln(S) = 8.06$, $s_u^A = 50$ kPa, and $G_{ur}/s_u^A = 900$)

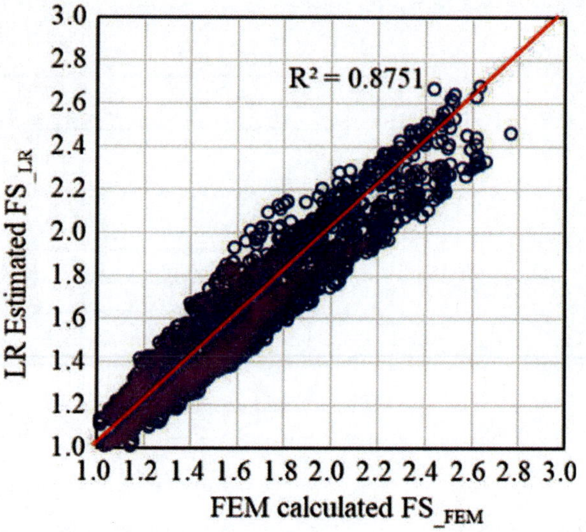

Fig. 2.34 Comparison of $FS_{\_FEM}$ and $FS_{\_LR}$

The parameter sensitivity is also evaluated and the results are shown in Fig. 2.35. The sensitivity index (SI) was used to assess the relative importance of each variable. The SI is calculated by changing one variable at a time over $\pm 10\%$ from the mean and determining the percent change in the FS. This simple indicator provides a good indication of parameter and model variability. The results indicate that the influence of the parameters s_u^A, γ, s_u^P/s_u^A, and B/H_e on the FS are more significant compared with the parameters D/H_e, $\ln(S)$, and G_{ur}/s_u^A.

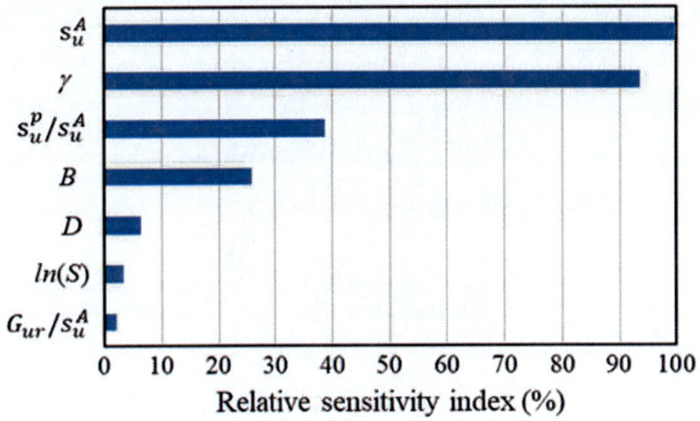

Fig. 2.35 Parameters sensitivity

2.6.5 Validation of the Predictive Model

This section presents the validation of the proposed LR model based on 10 Hangzhou case histories from Ying et al. [27] and three Taiwan cases from Hsieh et al. [5]. For the Hangzhou case, the depths of the excavations were in a range of 9.5–14.9 m and wall depths ranges from 19.5 to 35 m. The ratios of embedded depths to excavation depths varied from 0.91 to 1.46. The undrained shear strength obtained from field vane shear tests varies from 25 to 40 kPa, and the average total unit weight of the clays is about 17 kN/m^3. The support walls comprised of contiguous piles with diameters and spacing that ranged from 0.8 to 1.0 m and from 1.0 to1.40 m, respectively. The parameters of the case histories adopted in this study are listed in Table 2.8. The basal heave factor of safety $FS_{\_S2}$ is obtained by the simplified method S2 proposed by Hsieh et al. [5]. In the S2 method, the undrained shear strength $(s_u)_{avg}$ is derived from the average value of the CK0U-AC and CK0U-AE test results:

$$(s_u)_{avg} = \frac{s_u^e + s_u^c}{2} = \frac{s_u^P + s_u^A}{2} \tag{2.32}$$

And the $FS_{\_S2}$ are calculated by the slip circle method:

$$FS_{\_S2} = \frac{M_r}{M_d} = \frac{X' \int_0^{\frac{\pi}{2}+\rho} s_u \left(X' d\theta\right)}{W \frac{X'}{2}} \tag{2.33}$$

where s_u is the undrained shear strength of clay; *X'* is the radius of the failure arc; *W* is the total weight of the soil and surcharge above the excavation surface within an *X'* wide area outside the retaining construction; θ is the angle between the failure

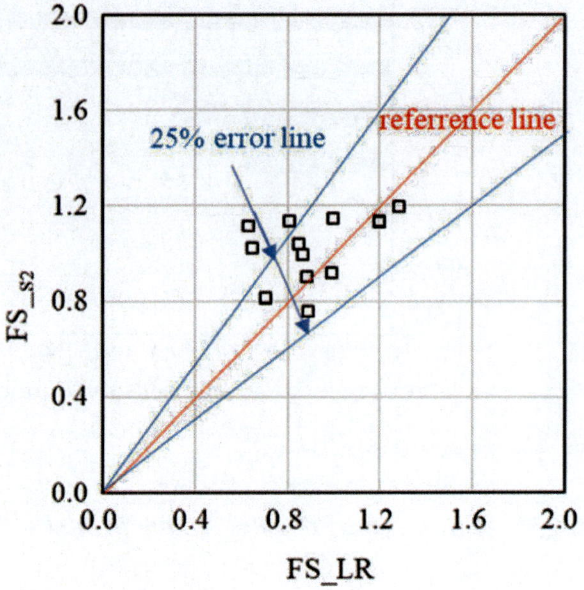

Fig. 2.36 Comparison of $FS_{\_S2}$ and $FS_{\_LR}$ for case histories

surface and the vertical direction; and ρ is the angle of failure arc in the excavation zone.

Figure 2.36 and Table 2.8 that present the comparison between $FS_{\_S2}$ and $FS_{\_LR}$ for 10 Hangzhou and three Taiwan cases indicate a good agreement between the two methods. It is shown that based on extensive field measurements or reliable data, it is feasible and economical to establish surrogate models to evaluate the basal heave stability of deep braced excavation [28].

2.6.6 Summary

In this study extensive FE analysis has been carried out to analyze the effects of the s_u^P/s_u^A,G_{ur}/s_u^A, s_u^A, the soil unit weight γ, the excavation width B, excavation depth H_e, system stiffness ln(S), and wall penetration D on the base stability of braced excavations with consideration of anisotropy of the soil undrained shear strength. The influence of the parameters s_u^A, γ, s_u^P/s_u^A , and B/H_e are found to be more significant on the FS, compared with the parameters D/H_e, ln(S), and G_{ur}/s_u^A. A simple logarithmic regression (LR) model was developed for preliminary assessment of the basal heave factor of safety for braced excavations in anisotropic clay. The proposed estimation model was validated by 13 well-documented case histories from Hangzhou and Taiwan.

Table 2.8 Documented case of Hangzhou excavation adopted from Ying et al. [27]

No.	$D/H_e(-)$	$B/H_e(-)$	$ln(S)$ (−)	s_u^A (kPa)	γ (kN/m^3)	$G_{ur}/s_u^A(-)$	$s_u^P/s_u^A(-)$	FS_LR	FS_S2	References
1	0.91	13.82	7.43	25	17	600	0.4	1.04	0.85	[27]
2	1.43	4.05	6.03	25	17	600	0.4	0.81	0.71	
3	1.04	4.85	5.59	35	17	600	0.4	1.13	1.20	
4	1.10	6.25	5.42	35	17	600	0.4	1.20	1.29	
5	1.21	4.64	6.15	35	17	600	0.4	1.14	1.20	
6	1.34	10.53	5.67	25	17	600	0.4	0.99	0.86	
7	1.19	3.13	5.67	25	17	600	0.4	0.76	0.89	
8	1.35	8.05	5.92	30	17	600	0.4	1.11	0.63	
9	1.16	12.40	5.84	25	17	600	0.4	1.02	0.65	
10	1.46	4.39	6.43	35	17	600	0.4	1.14	0.81	
11	0.78	1.93	6.89	40	18	600	0.61	1.15	1.00	[5]
12	0.66	1.32	7.5	35	18	600	0.56	0.90	0.88	
13	0.83	1.32	7.5	35	18	600	0.56	0.92	0.99	

References

1. Terzaghi, K. (1943). *Theoretical soil mechanics.* New York, USA: Wiley.
2. Bjerrum, L., & Eide, O. (1956). Stability of strutted excavations in clay. *Géotechnique, 6*, 32–47.
3. Eide, O., Aas, G., & Josang, T. (1972). Special applications of cast-in-place walls for tunnels in soft clay in Oslo (pp. 63–72). Norwegian Geotechnical Institute.
4. Goh, A. T. C. (1994). Estimating basal heave stability for braced excavations in soft clay. *Journal of Geotechnical Engineering, 120*, 1430–1436.
5. Hsieh, P. G., Ou, C. Y., & Liu, H. T. (2008). Basal heave analysis of excavations with consideration of anisotropic undrained strength of clay. *Canadian Geotechnical Journal, 45*, 788–799.
6. Goh, A., Kulhawy, F., & Wong, K. (2008). Reliability assessment of basal-heave stability for braced excavations in clay. *Journal of Geotechnical and Geoenvironmental Engineering, 134*, 145–153.
7. NAVFAC. (1986). Foundations and earth structures. Alexandria, USA: Design Manual. Department of the Navy Facilities Engineering Command.
8. Drucker, D. C., Prager, W., & Greenberg, H. J. (1952). Extended kungit design theorems for continuous media. *Quarterly of Applied Mathematics, 9*, 381–389.
9. Chen, W. (1975). *Limit analyses and soil plasticity.* Amsterdam, Netherlands: Elsevier.
10. Chang, M.-F. (2000). Basal stability analysis of braced cuts in clay. *Journal of Geotechnical and Geoenvironmental Engineering, 126*, 276–279.
11. Ying, H. W., Zhang, J. H., Zhou, J., Sun, W., & Yan, J. J. (2016). Analysis of stability against basal heave of excavation in anisotropic soft clay based on tests of hollow cylinder apparatus. *Rock and Soil Mechanics, 37*, 1237–1242. (in Chinese).
12. Tang, Z., Huang, M. S., & Yuan, J. Y. (2017). Upper bound analysis of basal stability of excavations based on continuous velocity fields. *Rock and Soil Mechanics, 38*(3), 833–839. (in Chinese).
13. Zou, G. D. (2004). Analysis of stability against upheaval of deep excavation by an upper limit method. In *Rock and soil mechanics* (pp. 1873–1878). (in Chinese).
14. Greco, V. R. (1996). Efficient Monte Carlo technique for locating critical slip surface. *Journal of Geotechnical Engineering-ASCE, 122*, 517–525.
15. Afiri, R., & Gabi, S. (2018). Finite element slope stability analysis of Souk Tleta dam by shear strength reduction technique. Innovative Infrastructure Solutions 3.
16. Do, T.-N., Ou, C.-Y., & Lim, A. (2013). Evaluation of factors of safety against basal heave for deep excavations in soft clay using the finite-element method. *Journal of Geotechnical and Geoenvironmental Engineering, 139*, 2125–2135.
17. Griffiths, D. V., & Lane, P. A. (1999). Slope stability analysis by finite elements. *Geotechnique, 49*, 387–403.
18. Zhang, W. G., & Goh, A. T. C. (2012). Reliability assessment on ultimate and serviceability limit states and determination of critical factor of safety for underground rock caverns. *Tunnelling and Underground Space Technology, 32*, 221–230.
19. Goh, A. T. C., Zhang, W. G., & Wong, K. S. (2019). Deterministic and reliability analysis of basal heave stability for excavation in spatial variable soils. *Computers and Geotechnics, 108*, 152–160.
20. Brinkgreve, L., Swolfs, W., & Engin, E. (2017). *Plaxis manual.* Netherlands: PLAXIS BV.
21. Cheng, C., & Likitlersuang, S. (2018). Underground excavation behaviour in Bangkok using three-dimensional finite element method. *Computers and Geotechnics, 95*, 68–81.
22. Vanmarcke, E. H. (1977). Probabilistic modeling of soil profiles. *Journal of the Geotechnical Engineering Division-Asce, 103*, 1227–1246.
23. Vanmarcke, E. (1986). Random fields: analysis and synthesis MIT Press, Cambridge (MA).
24. Vanmarcke, E. (2010). *Random fields: Analysis and synthesis* (2nd ed.). Hoboken (NJ): Wiley.

25. Zhang, R. H. (2021). Study on deep braced excavation-induced responses considering anisotropy of clay and parameter back analysis. Doctor thesis, Chongqing University, Chongqing China.
26. Grimstad, G., Andresen, L., & Jostad, H. P. (2012). NGI-ADP: Anisotropic shear strength model for clay. *International Journal for Numerical and Analytical Methods in Geomechanics, 36*(4), 483–497.
27. Ying, H. W., Cheng, K., Zhang, L. S., Ou, C. Y., & Yang, Y. W. (2020). Evaluation of excavation-induced movements through case histories in Hangzhou. *Engineering Computations, 37*(6), 1993–2016.
28. Zhang, W. G., Zhang, R. H., Wu, C. Z., Goh, A. T. C., & Wang, L. (2020). Assessment of basal heave stability for braced excavations in anisotropic clay using extreme gradient boosting and random forest regression. *Underground Space*. https://doi.org/10.1016/j.undsp.2020.03.001

Chapter 3
Lateral Earth Pressure and Strut Forces

3.1 Introduction

Problems of deep excavation, either stability analysis, stress analysis, or deformation analysis, entail the distribution of earth pressures. Though introductory books on soil mechanics or foundation engineering have discussed quite a few earth pressure theories along with many examples, systematic research is deficient. Some important points may not be sufficiently emphasized. In actual analyses, a wrong choice of earth pressure theory may lead to an uneconomical or unsafe design.

Other than the gravity retaining walls (the retaining wall alone can rarely resist the lateral pressure), supplementary strutting systems are also required. The selection of the strutting system depends on not only the magnitude of lateral pressure, but also on the period it will take to install the strutting system and the obstruction it may bring about on the construction. A strutting system comprises of either horizontal struts or anchors, which contribute to the resistance to the lateral earth pressure generated by excavation. The struts are the most important elements in a braced excavation. After preliminary estimation of deformation and stress calculated, the type and size of the struts or anchors are accordingly determined. The data on the strut load is necessary for the detailed design of struts or anchors. To design a strutting system, one first has to analyze the load on the strut during excavation.

When excavation proceeds, the pressure distributions during different excavation stages are shown in Fig. 3.1, in which the states follow:

Stage 1: The wall is subjected to an active earth pressure and the wall deforms.

Stage 2: A strut is installed and preloaded. Generally, the wall and soil will not be pushed back to their original position, but since the strut force is greater than the active pressure, this causes an increase in the wall contact pressure.

Stage 3: The excavation in stage 3 causes a new lateral displacement. The soil moves out of the zone behind the first strut into the displacement between b and c.

Stage 4: The installation of the second strut in stage 4 will result in similar changes to the earth pressures.

W. Zhang and H. Liu, *Design of Deep Braced Excavation and Earth Retaining Systems Under Complex Built Environment*,
https://doi.org/10.1007/978-981-16-5320-9_3

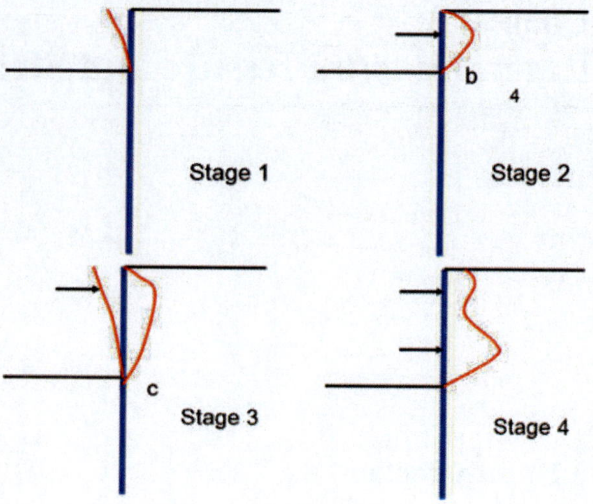

Fig. 3.1 Pressure distributions during different excavation stages

The strut load can be calculated following the finite-element method, the beam on elastic foundation method, or the apparent earth pressure method. This chapter will review the conventional earth pressure theory and apparent pressure diagrams method and then propose modified apparent pressure diagrams for excavations in different soils.

3.2 Review of Conventional Earth Pressure Theory

3.2.1 Lateral Earth Pressure at Rest

Horizontal (lateral) stress in soils are usually described and quantified by means of a lateral earth pressure coefficient, K:

$$K = \sigma_h^{'}/p_v^{'} \tag{3.1}$$

where

$p_v^{'}$ is the effective overburden pressure and $\sigma_h^{'}$ is the horizontal (lateral) effective stress at the same point within the soil mass.

The effective overburden pressure $p_v^{'}$ is given by

$$p_v^{'} = \int_0^z \gamma dz + q - u \tag{3.2}$$

where
γ is the bulk density, z is the depth below, u is the pore water pressure, and q is any uniform surcharge at the ground surface.

Unlike some other walls, embedded walls usually retain predominantly natural ground. So the pre-existing or in situ horizontal (lateral) earth pressure, as modified by wall installation, is potentially important. The symbol K_0 is used to denote the earth pressure coefficient describing the initial in situ stress state in the ground before the wall is installed.

The in situ earth pressure coefficient may be of particular concern in a clay deposit, in which K_0, like the specific volume, depends on the stress history. Deposition (or burial under a glacier) corresponds approximately to one-dimensional (1D) compression, during which the horizontal effective stress σ_h' increases in proportion to the effective overburden pressure p_v'. Clays may also become consolidated by desiccation (drying) on exposure to air, by vegetation, or by freezing, where the effective stress is increased through a reduction in pore water pressure while the total stress remains constant, as shown in Fig. 3.2.

On unloading path (e.g., due to the erosion of overlying soil, re-saturation after desiccation, the melting of an overlying glacier, or a rise in groundwater level), the horizontal effective stress,σ_h', tends to remain "locked-in", decreasing proportionately less quickly than the effective overburden pressure, p_v'. So the in situ earth pressure coefficient K_0 in an overconsolidated clay stratum is usually greater than unity. In heavily overconsolidated clays, a zone of soil extending to a depth of several meters from the surface may be be close to its limiting passive pressure because of geological unloading.

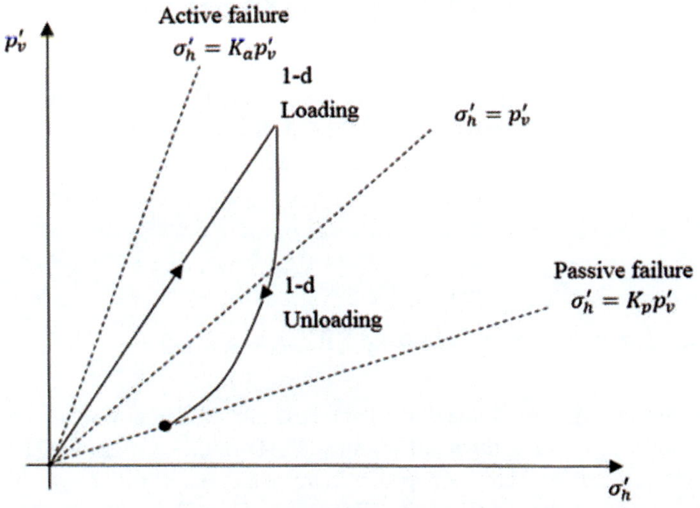

Fig. 3.2 Schematic stress history of an overconsolidated clay deposit

The earth pressure coefficient in granular material or weak rock is not uniquely related to the stress history of the deposit as it is in clay.

In situ lateral earth pressure may be estimated using an equation of the form

$$K_0 = \left(1 - sin\varphi^{'}\right).OCR^{\lambda}.(1 + sin\beta) \tag{3.3}$$

where $\varphi^{'}$ is the drained angle of shearing resistance, OCR is the over consolidation ratio, and β is the slope angle (to the horizontal) of the soil surface.

For normally consolidated deposit (OCR = 1) and level ground ($\beta = 0$), this reduces to Jaky's empirical formula $K_0 = \left(1 - sin\varphi'\right)$.

3.2.2 Rankine's Earth Pressure Theory

When soil is removed from in front of an embedded retaining wall, the wall will tend to move into the excavation. This will result in a reduction in the lateral stress in the ground behind the wall, eventually bringing it to the active condition in which the ground is at failure with the horizontal effective stress as small as it can be for the effective overburden pressure. In the ground that remains in front of the wall below formation level, the horizontal effective stress at failure is as large as it can be for the effective overburden pressure.

Approximations to these limiting pressures may be calculated by considering either the stresses in a zone of ground at failure or the equilibrium of an assumed sliding wedge. The first approach, following Rankine [1], is based on a stress state that can be in equilibrium without exceeding the limiting strength of the ground. In a uniform deposit, the limiting ratio of horizontal to vertical effective stresses is constant, i.e., if the vertical effective stress increases linearly with depth, so will the limiting horizontal effective stress. The limits calculated in this way ensure stability, but may be unnecessarily severe, and are inherently safe (assuming that the correct boundary conditions, ground strength parameters, and pore pressures have been identified) (Fig. 3.3).

3.2.3 Coulomb's Earth Pressure Theory

Following Coulomb [2], the force that must be exerted by a retaining wall to prevent a wedge of soil from sliding down an assumed slip surface is determined and is then usually assumed to arise from a stress that increases linearly with depth. The limits obtained will prevent sliding along the slip surface assumed in the calculation, but may not be sufficient to prevent failure from occurring in some other, unidentified mechanism, so they could be unsafe (Fig. 3.4).

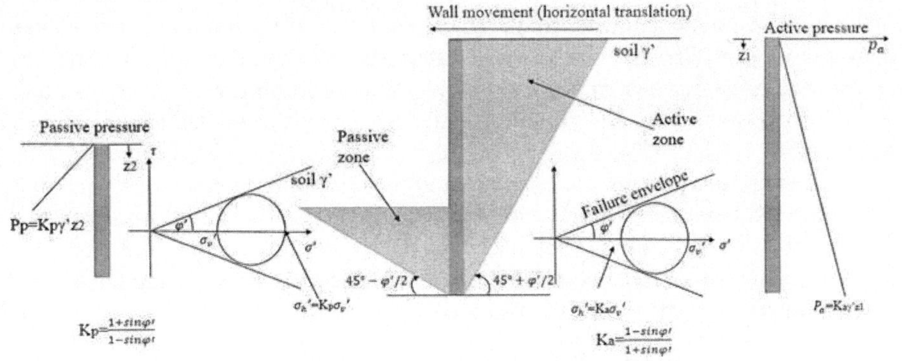

Fig. 3.3 Rankine's plastic equilibrium for a frictionless wall or soil interface translating horizontally

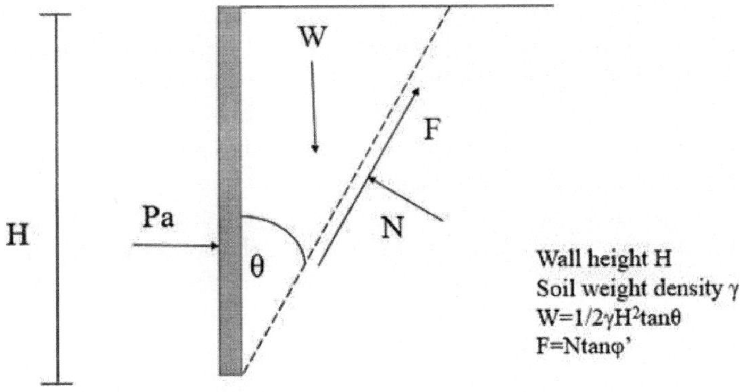

Res. Hor. $Pa = N\cos\theta - F\sin\theta$

$= N\ (\cos\theta - \sin\theta\tan\varphi')$

Res. vert. $W = 1/2\gamma H^2\tan\theta = N\ (\sin\theta + \cos\theta\tan\varphi')$

Eliminate N and minimize Pa

$\theta = 45° - \varphi'$ $\quad Pa = 1/2\gamma H^2 Ka$ where $Ka = (1-\sin\varphi')/(1+\sin\varphi')$

Fig. 3.4 Coulomb's method to calculate the limiting active force for a frictionless wall or soil interface translating horizontally

3.2.4 Comparison and Summary

The results obtained from Rankine's and Coulomb's earth pressure theories are identical under the same conditions (smooth wall surfaces, level grounds, and homogeneous cohesionless soils) though the two theories are quite differently based. The actual conditions, however, may not conform to the hypothetical terms: the wall surface may be rough and the ground surface in the back of the wall may be of irregular shape with a certain load. Rankine's theory can hardly apply under these conditions. Coulomb's theory can cope with these complicated conditions, though it is difficult to obtain a theoretical solution. However, experience overwhelmingly demonstrates that these more recent theories give accurate values of the limiting lateral stresses for use as a basis for design, assuming that variation in pore water pressures and strength properties, etc. in the ground around the wall have been identified and taken into account.

In summary, the conventional approach to the design of an embedded retaining wall is something of a hybrid, with earth pressure coefficients derived using a mechanism-based approach to take account of factors such as wall interface friction, the profile of the retained ground surface, and the presence of line loads used in an equilibrium analysis of the wall.

3.3 APD for Different Soils and Retaining Systems

The empirical methods of Peck [3] and Terzaghi and Peck [4] were obtained from field measurements of strut loads and are envelopes of the maximum measured pressures. They are considered to give conservative estimates of earth pressures and provide estimates of the maximum strut loads for design purposes. They are not the actual pressure distributions but correspond to the maximum values expected. These pressure envelopes are commonly referred to as apparent pressure diagrams (APDs).

Figure 3.5 illustrates the procedure of calculating the apparent earth pressure (by the method of Tributary Areas). That is the load on the struts can be approximately determined by the method of tributary areas, which consists of dividing into sections the pressure acting on the diaphragm wall. The division of these sections is performed in the middle of the distance between the struts, assigning the pressure value of each section to the strut that is in that section. The horizontal distance between struts here is considered to be a constant as b. From this way, strut load Q is also easy to calculate.

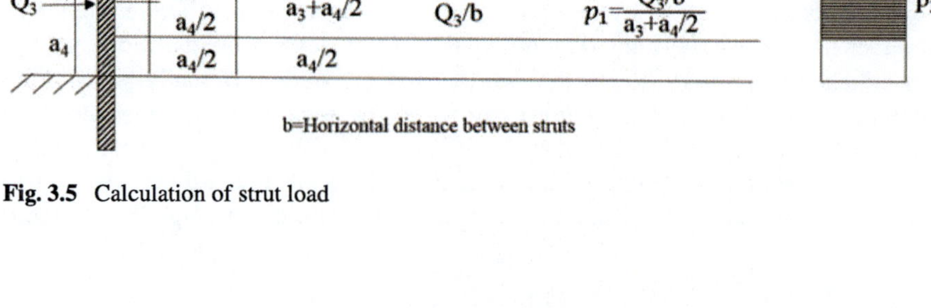

Fig. 3.5 Calculation of strut load

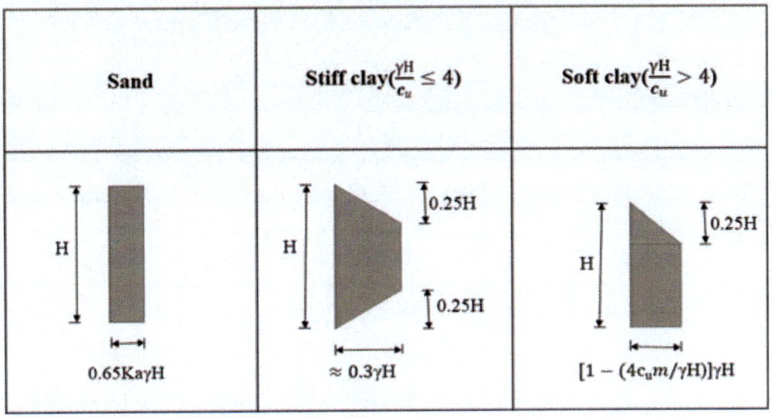

Fig. 3.6 Diagrams of the apparent earth pressure [3]

Figure 3.6 shows diagrams of the apparent earth pressure established by Peck [3]. As shown in the figure, when the soil in the back of the wall mainly consists of sandy soils, the apparent earth pressure P_a will be

$$P_a = 0.65 K_a \gamma H_e \tag{3.4}$$

where

γ = unit weight of sandy soils,

H_e = excavation depth, and

K_a = Rankine's coefficient of active earth pressure = tan $(45° - \varphi/2)$.

If the soil in the back of the wall is stiff clay ($\gamma H/c_u \leq 4$), the apparent earth pressure P_a would be

$$p_a = 0.2\gamma H \sim 0.4\gamma H \text{ (the average is } 0.3\gamma H) \tag{3.5}$$

If the soil in the back of the wall is soft to medium soft clay (i.e.,($\gamma H/c_u > 4$), the apparent earth pressure, P_a, would be the larger of

$$p_a = \left[1 - 4c_u m/\gamma H\right]\gamma H \text{ or } p_a = 0.3\gamma H \tag{3.6}$$

where

c_u = undrained shear strength of soil within the range of the excavation depth and.

m= an empirical coefficient; for deep deposit of soft clay, use m = 0.4, otherwise use m = 1.

3.3.1 ADP Determination for Braced Excavation in Clays

3.3.1.1 Introduction

Ou [5] summarized Peck's work on APD to estimate the magnitude and distribution of strut loads in different clays, where the Rankine's coefficient of lateral active earth pressure K_a is expressed as

$$K_{\mathrm{a}} = 1 - \mathrm{m}\frac{4c_{\mathrm{u}}}{\gamma H_{\mathrm{e}}} \tag{3.7}$$

where cu is the soil undrained shear strength (in kPa), γ is the soil unit weight (in kN/m^3), H_{e} is the depth of the excavation (in meter), and m is an empirical coefficient. Most of the commonly used empirical methods that were developed from either numerical results or field studies have been for excavations with flexible wall systems such as sheet pile walls. To date, there have been limited studies on the excavation performance for stiffer wall systems such as diaphragm walls and bored piles.

Chang and Wong [6] proposed a modified APD for diaphragm walls in deep clay deposits. The research was based on a case study and a parametric finite-element study. Their research showed that strut loads computed using Peck's APD underestimated the strut loads significantly. By introducing a strut force exceedance ratio α, as functions of the soil stiffness ratio and undrained shear strength, a modified APD was proposed. They commented that the amended APD was derived from the cases with *T*/*B* ratio greater than 1 (where T is the clay thickness below the final excavation level, and *B* is the excavation width, as illustrated in Fig. 3.7). If *T*/*B* is less than 1, they inferred that there would be a strong restraining effect from the hard stratum reducing the strut force.

Hashash and Whittle [7] compared Peck's conventional APD with their FE results which considered undrained strength anisotropy and strength non-homogeneity. Their research indicated that the conventional APD was smaller than the finite-element results for diaphragm wall, especially for deep excavations. Also, the wall stiffness plays an important role in the apparent earth pressures. As the wall stiffness decreased, the apparent earth pressure decreased.

Wong et al. [8] observed that for the construction of a major Singapore underground expressway project, most of the data were within the vertical boundary of the apparent earth pressure diagram proposed by Terzaghi and Peck [4]. However, they recommended that the vertical pressure diagram should extend to the ground surface instead of decreasing to zero to fit all the measured data.

Twine and Roscoe [9] enhanced Terzaghi and Peck's [4] work and introduced the distributed prop load (DPL) method based on 81 case histories and field measurements of prop loads. However, of the 81 case histories, 28 cases are for flexible walls in soft to medium clays (denoted as class AF) while only 2 cases are for stiff walls

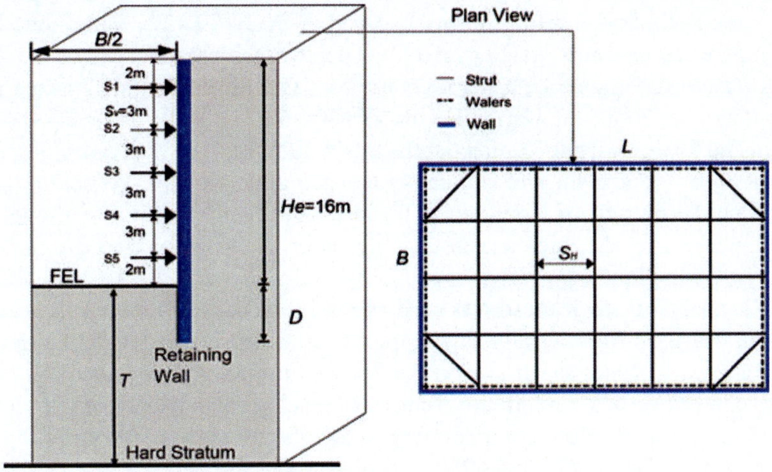

Fig. 3.7 Cross-section and plan view

(class AS). In addition, although there are 10 reported cases for stiff walls in stiff clays (class BS), five of them are singly propped while two cases have two strut levels and only the remaining three cases have three levels of struts. In view of these limited published data, it is, therefore, relevant to reassess the DPL method for the class AS and class BS excavation types.

All these studies outlined earlier generally indicated that Peck's APD under-predicted the apparent earth pressure of the braced excavations, especially for those involving diaphragm walls and large excavation depths. As various factors are likely to influence the APD such as the clay thickness, soil strength and stiffness, wall stiffness, excavation width, and strut stiffness, this chapter explores the performance of the strutting system, including the apparent earth pressure, through a series of the plane strain and three-dimensional finite-element analysis. Some differences were observed between the numerical results and Peck's APD.

As discussed previously, for the case of excavations in stiff wall systems, the proposed distributed prop loads (DPLs) by Twine and Roscoe [9] were based on only very limited measured data. The main contribution from this chapter is to propose updated APD for stiff wall systems based on extensive numerical analyses supplemented by additional measured data (eight cases in soft clays and eight cases in stiff clays, with up to five levels of struts).

3.3.1.2 Numerical Scheme

For this study, the finite-element analyses were carried out using the geotechnical software PLAXIS 2D and PLAXIS 3D. Figure 3.7 shows a typical cross-section and plan view for the cases considered. The parameters shown in the figure include the following: L = excavation length, B = excavation width, D = wall penetration depth, T = clay thickness below the final excavation level (FEL), S_H = horizontal strut spacing, S_V = vertical strut spacing, and H_e = depth of final excavation, all in meters. In this study, the geometrical properties considered include the following: excavation widths B fixed at 20 m; excavation lengths L of 44, 68, and 84 m; excavation depths H_e of 14 and 16 m; wall penetration depths D of 6 and 4 m; T = 32 m; horizontal strut spacings S_H of 4 and 5 m; and vertical strut spacings S_V of 3 m. Vertical retaining walls along the excavation boundary were installed together with a five-level strut and waling system. The thickness of the clay layers is varied for the different cases studied.

For 2D analyses, a half mesh was used due to geometrical symmetry. A very fine mesh size was used for 2D analysis to improve the accuracy of FE calculations. For 3D analyses, a medium mesh size in the horizontal direction and medium coarse mesh size in the vertical direction were used to reach a balance between processing time and accuracy. For brevity, only a typical 3D quarter mesh is shown in Fig. 3.8, comprising of 15,679 nodes and 4980 15-noded wedge elements.

For the 2D simulations, fourth-order 15-noded triangular elements, which are considered to be very accurate elements, were used to model the soil while the interface elements have five integration points. In 3D PLAXIS, the interface elements have 9-point Gauss integration with three translational degrees of freedom for each node. For the 2D analysis, the retaining wall is simulated using 5-noded elastic plate elements. The elastic behavior is defined by the following parameters: EA: normal stiffness, EI: bending stiffness, and m: Poisson's ratio. For the 3D analysis, the wall is simulated using 8-noded quadrilateral plate elements with six degrees of freedom per node.

In this study, three wall types, with five different stiffness values, were considered for each soil type, as listed in Table 3.1. Based on the approach adopted by Finno et al. [10], the wall thickness of 0.42 m was set to an arbitrary (constant) value so that the

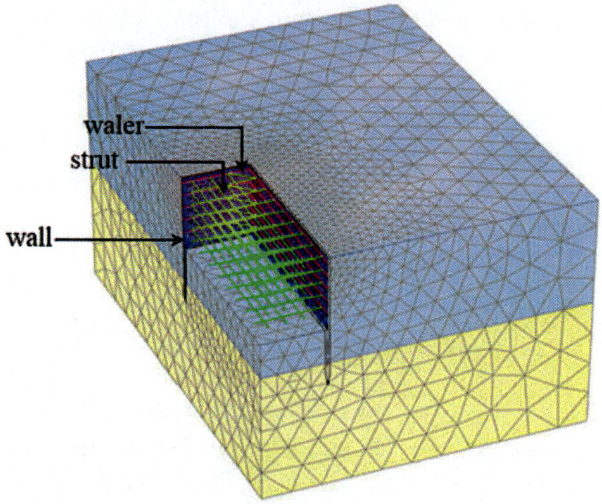

Fig. 3.8 Typical quarter 3D mesh of excavation

Table 3.1 Wall properties

Parameters		Wall types				
		Flexible		Medium	Stiff	
		Plane strain (2D) FE parameters				
α		0.1	0.2	1.0	2.0	10
System stiffness, S		32	64	320	320	3200
Wall stiffness EI ($\mathrm{kNm^2/m}$)		5.04×10^4	1.008×10^5	5.04×10^5	1.008×10^6	5.04×10^6
Compressive stiffness EA (kN/m)		3.427×10^6	6.854×10^6	3.427×10^7	6.854×10^7	3.427×10^8
Poisson's ratio, ν		0	0	0	0	0
		Three-dimensional (3D) FE parameters				
Young's modulus (kPa)	E_1	8.16×10^6	1.632×10^7	8.16×10^7	1.632×10^8	8.16×10^8
	E_2	4.08×10^5	8.16×10^5	4.08×10^6	8.16×10^6	4.08×10^7
	E_3	2.00×10^8	4.00×10^8	2.00×10^9	4.00×10^9	2.00×10^6
Shear modulus (kPa)	G_{12}	4.08×10^5	8.16×10^5	4.08×10^6	8.16×10^6	4.08×10^7
	G_{13}	4.00×10^5	8.00×10^5	4.00×10^6	8.00×10^6	4.00×10^7
	G_{23}	1.333×10^6	2.666×10^7	1.333×10^7	2.666×10^8	1.33×10^8
Poisson's ratio, ν		0	0	0	0	0

moment of inertia I and area A were kept constant, and only the wall elastic modulus E was varied. A wall stiffness coefficient a was introduced to represent walls with different rigidities [11] as shown in Table 3.1. The baseline bending stiffness EI for the analysis is 5.04×10^5 kN m^2/m, which refers to a wall of medium stiffness based on the databases of Long [12] and Moormann [13]. Therefore, $\alpha = 1.0$ for cases with this medium wall stiffness. For flexible walls, the baseline bending stiffness EI of 5.04×10^5 kN m^2/m was multiplied by smaller factors of 0.1 and 0.2, while for stiff walls, the baseline bending stiffness was multiplied by larger factors of 2 and 10. The system stiffness [14], S, in Table 3.1 is defined as

$$S = \frac{EI}{\gamma_w h_{avg}} \tag{3.8}$$

where EI = wall stiffness, γ_w = unit weight of water, and h_{avg} = average vertical strut spacing.

Meanwhile, based on the method adopted by Finno et al. [10], the wall thickness of 0.42 m was set to be constant so that the moment of inertia I and area A of

Table 3.2 Properties of the waling system

Parameter		Struts	Walers
Unit weight γ (kN/m^3)		78.5	78.5
Cross-sectional area A (m^2)		0.007367	0.008682
Young's modulus E (kPa)		2.1E8	2.1E8
Moment of inertia (m^4)	I_3	5.073E-5	1.045E-4
	I_2	5.073E-5	3.668E-4
	I_{23}	0	0

the plate were kept constant, and only elastic modulus E was varied. Besides, the soil–structure interactions are simulated through interfaces on both sides of the wall, which allow for the specification of reduced friction between the wall and the soil.

The struts were simulated using node-to-node anchor elements in the 2D analysis. For 3D analysis, the struts and walers were modeled as beam elements, which have six degrees of freedom per node. This is described in greater detail in Brinkgreve et al. [15]. For the braced excavations in this chapter, the struts were placed horizontally at a spacing of 4 or 5 m (for different case studies) in two directions to form a frame net. The walings were used to connect the excavation wall and the struts. The material properties are tabulated in Table 3.2.

The boundary conditions for 2D and 3D cases were as follows: (1) rollers at side boundaries to allow vertical displacements and (2) pinned at the base to restrain any movements. For both 2D and 3D cases, the lateral boundaries in the side directions were set at least 100 m away from the center of the excavation to eliminate the influence of the boundary restraints on the ground movements. This ensures that the lateral boundaries are beyond the settlement influence zone (which is typically two times the excavation depth) induced by the excavation as suggested by Hsieh and Ou [16]. In this study, the clay thickness below the final excavation level T is assumed as 32 m, which is regarded as fairly large. A typical staged construction simulation is shown in Table 3.3; the struts are installed when the excavation level reaches 1 m lower than the struts will be. The original groundwater table was assumed to be 2 m below the ground surface in the retained soil. The water table inside the excavation was progressively lowered with the excavation of the soil during each phase.

The properties of three different types of clays which were considered in this parametric study are similar to the properties assumed by Bryson and Zapata-Medina [11] and are tabulated in Table 3.4. The soils are assumed to follow the hardening soil (HS) model. The three soil types are as follows: soft clay, medium clay, and stiff clay. The clays are real soils whose properties have been extensively reported in the literature. The properties of the soft clay with average $c_u = 20$ kPa are based on the Upper Blodgett soft clay reported by Finno et al. [17]. The properties of the medium clay with average $c_u = 45 -$ kPa are based on the Taipei silty clay found at the Taipei Enterprise Center (TNEC) project [18]. The Gault clay at Cambridge [19] with average $c_u = 125$ kPa was used as the model for the stiff clay.

A series of 2D and 3D analyses were carried out covering various cases of wall stiffness, soil types, excavation geometries, and different strut levels and locations

Table 3.3 Typical construction sequence

Phases	Construction details
Phases 1	Install the retaining wall
Phases 2	Excavate to 3 m below the ground surface
Phases 3	Install strut system at 2 m below ground surface
Phases 4	Excavate to 6 m below the ground surface
Phases 5	Install strut system at 5 m below ground surface
Phases 6	Excavate to 9 m below the ground surface
Phases 7	Install strut system at 8 m below ground surface
Phases 8	Excavate to 12 m below the ground surface
Phases 9	Install strut system at 11 m below ground surface
Phases 10	Excavate to 15 m below the ground surface
Phases 11	Install strut system at 14 m below ground surface
Phases 12	Excavate to 16 m below the ground surface

Table 3.4 Soil parameters

Parameter	Unit	A: soft clay (Chicago clay)	B: medium clay (Taipei silty clay)	C: stiff clay (Gault clay)
γ_{unsat}	kN/m^3	18.1	18.1	20
γ_{sat}	kN/m^3	18.1	18.1	20
E_{50}^{ref}	kN/m^2	2350	6550	14,847
E_{oed}^{ref}	kN/m^2	2350	6550	14,847
E_{ur}^{ref}	kN/m^2	7050	19,650	44,540
C'	kN/m^2	0.05	0.05	0.05
φ	°	24.1	29	33
Ψ	°	0	0	0
ν_{ur}	[-]	0.2	0.2	0.2
p^{ref}	kN/m^2	100	100	100
M	[-]	1.0	1.0	1.0
K_0^{nc}	[-]	0.59	0.55	1.5
R_f	[-]	0.7	0.95	0.96
R_{inter} (interface friction)	[-]	1	1	1

using the hardening soil (HS) model. For brevity, only the main findings of the numerical results are presented in the following sections. For all the 2D and 3D cases, the height of wall H_w is fixed at 20 m, so the depth of wall penetration D decreases as He increases. Apart from slight differences with regard to the excavation depth H_e, identical construction procedures were applied as described in Table 3.3.

3.3.1.3 Results

Figure 3.9 depicts the plan view arrangement of the struts and walers (depicted schematically in red1) for the 3D analyses. The symbol x denotes the horizontal distance from the center of the excavation. There is no strut located at the middle section of the excavation, so the simulations of both the struts and the walers of 3D excavations are symmetrical. For each level of struts, the struts at the $x = -2$ m and $x = 2$ m section are identical because of symmetry. These struts which are the closest to the middle section are subjected to the largest force among all the struts at the same level [5]. Hence, only the strut forces at location $x = 2$ m section are presented and discussed. These values will be used in the subsequent section to assess the APD.

For brevity, only some of the results from the 2D and 3D analyses are presented. Tables 3.5 and 3.6 show some typical results of the strut forces from the 2D and 3D analyses. S1 denotes the top strut and S5 is the lowest strut level. Generally, the struts S2, S3, and S4 are subjected to larger axial forces. The general trend was for the forces in struts S2 to S5 to decrease with increasing soil strength. It can be observed from Tables 3.5 and 3.6 that the differences of maximum compressive strut forces depend on the clay types, *L/B* ratios, and the strut levels. Figure 3.10 shows the maximum axial force of different strut levels for various *L/B* ratios in stiff clay (a = 0.1, 1.0, and 10). Generally, the strut forces in struts S2 to S5 increased with increasing wall system stiffness which is similar to the trend observed by Hashash and Whittle [7]. It is obvious from this plot that the differences of maximum compressive strut forces for various *L/B* ratios (and 2D vs 3D) in stiff clay are less significant.

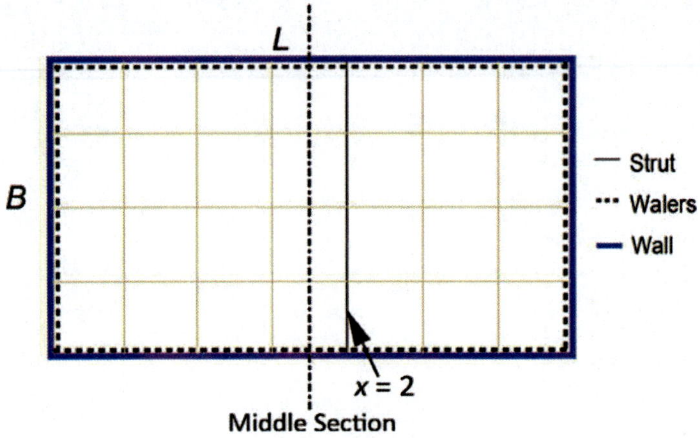

Fig. 3.9 Plan view: horizontal location of struts and walers

Table 3.5 Typical maximum compressive strut force for $H_e = 16$ m and $\alpha = 1.0$

Clay type	Struts	Maximum strut force (kN)				Max. difference between 2 and 3D (%)
		2D	3D			
			L/*B* = 2.2	*L*/*B* = 3.4	*L*/*B* = 4.2	
Medium clay	S1	605.6	645.2	631.1	–	6.4
	S2	16,899	1453.4	1591.1	–	14.0
	S3	2315.5	2094.7	2007.1	–	9.6
	S4	2886.1	2598.2	2204.5	–	10.0
	S5	1130.5	384.4	349.0	–	66.0
Stiff clay	S1	944.2	631.1	920.5	888.2	33.2
	S2	1670.0	1591.1	1609.5	1597	4.7
	S3	1791.4	2007.1	1699.7	1722.6	12.0
	S4	1487.1	2204.5	1524.3	1486.4	48.2
	S5	492.1	349.0	440.0	482.3	29.1

Table 3.6 Typical maximum compressive strut force for $H_e = 16$ m and $\alpha = 10.0$

Clay type	Struts	Maximum strut force (kN)				Max. difference between 2 and 3D (%)
		2D	3D			
			L/*B* = 2.2	*L*/*B* = 3.4	*L*/*B* = 4.2	
Soft clay	S1	1034.2	1074.6	1055.2	1046.4	3.9
	S2	2342.7	1923.1	1953.6	1934.1	17.9
	S3	5019.9	4096.8	4467.6	4529.3	18.4
	S4	3931.5	360.4	3957.1	4046.8	6.1
	S5	1028.3	1143.9	1010.0	971.8	11.2
Medium clay	S1	949.5	1023.6	993.4	992.3	7.8
	S2	1475.3	1292.1	1275.8	1359.1	13.6
	S3	3148	2554.7	2452.9	2899.1	22.1
	S4	3159.2	3005.6	2671.6	3065.6	15.4
	S5	915.7	902.9	767.9	492.3	46.2
Stiff clay	S1	1403.1	1430.2	1476.7	1349	3.9
	S2	1752.6	1630.9	1717.7	1577.2	10.0
	S3	1818.4	1685.9	1762.6	1643.4	9.6
	S4	1379.4	1303.5	1430.3	1262.6	8.5
	S5	418.7	377.6	357.5	373.0	14.8

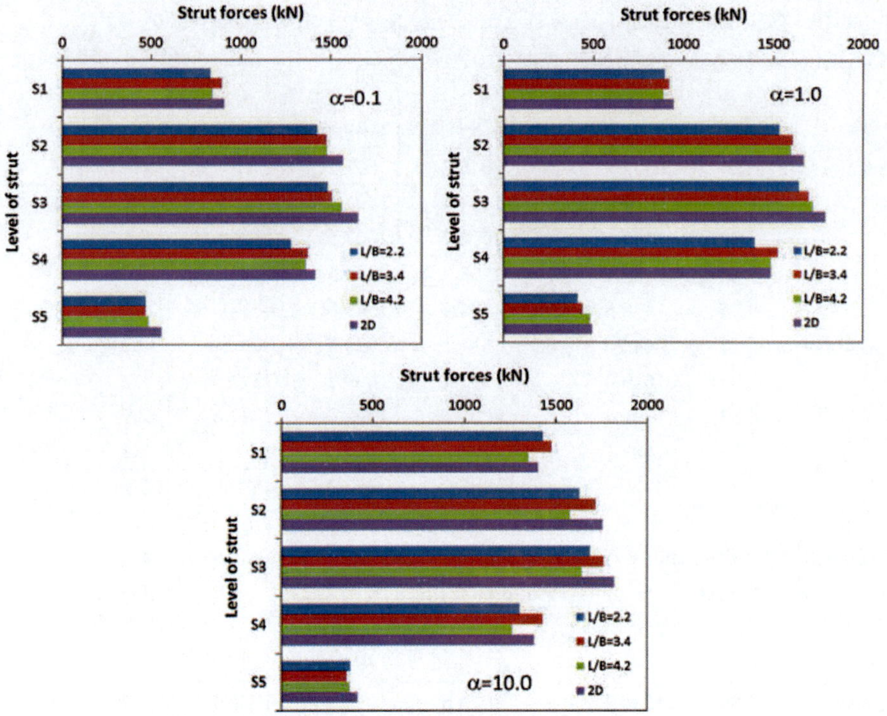

Fig. 3.10 Maximum compressive strut forces for various *L/B* ratios in stiff clay ($\alpha = 0.1$, 1.0, and 10)

3.3.1.4 Apparent Earth Pressure Envelope

For the design of braced deep excavation in soils, apparent earth pressure diagrams are usually used to calculate the minimum horizontal strut loads for each supporting level during vertical excavation. Herein the results from the 2D and 3D maximum strut forces for $H_e = 14$ m and $H_e = 16$ m were used to derive the apparent earth pressures for excavations in stiff walls. The tributary area load distribution procedure proposed by Peck [3] was used. The assumed loading area for each strut is plotted in Fig. 3.11 taking as an example the case with $H_e = 16$ m, where hsi represents the loading area depth for the corresponding strut.

For the force on the strut *P*, since the horizontal spacing of the struts S_H is 4 m, the apparent earth pressure σ_h can be obtained:

$$\sigma_h = \frac{P}{S_H h_{si}} \tag{3.9}$$

The normalized apparent earth pressure ($\sigma_h/\gamma H_e$) and the normalized depth (z/H_e) are then plotted. Only excavations in stiff walls for different clays are considered as

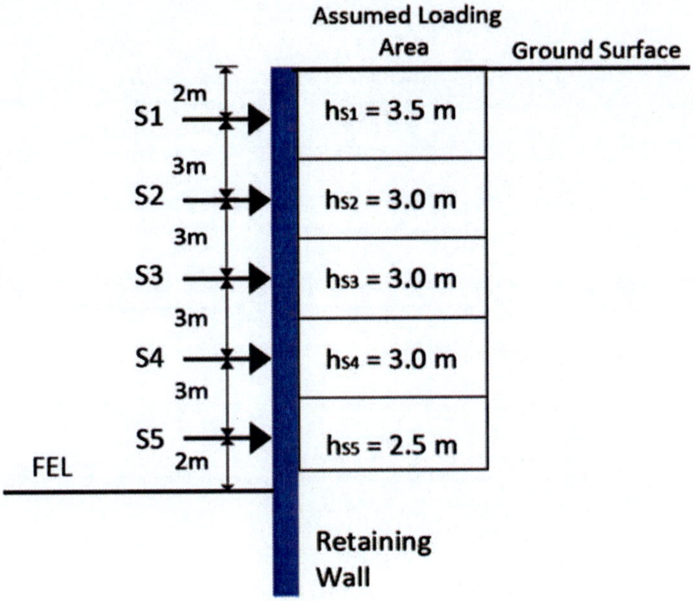

Fig. 3.11 Cross-section: assumed tributary loading areas of struts for $H_e = 16$ m

explained in the Introduction because of the limited studies that have been carried out so far and also because these walls are becoming more common for very deep excavations in congested urban environments. For comparison, Peck's APD and the CIRIA DPL recommendations [9], together with various more recent case histories from the literature as listed in Table 3.7, are also plotted alongside the FE results. For the soft clay, the K_a value is calculated using Eq. 3.7. Following the approach from CIRIA for excavations with flexible wall systems, this section provides an update for excavations with stiff walls. The proposed normalized apparent earth pressures for excavations with stiff walls in soft clays and stiff clays are shown in Fig. 3.12 and Fig. 3.13, respectively. Excavations in medium clays have been excluded due to limited field measurements from the literature. For comparison, Peck's APD for soft clays ($c_u = 20$ kPa) assuming m = 0.4 is also shown in Fig. 3.12. For stiff clays (Fig. 3.13), two dash lines are used to define the upper and lower bound of Peck's APD. Also shown in Fig. 3.13, represented by the blue dash line is the characteristic distributed prop load diagram of DPL = γH_e = 0:5 reported by CIRIA [9] for stiff walls in stiff to very stiff clays. Figure 3.13 also includes the proposal by Yau and Sum [20] who numerically analyzed the strut loads of typical excavation and lateral support work for two case histories in Hong Kong and proposed the maximum prop load line of $0.65\gamma H_e$.

Figure 3.12 shows that for excavations in soft clays with stiff walls, the FE results (represented by open symbols) indicate the apparent earth pressure is generally less than $0.3\gamma H_e$ for $z/H_e \leq 0.25$ while for $z/H_e \geq 0.25$, the apparent earth pressure is generally less than $0.9\gamma H_e$. Based on the measured strut forces from the case histories

Table 3.7 Summary of case histories

Case histories	Soil type	Unit weight (kN/m^3)	H_e (m)	Wall type	No. of strut levels	Horizontal strut spacing (m)	Vertical strut spacing (m)	References
BTG residual soils	Soft clay	18.5–19.2	19.5–24.0	D-wall/SBP wall (stiff wall)	4–6	4.2–8.5	4.0–6.0	[21]
Taipei (CIRIA report) AS1	Soft clay	17.5	14.4	D-wall (stiff wall)	4	5.5	4.0	[9]
						5.5	3.8	
						5.5	2.9	
						5.5	2.3	
Taipei (CIRIA report) AS2	Soft clay	19.2	14.1	D-wall (stiff wall)	4	6.0	2.7	
						6.0	2.9	
						6.0	3.4	
						6.0	3.4	
(CIRIA report) BS4	Stiff clay	19.6	19.5	D-wall (stiff wall)	2	4.5	6.0	
						4.5	2.9	
(CIRIA report) BS6	Stiff clay	20.0	10.0	D-wall (stiff wall)	3	1.7	2.6	
						1.7	3.0	
						1.7	3.0	
(CIRIA report) BS7	Stiff clay	19.5	12.4	D-wall (stiff wall)	3	1.7	4.3	
						1.7	3.0	
						1.7	3.3	
(CIRIA report) BS9	Stiff clay	18.9	15.3	D-wall (stiff wall)	3	3.6	6.3	
						3.6	4.1	
						3.6	3.1	

(continued)

Table 3.7 (continued)

Case histories	Soil type	Unit weight (kN/m^3)	H_e (m)	Wall type	No. of strut levels	Horizontal strut spacing (m)	Vertical strut spacing (m)	References
(CIRIA report) BS10	Stiff clay	18.8	16.6	D-wall (stiff wall)	2	4.2	4.3	
						4.2	5.3	
Singapore circle line projects	Soft clay	16.0	18.0	D-wall (stiff wall)	4	–	–	[22]
Singapore circle line projects	Stiff clay	20.0	22.0	D-wall (stiff wall)	4–5	–	–	
		22.0	20.0					
		22.0	21.0					
		22.0	18.0					
Singapore Old Alluvium soil sites	Soft clay	–	–	D-wall (stiff wall)	–	–	–	[23]
Singapore Old Alluvium soil sites	Stiff clay	–	–	D-wall (stiff wall)	–	–	–	
Shanghai CBD deep excavation	Soft clay	17.2	17.5	D-wall (stiff wall)	4	–	4.4	[24]
							4.6	
							4.0	
							4.6	
Singapore C907 site, CST 3A	Soft clay	18.0	16.0	D = 1.1 m CBP wall (stiff wall)	5	5	1.0	[25]
							3.0	
							3.0	

(continued)

Table 3.7 (continued)

Case histories	Soil type	Unit weight (kN/m^3)	H_e (m)	Wall type	No. of strut levels	Horizontal strut spacing (m)	Vertical strut spacing (m)	References
							3.0	
							3.0	
Singapore C907 site,	Soft clay	18.0	15.0	D-wall (stiff wall)	4	6	2.5	
							3.0	
							3.5	
							4.0	
Singapore central expressway Phases II	Soft soil overlying stiff clay	18.0	Up to 21.0	Arbed wall, CBP wall, and D-wall	3–5	5–6	3–5	[8]

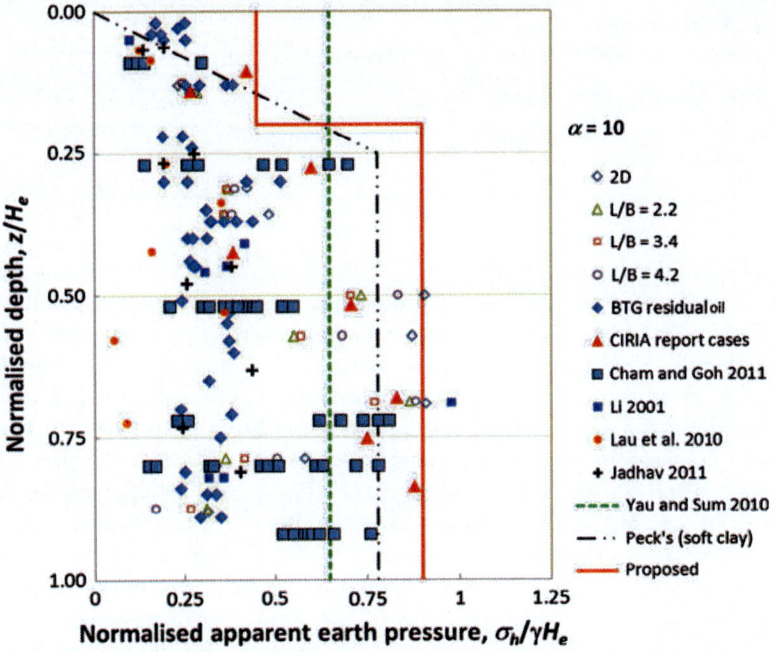

Fig. 3.12 Normalized apparent earth pressure versus normalized depth for stiff walls in soft clays

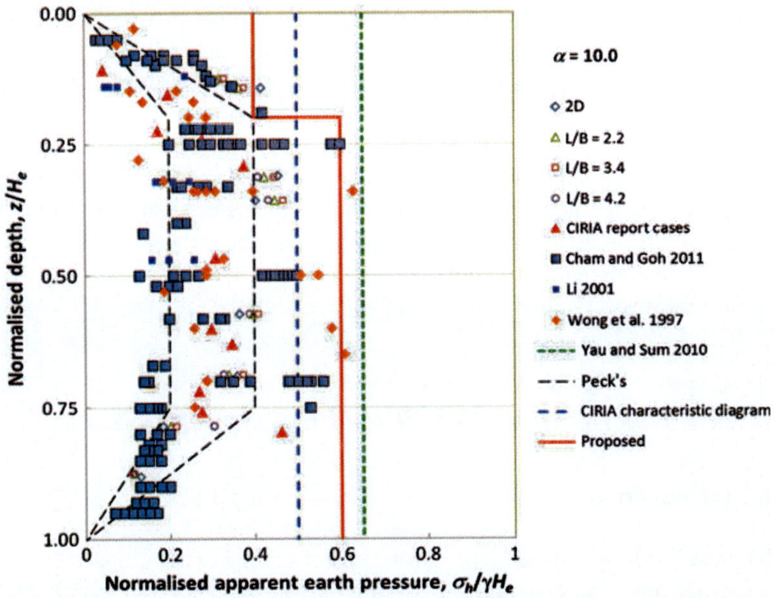

Fig. 3.13 Normalized apparent earth pressure versus normalized depth for stiff walls in stiff clays

Table 3.8 Recommended K_a values for stiff walls in clays

Clay type	Ka
Soft clay	0.45 ($z/H_e \leq 0.20$)
	0.9 ($0.2 \leq z/H_e \leq 1.0$)
Stiff clay	0.4 ($z/H_e \leq 0.20$)
	0.6 ($0.2 \leq z/H_e \leq 1.0$)

(denoted as solid symbols), for $z/H_e \geq 0.25$, the apparent earth pressure is generally less than $0.4\gamma H_e$, while for $z/H_e \geq 0.25$, the apparent earth pressure is generally less than $1.0\gamma H_e$. Numerical and field measurements indicate that Peck's APD (which was developed solely based on flexible walls) can sometimes underestimate the apparent earth pressure for soft clays with stiff walls. From Fig. 3.12, it can also be observed that the line proposed by Yau and Sum [20] overestimates the apparent earth pressure for $z/H_e \leq 0.25$ and underestimates the apparent earth pressure for $z/H_e \geq 0.25$. Thus, for excavations in soft clays with stiff walls, following the similar approach adopted by CIRIA for excavations with flexible wall systems, a threshold apparent earth pressure of $0.45\gamma H_e$ for $z/H_e \leq 0.2$ and threshold apparent earth pressure of $0.9\gamma H_e$ for $0.2 < z/H_e \leq 1.0$ is proposed, as indicated in Fig. 3.12, by the red solid lines.

For excavations in stiff clays with stiff wall systems, for the numerical results the apparent earth pressure is generally less than $0.4\gamma H_e$ for $z/H_e \leq 0.25$, while for $z/H_e \geq 0.25$, the apparent earth pressure is generally less than $0.5\gamma H_e$, which fits the CIRIA characteristic DPL of $0.5\gamma H_e$ well. Based on the measured strut forces from the case histories, for $z/H_e \leq 0.25$, most of the data are less than $0.5\gamma H_e$, while for $z/H_e \geq 0.25$, the apparent earth pressure is also generally less than $0.6\gamma H_e$, except for two points (with $0.63\gamma H_e$ and $0.61\gamma H_e$, respectively), which are slightly larger than $0.6\gamma H_e$ but within the limit of $0.65\gamma H_e$ as proposed by Yau and Sum [20]. Thus, for excavations in stiff clays with stiff walls, a threshold apparent earth pressure of $0.4\gamma H_e$ for $z/H_e \leq 0.2$ and threshold apparent earth pressure of $0.6\gamma H_e$ for $0.2 < z/H_e \leq 1.0$ are proposed, as indicated in Fig. 3.13, by the red solid lines.

Table 3.8 summarizes the recommended APD for excavations using stiff wall systems. It is expressed in the form of $K_a \gamma H_e$.

3.3.2 ADP Determination for Braced Excavation in Sands

3.3.2.1 Introduction

Most previous studies focused on performances of braced excavations in soft to stiff clays; there are only limited studies involving braced excavation in granular soils.

Juran and Elias [26] proposed a modified apparent pressure diagram based on in situ construction data in the sand. Their study indicated that the shape of the APD is almost the same as the stiff clay proposed by Terzaghi and Peck [4], as shown in Fig. 3.14a. Other APDs for dense sand and loose sand provided by Terzaghi and Peck [4] are shown in Fig. 3.14b and c, respectively.

Hsu et al. [27] summarized a number of twenty-five deep excavation cases in gravel formation in Taichung and obtained the distributions of APD at excavation depths of 15, 20, and 27 m. His research demonstrated that the apparent earth pressure is much less than that of Peck's APD. The coefficient of lateral earth pressure for H less than 15 m, between 15 and 20 m, and greater than 20 m are 0.07, 0.09, and 0.11, respectively.

Hsiung et al. [28] carried out FE analysis for the performance of braced excavations in loose to medium dense sand. The consistency of the results between measured data and the predicted values validates the reliability of FE analysis. It has been pointed out that the findings in their study generally agree well with the distributed prop load (DPL) suggested by Twine and Roscoe [9] as well as the APD for sand suggested by Peck [3] at depths from 3 to 9 m.

Chee [29] firstly compared the strut forces obtained from numerical analysis with the semi-empirically derived APD for braced excavations in the sand and investigated the influences of sand friction angle and stiffness on strut forces. The results indicate that the APD for sand generally overestimates strut forces for the top two levels while underestimates the loads of the lower levels. Increasing the sand stiffness alone did not lead to significantly lower strut forces. Nevertheless, increasing the sand friction angle, which represents a higher stiffness, can result in significantly lower strut forces.

Apart from the numerical analysis, a model test has been conducted by Nakai et al. [30], and the results were compared with FE analysis. It was found that the computed results agreed well with the model test results.

As discussed previously, the determination of strut forces for braced excavation in granular soils is still a hot issue due to limited research and the scarceness of reliable in situ monitoring data [31]. The main purpose of this section is to propose

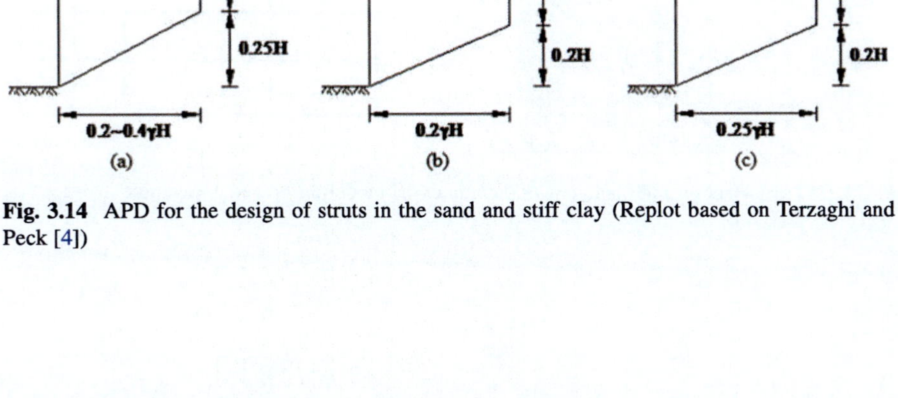

Fig. 3.14 APD for the design of struts in the sand and stiff clay (Replot based on Terzaghi and Peck [4])

the updated APD for medium rigid wall systems based on extensive plane strain and three-dimensional numerical analyses supplemented by additional measured data from case histories. The influential factors on APD include the granular thickness, soil strength and stiffness, wall embedded depth, and wall rigidity. Consequently, three types of granular soil types, i.e., medium dense sand, dense sand, and gravel with different geometrical properties are considered in this study. Some differences were observed between the numerical results and estimations from Peck's APD.

3.3.2.2 Numerical Scheme

The FE software PLAXIS 2D and PLAXIS 3D were used to perform the excavation simulations. Figure 3.15 shows a typical cross-section and plan view for the cases considered. The symbol x represents the horizontal distance from the center of the excavation. For each level of struts, the struts at $x = -2$ m and $x = 2$ m sections (when horizontal strut spacing was 4 m) are identical due to the symmetry of the struts and walers. Figure 3.15 also plots the embedded retaining wall together with a five-level strut system for $H_e = 17$ m. Hp refers to the wall's penetration depth, and other relevant design parameters shown in Fig. 3.2 are summarized in Table 3.9.

Considering that the relatively flexible wall was generally designed and constructed for braced excavations in the competent granular soil in engineering practice, two types of wall with four different stiffness values were considered for each soil type, as listed in Table 3.2. A stiffness coefficient α was utilized to represent walls with varying rigidities. The baseline bending stiffness *EI* used to develop the different models was 540,675 kNm2/m, which refers to a wall of medium rigidity

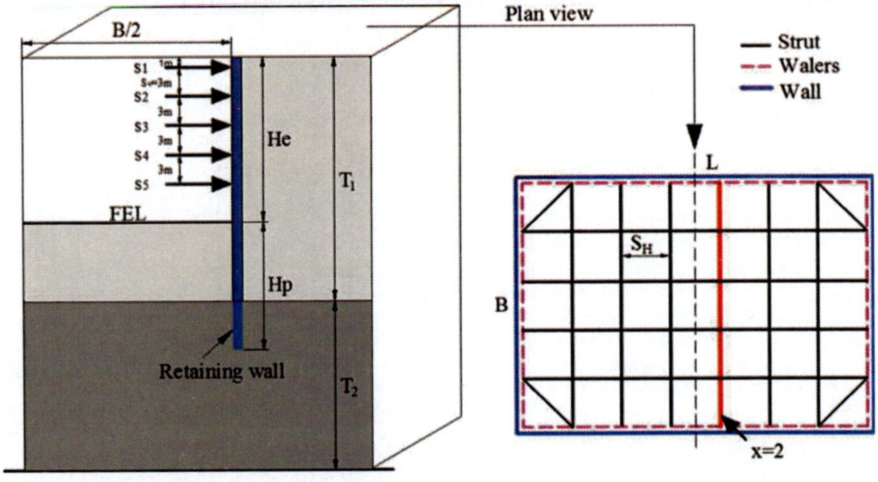

Fig. 3.15 Cross-section and plan view of the numerical model for braced excavation

Table 3.9 Geometric parameters

Parameter	Name	Geometrical properties
Excavation length (m)	L	36,60,84*
Excavation width (m)	B	30
Thickness of granular layer (m)	T_1	25
Thickness of stiff clay layer (m)	T_2	30
Horizontal strut spacing (m)	S_H	4,5
Vertical strut spacing (m)	S_V	3
Depth of final excavation (m)	H_e	14,17

*L = 36 m, 60 m, 84 m refers to L/B = 1.2, 2.0, 2.8, respectively. For each L/B, two types of wall with four different stiffness values are considered

based on the databases of Long [12] and Moormann [13]. Therefore, $\alpha = 1.0$ represents the cases with the medium rigid wall in this study. For flexible walls in granular soils, the baseline bending stiffness EI of 540,675 kNm^2/m was multiplied by smaller α values of 0.06, 0.1, and 0.2.

The shoring system of the braced excavation comprised of struts and walers; the details are the same as those in Sect. 3.3.1.2 and listed in Table 3.2.

The construction sequence is also similar to that in Sect. 3.3.1.2. For brevity, please refer to Table 3.3, although there are some differences in excavation levels (Table 3.10).

The hardening soil (HS) model was used to simulate the constitutive behavior of the granular soils. This model involves frictional hardening characteristics to model plastic shear strain when subjected to primary deviatoric loading, and cap hardening

Table 3.10 Wall properties

Parameters		Wall types			
		Flexible			Medium
Plane strain (2D) FE parameters					
		0.06	0.1	0.2	1.0
Wall stiffness EI (kNm^2/m)		3.224×10^4	5.407×10^4	1.081×10^5	5.407×10^5
Compressive stiffness EA (kN/m)		2.206×10^6	3.678×10^6	7.356×10^6	3.678×10^7
Poisson's ratio, ν		0.15	0.15	0.15	0.15
Three-dimensional (3D) FE parameters					
Young's Modulus (kPa)	E_1	5.252×10^6	8.754×10^6	1.751×10^7	8.754×10^7
	E_2	2.626×10^5	4.377×10^5	8.754×10^5	4.377×10^6
Shear Modulus (kPa)	G_{12}	2.627×10^5	4.378×10^5	8.757×10^5	4.379×10^6
	G_{13}	8.754×10^5	1.459×10^6	2.918×10^6	1.459×10^7
	G_{23}	2.626×10^5	4.377×10^5	8.754×10^5	4.377×10^6
Poisson's ratio	ν	0	0	0	0

Table 3.11 Hardening soil parameters used for finite-element modeling

Hardening soil parameter	Unit	Medium dense sand (drained)	Dense sand (drained)	Gravel (drained)	Stiff clay (undrained)
γ_{unsat}	kN/m^3	17	18.2	16	20
γ_{sat}	kN/m^3	19.8	20.3	20	20
$k_x = k_y = k_z$	m/day	1×10^{-8}	1×10^{-8}	1×10^{-4}	1×10^{-8}
E_{50}^{ref}	kN/m^2	30,000	48,000	60,000	14,847
E_{oed}^{ref}	kN/m^2	30,000	48,000	60,000	14,847
E_{ur}^{ref}	kN/m^2	90,000	144,000	250,000	44,540
C	kN/m^2	0	0	0	0.05
φ	°	34.3	38	44	33
Ψ	°	4.3	8	14	0
ν_{ur}	[-]	0.3	0.35	0.25	0.2
p^{ref}	kN/m^2	100	100	100	100
m	[-]	0.544	0.45	0.5	1
Knc 0	[-]	0.436	0.384	0.305	0.455
R_f	[-]	0.938	0.9	0.9	0.96
R_{inter}	[-]	0.8	0.8	0.8	1

to model plastic volumetric strain in triaxial compression loading conditions. Failure is still defined by the MC failure criteria. It should be noted that the hardening soil parameters for the sand and gravel considered in this parametric study were based on data extracted from Halilm [32], respectively. The stiff clay with average undrained shear strength c_u = 125 kPa was based on the Gault clay found in Cambridge[19]. The HS parameters for different soils are summarized in Table 3.11.

A total of 113 FE simulations using the hardening soil (HS) model were carried out to investigate the influences of soil types, excavation geometries, wall stiffness α, and strut level on the strut forces induced by excavation. The influence of the wall's embedded depth on strut forces was investigated with typical FE models forehand to choose the proper embedded-ratio H_p/H_e. For brevity, the numerical results of the study are presented with the main findings, as described in the next sections.

3.3.2.3 Results

To illustrate the influence of the wall's embedded depth on values of strut forces, sensitive analysis was performed with different ratios between wall embedded depth to excavation depth (H_p/H_e). Figure 3.16 shows the typical cases in medium dense sand with α = 1, L = 36 m, B = 30 m, T1 = 25 m, T2 = 30 m, SH = 4 m, and H_e = 17 m; Hp was assumed to be 17 m, 15 m, 13 m, 11 m, 9 m,7 m, and 5 m, where corresponding embedded-ratio H_p/H_e equals 1.0, 0.88, 0.76, 0.65, 0.53, 0.41, and

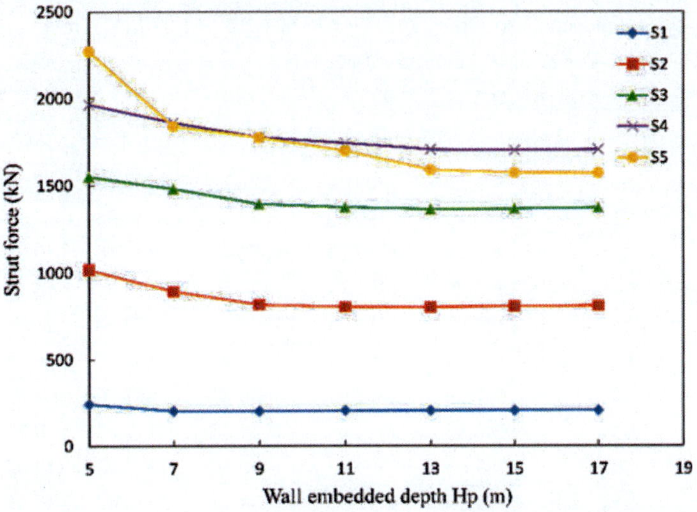

Fig. 3.16 Variations of strut forces versus increasing the wall embedded depth in medium dense sand

0.29, respectively. S1 denotes the first strut level and S5 represents the fifth level. As can be inferred from the diagrams illustrated in the figure, the general trend is that strut forces in strut S2 ~ S4 decrease markedly when H_p varies from 29%H_e to 53%H_e; strut forces in strut S5 are noticeably varied when H_p/H_e rises from 0.29 to 0.76 and slightly varies when H_p/H_e greater than 0.76. Strut forces in strut S1 change mildly when wall embedded depth varies from 29%H_e to 100%H_e, which is similar to the trend observed by Bahrami et al. [33]. In this section, H_p was fixed at the length of 13 m (where H_p/H_e is equal to 0.76) to study the effect of various wall rigidity and *L*/*B* ratios of strut forces in three types of soils.

The general trends of strut forces with varying wall rigidity and *L*/*B* ratios for the three granular soils are also presented. As Fig. 3.15 depicted, the struts that are the closest to the center of the excavation among all the struts at the same level are subjected to the greatest force. Therefore, only the strut forces at location $x = 2$ m (depicted schematically in red in Fig. 3.15) are presented and analyzed.

For brevity, only some of the main results from the 2D and 3D analyses are presented in Tables 3.12 and 3.13. Generally, the struts S3, S4, and S5 are subjected to larger compressive strut forces. The general trend was that the axial forces in struts S2 to S5 decrease with the increase of soil strength while increase with decreasing wall stiffness. It can be also observed from Tables 3.12 and 3.13 that the differences of maximum strut forces depend on the granular soil types, *L*/*B* ratios, and strut levels.

Figure 3.17 presents the maximum strut forces of different strut levels for various *L*/*B* ratios in dense sand (α = 0.06, 0.1, 0.2, and 1). In general, the strut forces calculated by 3D simulations are greater than results from 2D simulations. It can be found that there are discrepancies between 2D simulations and 3D simulations

Table 3.12 Typical maximum strut force for $H_e = 17$ m and $\alpha = 1.0$ ($S_H = 4$ m)

Soil type	Strut level	Maximum strut force (kN)				Max. difference between 2 and 3D (%)
		2D	3D			
			$L/B = 1.2$	$L/B = 2.0$	$L/B = 2.8$	
Medium dense sand	S1	186.22	203.24	191.22	203.15	9.14
	S2	740.53	799.82	673.24	769.94	8.01
	S3	1228.64	1363.31	1052.42	1321.16	10.96
	S4	1491.76	1706.41	1397.50	1678.91	14.39
	S5	1324.08	1589.24	1367.19	1537.54	20.03
Dense sand	S1	141.88	165.51	165.95	163.61	16.97
	S2	667.89	683.97	698.44	687.96	4.57
	S3	1046.53	1071.64	1144.05	1117.87	9.32
	S4	1247.63	1304.53	1410.36	1374.93	13.04
	S5	1129.53	1252.95	1338.17	1280.45	18.47
Gravel	S1	84.16	128.45	118.44	114.55	52.63
	S2	632.34	592.62	587.56	584.41	6.28
	S3	986.49	935.02	956.31	942.38	3.06
	S4	1188.79	1135.78	1174.26	1136.67	1.22
	S5	1051.24	917.33	1046.82	1003.39	0.42

Table 3.13 Typical maximum strut force for $H_e = 17$ m and $\alpha = 0.06$ ($S_H = 4$ m)

Soil type	Strut level	Maximum strut force (kN)				Max. difference between 2 and 3D (%)
		2D	3D			
			$L/B = 1.2$	$L/B = 2.0$	$L/B = 2.8$	
Medium dense sand	S1	148.89	176.42	181.17	182.97	22.89
	S2	928.36	1027.06	999.60	980.74	10.63
	S3	1372.98	1361.79	1404.45	1450.58	5.65
	S4	1611.11	1704.55	1703.64	1723.25	6.96
	S5	1698.13	1906.37	1918.09	1977.13	16.43
Dense sand	S1	174.69	220.79	219.29	224.22	28.35
	S2	877.85	971.27	939.49	898.32	10.64
	S3	1233.71	1263.22	1308.61	1248.18	6.07
	S4	1359.24	1402.10	1520.28	1436.74	11.85
	S5	1382.82	1850.74	1699.90	1567.95	33.83
Gravel	S1	125.72	174.92	176.16	177.04	40.82
	S2	751.95	693.12	679.06	691.98	7.82
	S3	1079.90	948.65	1023.13	1024.65	5.12
	S4	1230.48	1171.73	1211.93	1232.38	0.15
	S5	1605.14	1528.05	1424.34	1358.03	4.80

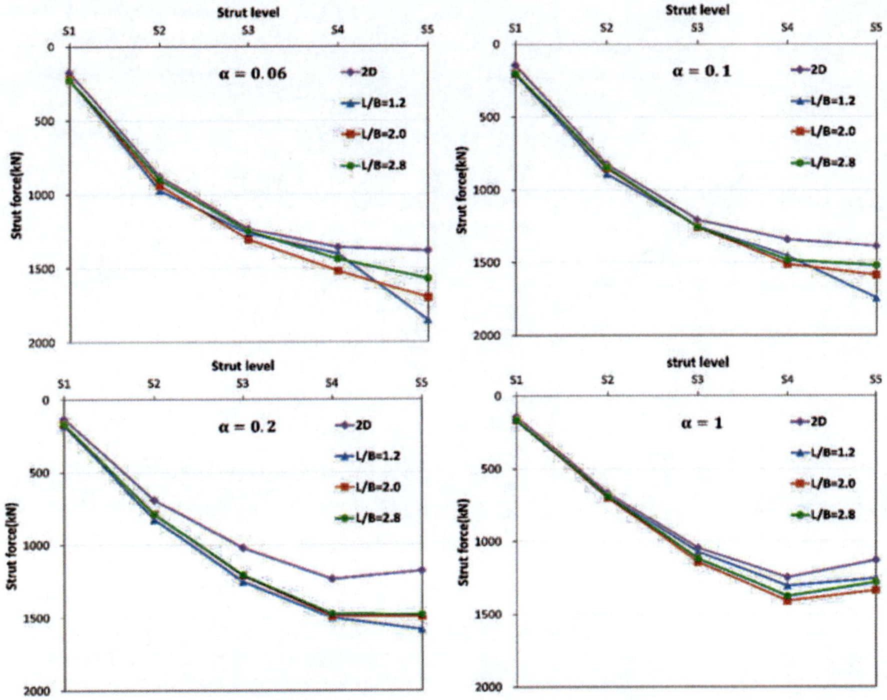

Fig. 3.17 Maximum strut forces for various *L/B* ratios in dense sand ($\alpha = 0.06$, 0.1, 0.2, and 1)

at each strut level. However, the distribution patterns of strut force obtained from different models are almost similar, i.e., the forces in struts S1 to S5 decrease with the increase of wall stiffness. Meanwhile, as the wall stiffness increases from $\alpha =$ 0.1. to $\alpha = 0.2$ and then to 1.0, the strut level with the greatest prop force changes from level S5 to S4. It is obvious from Fig. 3.18 that for a particular soil type, the maximum strut forces from S1 to S5 decrease with the increase of wall system stiffness. In addition, for the same wall rigidity, the maximum strut force decreases significantly from over 2000 kN for $\alpha = 0.06$ of dense sand to less than 1400 kN for $\alpha = 0.06$ of gravel. This again confirms that the strut forces depend on the soil type, the excavation geometries, the strut level, and the wall rigidity, and the resultant influences from these factors are complex.

3.3.2.4 Apparent Earth Pressure Envelope

The results of the 2D and 3D maximum strut forces for $H_e = 14$ m and 17 m were normalized to derive the APD for excavations in medium rigid walls in this study. The tributary area load distribution procedure proposed by Peck [3] was used. As an example to the case, Fig. 3.19 shows the assumed loading area of struts for $H_e =$ 17 m, where hsi represents the loading area depth for the corresponding strut.

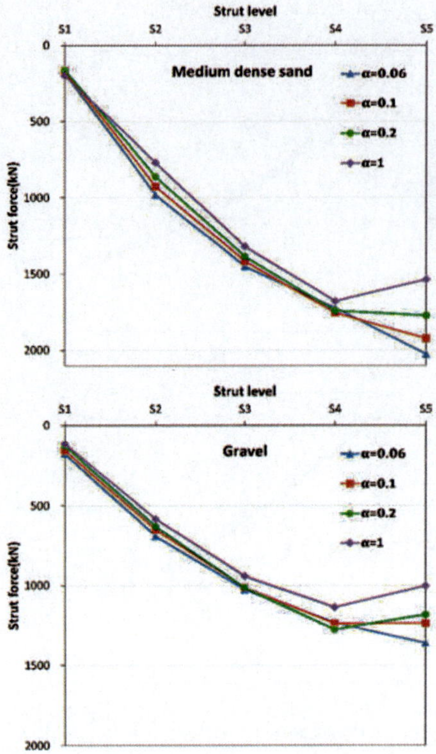

Fig. 3.18 Maximum strut forces for various in different soils (*L*/*B* = 2.8)

Fig. 3.19 Cross-section: assumed tributary loading areas of struts for $H_e = 17$ m

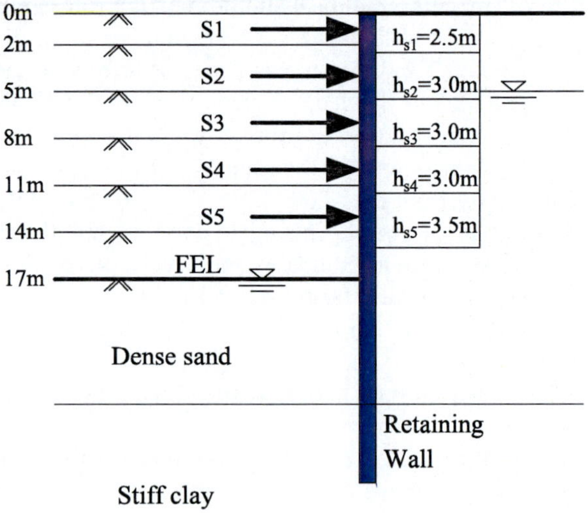

Only excavations in medium rigid walls ($\alpha = 1$) for different granular soil types are considered in view that the common use of these walls is for deep excavations in such soil types in actual projects. For comparison, Peck's APD and the CIRIA DPL recommendations, as well as various case histories from the literature as tabulated in Table 3.14, are plotted together with the numerical results, for better comparison.

The apparent earth pressure for medium rigid walls ($\alpha = 1$) is shown in Fig. 3.20 for dense sand and Fig. 3.21 for gravel, respectively. For cases with medium dense sand or flexible walls, due to the limited case histories reported, the calculation results are not included in the plot. For dense sand (Fig. 3.20), the dash line defines the upper bound of Peck's APD assuming the coefficient K_a is equal to 0.2. Represented by the dark blue solid line is the characteristic distributed prop load diagram reported by CIRIA. For gravel (Fig. 3.21), due to the limited studies on strut forces in gravel formations, the line proposed by Hsu [27] was plotted, where the maximum prop load line was $0.11\gamma H_e$, as illustrated by the green dash line.

For excavations in dense sand with medium stiff walls, as shown in Fig. 3.20, the numerical results (represented by open symbols) indicate that the apparent earth pressure is generally less than $0.1\gamma H_e$ for $z/H_e \leq 0.2$ while for $z/H_e \geq 0.2$, the apparent earth pressure is generally less than $0.35\gamma H_e$. Based on the measured strut forces from the case histories (denoted as solid symbols) listed in Table 3.8, for $z/H_e \leq 0.2$, the apparent earth pressure is generally less than $0.15\gamma H_e$, while for $z/H_e \geq 0.2$, it is less than $0.3\gamma H_e$. Based on the numerical results and field measurements, Peck's APD (which was developed based on medium rigid walls as well) underestimates the strut forces, especially for the normalized depth from $z/H_e = 0.2$ to $z/H_e = 0.8$.

It can also be observed in Fig. 3.20 that the diagram proposed by Twine and Roscoe [9] overestimates the apparent earth pressure for $z/H_e \leq 0.2$ while underestimates it for $z/H_e \geq 0.2$. Therefore, for braced excavations in dense sand with medium rigid walls, a threshold apparent earth pressure of $0.15\gamma H_e$ for $z/H_e \leq 0.2$ and a threshold apparent earth pressure of $0.3\gamma H_e$ for $z/ H_e \geq 0.2$ are proposed in this study, as indicated in Fig. 3.20 by the red solid lines.

For excavations in gravel with medium rigid walls, as shown in Fig. 3.21, the numerical results indicate that the apparent earth pressure is generally less than $0.1\gamma H_e$ for z/He ≤ 0.2, while for $z/H_e \geq 0.2$, the apparent earth pressure is generally less than $0.35\gamma H_e$. Based on the measured strut forces from the case histories in Table 3.14, for $z/H_e \leq 0.2$, most of the data are less than $0.1\gamma H_e$ except for one point (which is $0.25\gamma H_e$). While for $z/H_e \geq 0.2$, the apparent earth pressure is also generally less than $0.35\gamma H_e$, while two points (with $0.39\gamma H_e$ and $0.40\gamma H_e$, respectively.) are slightly over $0.35\gamma H_e$.

Considering that the case histories summarized by Twine and Roscoe [9] are not based on the pure gravel formations, the measured strut forces from these cases may sometimes deviate from the mean values of the strut forces in the pure gravel formations. Thus, for multi-propped excavations in gravel formations with medium rigid walls, a conservatively threshold distributed prop load of $0.2\gamma H_e$ for $z/H_e \leq 0.2$ and a threshold distributed prop load of $0.35\gamma H_e$ for $z/H_e \geq 0.2$ are proposed, as shown in Fig. 3.21 by the red solid lines.

Table 3.14 Summary of case histories

Case histories	Soil type	Unit weight (kN/m^3)	H_e (m)	Wall type	No. of strut levels	Horizontal strut spacing (m)	Vertical strut spacing (m)	References
New York (CIRIA report) C2	Gravel	17.7	8.6	KP (medium stiff wall)	3	3 6.1 6.1	3 2.2 2.2	[9]
Tokyo (CIRIA report) C9	Gravel	16.9	20.1	SP (medium stiff wall)	6	–	4.4 2.5 2.7 3 3 3.5	[9]
(CIRIA report) D13	Dense sand	19.6	21.3	KP (medium stiff wall)	4	3.6	5.3 4.6 4.6 4.5	[9]
Sao Paulo (CIRIA report) C8	Dense sand	20	18	KP (medium stiff wall)	5	–	3.5 3.2 3.1 3.4 3.3	[9]
Washington (CIRIA report) C7	Dense sand	18.9	18.3	KP (medium stiff wall)	5	4	2.9 3.4 3.3 3 3.6	[9]
Oakland (CIRIA report) C10	Dense sand	19.6	20.7	KP (medium stiff wall)	5	3.6	6.6 3.2 3 3.2 3.2	[9]
Munich (CIRIA report) C3	Gravel	19.6	11.9	KP (medium stiff wall)	4	2	2.4 3.1 2.7 2.5	[9]
(CIRIA report) D14a	Gravel	19.4	20	CC (medium stiff wall)	3	8.3	5.8 6.8 5.5	[9]
Toronto Transit Commission (TTC) subway structure	Dense sand	18–22	9–20	SP (medium stiff wall)	–	–	2.4–5.8	[34])
Taiwan, Taichung	Gravel	22.5	15–27		–	–	–	[27]

(continued)

Table 3.14 (continued)

Case histories	Soil type	Unit weight (kN/m^3)	H_e (m)	Wall type	No. of strut levels	Horizontal strut spacing (m)	Vertical strut spacing (m)	References
(CIRIA report) C4	Dense sand	16.7–16.9	10.7–11.8	KP (medium stiff wall)	4	2	2.3 3.2 2.6 2.0	[9]

Notations: KP = King post wall; SP = Sheet pile wall; CC: Concrete contiguous pile wall

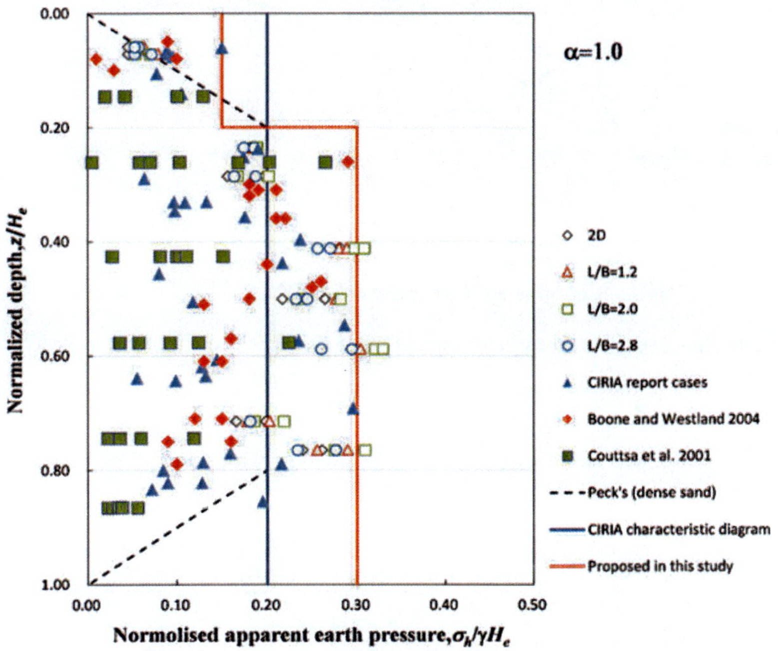

Fig. 3.20 Normalized apparent earth pressure versus normalized depth for medium rigid walls in dense sand

Noted that APDs for braced excavations in medium dense sand are not discussed in the present study for lack of relevant case histories. Table 3.15 summarizes the recommended APD in this study for braced excavations in dense sand and gravel using medium rigid wall systems, in the form of $K_a \gamma H_e$. The threshold distributed prop load of gravel is greater than that of dense sand for the scarceness of the pure gravel formation in the case histories considered.

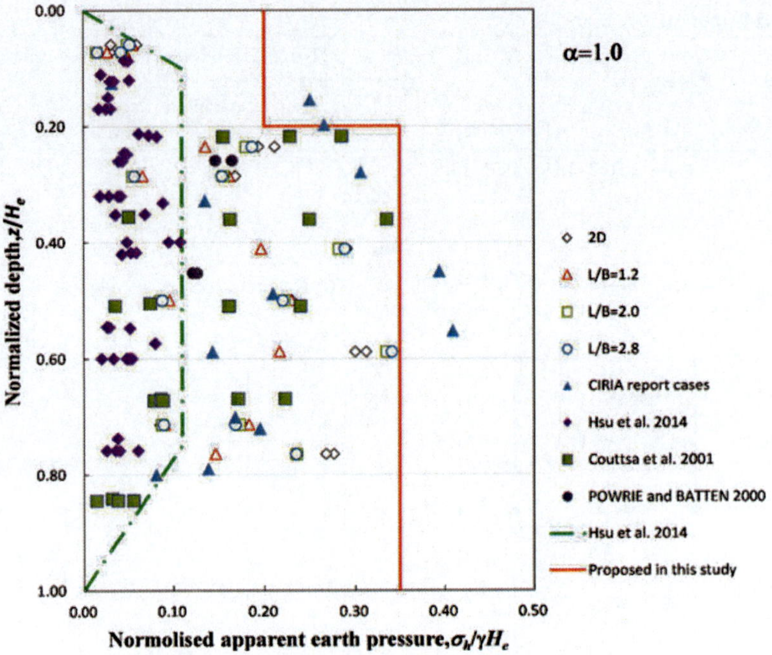

Fig. 3.21 Normalized apparent earth pressure versus normalized depth for medium rigid walls in gravel

Table 3.15 Recommended K_a values for medium rigid walls in granular soils

Soil type	K_a
Dense sand	0.15 ($z/H_e < 0.2$)
	0.30 ($0.2 \leq z/H_e \leq 1.0$)
Gravel	0.20 ($z/H_e < 0.2$)
	0.35 ($0.2 \leq z/H_e \leq 1.0$)

3.3.3 Summary and Conclusions

In this section, series of 2D and 3D FE simulations using the HS model for braced excavations in different soils (clays and granular soils) were carried out. It is concluded that for a particular soil type, the strut forces calculated by 3D simulations are generally greater than those from 2D. The general trend for strut forces is to decrease with increasing soil strength and increase with decreasing wall system rigidity. Combining the field measurements from a number of reported case histories and the numerical results, empirical charts for different soils have been proposed for braced excavations with various rigid wall systems as shown in Figs. 3.12, 3.13, 3.20, and 3.21. Recommended K_a values are also listed in Tables 3.8 and 3.15.

3.4 ADP for Braced Excavation in Anisotropic Clay

To analyze the strut forces and the apparent earth pressures for braced excavations in more realistic clay deposits, the complex characteristic of natural clay should be considered [35]. The NGI-ADP soil constitutive model was adopted in the finite-element analyses to assess the influence of various soil properties, as well as excavation geometrical and structural properties on the strut forces. Based on the numerical results, modified design envelopes (APDs) that consider the anisotropic behavior of clays are proposed for flexible and stiff walls.

3.4.1 Finite-Element Modeling

The plane strain finite-element method was adopted to analyze the braced excavation in anisotropic clay. The FE model comprises of supporting structures including a retaining wall and four levels of struts, and a 40 m thick soft clay layer overlying a 20 m thick stiff clay layer. Figure 3.22 shows the schematic cross-section of the excavation system. The wall was strutted at depths of 1, 3, 5, and 7 m below the original ground surface. The NGI-ADP constitutive model was used for the soft clay, while the Mohr–Coulomb (Undrained C) model was used for modeling the stiff clay. The soils are assumed to be subjected to undrained shearing during excavation. The excavation width B, the wall width b, the final excavation depth H_e, and the penetration depth of the wall below the formation level D are shown schematically in Fig. 3.22. The strut stiffness per meter *EA* is 6.1×10^5 kN/m. For simplicity, the elastic modulus of the wall E_{wall} is assumed to be constant and equal to 2.8×10^7 kPa, and the rigidity (stiffness) of the wall is studied by varying the wall width b. The K_0-value for both the soft clay layer and the stiff clay layer is 1.0. The ranges of

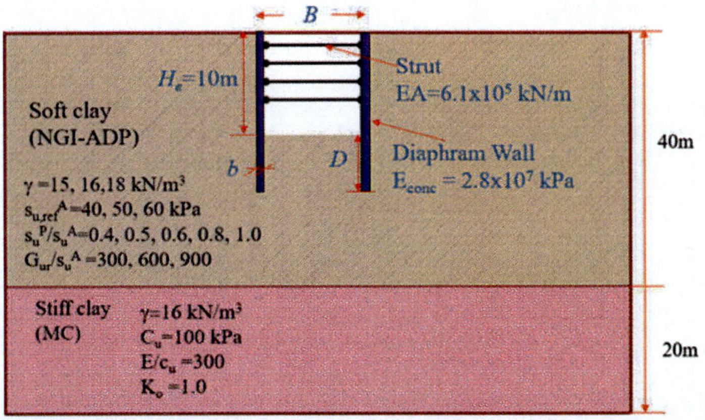

Fig. 3.22 Cross-sectional profile

three key varying soil parameters considered in this study including s_u^A (kPa), G_{ur}/s_u^A (−), and s_u^P/s_u^A (−) are also shown in Fig. 3.1. The range of s_u^P/s_u^A considered in this study is 0.4, 0.5, 0.6, 0.8, and 1.0, with $s_u^P/s_u^A = 1.0$ denoting that the clay is isotropic, and a smaller value of s_u^P/s_u^A indicate a higher degree of anisotropy of clay. The s_u^P/s_u^A ratio varies from 0.3 to 0.81. Therefore, this study looks at the range from 0.4 to 0.8 for anisotropic clay.

From symmetry, only half of the cross-section is considered. The right vertical boundary extends 60 m from the excavation to minimize the effects of the boundary restraints. The nodes along the left and right boundaries were constrained horizontally and the bottom boundary was constrained both vertically and horizontally. The soil was modeled using 15-noded triangular elements, the structural elements of the wall were assumed to be linearly elastic and modeled by 5-noded beam elements, and the struts were represented by 3-noded bar elements. The interface simulated between the wall and the soil was represented by 15-noded triangular elements and $R_{inter} = 1.0$.

The parametric study was carried out with an emphasis on the strut forces and the apparent earth pressure for braced excavation in anisotropic clay, using the NGI-ADP model for the soft clay layer. The apparent earth pressure is determined by dividing the strut force by the tributary area, using the approach proposed by Terzaghi and Peck [4]. The properties of the soft and stiff clay are listed in Tables 3.16 and 3.17, respectively. In this study, based on the recommendations of Brinkgreve et al. [15], the $s_u^{C,TX}/s_u^A$ and τ_0/s_u^A ratios are set at their default values of 0.99 and 0.7, respectively. The structural properties of the support system are shown in Table 3.18. A total of 270 different finite-element analyses were carried out in this study.

In this study, two different wall rigidities are considered, a flexible wall with $\ln(S) = 4.76$ and a stiff wall with $\ln(S) = 10.13$. A wall stiffness coefficient α is introduced to represent the walls with different rigidities. The baseline bending stiffness $(EI)_{\text{wall}}$ for the analysis is 5.04×10^5 kNm^2/m. Therefore, $\alpha = 1.0$ represents

Table 3.16 NGI-ADP soil model properties of soft clay (undrained C)

Parameters and units	Values
$s_u^{C,TX}/s_u^A(-)$	0.99
$\gamma_f{}^C$ (%)	0.75
$\gamma_f{}^E$ (%)	3.5
$\gamma_f{}^{DSS}$ (%)	1.735
R_{inter} (−)	1.0
$s_u^{DSS}/s_u^A(-)$	$(1 + s_u^P/s_u^A)/2$
τ_0/s_u^A (−)	0.7
υ_u (−)	0.495
$s_u^P/s_u^A(-)$	0.4, 0.5, 0.6, 0.8, 1.0
s_u^AkPa	40, 50, 60
$G_{ur}/s_u^A(-)$	300, 600, 900
γ(kN/m^3)	15, 16, 18

Table 3.17 Stiff clay properties (undrained C)

Parameters	Values	Units
γ	16	(kN/m^3)
c_u	100	(kPa)
E/c_u	300	(-)
ϕ_u	0	(°)
$K_O = 1 - \sin\phi$	1.0	(-)
υ_u	0.495	(-)
R_{inter}	1.0	(-)

Table 3.18 Parameters of the support system

Parameter and units	Ranges
Horizontal strut spacing $L_{spacing}$ (m)	4
Strut stiffness per meter EA (kN/m)	6.1×10^5
Elastic modulus of wall E_{wall} (kPa)	2.8×10^7
System stiffness ratio ln(S)*	4.76, 10.13
Bending stiffness of wall $(EI)_{wall}$ (kN·m^2)	1.90×10^4, 4.03×10^6
Wall width b (m)	0.2, 1.2
Wall stiffness coefficient $\alpha^{\#}$ (-)	0.04, 8
Excavation width B (m)	20
Final excavation depth H_e (m)	10
Wall penetration depth D (m)	5, 10, 15

*in which $S = (EI)_{wall}/\gamma_w h_{avg}{}^4$, where γ_w is the unit weight of water and h_{avg} is the average vertical strut spacing
#$\alpha = (EI)_{wall}/(5.032 \times 10^6)$, in which 5.032×10^6 is the baseline of bending stiffness

a wall with medium stiffness. Consequently, the finite-element analyses with $b =$ 1.2 m correspond to a stiff wall with α= 8, and flexible walls are modeled with $b =$ 0.2 m and $\alpha = 0.04$.

3.4.2 *Results and Analyses*

3.4.2.1 Strut Forces and APD for Excavation with Stiff Walls

The numerical study by Goh et al. [21] indicated that for excavations in isotropic soft clays with stiff walls, the normalized apparent earth pressure (i.e., the horiz.pressure/γH_e) is generally less than $0.3\gamma H_e$ for $z/H_e \leq 0.25$, where z is the depth of the wall below the top of the wall while for $z/H_e \geq 0.25$, the apparent earth pressure is less than $0.9\gamma H_e$. As mentioned previously, the apparent earth pressure is determined by dividing the strut force by the tributary area, using the approach proposed by Terzaghi and Peck [4]. Similar trends were observed for measured strut forces from the case histories [22], where for $z/H_e \leq 0.25$, the apparent earth pressure is generally less than $0.4\gamma H_e$, while for $z/H_e \geq 0.25$, the apparent earth pressure is less than $1.0\gamma H_e$. Figure 3.23 illustrates that the s_u^P/s_u^A ratio does affect the apparent earth pressure. The FE results of apparent earth pressure for $s_u^P/s_u^A = 0.4, 0.5, 0.6, 0.8,$ and 1.0 are consistent with the conclusions of above literatures.

Figure 3.24 shows the influence of the magnitude of s_u^A on the apparent earth pressures. As expected, the larger s_u^A results in smaller apparent earth pressures. The same trends as for Fig. 3.4 are observed. The apparent earth pressures for the isotropic clay with $s_u^P/s_u^A = 1.0$ are smaller than for the anisotropic clay with $s_u^P/s_u^A = 0.5$. For $s_u^A = 40$ kPa, the distribution of apparent earth pressures for cases with $s_u^P/s_u^A = 1.0$ and 0.5 is quite different. For $s_u^P/s_u^A = 0.5$, as a consequence of the significantly larger wall deflections, the force acting on the two lower struts is significantly larger compared to the other cases.

Figure 3.25 shows the influence of the soil stiffness ratio G_{ur}/s_u^A on the apparent earth pressures for $s_u^A = 40$ kPa. The range of G_{ur}/s_u^A considered is 300, 600, and 900, and as expected, the apparent earth pressures decrease with increasing G_{ur}/s_u^A. Comparing the isotropic and anisotropic cases, the same trends are observed where the force acting on the strut is significantly larger for $s_u^P/s_u^A = 0.5$, compared to the isotropic cases, because of the significantly larger wall deflection.

The penetration depth has little influence on the strut forces for stiff walls and flexible walls. The results of $s_u^P/s_u^A = 0.5$ and 1.0 cases with $D = 5$, 10, and 15 m show that the differences between the strut forces are minimal.

The apparent pressure diagrams (APDs) proposed by Terzaghi and Peck have been widely used in the design of the supporting strut system of braced excavation.

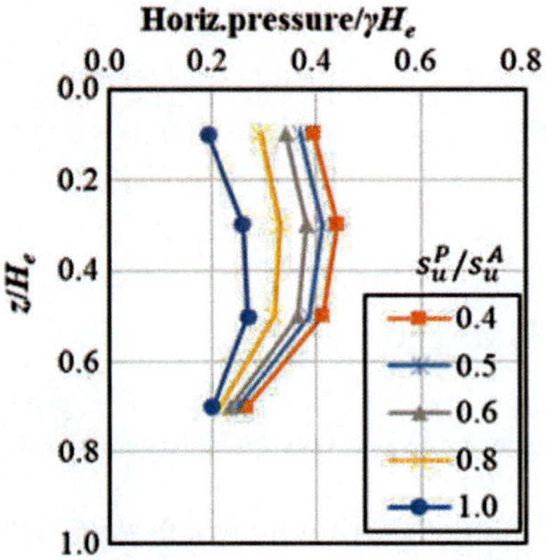

Fig. 3.23 Apparent earth pressures for $s_u^P/s_u^A = 0.4,\ 0.5,\ 0.6,\ 0.8,$ and 1.0 ($B/H_e = 2$, $D/H_e = 1.5$, $\gamma = 16$ kN/m^3, $s_u^A = 50$ kPa, $G_{ur}/s_u^A = 600$, and $\alpha = 8$)

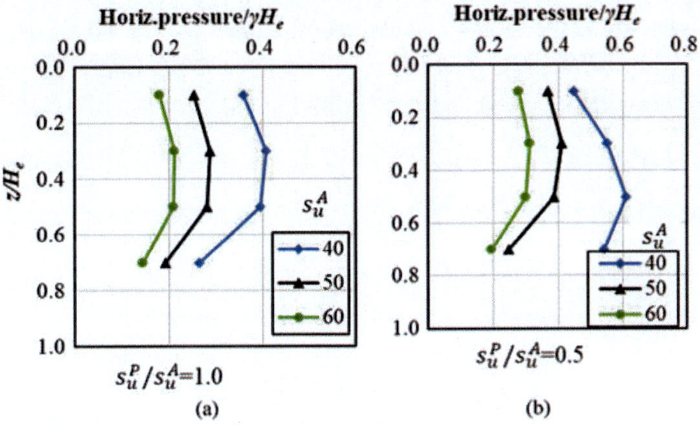

Fig. 3.24 Apparent earth pressures for $s_u^A = 40$, 50, and 60 kPa ($B/H_e = 2$, $D/H_e = 1.5$, $\gamma = 16$ kN/m^3, $G_{ur}/s_u^A = 600$, and $\alpha = 8$)

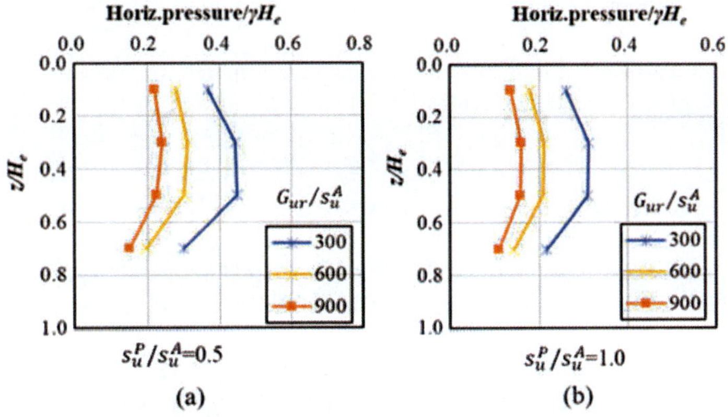

Fig. 3.25 Apparent earth pressures for $G_{ur}/s_u^A = 300$, 600, and 900 ($B/H_e = 2$, $D/H_e = 1.5$, $\gamma =$ 16 kN/m^3,$s_u^A = 60$ kPa, and $\alpha = 8$)

However, the study by Goldberg et al. [36] suggested that Peck's APD underestimated the apparent earth pressures for braced excavations. Following Goldberg et al. [36], numerous attempts have been made to evaluate these APDs from field measurements and finite-element simulations for different excavation support systems, particularly for those involving diaphragm walls [21].

In this study, the APD results of the numerical analyses for anisotropic clays will be compared with the distributed prop load (DPL) design envelope which is illustrated in Fig. 3.26, where γ_{ave} is the average unit weight of the soil over the retained height of the excavation H_e. Although various factors are likely to influence the APD, such as the excavation geometry, wall and strut stiffness, clay thickness, and clay strength and stiffness, this study only explores the influence of the anisotropy of clays on the

Fig. 3.26 Characteristic DPL Diagrams for soft clay with **a** stable base = adequate basal heave FS, **b** unstable base = wall needs to penetrate the competent layer to improve basal heave stability (*T/B* < 0.5; *T/H* < 0.8)

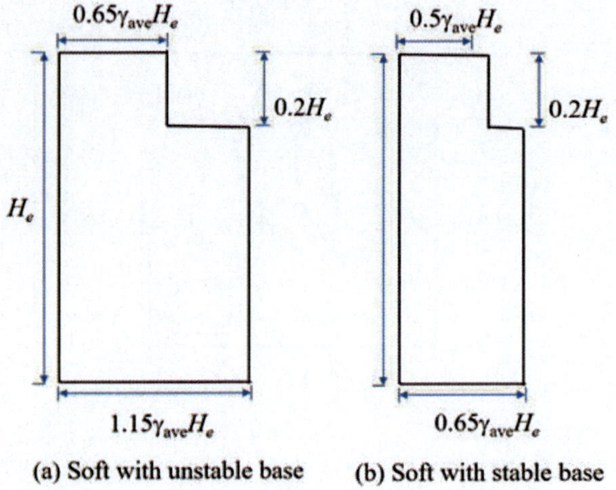

APD through a series of plane strain finite-element analysis. The numerical results which are plotted alongside the DPL design envelopes in Fig. 3.27 indicate that in some of the anisotropic clay cases the DPL underestimates the forces acting on the struts for excavations with stiff walls.

3.4.2.2 Strut Forces and APD for Excavation with Flexible Walls

Previous studies have demonstrated that the forces acting on the struts are affected by the wall stiffness. As the wall stiffness is reduced, the strut forces and apparent earth pressures decrease. For example, the study of Hashash and Whittle [7] indicated that increased wall flexibility causes a significant reduction in apparent earth pressure and was in close agreement with the design envelope of Peck [3]. Figure 3.28 shows the influence of the s_u^P/s_u^A on the apparent earth pressures. The apparent earth pressures are reduced significantly for the flexible wall when compared with the stiff wall and are consistent with the conclusion of Hashash and Whittle [7] For Fig. 3.28, only cases with s_u^A = 50 kPa and $s_u^P/s_u^A \geq 0.4$ are presented, as the numerical analyses indicated collapse (instability) had occurred for the case with s_u^A = 40 kPa and s_u^P/s_u^A = 0.4. As shown in Fig. 3.28, the apparent earth pressures are increased with s_u^P/s_u^A increases. As in the cases for stiff walls discussed previously, these results indicate that the use of isotropic APD can result in an underestimation of the struts forces for excavation in anisotropic clays for flexible walls.

Figure 3.29 shows the effects of s_u^A on the apparent earth pressures. As in the case of the stiff wall, the larger s_u^A results in smaller apparent earth pressures, and as the degree of anisotropy increases (i.e., smaller s_u^P/s_u^A ratio), the force acting on the struts increases.

Figure 3.30 shows the influence of the soil stiffness ratio G_{ur}/s_u^A on the apparent earth pressures for s_u^A = 40 kPa. The results indicate the same trends as for the stiff

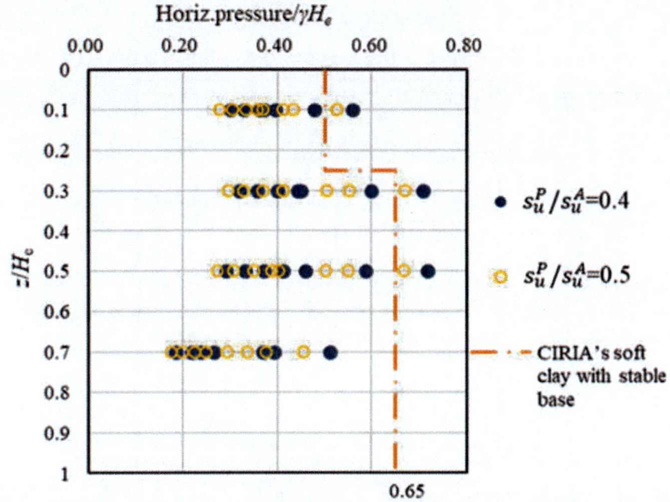

(a) s_u^P/s_u^A=0.4 and 0.5

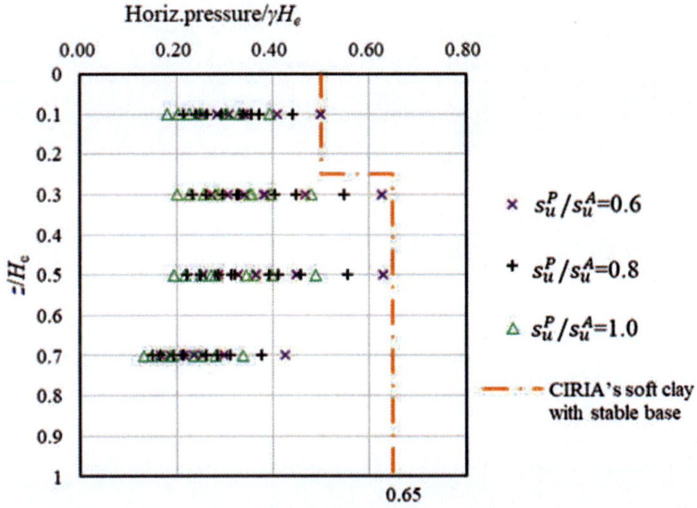

(b) s_u^P/s_u^A≥0.6

Fig. 3.27 Comparision of the CIRIA's DPL and APD for anisotropic soft clay for stiff wall ($B/H_e = 2$; $D/H_e = 1.5$; $s_u^A = 50$ kPa; $\gamma = 15$, 16, and 18 kN/m^3; $G_{ur}/s_u^A = 300$, 600, and 900; and $\alpha = 8$)

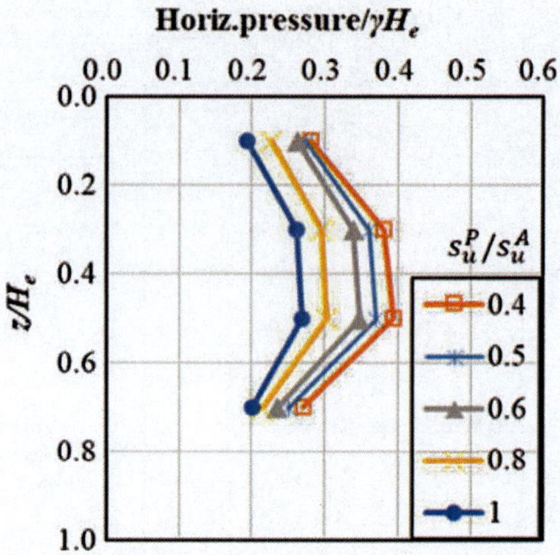

Fig. 3.28 Apparent earth pressures for $s_u^P/s_u^A = 0.5, 0.6, 0.8$, and 1.0 ($B/H_e = 2$, $D/H_e = 1.5$, $\gamma = 16$ kN/m^3, $s_u^A = 50$ kPa, $G_{ur}/s_u^A = 600$, and $\alpha = 0.04$)

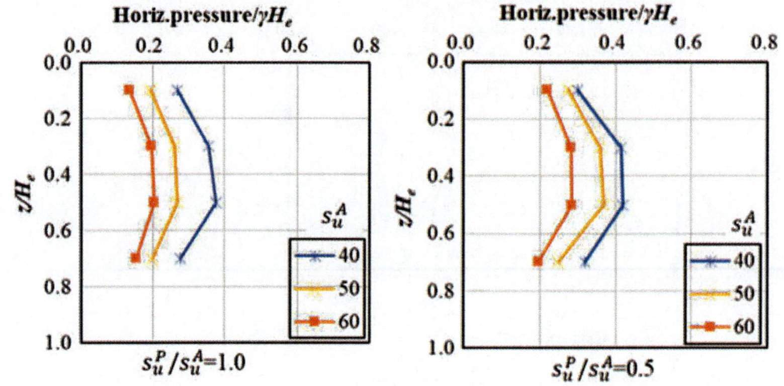

Fig. 3.29 Apparent earth pressures for $s_u^A = 40$, 50, and 60 ($B/H_e = 2$, $D/H_e = 1.5$, $\gamma = 16$ kN/m^3, $G_{ur}/s_u^A = 600$, and $\alpha = 0.04$)

wall with an increase in the apparent earth pressures as G_{ur}/s_u^A decreases, due to the corresponding larger wall deflections. The penetration depth has little influence on the strut forces for flexible walls as the same trends with the stiff walls.

The numerical results of the apparent pressure for flexible walls in Fig. 3.31 show the same trends for the stiff wall. The results indicate that in some of the anisotropic clay cases the DPL underestimates the forces acting on the struts.

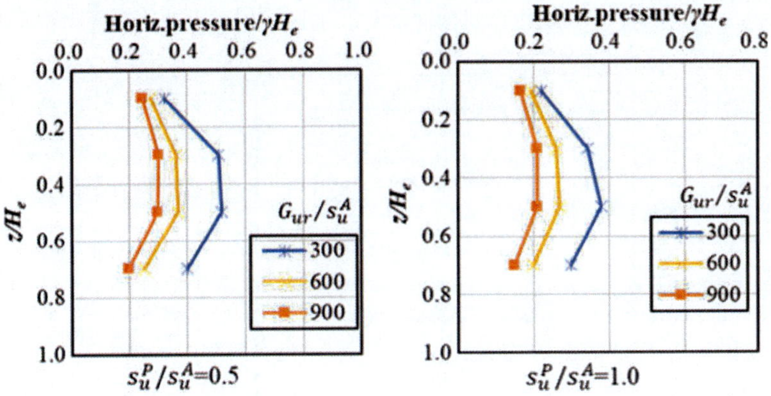

Fig. 3.30 Apparent earth pressures for $G_{ur}/s_u^A = 300$, 600, and 900 ($B/H_e = 2$, $D/H_e = 1.5$, $\gamma = 16$ kN/m^3, $s_u^A = 40$ kPa, and $\alpha = 0.04$)

3.4.2.3 Proposed APD with Consideration of Anisotropy

Based on the previous findings in Figs. 3.27 and 3.31, to consider clay anisotropy, it is proposed that the original DPL design envelopes be modified as shown in Fig. 3.32. Different values of k_1 and k_2 for stiff walls and flexible walls are proposed as shown in Fig. 3.33 and Table 3.19. The k values decrease with an increase in the undrained shear strength ratio s_u^P/s_u^A. Besides, smaller k_1 and k_2 values are recommended for flexible walls compared with stiff walls.

3.4.3 Summary and Conclusions

This part presents the results of plane strain finite-element analyses using the NGI-ADP soil constitutive model to assess the strut forces and apparent earth pressures for excavations in soft clays. The influence of soil strength and stiffness properties were also considered. The results show that the soil parameters s_u^P/s_u^A, G_{ur}/s_u^A, and s_u^A have a significant influence on the strut forces and apparent earth pressures. The results also indicate that the clay anisotropy has a more significant influence on the stiff walls than the flexible walls, because of the higher wall stiffness. This study analyzed and compared apparent earth pressures for flexible and stiff walls and proposed modified APDs for flexible and stiff walls with consideration of the anisotropic behavior of the clays.

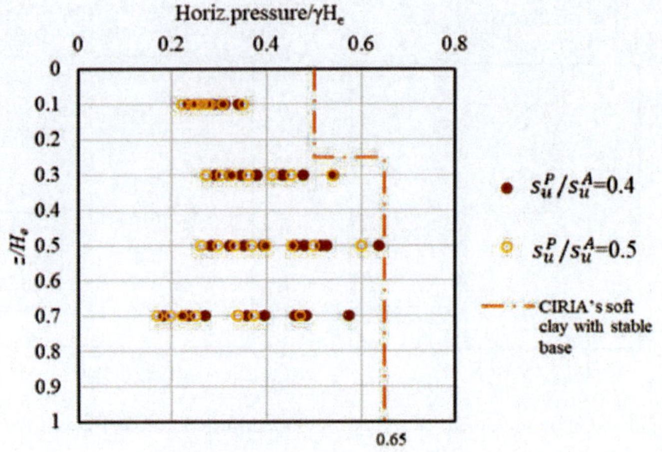

(a) s_u^P/s_u^A=0.4 and 0.5

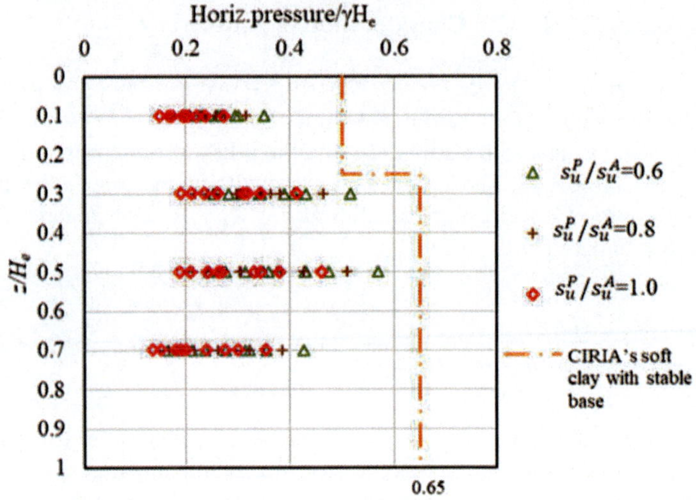

(b) $s_u^P/s_u^A \geq 0.6$

Fig. 3.31 Comparison of the CIRIA's DPL and APD for anisotropic soft clay for flexible wall ($B/H_e = 2$; $D/H_e = 1.5$; $s_u^A = 50$ kPa; $\gamma = 15$, 16, and 18 kN/m^3; $G_{ur}/s_u^A = 300$, 600, and 900; and $\alpha = 0.04$)

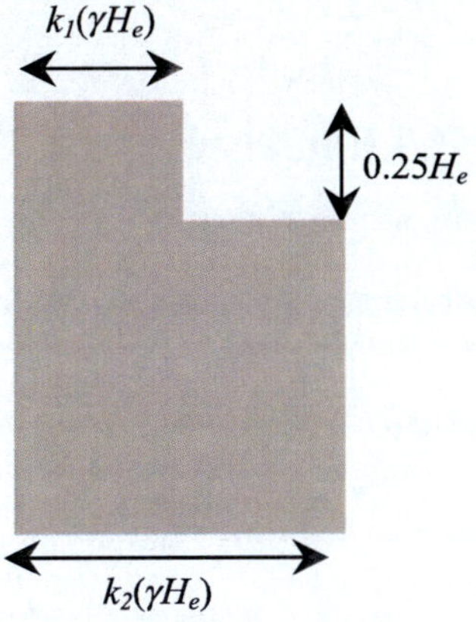

Fig. 3.32 The geometry of the proposed APD

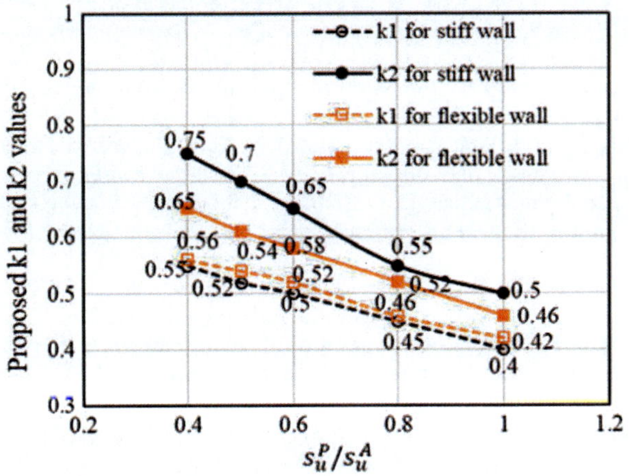

Fig. 3.33 Proposed k_1 and k_2 values of APD for anisotropic soft clay ($B/H_e = 2$; $D/H_e = 1.5$; $s_u^A = 40,60$, and 80 kPa; $\gamma = 15$, 16, and 18 kN/m^3; $G_{ur}/s_u^A = 300$, 600, and 900; $\alpha = 0.04$ and 8.0)

Table 3.19 Mathematical equations for k_1 and k_2

	k_1	k_2	s_u^P/s_u^A
Stiff wall	$k_1 = -0.25(s_u^P/s_u^A) + 0.65$	$k_2 = -0.43 (s_u^P/s_u^A) + 0.91$	0.4 – 1.0
Flexible wall	$k_1 = -0.24(s_u^P/s_u^A) + 0.66$	$k_2 = -0.31(s_u^P/s_u^A) + 0.77$	0.4 – 1.0

References

1. Rankine, W. M. J. (1857). On stability on loose earth. Philosophic Transactions of Royal Society.
2. Coulomb, C. A. (1776). Essaisurune application des regles de maximis et minimis a quelques problemes de statique. Mem. Roy. des Sciences.
3. Peck, R. B. (1969). Deep excavation and tunneling in soft ground. In: *Proceedings of the 7th International Conference on soil Mechanics and Foundation Engineering* (pp. 225–290). Mexico City.
4. Terzaghi, K., & Peck, R. B. (1967). *Soil Mechanics in engineering practice.* John Wiley & Sons.
5. Ou, C. Y. (2006). *Deep excavation: Theory and practice.* Taylor and Francis.
6. Chang, J. D., & Wong, K. S. (1996). Apparent pressure diagram for braced excavations in soft clay with diaphragm wall. In: *Proceedings of International Symposium on Geotechnical Aspects of Underground Construction in Soft Ground* (pp. 87–92). A.A. Balkema.
7. Hashash, Y. M. A., & Whittle, A. J. (2002). Mechanism of load transfer and arching for braced excavation in clay. *Journal of Geotechnical and Geoenvironmental Engineering, 128*, 198–197.
8. Wong, I. H., Poh, T. Y., & Chuah, H. L. (1997). Performance of excavations for depressed expressway in Singapore. *Journal of Geotechnical and Geoenvironmental Engineering, 123*, 617–625.
9. Twine, D., & Roscoe, H. (1999). Temporary propping of deep excavations-guidance on design. CIRIA C517, CIRIA London.
10. Finno, R. J., Blackburn, J. T., & Roboski, J. F. (2007). Three-dimensional effects for supported excavations in clay. *Journal of Geotechnical and Geoenvironmental Engineering, 133*, 30–36.
11. Bryson, L. S., & Zapata-Medina, D. G. (2012). Method for estimating system stiffness for excavation support walls. *Journal of Geotechnical and Geoenvironmental Engineering, 138*, 1104–1115.
12. Long, M. (2001). Database for retaining wall and ground movements due to deep excavations. *Journal of Geotechnical and Geoenvironmental Engineering, 127*, 203–224.
13. Moormann, C. (2004). Analysis of wall and ground movements due to deep excavations in soft soil based on a new worldwide database. *Soils and Foundations, 44*, 87–98.
14. Clough, G. W., Smith, E. M., & Sweeney, B. P. (1989). Movement control of excavation support systems by iterative design. In: *Proceedings of Foundation Engineering Congress on Current Principles and Practices* (pp. 869–884). New York.
15. Brinkgreve, L., Swolfs, W., & Engin, E. (2017). *Plaxis manual.* PLAXIS BV.
16. Hsieh, P. G., & Ou, C. Y. (1998). Shape of ground surface settlement profiles caused by excavation. *Canadian Geotechnical Journal, 35*, 1004–1017.
17. Finno, R. J., Bryson, S., & Calvello, M. (2002). Performance of a stiff support system in soft clay. *Journal of Geotechnical and Geoenvironmental Engineering, 128*, 660–671.
18. Ou, C. Y., Liao, J. T., & Lin, H. D. (1998). Performance of diaphragm wall constructed using Top-Down method. *Journal of Geotechnical and Geoenvironmental Engineering, 124*, 798–808.
19. Ng, C. W. W. (1992). An evaluation of soil - structure interaction associated with a multi - propped excavation. Ph D Thesis University of Bristol 13, 31–35.
20. Yau, P. K. F., Sum, A. H. L. (2010). An analytical review of excavation and lateral support, case history in Hong Kong. In: *Proceedings of the 30th annual seminar geotechnical division. The Hong Kong Institution of Engineers* (pp. 47–53). Hong Kong.
21. Goh, A. T. C., Zhang, F., Zhang, W. G., & Chew, O. Y. S. (2017). Assessment of strut forces for braced excavation in clays from numerical analysis and field measurements. *Computers & Geotechnics, 86*, 141–149.
22. Cham, W. M., Goh, K. H. (2011). Observed apparent pressure diagrams from actual strut monitoring of excavations in Circle Line project, In: *Proceedings of Underground Singapore* (pp. 289–297). Singapore,.

23. Li, W. (2001). *Braced excavation in old alluvium in Singapore*. Nanyang Technological University.
24. Lau, C. S., Chiu, S. L., Lo, K. L., Chu, K. K. N. (2010). Ground response in deep excavation in soft soil in Shanghai, In: *Proceedings of the 30th annual seminar geotechnical division. The Hong Kong Institution of Engineers* (pp. 149–161). Hong Kong.
25. Jadhav, A. S. (2003). Field measurements of strut loads in LTA contract C907. In: *Proceedings of underground singapore* (pp. 267–276). Singapore.
26. Juran, I., & Elias, V. (1987). Soil nailed retaining structures: Analysis of case histories. Soil Improvement.
27. Hsu, S. C., Huang, Y. P., & Cheng, T. M. (2014). Earth Pressure Distribution for Deep Excavations in Gravel Formations.
28. Hsiung, B. C. B., Yang, K. H., Aila, W., & Hung, C. (2016). Three-dimensional effects of a deep excavation on wall deflections in loose to medium dense sands. *Computers and Geotechnics, 80*, 138–151.
29. Chee, B. J. S. (2014). Analysis of strut forces for braced excavations in sand. CEE Student Reports (FYP/IA/PA/PI), Nanyang Technological University.
30. Nakai, T., Kawano, H., Murat, K., Banno, M., & Hashimoto, T. (1999). Model tests and numerical simulation of braced excavation in sandy ground: influences of construction history, wall friction, wall stiffness, strut position and strut stiffness. *Soils and Foundations, 39*, 1–12.
31. Zhang*, W., Hou, Z., Anthony T.C. Goh, & Zhang, R. (2019). Estimation of strut forces for braced excavation in granular soils from numerical analysis and case histories. Comput Geotech 106, 286–295.
32. Halim, D. (2008). *Effect of excavation on performance of adjacent buildings*. Nanyang Technological University.
33. Bahrami, M., Khodakarami, M. I., & Haddad, A. (2018). 3D numerical investigation of the effect of wall penetration depth on excavations behavior in sand. *Computers and Geotechnics, 98*, 82–92.
34. Boone, S. J., & Westland, J. (2004). Design, construction, and performance of a deep braced excavation. In: *International conference on Case Histories in Geotechnical Engineering* 29.
35. Zhang, R. H., Goh, A. T. C., Li, Y. Q., Hong, L., & Zhang, W. G. (2021). Assessment of apparent earth pressure for braced excavations in anisotropic clay. *Acta Geotechnica*. https://doi.org/10.1007/s11440-020-01129-x
36. Goldberg, D. T., Jaworski, W. E., & Gordon, D. (1976). Lateral support systems and underground. volumn II. Design Fundamentals Construction.

Chapter 4
Retaining Wall Deflection

4.1 Introduction

One of the key concerns in constructing an underground facility within a built-up environment is the impact of the associated ground movements on adjacent buildings [1]. For excavations in ground that comprises of thick soft clays overlying stiff clay, braced walls are usually used to minimize ground movements. It is common to extend the wall length into the stiff-clay layer to prevent basal heave failure and to reduce the movement of the wall toe. On the other hand, wall deflection is a critical factor used to estimate safety during construction. To ensure the serviceability limit state is satisfied, a common design criterion is to limit the maximum wall deflection to a fraction of the excavation depth H_e, typically in the range of 0.5–1.5% of H_e. Unnecessary severe restrictions may lead to uneconomic design. Therefore, reliable estimates of wall deflections under working conditions are essential [2, 3].

This chapter will review the factors influencing on retaining wall deflection, as well as analysis methods. Then, some predictive models are developed.

4.2 Triggering Factors

Retaining wall deformation arises from:

(1) excavation in front of the wall (Sect. 4.2.1)
(2) groundwater flow resulting in loss of ground and consolidation caused by changes in water pressures due to seepage through and/or around the wall (Sect. 4.2.2)
(3) other site-specific sources of movements, such as:

- ground improvement installations such as grouting (compensation or permeation) in or around the site

W. Zhang and H. Liu, *Design of Deep Braced Excavation and Earth Retaining Systems Under Complex Built Environment*,
https://doi.org/10.1007/978-981-16-5320-9_4

- installation of excavation access ramps
- removal of temporary sheet piles
- installation of temporary anchors, etc.

4.2.1 Excavation in Front of the Wall

For any given applied loading conditions (actions) behind an embedded retaining wall, ground movements arising from the excavation in front of the wall are influenced by:

(1) stress changes due to wall installation and excavation
(2) excavation geometry
(3) ground stiffness and strength
(4) changes in groundwater conditions
(5) type and stiffness of the wall and its support system
(6) construction sequence and methods
(7) quality of workmanship

Although any one of these factors may control the overall movement of a supported excavation, this complex interaction of all of them in three dimensions (3D). For this reason and the fact that it is impossible to quantify many of these factors, structural modeling and analysis of excavation support systems seldom results in reliable prediction of movement. So, the estimation of ground movements associated with supported excavations is generally based on a combination of empirical and analytical methods in conjunction with the application of sound judgment, experience as well as building some surrogate models [4–7].

However, before considering in detail how such estimates can be made quantitatively, it is appropriate to first consider each of these factors in more detail.

4.2.1.1 Stress Changes Due to Wall Installation and Excavation

Stress changes at four locations beneath, adjacent to, and remote from an excavation in an overconsolidated clay supported by a diaphragm wall during excavation.

It is assumed that pore water pressures are initially hydrostatic below an in situ groundwater level. After excavation, steady-state seepage eventually develops from the initial groundwater level behind the wall to a groundwater level at the formation level in front.

The geological stress history of the clay comprises deposition followed by the removal of overburden, resulting in an overconsolidated material with $\sigma_h' > \sigma_v'$ in situ. This is represented on the indicative stress path 0'–0 in Fig. 4.1. There may also be reloading by superficial deposits, which is not shown.

During wall installation, both elements A and B will be subjected to a reduction in lateral total stress as the excavation is made under support fluid, followed by an

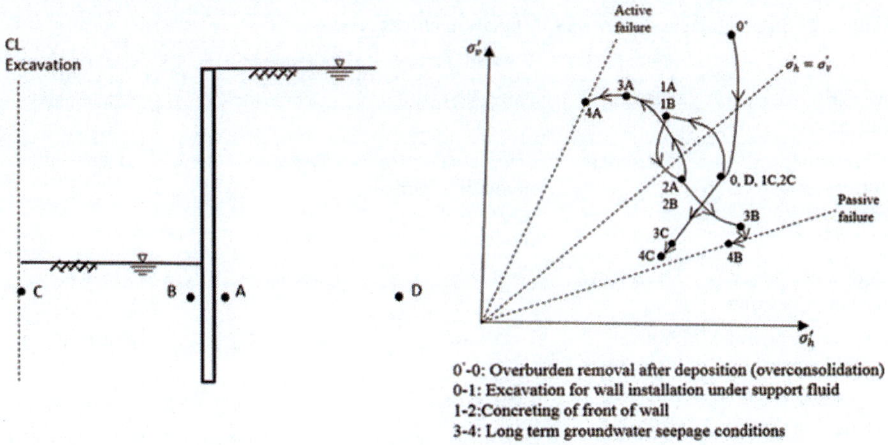

Fig. 4.1 Stress changes due to wall installation and excavation

increase in lateral total stress as the concrete is poured (assuming uncased bores or panels). Field measurements [8] and centrifuge model tests [9] show that pore water pressures will fall during excavation under bentonite and recover back to approximately their in situ values during concreting Indicative effective stress paths are shown in Fig. 4.1 (0–1 under bentonite, 1–2 during concreting).

During excavation in front of the wall, the wall is likely to move forward into the excavation resulting in a reduction in horizontal total stress for element A behind the wall. This will cause a reduction in pore water pressure behind the wall. In an overconsolidated clay, the shear following yield will also generate negative excess pore water pressures (2–3 in Fig. 4.1). The long-term steady-state seepage pore water pressure behind the wall is less than the initial in situ hydrostatic value, but probably greater than the pore water pressure immediately following excavation so that the pore water pressure will probably increase in the long term as steady-state conditions are approached (3–4). Overall (2–4) element A will experience a reduction in pore water pressure and a reduction in horizontal total stress at constant vertical total stress. These changed in pore water pressures and boundary stresses will result in an increase in vertical effective stress and a decrease in horizontal effective stress, bringing element A toward the active condition.

During excavation, element B will experience a large reduction in vertical total stress, which will result in a large reduction in pore water pressure. Movement of the wall below formation level into the soil in front will tend to increase the horizontal total stress. These changes are likely to result in an increase in horizontal effective stress and a reduction in vertical effective stress during excavation (2–3). In the long term, pore water pressure will increase again as steady-state seepage develops, reducing both vertical and horizontal effective stress and bringing the soil toward passive failure (3–4).

Table 4.1 Stress changes due to wall installation and excavation

	Element A	Element B	Element C	Element D[a]
Vertical total stress during excavation	Constant	Decreases	Decreases	Unchanged
Horizontal total stress during excavation	Decreases	Decreases due to unloading Increase due to wall movement	Decreases	Unchanged
Pore water pressure during excavation	Decreases	Decreases	Decreases	See note[b]
Pore water pressure in the long term	Probably increases	Increases	Decreases	See note[b]
Undrained shear strength in the long term	Probably increases	Decreases	Decreases	Unchanged
Strain during excavation	Vertical compression	Vertical extension	Vertical extension	Unchanged
Strain in the long term	Vertical compression	Vertical extension	Vertical extension	Unchanged

Notes

[a]assumed to be located sufficiently remotely from the wall so as not to be affected by changes in soil stress due to excavation in front of the wall

[b]depends on ground permeability

These changes in stress from the initial, prior to excavation, stage are summarized in Table 4.1.

In summary, the proximity of element B to the failure envelope is the most significant factor that influences the horizontal movement of the soil below excavation level. Yield will be small and both heave and horizontal movements will also be small if stress path 2 to 4 is well within the passive failure envelope. If the effective stress points for element B are close to the failure envelope this is indicative of significant yield and local passive failure resulting in relatively high magnitude lateral movements.

4.2.1.2 Excavation Geometry

The geometrical shape and area in plan and the excavation depth all have a critical influence on the magnitude and distribution of ground movements around an excavation in any given ground conditions.

Ou et al. [10] suggests that the deformation of a retaining wall deteriorates with the increase of the excavation depth. The maximum deflection can be estimated to be 0.2%–0.5% excavation depth H_e (Fig. 4.2).

In many situations, embedded retaining walls form a closed box, or a more complex shape. In such cases, the distribution of movement will be complex and

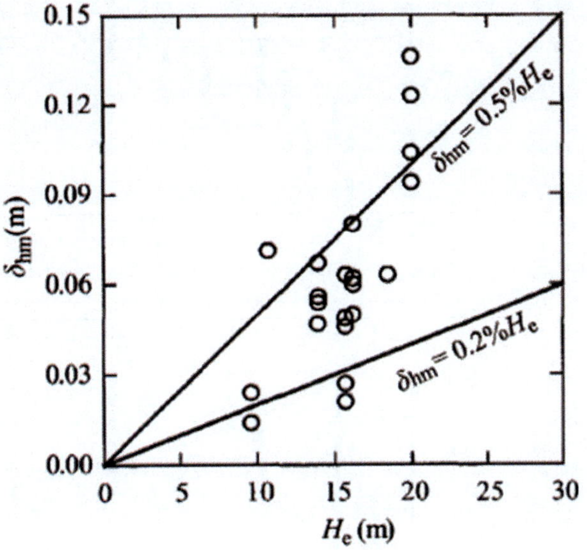

Fig. 4.2 Relationships between maximum wall deflections and excavation depths (Ou et al. [10])

its magnitude difficult to estimate without the benefit of comparable experience. The corners of an excavation tend to restrict movement.

Analyses undertaken to date allowing for corner effects are site specific and the extrapolation of the results of such studies for general application is limited only to reasonably comparable situations. Further work is necessary before reliable correlations can be established for general application. However, St [11] found that plane strain and axisymmetric analyses gave similar vertical movements. Horizontal movements from the axisymmetric analysis were some 50% of those computed in the plane strain model. Simic and French [12] found that steel quantities in diaphragm wall reinforcement could be reduced by about 25% at the corners of the excavation they studied. Temporary Prop loads measured across the corners of the Mayfair car park excavation in London indicate a 40% reduction within a horizontal distance from the corner equal to the excavation depth, compared to the central props where plane strain conditions predominated [13]. Significant economies are possible.

In addition, the effects of the existence of a corner on the deflection behavior of an excavation wall as represented by the 3D response of excavations have been reported in the literature by a number of researchers [14]. For a typical rectangular-shaped excavation, Ou et al. [14] defined the plane strain ratio (PSR) as the maximum lateral wall displacement computed from the results of 3D analysis normalized by that derived from a plane strain analysis. Finno et al. [15] extended this work and, based on field and parametric numerical studies, concluded that the PSR for excavations in clays was strongly influenced by the ratio of the length of the excavation in relation to its width (L/B) and the ratio of the length of the wall to the excavation depth (L/H). Their findings are presented in Fig. 4.3 and show that PSR is less than 1 for L/B and L/H ratios of less than 2 and 6, respectively. In general, greater corner effects are

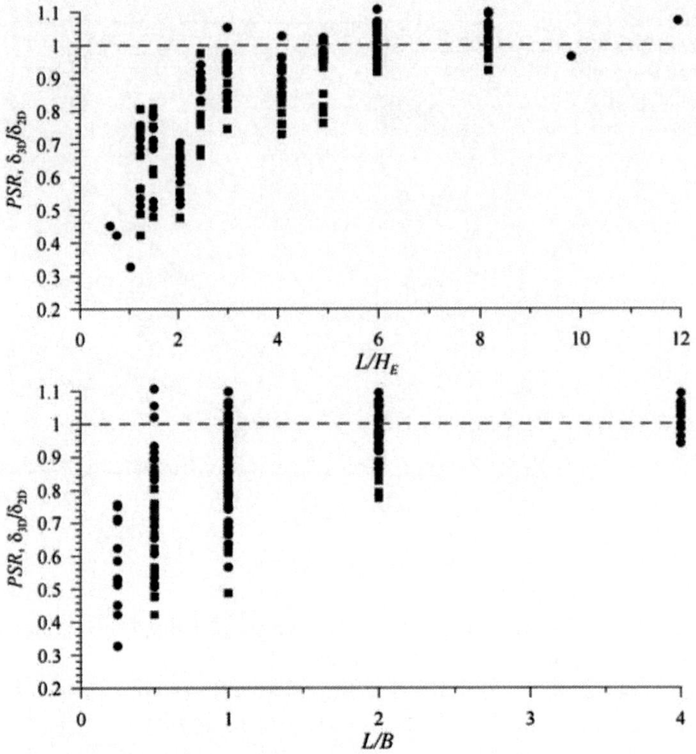

Fig. 4.3 Relationship between PSR and excavation geometry (Adapted from [15])

observed for relatively deep excavations (as evidenced by small L/H ratios), i.e., for any given geometry, higher reductions in movements occur near the corners of deep excavations as compared to shallower excavations in similar soil conditions with similar support systems. When L/H is larger than 6, plane strain and 3D analyses yield similar wall movements along the perimeter wall at the center of the excavation.

4.2.1.3 Ground Stiffness and Strength

Overall ground movements (heave beneath the excavation, wall deflections, and ground movements around the excavation) are strongly influenced by the stiffness of the ground and in weaker ground soil strength. Movements in competent ground conditions, such as dense coarse-grained soils, stiff clays, and weak rocks are generally much smaller than in less competent ground conditions such as soft and firm clays and loose sands. Soft and firm clays can be susceptible to large movements even at small excavation depths, owing to the potential for undrained bearing capacity failure at the toe of the wall. The requirement that the difference in vertical stress in the soil on either side of the wall at the toe should not exceed the bearing capacity of the soil gives a theoretical upper limit to the excavation depth, irrespective of the nature of the support system provided.

4.2.1.4 Changes in Groundwater Condition

If the retaining wall does not provide a hydraulic cut-off at depth, a steady-state groundwater seepage condition will develop involving flow around the toe of an impermeable retaining wall and subsequently upwards to the excavation formation level unless recharged from another source, such groundwater flow will result in drawdown behind the wall (and related decreased pore water Pressures) and reduced pore water pressures within the excavation compared to the initial hydrostatic condition. Establishment of a steady-state groundwater flow regime may result in both vertical ground movements, owing to the consolidation or swelling of fine-grained soils, and horizontal ground movements associated with the lateral movement of the wall into the excavation.

4.2.1.5 Type and Stiffness of the Wall and Its Support System

The type of wall selected and the excavation construction method are major influences on the magnitudes of wall and ground movements. It is well known that stiffer walls attract larger bending moments than more flexible walls with the same ground and support conditions [16]. Wall flexure redistributes the stresses imposed by the soil away from the linear, limit equilibrium ideal, which reduces the structural stress resultants, but at the expense of larger wall and ground movements. For a propped or anchored embedded wall of given overall height H and flexural stiffness EI (where E is the Young's modulus and I is the second moment of the cross-sectional area of the wall), bending effects are most significant when the wall is supported at the top. In general, wall deformation occurs partly due to rigid body rotation (in the case of a propped or anchored wall, about the position of the prop or anchor) and bending (Fig. 4.4).

Strut stiffness (spacing) and preload are also influential. As shown in Fig. 4.5a, with the start of the first stage of excavation, wall movement will be produced and form a cantilever shape. The second stage of excavation starts after the installation of the first level of struts. If the stiffness of the struts is high enough, the compression of the struts will be rather small, so that the retaining wall will rotate about the contact point between the struts and the wall, and wall deformation is thus generated. The maximum wall deformation will occur near the excavation surface as shown in Fig. 4.5b. With the comp letion of the second level of struts, the third stage of excavation starts. Suppose the stiffness of the second level of struts is also strong enough, the retaining wall will continue rotating about the contact point with the second level of the struts and wall deformation is produced again. The location of the maximum deformation will be near the excavation surface (Fig. 4.5c). If the soil below the excavation surface is soft soil, the resisting force to prevent the retaining

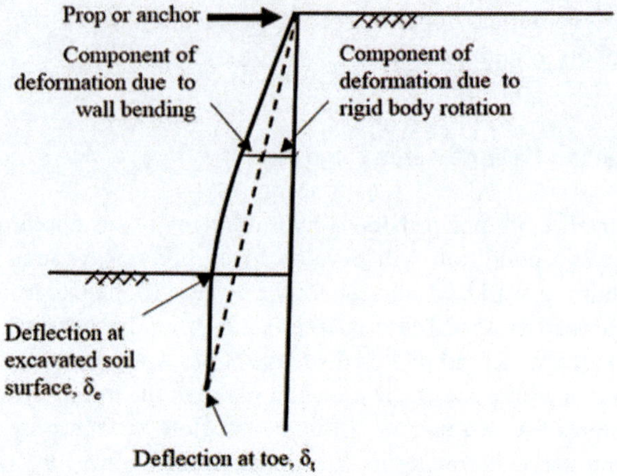

Fig. 4.4 Component of wall displacements and definition of a stiff wall

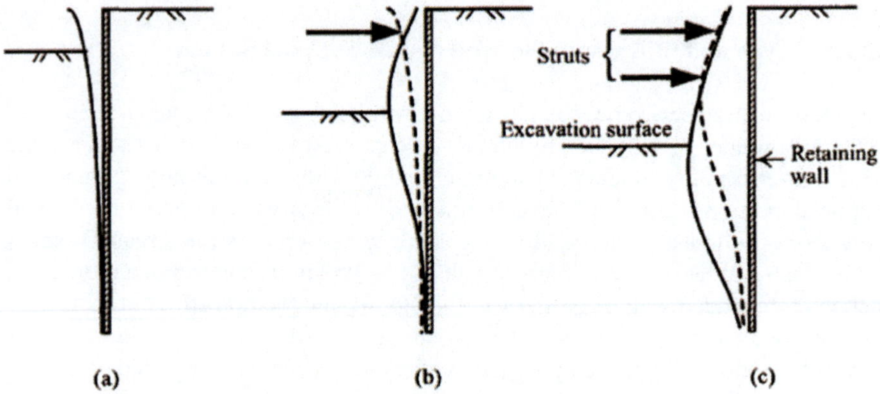

Fig. 4.5 Relationship between the shape of wall deformation and high strut stiffness: **a** first stage of excavation, **b** second stage of excavation, and **c** third stage of excavation (from Ou [17])

wall from pushing in will be weak and the location of the maximum deformation will be mostly below the excavation surface. Inferred from the same extrapolation, excavation in stiff soils (such as sand) will mostly produce the maximum deformation above the excavation surface.

As shown in Fig. 4.6, with the stiffness of struts not high, the compression of the struts should be quite large. There will be larger wall displacement around the contact points during the second and the third stages of excavation. The final deformation pattern of the retaining wall will be close to that of the cantilever type and the maximum deformation will be produced at the top of the retaining wall.

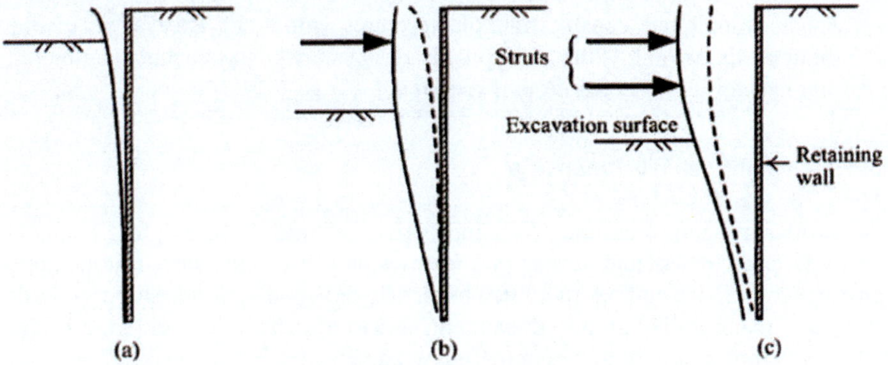

Fig. 4.6 Relationship between the shape of wall deformation and high strut stiffness: **a** first stage of excavation, **b** second stage of excavation, and **c** third stage of excavation (from Ou [17])

4.2.1.6 Construction Sequence and Methods

The choice of wall type and its installation and the overall excavation construction sequence and duration significantly influence ground movements around an excavation for any given ground and groundwater conditions and excavation geometry and dimensions construction sequence (e.g., top-down versus bottom-up) appropriate for particular circumstances is provided in Gaba et al. [18] where the respective advantages and disadvantages of each approach are discussed in detail. It is generally assumed that the top-down construction method helps to minimize ground movements because this applies early stiff support (typically via a structural slab) near the top of the wall ahead of any significant excavation in front of the wall. However, the regularity of support provided by subsequent floors at each storey height of the structure may not necessarily provide support to the wall at optimum levels (e.g., where external surcharge loading may be applied adjacent to the excavation from existing foundations, or where lateral support to the wall maybe referable at locations other than where the structural slabs are located due to, for example, the need to minimize the vertical height from the penultimate support to excavation formation level where weak strata exist immediately below final formation level). CGS [19] reports that movements of anchored walls can be less than those that are supported struts for the following reasons:

(1) anchors can be stressed to a proportion of the design load to the completed excavation before excavation below the anchors level
(2) typically, little excavation occurs below the anchor level to enable anchor installation equipment to operate
(3) the connection between struts, waler beams, and the wall is often imperfect and without significantly pre-loading the struts, compression of these connections can lead to additional deformation

(4) when completing construction of structures within the excavation by the bottom-up method, struts are typically removed, whereas anchors can be left in place for a longer duration, if required.

4.2.1.7 Quality of Workmanship

Not surprisingly, unprofessional or inappropriate attitudes or failure to effectively apply risk management and mitigation measures on-site as a matter of routine good construction practice and not to adhere to recognized good workmanship standards can lead to uncontrolled ground movements and even local failure and progressive collapse. In this regard, there are numerous examples of poor site practices such as:

(1) late installation of supports
(2) unplanned over-excavation
(3) poor pile driving and caisson construction
(4) water seepage and loss of ground through holes for tie backs and at joints or sheet pile interlocks or diaphragm wall joints
(5) remolding and undercutting of clay berms
(6) excessive surcharge loads from spoil heaps and construction equipment
(7) lack of rigidity and tightness of shores and braces.
(8) control of ground movements

4.2.1.8 Movement Due to Water

Figure 4.7 shows some potential water flow situations that can result in ground movements, water movements in and around an excavation can occur:

(1) Through flaws in an impervious wall
(2) By flow through a wall (e.g., contiguous bored pile wall, leakage through sheet pile interlock, and king post wall lagging)
(3) By flow under the wall

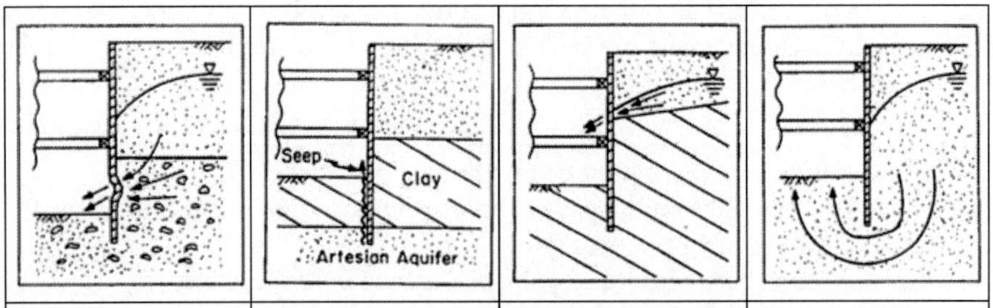

Flow through wall flaw	Flow along wall interface	Flow from perched water	Flow beneath wall

Fig. 4.7 Movements potentially associated with water flow

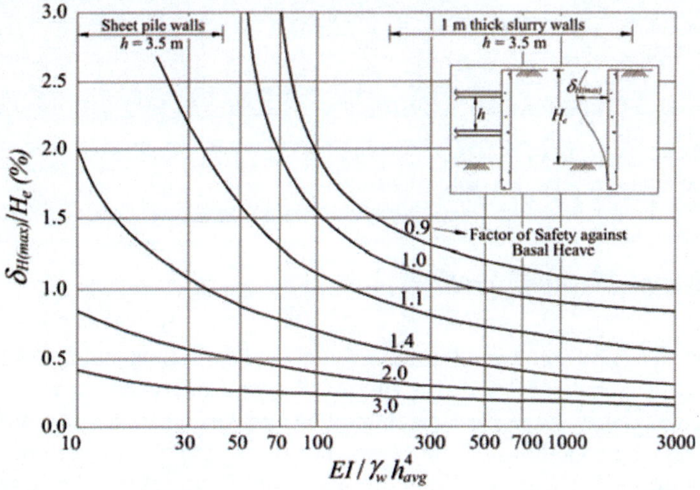

Fig. 4.8 Clough and O'Rourke design chart [20]

(4) By flow along boundaries between soils of different permeability
(5) By flow along the wall if the wall penetrates an underlying aquifer
(6) By dewatering

If piezometric pressures in an aquifer underlying an excavation are not properly reduced, heaving of the base can occur, leading to loss of passive restraint to the wall. Piezometric level outside the excavation zone drawdown results less lateral pressure on the wall, leading to less deflection.

4.3 Wall Deflections Induced by Deep Braced Excavation

4.3.1 Empirical Methods

Generally speaking, empirical methods employ the monitoring results of excavation case histories and then sort them into the stress and deformation characteristics of retaining walls and soils. The characteristics are useful not only to help understand the actual excavation behavior, but also to offer information for excavation-induced stress and deformation analyses.

The strut-retaining system of excavation is, in nature, a highly static indeterminate structure and thereby is difficult to analyze by hand calculation unless the loading pattern, boundary conditions, and analysis method are simplified.

The most well-known one is the Clough and O'Rourke [20] model, which related nomarlized maximum deflections of walls $\delta_{\mathrm{hm}}/H_{\mathrm{e}}$ to factors of safety against basal heave F_{b} and retaining system stiffness S. The retaining system stiffness S is defined as

$$S = EI/(\gamma_w h^4) \tag{4.1}$$

where:

γ_w is the unit weight of water;
h is the average vertical space of struts.

4.3.2 Finite Element Method

Under normal excavation conditions, excavation-induced stress and deformation are engendered by unbalanced forces acting on the wall due to the removal of soils within the excavation zone. The magnitude of unbalanced forces is influenced by many factors: the conditions of soil layers, the table and pressures of groundwater, the excavation depth, the excavation width, and so on. Theoretically, the finite element method is capable of simulating these factors, and therefore, the results derived from the method would be more accurate than those derived from simplified methods or the beam on elastic foundation method. The theories on which the finite element method is based, however, are rather complicated and the data to be processed both before and after analysis are enormous. What's more, some of the theories are not fully developed. To apply a program based on the finite element method, analysts are required to be well equipped with comprehensive geotechnical knowledge and experience. All this adds confusion and trouble for analysts.

4.3.3 Beam on Elastic Foundation Method

Theories concerning the finite element method are quite complicated and some of them are not fully developed. The finite element analysis normally requires enormous preprocessor and postprocessor time, computation time, and analysts have to be well equipped with comprehensive geotechnical knowledge and experience. Thus, the method is not widely adopted in the analysis and design of excavations. The beam on elastic foundation method, on the other hand, is simpler in its analysis model. With succinct input parameters, it does not take much time for processing, and therefore, is favored by most engineers. Nevertheless, the simplicity of the beam on elastic foundation method requires more delicacy and prudence when dealing with complicated excavation problems so that mistakes can be avoided.

In foundation engineering, the soil-structure interaction problem is often simulated as a series of springs to simplify analysis. Among them, Winkler's model [21] is the most widely applied.

As shown in Fig. 4.9, the basic assumption of the Winkler model is given that the foundation is a structure with stiffness and soils are of elastic foundation, their

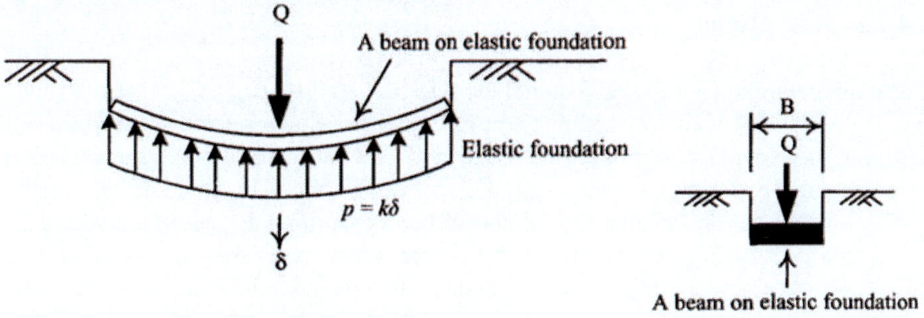

Fig. 4.9 Winkler's model [21]

interaction can be simulated as a series of individual springs. The spring constant is the ratio of stress (p) to displacement (δ), which can be expressed as follows:

$$k_s = \frac{p}{\delta} \tag{4.2}$$

where the constant k_s is called the coefficient of subgrade reaction, the modulus of subgrade reaction, or the soil spring constant, the unit of which is (force) × (dimension)$^{-3}$. The strength of the Winkler model is that it greatly simplifies analysis, for it assumes the elements are individually acting without interaction.

4.4 Surrogate Models for Maximum Retaining Wall Deflections

4.4.1 A Simple Prediction Model (2D)

4.4.1.1 Introduction

Parametric studies were carried out using the plane strain finite element (FE) software Plaxis 2D in which the soft clay stress–strain behavior was modeled using the hardening small strain (HSS) constitutive relationship that considers the small strain effect. Analyses were carried out to evaluate the behavior of excavations with braced walls in soft clays. Based on these results, this section describes the use of a simple Polynomial Regression (PR) model for relating the maximum wall deflection to various parameters such as the excavation geometry, soil strength and stiffness parameters, and the wall stiffness. Validations of the proposed PR model were also carried out through comparisons with well-documented excavation case histories.

4.4.1.2 Soil Model

The main parameters of the HSS model include G_0^{ref}, φ, and E_{50}^{ref}. G_0^{ref} is a reference initial shear stiffness corresponding to the reference pressure pref and shear strain $\gamma_{0.7}$ at which the secant shear modulus is reduced to 70% of G_0. Following the approach recommended by Brinkgreve et al. [22], G_0^{ref} ref was obtained by first determining the E_0/E_{ur} ratio based on the chart by Alpan [23] and assuming $E_{ur} = 3E_{50}$, where E_0 is the small strain Young's modulus, and subsequently using the expression $G_0^{ref} = E_0^{ref}/(2(1 + \nu_{ur}))$ with tur assumed as a constant. Since the chart for estimating the parameter $\gamma_{0.7}$ reported in Brinkgreve et al. [22] shows that $\gamma_{0.7}$ only varies within a narrow range between 1×10^4 and 4×10^4, in this study $\gamma_{0.7} = 2 \times 10^{-4}$ was assumed. The G_0 is defined as

$$G_0 = G_0^{ref}\left(\frac{c' \cos\varphi - \sigma_3' \sin\varphi}{c' \cos\varphi + p^{ref} \sin\varphi}\right)^m \tag{4.1}$$

where σ_3' is the effective confining stress (assuming compressive stress is negative). The effective friction angle φ is computed using the correlation as shown in Eq. 4.2.

$$\frac{c_u}{\sigma_v'} = 0.5743\frac{3\sin\phi'}{3 - \sin\phi'} \tag{4.2}$$

in which c_u is the undrained shear strength and σ_v' is the vertical effective stress. When the ground water table is at the ground surface and assuming $m = 1$, $c_u/\sigma_v' = \alpha$, soil stiffness ratio $E_{50}/c_u = \beta$ and $\sigma_3' = K_0\sigma_1'$ in the HSS model, E_{50}^{ref} can be expressed as:

$$E_{50}^{ref} = \frac{E_{50}}{\left(\frac{\sigma_3'}{p^{ref}}\right)^m} \tag{4.3}$$

The HSS model accounts for the increased stiffness of soils at small strains. At low strain levels, most soils exhibit a higher stiffness than at engineering strain levels, and this stiffness varies non-linearly with strain. In the TNEC case history back analysis, Kung et al. [24] used a small-strain constitutive model, as well as a Modified Cam Clay (MCC) model, for soft/medium clay. Their results indicated that the small-strain model was able to predict the wall lateral deflection and ground surface settlement fairly well, but that the MCC model could not predict accurately the surface settlement. Other publications in which small strain has been used to model excavation in soft/medium clay are Kung [25], Kung et al. [26], and Clayton [27].

The Plaxis default values are used to define the power for stress-level dependency of the stiffness m, the coefficient of earth pressure at-rest K_0^{nc}, and the Poisson's ratio ν_{ur} and E_{ur} with $m = 1$, $K_0^{nc} = 1 - \sin\varphi$, and $\nu_{ur} = 0.2$ and $E_{ur} = 3E_{50}$.

4.4.1.3 Numerical Scheme

Parametric studies have been carried out using the HSS model for the soft clay with emphasis on the maximum wall deflection predictions. Figure 4.10 shows schematically the cross-section of the excavation system, with a slightly simplified soil profile comprising of a thick normally consolidated soft clay layer overlying a stiff clay layer, typical of soil conditions in many coastal areas. The MC constitutive relationship was used to model the stiff clay ($\gamma = 20$ kN/m3, $c_u = 500$ kPa, $E_u = 250$ MPa) underlying the soft clay deposit. The soft-clay thickness is denoted as T in Fig. 4.10. The penetration depth of the wall into the stiff layer was varied between 3 and 5 m. Results indicated minimal differences in the wall deflections for these different penetration depths.

The analyses considered a plane strain excavation supported by a retaining wall system. Considering symmetry, only half the cross-section was considered. The soil was modeled by 15-noded triangular elements. The structural elements were assumed to be linear elastic with the wall represented by 5-noded beam elements and 3-noded bar elements were used for the 7 levels of struts located at depths of 1 m, 4 m, 7 m, 10 m, 13 m, 16 m, and 19 m below the original ground surface. The nodes along the side boundaries of the mesh were constrained from displacing horizontally while the

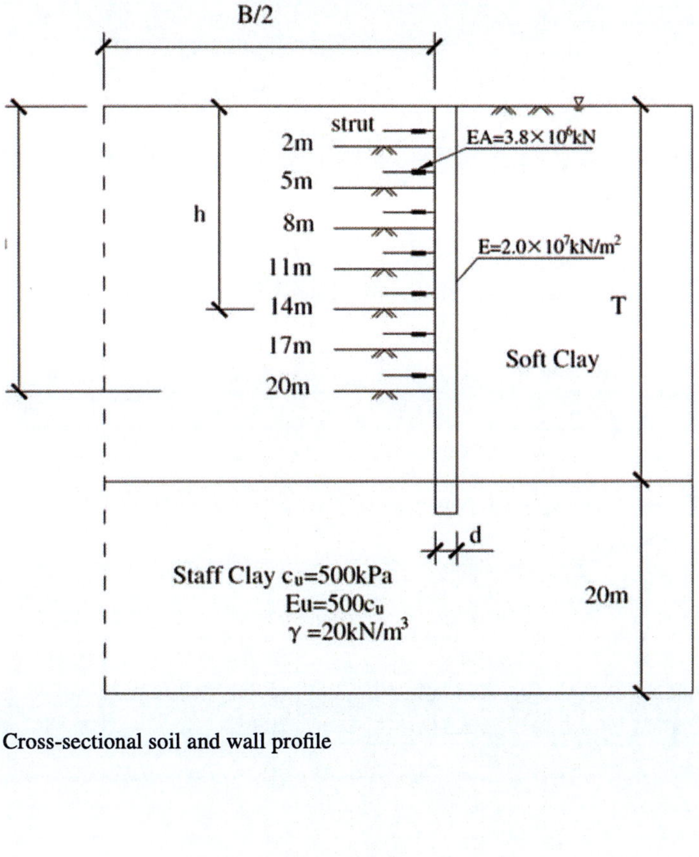

Fig. 4.10 Cross-sectional soil and wall profile

nodes along the bottom boundary were constrained from moving horizontally and vertically. The right vertical boundary extends far from the excavation to minimize the effects of the boundary restraints. The ranges of properties varied are shown in Table 4.2. The various φ, K_0, and E_{50}^{ref} values derived from empirical Eqs. 4.1–4.3, listed in Tables 4.3 and, 4.4 respectively.

The influence of the wall stiffness was studied by varying the wall thickness d while keeping the Young's modulus of the wall constant ($E = 2.0 \times 10^7$ kN/m²). The corresponding natural logarithm of the system stiffness $\ln\left(EI/\gamma_w h_{avg}^4\right)$ for the wall thickness of 0.6, 0.9, 1.2, and 1.5 m with average vertical strut spacing h_{avg} = 3 m are 6.097, 7.313, 8.176, and 8.846, respectively.

The strut stiffness per meter EA is assumed as a constant at 1.0 _ 106 kN/m since the influence of strut stiffness on wall deflection is not very significant when the strut is stiff [28, 29]. A total of 1120 hypothetical cases were analyzed.

The construction sequence comprised the following steps: (1) the wall is installed ("wished into place") without any disturbance in the surrounding soil; (2) the soil is

Table 4.2 Range of parameters

Parameter	Range
Relative soil shear strength ratio c_u/σ_v'	0.21, 0.25, 0.29, 0.34
Relative soil stiffness ratio E_{50}/c_u	100, 200, 300, 400
Soil unit weight (kN/m³)	15, 17, 19
Soft clay thickness (m)	25, 30, 35
Excavation width (m)	20, 30, 40, 50, 60
Excavation depth He (m)	11, 14, 17, 20
Wall stiffness EI ($\times 10^6$ kNm²/m)	0.36, 1.21, 2.88, 563

Table 4.3 Corresponding φ and K_0 values considering various c_u/σ_v' values

c_u/σ_v'	0.21	0.25	0.29	0.34
φ	19	22.3	25.6	29.6
K_0	0.674	0.621	0.568	0.506

Table 4.4 E_{50}^{ref} values for different combinations of strength and stiffness ratio

c_u/σ_v'	E_{50}^{ref} (kPa)			
	$E_{50}/c_u = 100$	$E_{50}/c_u = 200$	$E_{50}/c_u = 300$	$E_{50}/c_u = 400$
0.21	3114	6228	9342	12,456
0.25	4031	8062	12,093	16,124
0.29	5105	10,210	15,315	20,420
0.31	6721	13,442	20,163	26,884

excavated uniformly 1 m below each strut level prior to adding the strut support with struts at 3 m vertical spacing until the final depth He is reached. The soil is assumed to be subjected to undrained shearing (undrained analysis method A in Plaxis) during excavation. Analyses were performed assuming the groundwater table to be at the original ground surface.

Figure 4.11 presents the wall deflections corresponding to different excavation stages for the case of B = 40 m, H_e = 20 m, $c_u/\sigma_v^{'}$ = 0.29, E_{50}/c_u = 200, γ = 17 kN/m^3, $\ln\left(EI/\gamma_w h_{avg}^4\right)$ = 7.313, and T = 30 m with a penetration depth of 5 m. The cantilever wall deflection profile can be observed at the first excavation stage (prior to installation of the top-level strut). The diaphragm wall then displays deep inward movements at subsequent stages. The maximum wall deflection dhm increases as excavation proceeds.

For brevity, only some general trends are highlighted. The influence of the soil stiffness ratio E_{50}/c_u is shown in Fig. 4.12a–c for cases with γ = 17 kN/m^3, B = 30 m, and T = 30 m for $c_u/\sigma_v^{'}$ = 0.25, 0.29, and 0.34, respectively. It is obvious that the wall deflection decreases with the increase of the relative soil stiffness ratio E_{50}/c_u. In addition, the influence of E_{50}/c_u is more significant for lower wall thickness d. The influence of the soil shear strength ratio $c_u/\sigma_v^{'}$ for the cases with γ = 17 kN/m$_3$, B = 40 m, d = 1.2 m, and T = 30 m is presented in Fig. 4.13a and b for E_{50}/c_u = 200 and 300, respectively. The results show the maximum wall deflection dhm decreases with the increase of the soil shear strength ratio.

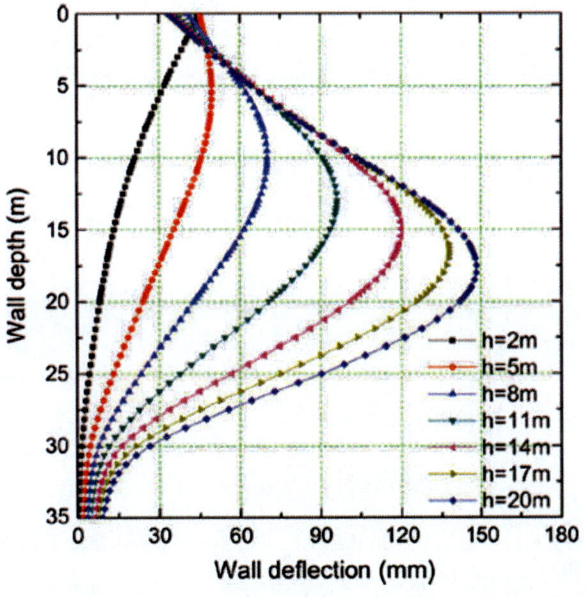

Fig. 4.11 Wall deflection profile for different excavation stage

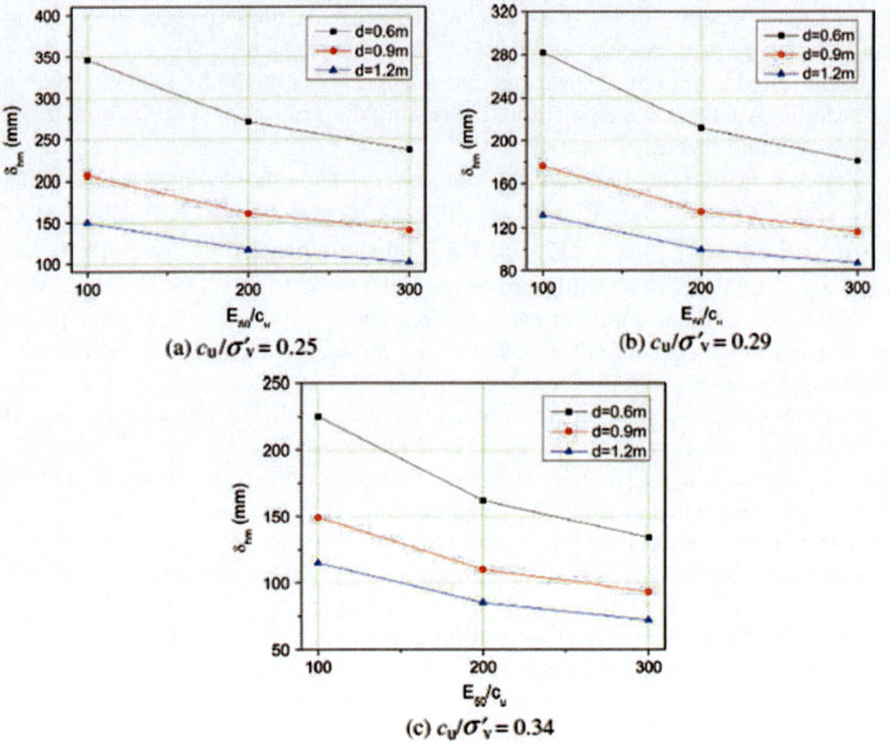

Fig. 4.12 Effect of soil stiffness on wall deflection for $H_e = 20$ m ($\gamma = 17$ kN/m^3, B = 30 m, T = 30 m)

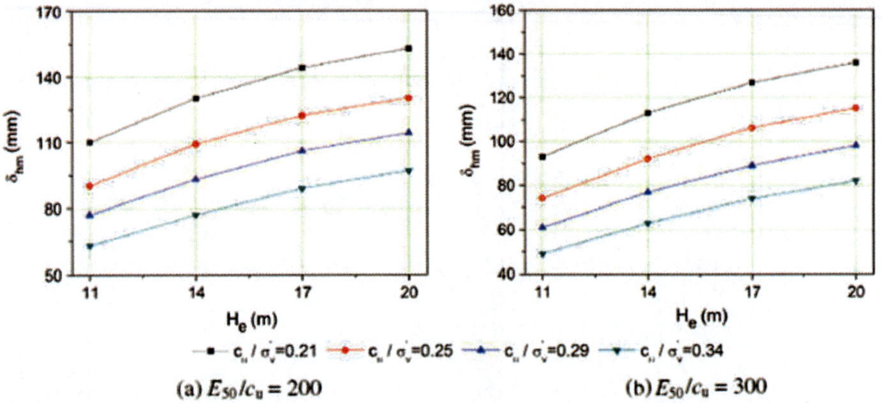

Fig. 4.13 Effect of soil shear strength ratio on wall deflection for $d = 1.2$ m ($\gamma = 17$ kN/m^3, B = 40 m, T = 30 m)

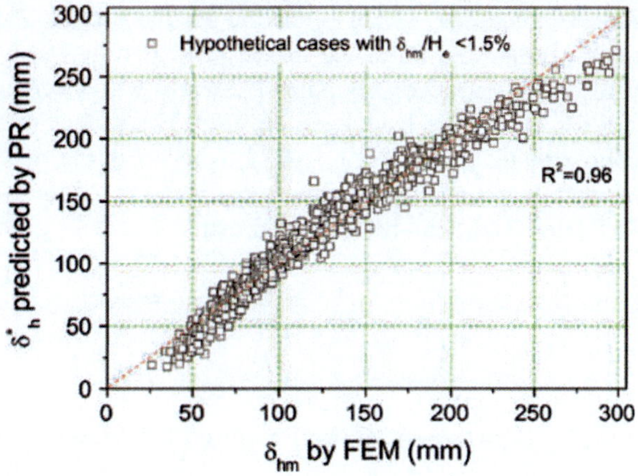

Fig. 4.14 Predicted maximum wall deflection δ_h^* versus δ_{hm} by FEM

4.4.1.4 Polynomial Regression (PR) Model

Based on the results, a Polynomial Regression (PR) model has been developed for estimating the maximum wall deflection dhm as a function of seven input parameters: γ, B, c_u/σ_v', E_{50}/c_u, system stiffness in logarithmic scale $\ln\left(EI/\gamma_w h_{avg}^4\right)$, and T. Only those cases with $\delta_{hm}/H_e \leq 1.5\%$ (a total of 1032 of the original 1120 cases) were considered in developing the regression model, as the serviceability limit state is likely to be exceeded for $\delta_{hm}/H_e > 1.5\%$. The optimal regression equation for wall deflection takes the following form:

$$\begin{aligned}\delta_h^* = {} & a_0 + a_1 B + a_2 B^2 + a_3 T + a_4 T^2 + a_5 H_e + a_6 H_e^2 + a_7(c_u/\sigma_v') + a_8(c_u/\sigma_v')^2 + \\ & a_9(E_{50}/c_u) + a_{10}(E_{50}/c_u)^2 + a_{11}\ln\left(EI/\gamma_w h_{avg}^4\right) + a_{12}\left(\ln\left(EI/\gamma_w h_{avg}^4\right)\right)^2 + a_{13}\gamma + a_{14}\gamma^2 + \\ & a_{15}\gamma\ln\left(EI/\gamma_w h_{avg}^4\right) + a_{16}(c_u/\sigma_v')\ln\left(EI/\gamma_w h_{avg}^4\right) + a_{17}H_e\ln\left(EI/\gamma_w h_{avg}^4\right) + a_{18}H_e T\end{aligned} \tag{4.4}$$

The values of the coefficients are shown in Table 4.5. Figure 4.14 shows the plot of the maximum wall deflection estimations using Eq. 4.4 versus the FEM values for the 1032 hypothetical cases. Equation 4.4 is reasonably accurate with a high coefficient of determination (R^2) of 0.96.

In all the previous analyses, the ground water table was assumed at the ground surface, which is the most unfavorable condition. In many situations with soft clay,

Table 4.5 Response surface coefficient for δ_{hm}

a_0	2816.7	a_4	−0.1515	a_8	1691.0	a_{12}	5.1584	a_{16}	213.27
a_1	3.2659	a_5	11.634	a_9	−0.7102	a_{13}	−136.76	a_{17}	−2.5553
a_2	−0.0255	a_6	−0.2348	a_{10}	0.00098	a_{14}	2.1079	a_{18}	0.6817
a_3	2.7198	a_7	−3089.0	a_{11}	−259.91	a_{15}	7.0204		

the water could be 1–2 m below the ground surface. Additional analyses carried out to investigate the influence of the ground water table indicate that the maximum wall deflection decreases almost linearly with decreasing ground water level. For brevity, these plots have been omitted. The water table correction factor μ_w can be approximated as $\mu_w = 1 - 0.1l$, where l is the depth of the ground water table below the ground surface (in metres) and $l \leq 2$. Thus, the predicted maximum wall deflection $\delta_{h,\mathrm{PR}}$, can be estimated using

$$\delta_{h,\mathrm{PR}} = \mu_w \delta_h^* \tag{4.5}$$

4.4.1.5 Assessment of the Proposed PR Model

To validate the proposed PR model, a total of 21 well documented excavation case histories from various countries as listed in Table 4.6 were analyzed [30]. Also, listed in Table 4.6 are the upper and lower bound values of the predicted $\delta_{h,\mathrm{PR}}$ ($\delta_{h,\mathrm{PR\_u}} and \delta_{h,\mathrm{PR\_l}}$, respectively), assuming that the soil parameter E_{50}/c_u follows a normal distribution with Coefficient of Variation (COV) of 0.15.

Figure 4.15 shows the predicted wall deflections (with error bars depicting the upper and lower bounds) versus the measured wall deflection for the 21 cases. The plot indicates that the PR model is able to predict reasonably well (within a factor of 2) the excavation induced wall deflections for the case histories considered.

4.4.2 A Simple Prediction Model (3D)

4.4.2.1 Typical 2D Analyses and Results

Figure 4.16 shows the wall deflection profiles from the 2D analyses at different excavation depths (He = 6 m, 9 m, 12 m, and 16 m) in stiff clays for wall stiffness of a = 1.0 and 10. For brevity, the wall deflections in soft and medium clays are omitted. The profiles follow a general trend that increasing wall stiffness leads to smaller maximum wall deflection. The strut installation generally results in lateral restraint of the wall movement above the level of the installed strut, especially for flexible and medium walls, so that the wall displacement profiles at different excavation stages almost coincide with each other above the installed struts. This agrees with the previous research by Finno et al. [31]. Generally, for the flexible walls, the wall deflection profile has a bulging shape with the maximum wall deflection between the excavation level and the toe of the wall.

For the 3D rectangular braced excavation, the maximum lateral deflection of the wall occurs at the center line due to the symmetry as shown in Fig. 4.17 from the plan view perspective. Therefore, for the 3D analyses, only the profiles of the horizontal wall deflection at this particular center line location are presented.

Table 4.6 Summary of excavation case histories

Case no	Case name	B (m)	T (m)	H_e (m)	c_u/σ'_v	E_{50}/c_u	ln(S)	γ (kN/m^3)	μ_w	δ_{h,PR_u} (mm)	δ_{h,PR_l} (mm)	Δh,M (mm)
1	Form case	35	27	18.5	0.34	200	7.3	19.0	0.8	63	79	56
2	Far east enterprise	70	24	20	0.31	325	6.44	19.0	0.9	90	103	89
3	Song-san	42	20	9.31	0.25	150	6.24	17.6	1.0	139	157	131
4	NTUH	140	40	15.7	0.25	100	7.80	20.0	0.8	47	59	41
5	Ease Taipei basin	68	23.5	14.1	0.33	100	6.51	16.5	0.8	136	147	129
6	Taiwan Power Company	60	13.5	14.7	0.3	150	6.65	19	0.9	46	53	29
7	Shaodao temple	21.5	26.5	18.5	0.3	250	7.82	18.7	0.9	35	63	29
8	Bugis MRT	21	35	18	0.25	150	8.18	16.5	0.9	125	142	117
9	Lavender	24	16	15.7	0.25	150	7.96	17.0	0.8	35	50	28
10	Syed alwi	28	16	7.8	0.25	150	6.43	16.0	0.9	120	138	112
11	MRT line in Singapore	20	20	16	0.25	150	8.11	17.6	0.8	31	46	24
12	Farrer park	21	22	17.5	0.25	150	7.3	17.3	0.8	76	92	69
13	Rochor complex	95	24	8.3	0.25	150	4.02	16.0	0.8	248	264	241
14	HDR-4 subway	12.2	22	12.2	0.25	150	6.3	19.1	0.9	103	121	95
15	Flagship wharf	34	8	14	0.22	110	6.8	18.5	0.9	72	86	65
16	Muni metro turnback	16	20.5	13.1	0.2	250	7.31	16.5	0.8	59	75	54
17	Luire	64	7.4	11.8	0.22	165	5.85	18.9	0.8	61	77	54

(continued)

Table 4.6 (continued)

Case no	Case name	B (m)	T (m)	H_e (m)	c_u/σ'_v	E_{50}/c_u	ln(*S*)	γ (kN/m^3)	μ_w	δ_{h,PR_u} (mm)	δ_{h,PR_l} (mm)	$\Delta h,M$ (mm)
18	Shanghai bank	43	19.3	15.2	0.25	190	6.57	18.6	0.9	70	89	62
19	Yanchang road	18.1	15.5	15.2	0.3	190	5.77	18.0	0.9	45	63	37
20	Pudian road	20.4	15.5	16.5	0.3	190	6.12	18.0	0.9	34	52	26
21	kotoku	30	30	17	0.34	100	7.3	18.0	0.8	128	140	122

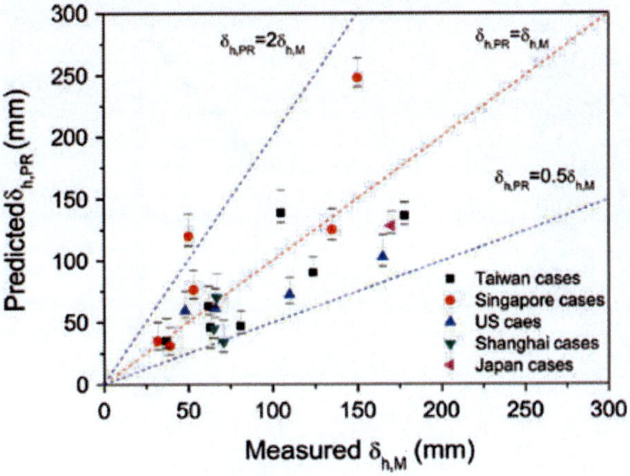

Fig. 4.15 Measured $\delta_{h,\mathrm{M}}$ versus predicted $\delta_{h,\mathrm{PR}}$

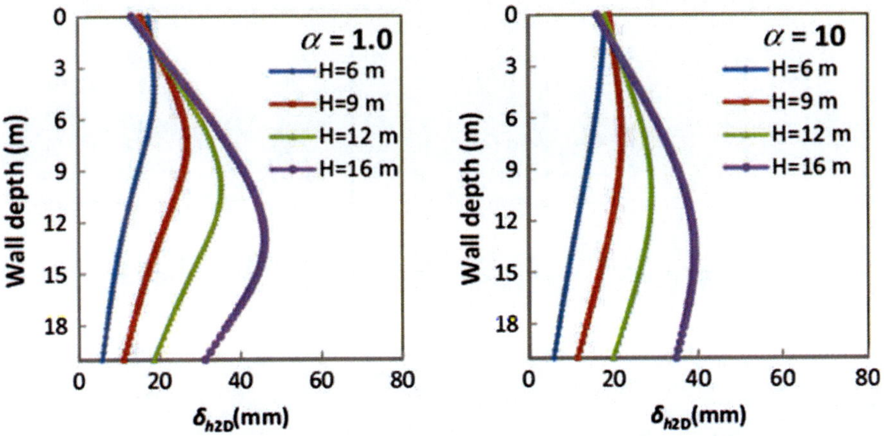

Fig. 4.16 Wall deflection profiles from 2D analyses for three wall stiffness in stiff clay

For brevity, only wall deflections with L/B = 1.0 for α = 1.0 and 10 are plotted for stiff clays in Fig. 4.18. In general, the increase of α leads to a reduction in the horizontal wall deflection. Similar to the trends of the 2D results, the wall deflection patterns from the 3D analyses show that the strut installation generally results in lateral restraint of the wall movement above the level of the installed strut, so that the wall deflection profiles at different excavation stages almost coincide with each other above the installed struts.

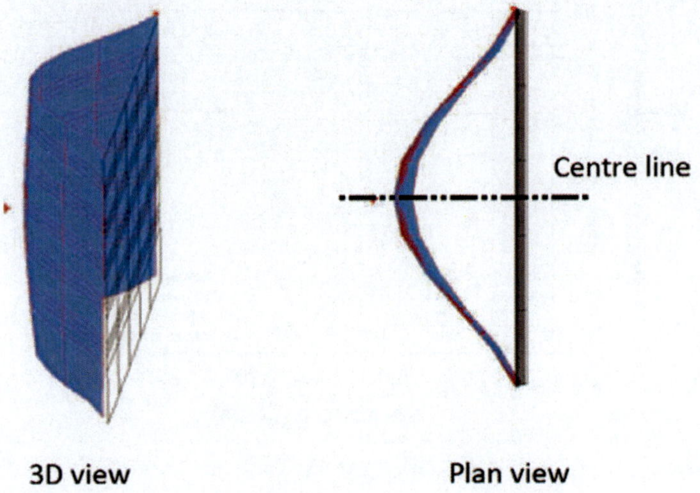

Fig. 4.17 Deformed shape of wall deflection in three dimensions from 3D view and plan view

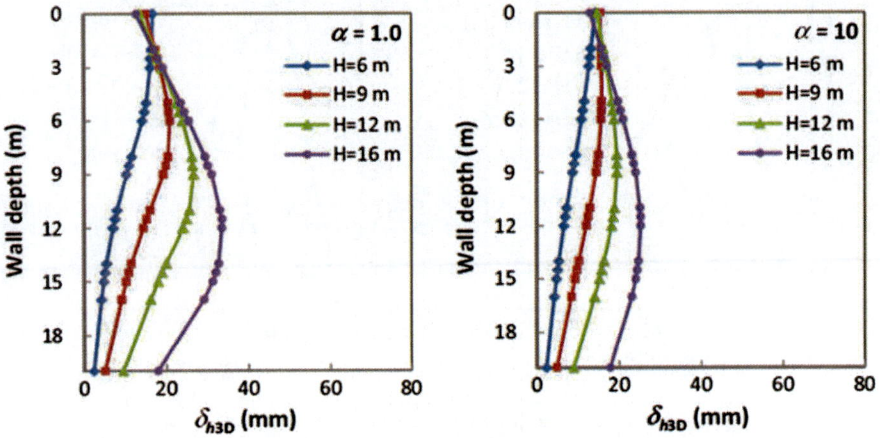

Fig. 4.18 Wall deflections from 3D analyses at different excavation stages in stiff clay (L/B = 1.0)

4.4.2.2 A Summary of 2D and 3D Results and the Estimation Model

The maximum wall deflections for all the cases considered are summarized in Table 4.7. A comparison of the 2D and 3D results indicates that the differences in the maximum wall deflections are minimal for $L/B \geq 2.2$.

Based on the results in Table 4.7, and assigning $L/B = 6.0$ for the 2D cases [], a simple logarithmic regression model was developed for estimating δ_{hm} in terms of α, c_u, H_e, D, L/B and D/B, as shown in Eq. 4.6

Table 4.7 Summary of maximum wall deflections for various L/B ratios

Cases		Max. wall deflection, δ_{hm} (mm)				
		$L/B = 1.0$	$L/B = 2.2$	$L/B = 3.4$	$L/B = 4.2$	2D
$H = 9$ m $D = 11$ m	Soft, α = 1.0	92.6	141.4	148.2	150.0	145.2
	Soft, α = 10	49.0	110.9	111.5	110.9	1296.1
	Medium, α = 1.0	45.5	65.5	65.0	63.3	63.2
	Medium, α = 10	26.5	46.1	46.2	47.3	45.7
	Stiff, α = 0.1	25.9	35.0	36.8	35.4	31.4
	Stiff, α = 1.0	20.9	28.7	330.4	29.5	26.8
	Stiff, α = 10	15.7	23.3	24.6	23.7	22.0
$H = 12$ m $D = 8$ m	Soft, α = 1.0	123.5	209.9	220.7		273.3
	Soft, α = 10	75.4	192.3	202.0	225.4	247.5
	Medium, α = 1.0	62.1	90.3	88.5	205.6	93.9
	Medium, α = 10	35.4	64.3	74.9	-	90.9
	Stiff, α = 0.1	32.6	44.0	45.6	77.6	39.9
	Stiff, α = 1.0	26.7	36.8	38.5	38.2	35.3
	Stiff, α = 10	19.5	29.4	31.0	30.7	29.0
$H = 16$ m $D = 4$ m	Soft, α = 1.0	254.2	419.1	446.8	-	538.0
	Soft, α = 10	126.4	290.7	305.6	314.8	348.0
	Medium, α = 1.0	94.6	143.7	123.5	-	211.1
	Medium, α = 10	66.7	134.9	156.5	151.1	166.8
	Stiff, α = 0.1	39.0	54.4	57.0	56.7	50.1
	Stiff, α = 1.0	33.4	47.0	49.9	49.9	46.0
	Stiff, α = 10	25.2	38.6	41.3	40.4	39.2

$$\delta_{hm} = \alpha^{-0.0617} c_u^{-3.0899} H_e D^{4.5997} (L/B)^{0.3024} (D/B)^{0.3024} \tag{4.6}$$

A plot of the logarithmic regression model estimated δ_{hm_LR} versus the FEM calculated maximum wall deflection δ_{hm_FEM} shown in Fig. 4.19 indicates that the LR predictions are generally in agreement with the target FEM δ_{hm}. It should be noted that for the large wall deflection cases (δ_{hm} > 400 mm), the regression model underestimates the magnitude of the wall deflection compared with the FEM predictions. The three data points with these large wall deflections and correspondingly low basal heave factor of safety are for the cases with $H_e = 16$ m, soft clay, α = 1.0, and L/B = 3.4, 4.2, and 6.0, respectively. Consequently, the developed logarithmic regression model should be used with caution for such large wall deflections where the serviceability limit state is likely to have been exceeded.

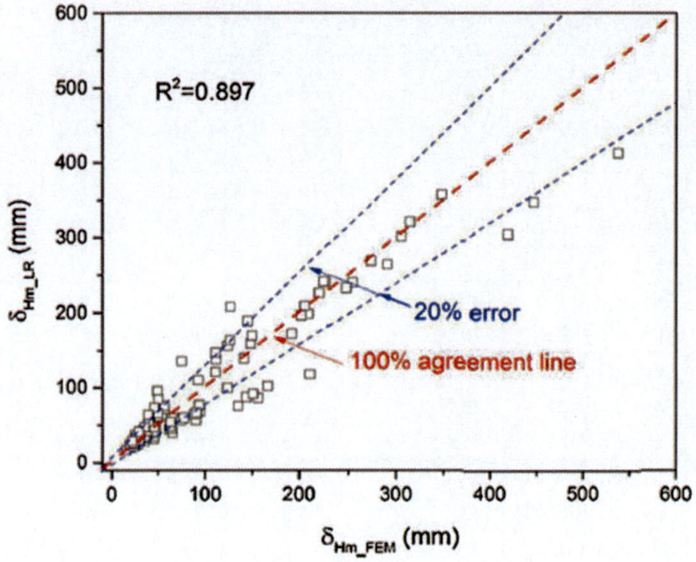

Fig. 4.19 Comparison between dHm FEM and $\delta_{\mathrm{hm\_FEM}}$ $\delta_{\mathrm{hm\_LR}}$

4.4.2.3 Comparisons with Previous Semi-Empirical Design Charts

The FE results are compared with two commonly used semi-empirical design charts from (i) Clough and O'Rourke [20] denoted as the Clough chart and (ii) Bryson and Zapata-Medina [33], denoted as the Bryson chart. It should be noted that the original Clough chart is based on the basal heave FS from Terzaghi's method [34] (denoted as FS_{T1} in Eq. 2.6), while for the Bryson chart, the FS values are based on a modified Terzaghi's method [35] (denoted as FS_{T2} in Eq. 4.7.)

$$FS_{T2} = \frac{5.14c_u + \sqrt{2}(\mathrm{H}_w/B) + 2c_u(D/B)}{\gamma \mathrm{H}} \tag{4.7}$$

Figure 4.20, the data points from this study are plotted on the Clough chart for comparison. The plot shows obvious differences between FE results and predictions from the Clough chart

Instead of using the wall system stiffness factor S, Bryson and Zapata-Medina [33] introduced a relative stiffness ratio R (Eq. 4.8) and a closed form equation relating $\delta_{\mathrm{hm}}/H_{\mathrm{w}}$ ratio (Eq. 4.9) with R and FS_{T2} (Eq. 4.7)

$$R = \frac{E_s}{E}\frac{S_H S_V H_W}{I}\frac{\gamma_s H_e}{s_u} \tag{4.8}$$

where E_s = Young's modulus of the soil, E = Young's modulus of the wall, I = moment of inertia per unit length of the wall, S_H = average horizontal support

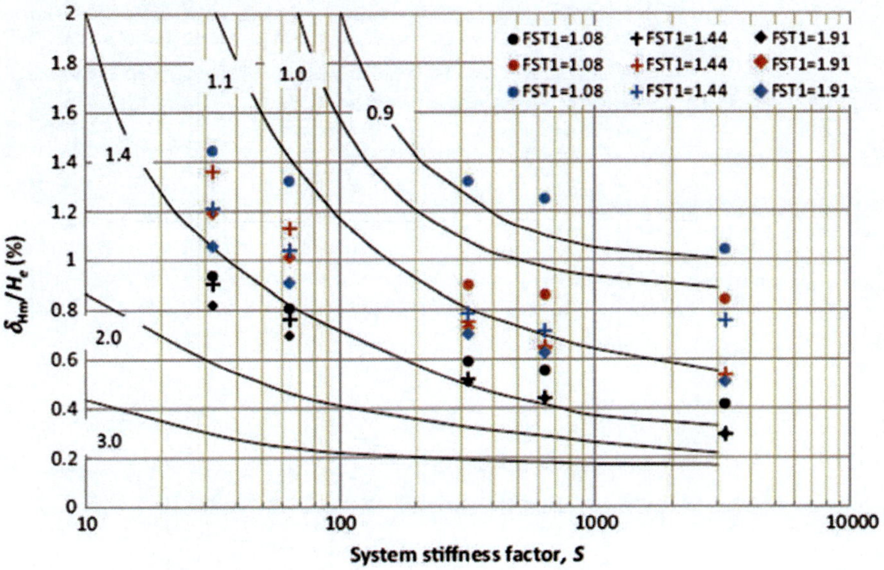

Fig. 4.20 Comparison of the 3D FE results with the Clough chart

spacing, S_V = average vertical support spacing, H_w = total height of the wall, H_e = excavation depth, γ_s = average unit weight of the soil, s_u = undrained shear strength of the soil at the bottom of the excavation.

$$\frac{\delta_{hm}}{H_W}(\%) = 0.275 \times FS^{-0.9322} R^{(0.2585-0.0351 \times FS)} \tag{4.9}$$

where H_w is defined as the total height of the wall, which is 20 m in this study. In this current study, the soil parameters, the constitutive soil model, and the finite element software employed are identical to the Bryson analyses. Only the excavation and wall geometries are different. But as plotted in Fig. 4.21, the 3D results from this study differ from the design curves for the same FS values. For L/B = 1.0, the data points are slightly below or coincide with the corresponding curves. For L/B ratio = 2.2 and 3.4, the 3D results from this study are generally above the corresponding curves, indicating larger δ_{hm}/H_w values compared with Bryson's study (L/B = 1.7). This is probably because the L/B ratio was not taken into consideration in Bryson's assessment in predicting the maximum wall deflection.

Another possible reason why larger δ_{hm}/H_w values are obtained in this study is the difference of embedment depth D, which affects the horizontal wall deflection, particularly for stiff walls. In Fig. 4.22, two groups of data points with two different D/B ratios are plotted in the Bryson chart. The plot shows slight differences of up to 5% for the two different D/B ratios. In Bryson's finite element study, the D/B ratio was 0.27.

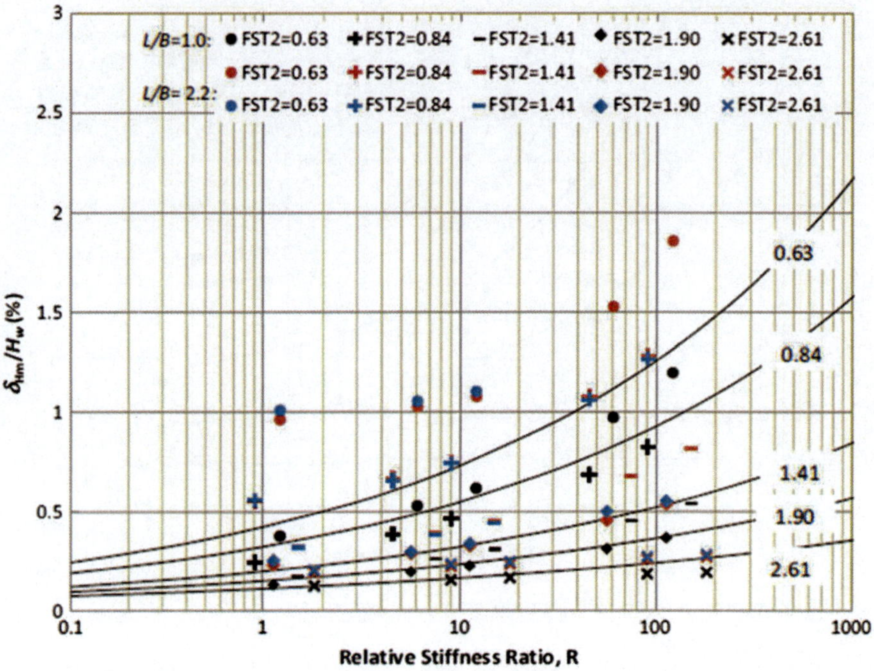

Fig. 4.21 Comparison of the 3D FE results with Bryson chart: variation of L/B ratio (D = 4 m)

4.4.3 A Multivariate Adaptive Regression Splines (MARS) Model

4.4.3.1 MARS Methodology

Friedman [36] introduced MARS as a statistical method for fitting the relationship between a set of input variables and dependent variables. It is a nonlinear and nonparametric regression method based on a divide and conquer strategy in which the training data sets are partitioned into separate piecewise linear segments (splines) of differing gradients (slope). No specific assumption about the underlying functional relationship between the input variables and the output is required. The end points of the segments are called knots. A knot marks the end of one region of data and the beginning of another. The resulting piecewise curves (known as basis functions), give greater flexibility to the model, allowing for bends, thresholds, and other departures from linear functions.

MARS generates basis functions by searching in a stepwise manner. An adaptive regression algorithm is used for selecting the knot locations. MARS models are constructed in a two-phase procedure. The forward phase adds functions and finds potential knots to improve the performance, resulting in an overfit model. The

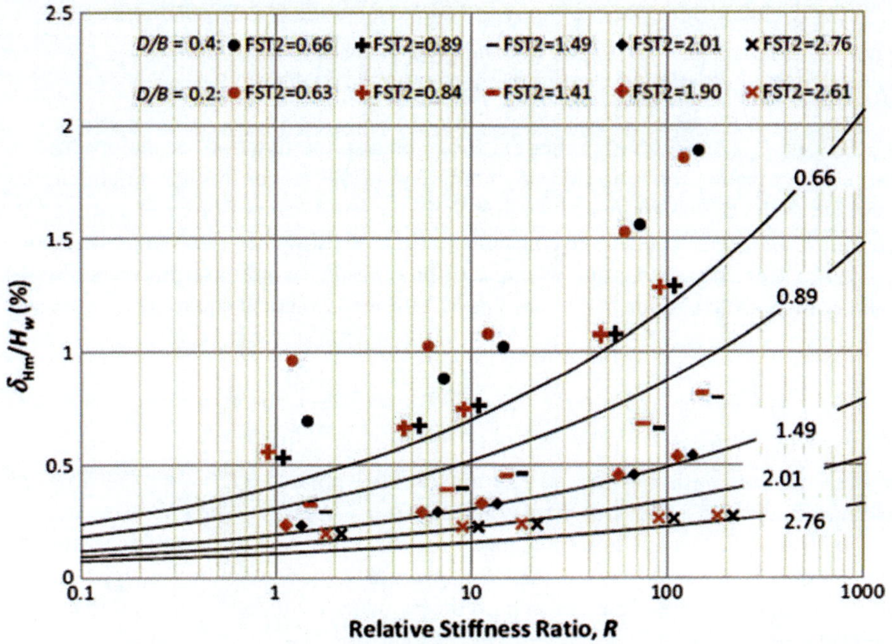

Fig. 4.22 Comparison of the 3D FE results with Bryson chart: variation of D/B ratio (L = 44 m)

backward phase involves pruning the least effective terms. An open source code on MARS from Jekabsons [37] is used in carrying out the analyses presented in this book.

Let y be the target output and $X = (X_1, \ldots, X_P)$ be a matrix of P input variables. Then it is assumed that the data are generated from an unknown "true" model. In case of a continuous response, this would be

$$y = f\left(X_1, \ldots, X_p\right) + e = f(X) + e \tag{4.10}$$

in which e is the distribution of the error. MARS approximates the function f by applying basis functions (BFs). BFs are splines (smooth polynomials), including piecewise linear and piecewise cubic functions. For simplicity, only the piecewise linear function is expressed. Piecewise linear functions are of the form max $(0, x - t)$ with a knot occurring at value t. The equation max(.) means that only the positive part of (.) is used otherwise it is given a zero value. Formally,

$$\max(0, x - t) = \begin{cases} x - t, if\, x \geq t \\ 0, otherwise \end{cases} \tag{4.11}$$

The MARS model $f(X)$ is constructed as a linear combination of BFs and their interactions, and is expressed as

$$f(X) = \beta_0 + \sum_{m=1}^{M} \beta_m \lambda_m(X) \tag{4.12}$$

where each $\lambda_m(\boldsymbol{X})$ is a basis function. It can be a spline function, or the product of two or more spline functions already contained in the model (higher orders can be used when the data warrants it; for simplicity, at most second-order is assumed in this part). The coefficients β are constants, estimated using the least-squares method.

Figure 4.23 shows a simple example of how MARS would use piecewise linear spline functions to attempt to fit data. The MARS mathematical equation is expressed as

$$y = 4.467 + 1.104 \times BF1 - 3.997 \times BF2 + 1.967 \times BF3 \tag{4.13}$$

where BF1 = max(0, x − 16), BF2 = max(0, 16 − x), and BF3 = max(0, 25 − x) and max is defined as: max (a, b) is equal to a if $a > b$, else b. The knots are located at $x = 16$ and 25. They delimit three intervals where different linear relationships are identified. It is obvious that the MARS approach is good at dealing with problems in which there are great scatters in both the explanatory independent variables and the target responses.

The MARS modeling is a data-driven process. To fit the model in Eq. 4.12, first, a forward selection procedure is performed on the training data. A model is constructed with only the intercept, β_0, and the basis pair that produces the largest decrease in the training error is added. Considering a current model with M basis functions, the next pair is added to the model in the form

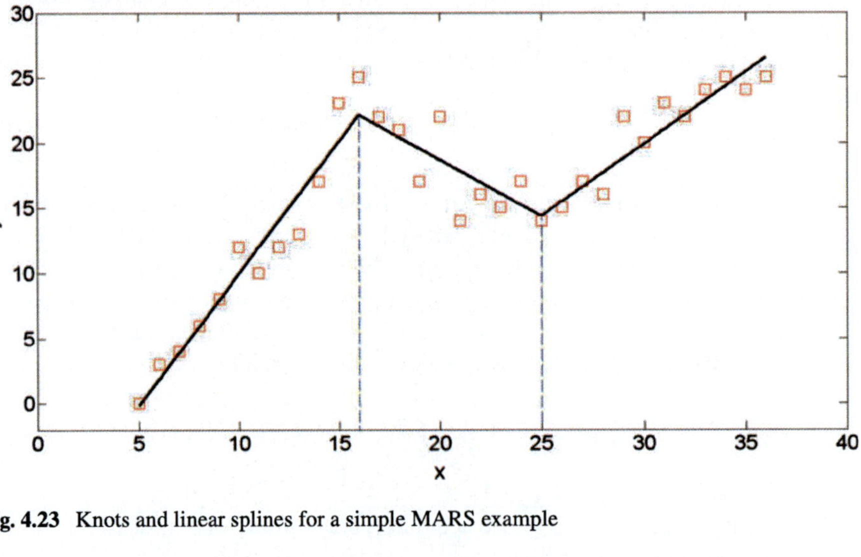

Fig. 4.23 Knots and linear splines for a simple MARS example

$$\hat{\beta}_{M+1}\lambda_m(X)\max(0, X_j - t) + \hat{\beta}_{M+2}\lambda_m(X)\max(0, t - X_j) \tag{4.14}$$

with each β being estimated by the method of least squares. As a basis function is added to the model space, interactions between BFs that are already in the model are also considered. BFs are added until the model reaches some maximum specified number of terms leading to a purposely overfit model.

To reduce the number of terms, a backward deletion sequence follows. The aim of the backward deletion procedure is to find a close to optimal model by removing extraneous variables. The backward pass prunes the model by removing the BFs with the lowest contribution to the model until it finds the best sub-model. Thus, the BFs maintained in the final optimal model are selected from the set of all candidate BFs, used in the forward selection step. Model subsets are compared using the less computationally expensive method of Generalized Cross-Validation (GCV). The GCV equation is a goodness of fit test that penalizes large numbers of BFs and serves to reduce the chance of overfitting. For the training data with N observations, GCV for a model is calculated as follows [38]:

$$\text{GCV} = \frac{\frac{1}{N}\sum_{i=1}^{N}[y_i - f(x_i)]^2}{\left[1 - \frac{M + d\times(M-1)/2}{N}\right]^2} \tag{4.15}$$

in which M is the number of BFs, d is the penalizing parameter, N is the number of observations, and $f(x_i)$ denotes the predicted values of the MARS model. The numerator is the mean squared error of the evaluated model in the training data, penalized by the denominator. The denominator accounts for the increasing variance in the case of increasing model complexity. Note that $(M - 1)/2$ is the number of hinge function knots. The GCV penalizes not only the number of the model's basis functions but also the number of knots. A default value of 3 is assigned to penalizing parameter d. At each deletion step, a basis function is removed to minimize Eq. 4.12, until an adequately fitted model is found. MARS is an adaptive procedure because the selection of BFs and the variable knot locations are data-based and specific to the problem at hand.

After the optimal MARS model is determined, by grouping together all the BFs that involve one variable and another grouping of BFs that involve pairwise interactions (and even higher level interactions when applicable), the procedure known as analysis of variance (ANOVA) decomposition can be used to assess the contributions from the input variables and the BFs through comparing (testing) variables for statistical significance. Previous applications of the MARS algorithm in civil engineering can be found in Goh and Zhang [39], Goh et al. [40], Zhang and Goh [41], Zhang et al. [42], Adoko et al. [43], Xiang et al. [44].

4.4.3.2 The Developed MARS Model

The data used to build MARS model is also extracted from the numerical analysis described in Sect. 4.4.1. The strut stiffness per meter (*EA*)strut is assumed as a constant at 1.0 × 10^6kN/m since the influence of (*EA*)strut on wall deflection is not very significant when the strut is stiff [45]. However, taking into account the excavation width, the strut stiffness parameter k_{strut} is introduced as

$$k_{strut} = \frac{(EA)_{strut}}{ls} = \frac{(EA)_{strut}}{(B/2)s} = \frac{2(EA)_{strut}}{Bs} \tag{4.18}$$

where k_{strut} = stiffness of strut; $(EA)_{strut}$ is the strut stiffness per meter, l = half-length of each strut; and s = horizontal spacing of each strut. For l = 10 m (i.e., if the excavation is symmetrical with width B = 20 m).

The influence of the soil stiffness ratio E_{50}/c_u and system stiffness ln(S) (denoted by wall thickness) is shown in Fig. 4.24 for cases with $\gamma = 17$ kN/m^3, $k_{strut} = 42{,}222$ kN/m2, and T = 30 m for $c_u/\sigma_v' = 0.25$, 0.29, and 0.34, respectively. It is obvious that the normalized wall deflection decreases with the increase of the relative soil stiffness ratio E_{50}/c_u. In addition, the influence of E_{50}/c_u is more significant for lower wall thickness *d*. For the same system stiffness (same *d*), the normalized wall deflection decreases with the increase of relative soil shear strength ratio c_u/σ_v'.

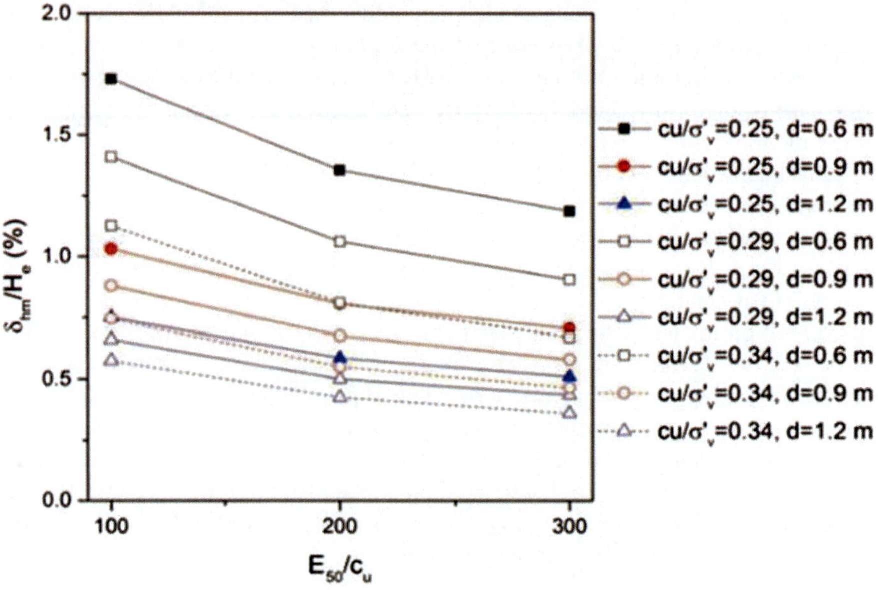

Fig. 4.24 Effect of soil stiffness on normalized wall deflection for $H_e = 20$ m,$\gamma = 17$ kN/m^3, B = 30 m and (T = 30 m)

The influence of strut stiffness *kstrut* for the cases with $H_e = 20$ m, $c_u/\sigma_v' = 0.34$, $d = 0.9$ m is presented in Fig. 4.25 for E_{50}/c_u=100, 200 and 300, respectively. The results show the normalized maximum wall deflection decreases with the increase of the strut stiffness. It is also obvious that the normalized wall deflection increases with the increase of the soft-clay thickness T. In addition, the influence of soil the influence of strut stiffness *kstrut* for the cases with $H_e = 20$ m, $c_u/\sigma_v' = 0.34$, $d = 0.9$ m is presented in Fig. 4.25 for E_{50}/c_u=100, 200 and 300, 100, 200 and 300, respectively.

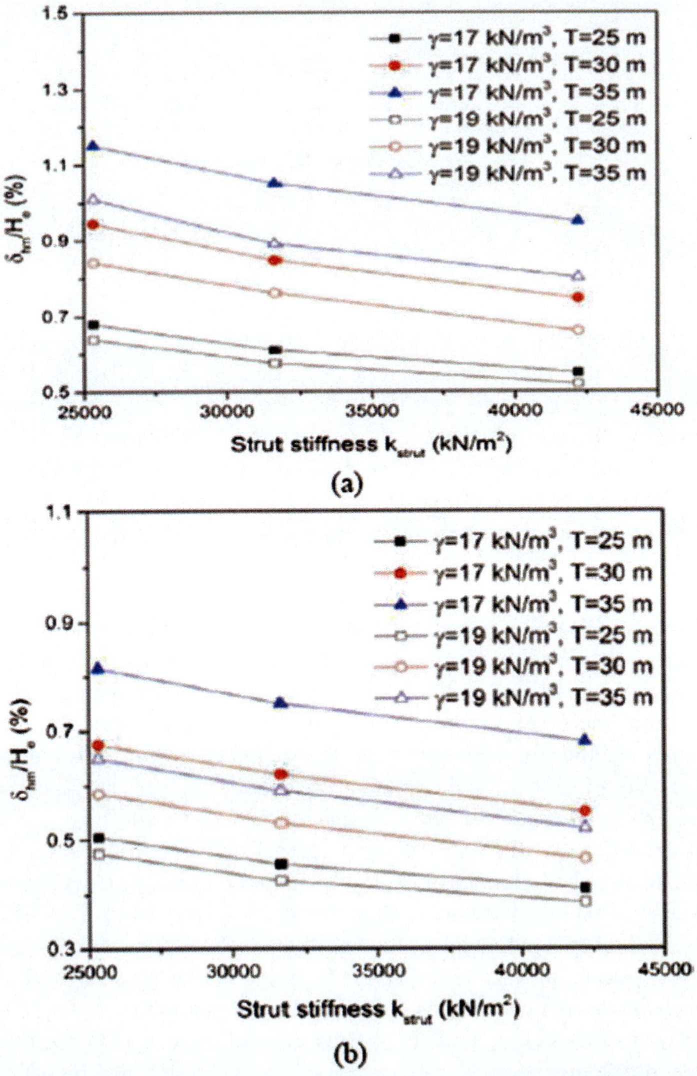

Fig. 4.25 Effect of strut stiffness k_{strut} on normalized wall deflection for **a** $E_{50}/c_u = 100$, **b** $E_{50}/c_u = 200$, and **c** $E_{50}/c_u = 300$ ($c_u/\sigma_v' = 0.34$, $d = 0.9$ m, and $H_e = 20$ m)

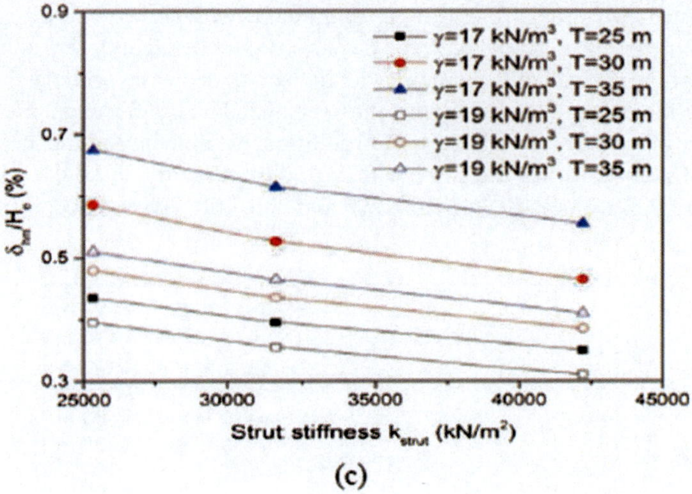

Fig. 4.25 (continued)

The results show that the normalized maximum wall deflection decreases with an increase in the strut stiffness. It is also obvious that the normalized wall deflection increases with the increase of the soft-clay thickness *T*. In addition, the influence of soil unit weight γ is more significant for larger soft-clay thickness *T*.

The influence of both the wall stiffness and the strut stiffness for the cases with $\gamma = 17$ kN/m3, $E_{50}/c_u = 200$, $c_u/\sigma_v^{'} = 0.29$ is presented in Fig. 4.26a–c for $T = 25$ m, 30 m and 35 m, respectively. Figure 4.26a shows the normalized wall deflection δ_{hm}/H_e for $k_{strut} = 25{,}333$ and 42,222 (strut moderately stiff and very stiff). It is obvious from Fig. 4.26a that δ_{hm}/H_e decreases with the increase in the system stiffness ln(*S*) and the strut stiffness k_{strut}. Figure 4.26b shows the normalized wall deflection δhm/*H*e for $k_{strut} = 21{,}111$ and 42,222 (strut less stiff and very stiff). It is obvious from Fig. 4.26b for less stiff strut, to keep the δ_{hm}/H_e within a reasonable limit, the system stiffness ln(S) must increase accordingly. In addition, it can be observed that for very stiff strut, the influence of excavation depth H_e on δ_{hm}/H_e is less significant while the influence of system stiffness is still considerable. Figure 4.26c shows the normalized wall deflection δ_{hm}/H_e for $k_{strut} = 25{,}333$ and 31,667 (strut moderately stiff and stiff) with a great thickness of soft soil. It is obvious from Fig. 4.26c that the influence of excavation depth H_e on δ_{hm}/H_e is becoming less significant as the system stiffness increases for $H_e = 11$ and 14 m. The difference of δ_{hm}/H_e is marginal even when the system stiffness ln(S) is at a small value of 6.097.

Based on Figs. 4.24, 4.25 and 4.26, it is obvious that δ_{hm}/H_e is significantly influenced by excavation geometries *T*, *B*, and *H*e, soil parameters $c_u/\sigma_v^{'}$, E_{50}/c_u, and γ, strut and wall stiffness k_{strut} and ln(S). Thus, estimation of δ_{hm}/H_e is a multivariate geotechnical problem.

Therefore, a MARS model has been developed for estimating the normalized maximum wall deflection $\delta_{hm}/H_e(\%)$ as a function of six input parameters: γ, k_{strut},

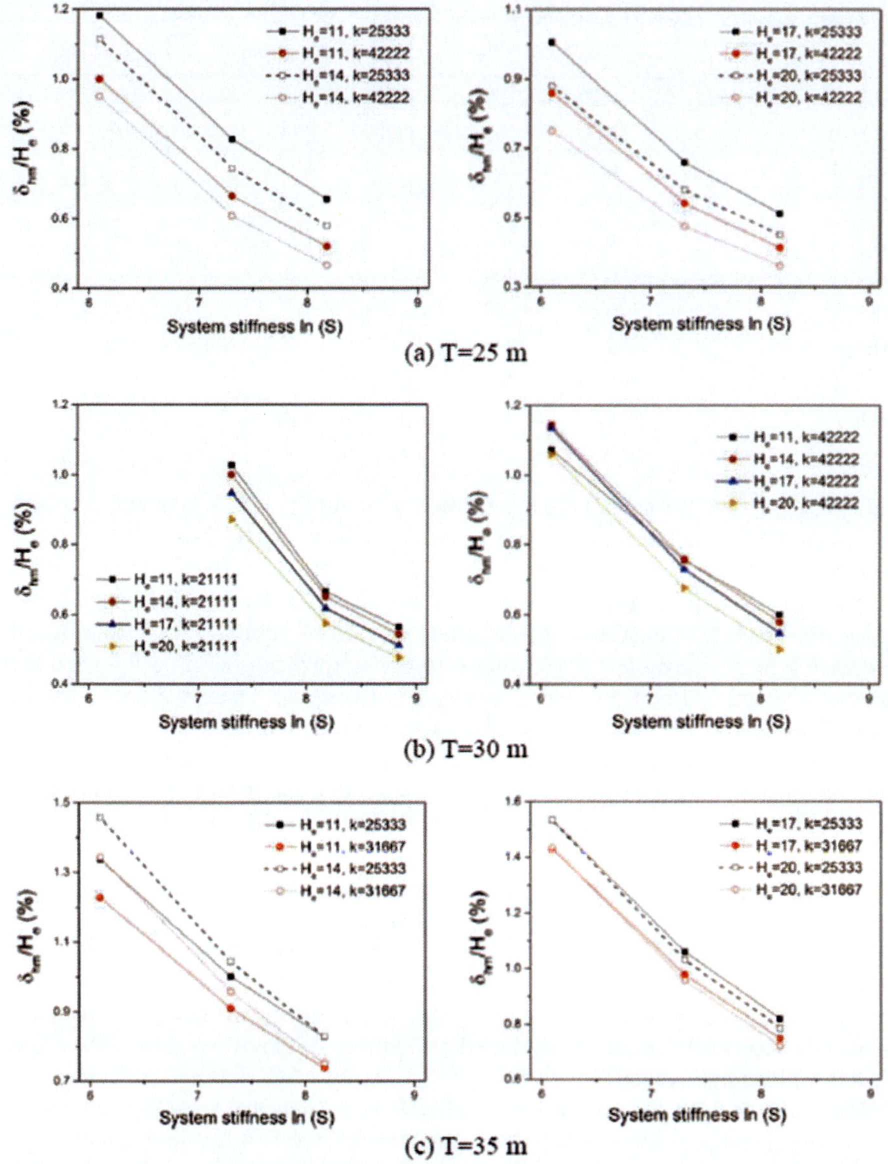

Fig. 4.26 Effect of excavation depth H_e on normalized wall deflection for **a** T = 25 m, **b** T = 30 m, and **c** T = 35 m ($c_u/\sigma_v^{\prime} = 0.29$, $E_{50}/c_u = 200$, and $\gamma = 17$ kN/m^3)

Table 4.8 Sample data sets of the testing patterns

k_{strut} (kN/m^2)	T (m)	c_u/σ'_v	E_{50}/c_u	ln(S)	γ (kN/m^3)	$\delta_{hm}/H_e(\%)$
31,667	35	0.29	100	7.313	19	1.191
31,667	30	0.34	200	6.097	15	1.407
25,333	25	0.29	200	7.313	17	0.827
42,222	30	0.34	300	7.313	19	0.382
31,667	35	0.34	300	8.176	19	0.405
42,222	30	0.34	400	8.846	19	0.241
25,333	35	0.34	100	7.313	17	1.291
31,667	30	0.34	200	8.176	19	0.509
25,333	35	0.25	200	7.313	17	1.273
31,666	25	0.29	200	6.096	19	0.900

c_u/σ'_v, E_{50}/c_u, system stiffness in logarithmic scale $\ln(S)$, and T. It should be noted that Sect. 4.4.1.4 had developed a polynomial regression model using those cases with $\delta_{hm}/H_e(\%) \leq 1.5\%$ (a total of 1032 of the original 1120 cases) considering that the serviceability limit state is likely to be exceeded for $\delta_{hm}/H_e(\%) \geq 1.5\%$. However, a review of measured diaphragm from various published case histories of successful deep excavations shows that wall deflections can be up to 3% time the excavation depth, without any serviceability problems [46]. Therefore, the results of all 1120 hypothetical cases were used for MARS model building.

Of the 1120 cases, 840 patterns were randomly chosen as the training data sets and the remaining as the testing sets. The criterion of data pattern selection used in this study was based on ensuring that the statistical properties including the mean and standard deviations of the training and testing subsets were similar to each other. Table 4.8 Sample data sets of the testing patterns lists sample data sets of the testing patterns.

The optimal MARS model consists of 32 BFs of linear spline functions with second-order interaction. A plot of the MARS predicted wall deflection values versus the FEM calculated values for the training and testing patterns is shown in Fig. 4.27. It can be observed that most of the MARS estimations of the data patterns fell within $\pm$ 20% of the target values.

Table 4.9 ANOVA decomposition for MARS model lists the ANOVA decomposition of the developed model. The first column lists the ANOVA function number. The second column gives an indication of the importance of the corresponding ANOVA function, by listing the GCV score for a model with all BFs corresponding to that particular ANOVA function removed. The third column provides the standard deviation of the function. It gives an indication of its relative importance to the overall model and can be interpreted in a manner similar to the standardized regression coefficient in a linear model. The fourth column gives the number of BFs comprising the ANOVA function. The last column gives the particular input variables associated with the ANOVA function.

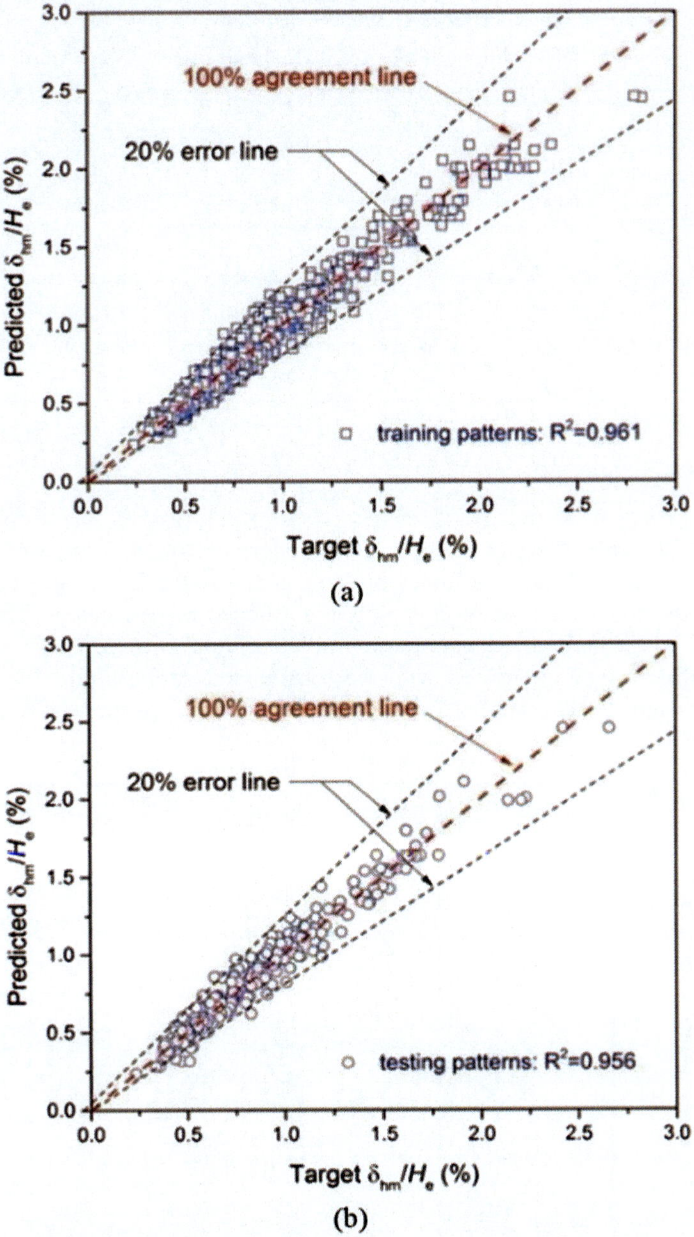

Fig. 4.27 Comparison between target and MARS predicted δ_{hm}/H_e, **a** training data and **b** testing data

Table 4.9 ANOVA decomposition for MARS model

Function No	GCV	STD	#basis	Variable(s)
1	0.011	0.052	2	k_{strut}
2	0.028	0.129	2	T
3	0.020	0.075	2	c_u/σ_v'
4	0.049	0.181	2	E_{50}/c_u
5	0.109	0.289	2	$\ln(S)$
6	0.029	0.133	2	γ
7	0.011	0.052	2	$T, c_u/\sigma_v'$
8	0.008	0.028	2	$T, E_{50}/c_u$
9	0.010	0.043	2	T, γ
10	0.009	0.031	2	$c_u/\sigma_v', E_{50}/c_u$
11	0.027	0.113	3	$c_u/\sigma_v', \ln(S)$

Figure 4.28 gives the plot of the relative importance of the input variables for the MARS model, which is evaluated by the increase in the GCV value caused by removing the considered variables from the developed MARS model. The results indicate that the normalized maximum wall deflection is more sensitive to the system stiffness $\ln(S)$ compared with the relative soil stiffness ratio E_{50}/c_u and clay unit weight γ. k_{strut} and $\ln(S)$ represent the structural stiffness of the bracing

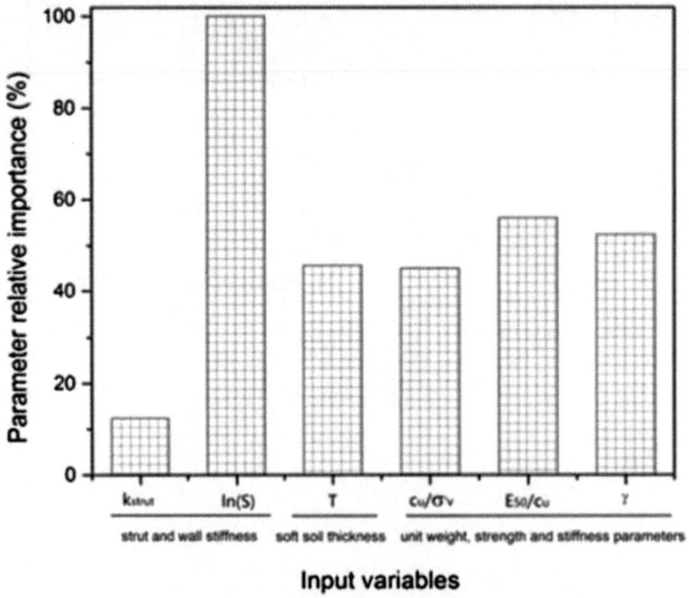

Fig. 4.28 Relative importance of the input variables in the MARS model

Table 4.10 Expressions of BFs for MARS model

BF	Equation	BF	Equation
BF1	$\max(0, \ln(S) - 7.31332)$	BF17	$BF2 \times \max(0, E_{50}/c_u - 200)$
BF2	$\max(0, 7.31332 - \ln(S))$	BF18	$BF2 \times \max(0, 200 - E_{50}/c_u)$
BF3	$\max(0, E_{50}/c_u - 200)$	BF19	$BF9 \times \max(0, \gamma - 17)$
BF4	$\max(0, 200 - E_{50}/c_u)$	BF20	$BF9 \times \max(0, 17 - \gamma)$
BF5	$\max\left(0, c_u/\sigma'_v - 0.25\right)$	BF21	$BF5 \times \max(0, T - 30)$
BF6	$\max\left(0, 0.25 - c_u/\sigma'_v\right)$	BF22	$BF5 \times \max(0, 30 - T)$
BF7	$\max(0, \gamma - 17)$	BF23	$BF4 \times \max(0, \gamma - 17)$
BF8	$\max(0, 17 - \gamma)$	BF24	$BF4 \times \max(0, 17 - \gamma)$
BF9	$\max(0, T - 30)$	BF25	$BF4 \times \max(0, T - 30)$
BF10	$\max(0, 30 - T)$	BF26	$BF4 \times \max(0, 30 - T)$
BF11	$BF2 \times \max(0, \gamma - 17)$	BF27	$BF2 \times \max\left(0, 0.29 - c_u/\sigma'_v\right)$
BF12	$BF2 \times \max(0, 17 - \gamma)$	BF28	$BF5 \times \max(0, \gamma - 17)$
BF13	$BF5 \times \max(0, \ln(S) - 8.1763)$	BF29	$BF5 \times \max(0, 17 - \gamma)$
BF14	$BF5 \times \max(0, 8.1763 - \ln(S))$	BF30	$BF5 \times \max(0, E_{50}/c_u - 300)$
BF15	$\max(0, k_{strut} - 31667)$	BF31	$BF5 \times \max(0, 300 - E_{50}/c_u)$
BF16	$\max(0, 31667 - k_{strut})$	BF32	$BF5 \times \max(0, 200 - E_{50}/c_u)$

and supporting systems. *T*, the thickness of the clay, can be individually categorized into the excavation geometry since it defines the soil profile. Similarly, unit weight γ, relative soil stiffness ratio E_{50}/c_u and relative shear strength ratio c_u/σ'_v are soil physical and mechanical parameters. The three categories: the total relative importance values for structural stiffness, excavation geometries, and clay physical and mechanical parameters are 112, 46, and 153%, respectively.

Table 4.10 Expressions of BFs for MARS model lists the BFs and their corresponding equation for the developed MARS model. It is observed from Table 6 that interactions have occurred between BFs (20 of the 32BFs are interaction terms). These terms include interaction between(S) and γ, ln(S) and E_{50}/c_u, T and γ, T and c_u/σ'_v, T and c_u/σ'_v, T and E_{50}/c_u, γ and c_u/σ'_v, E_{50}/c_u and c_u/σ'_v . The presence of various interactions suggests that the built MARS model is not simply additive and that interactions play a significant role in building an accurate model for normalized maximum wall deflection δ_{hm}/H_e predictions. This again indicates that MARS is capable of capturing the nonlinear and complex relationships between δ_{hm}/H_e and a multitude of soil parameters, excavation geometries, and structural stiffness with interactions among each other without making any specific assumption about the underlying functional relationship between the input variables and the dependent response. The equation of MARS normalized maximum wall deflection model is given by

$$(\delta_{\rm hm}/H_{\rm e})^*_{MARS} = 0.987 - 0.296 \times \text{BF1} + 0.373 \times \text{BF2} - 0.003 \times \text{BF3} + 0.0032 \times \text{BF4} - 1.68 \times BF5 + 4.35 \times BF6 - 0.069 \times BF7 + 0.113 \times BF8 + 0.045 \times BF9 - 0.042 \times BF10 - 0.054 \times BF11 + 0.13 \times BF12 + 1.9 \times BF13 - 1.7 \times BF14 - 9 \times 10^{-6} \times \text{BF15} + 7 \times 10^{-6} \times \text{BF16} - 0.0007 \times \text{BF17} + 0.001 \times \text{BF18} - 0.012 \times BF19 + 0.011 \times BF20 - 0.3 \times BF21 + 0.3 \times BF22 + 0.037 \times BF23 - 0.051 \times \text{BF24} + 0.00024 \times \text{BF25} - 0.00012 \times \text{BF26} + 2.52 \times \text{BF27} + 0.265 \times BF28 - 0.427 \times BF29 + 0.009 \times BF30 - 0.005 \times BF31 - 0.008 \times \text{BF32} \quad (4.16)$$

In all the previous numerical analyses, the ground water table was assumed at the ground surface, which is considered to be the most unfavorable condition. In many situations with soft clay, the water could be 1–2 m below the ground surface. Additional analyses carried out to investigate the influence of the groundwater table indicate that the maximum wall deflection decreases almost linearly with decreasing groundwater level. For brevity, these plots have been omitted. The water table correction factor μw can be approximated as $\mu_w = 1 - 0.1l$, where l is the depth of the ground water table below the ground surface (in metres) and $l \le 2$. Thus, the predicted maximum wall deflection $\delta_{\rm h,PR}$ can be estimated using.

Equations 4.16 and was obtained from multivariate regression analysis of numerical data for excavations in the following ranges: $k_{strut} = 21{,}111$ to 42222 KN/m^2, $T = 25$ to 35 m, $c_u/\sigma'_v = 0.21$ to 0.34, $E_{50}/c_u = 100$ to 300, $\ln(S) = 6.097$ to 8.846, and $\gamma = 15$ to 19 kN/m^3. Therefore, when predicting the normalized maximum wall deflection using the above approaches, it is better that the excavation geometry and soil properties conform to the ranges used in the parametric study.

4.4.3.3 Parametric Sensitivity Analysis

To validate the MARS normalized maximum wall deflection model, a parametric analysis was performed, aiming to find the effect of each input variable on $\delta_{\rm hm}/H_{\rm e}$. This parametric sensitivity analysis investigates the response of δhm/*H*e predicted by the MARS model to a set of hypothetical input data generated over the ranges of the minimum and maximum data sets. One input variable was changed each time within its range while the others were kept at the average values of their entire data sets. As suggested by Alavi et al. [47], a set of synthetic data for the single varying parameter was generated by increasing the value of this in increments. These values were presented to the MARS prediction model and $\delta_{\rm hm}/H_{\rm e}$ was calculated. This procedure was repeated using another variable until the responses of the models were tested for all of the predictor variables. To better illustrate the effects of all parameters on normalized maximum wall deflection, the range of all parameters is considered on a scale of 0 to 1. An example for changing the range of clay thickness from normal scale to $\alpha = 0$ to 1 scale is given in Eq. 4.17

$$\alpha_T = \frac{actual \text{ clay thickness} - \text{lower value (25m)}}{\text{upper value (35m)} - \text{lower value (25m)}} \quad (4.17)$$

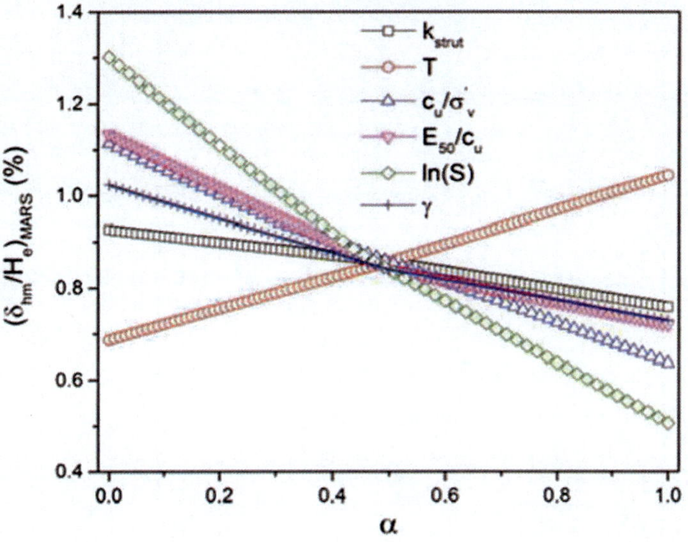

Fig. 4.29 Effect of all parameters on normalized maximum wall deflection at 0 to 1 scale

The effects of all parameters on δ_{hm}/H_e on a zero to 1 scale is shown in Fig. 4.29, plotting the influence of the δ_{hm}/He predictions to the variations of k_{strut}, T, c_u/σ'_v, E_{50}/c_u, ln(S), and γ, respectively. It is obvious that ln(S) influences δ_{hm}/H_e most, followed by E_{50}/c_u and γ, while k_{strut} is the least influential factor, which is consistent with the finding in Fig. 4.28.

4.5 Predictive Models for Wall Deflection Profiles

Most studies aim at solving for the lateral maximum wall deflection, without any attention to the lateral wall deflection profiles, which is thought to be of the same significance with the maximum value [48]. This section utilized MARS to derive another estimation model for the lateral wall deflection profiles induced by excavations in clays, using some well-documented case histories.

4.5.1 Case Histories

Zapata-Medina [49] compiled the expended database with case histories distinguished by soil type defined by the undrained shear strength found at the dredge level of the excavation. Table 4.11 lists the locations and the references of these case histories. Note that 10 case histories are presented for each soil type, giving a

Table 4.11 Case histories of the expended database (adapted from [49])

Case No	Soil type	Location	H_w (m)	H_e (m)	B (m)	S	$\gamma_{s,avg}$ (kN/m^3)	s_u (kPa)	δ_{hm} (mm)	References
1	Stiff	Lion Yard Development, Cambridge	16.3	9.6	45.0	543	20.0	120	17.7	[50]
2		New Palace Yard Park Project, London	30.0	18.5	18.5	1632	20.0	170	24.1	[51]
3		Far-East Enterprise Center Project, Taipei	33.0	20.0	63.8	1443	19.0	76.5	124.8	[52]
4		Oxley Rise Development,	14.0	11.1	33.0	149	20.8	80	10.0	[28]
5		Insurance Building, Taipei	23.0	11.4	33.7	186	19.7	50	44.5	[53]
6		Post Office Square Garage, Boston	25.6	20.2	61.0	1760	20.2	91	53.6	[54]
7		Taiwan University Hospital, Taiwan	27.0	15.7	140.0	2433	20.0	77.5	81.3	[55]
8		Taipei County Administration Center, Taiwan	38.0	20.0	93.0	13,760	20.0	65	54.3	[55]
9		75 State Street, Boston	26.0	20.0	45.7	660	18.0	70	47.3	[56]
10		Smith Tower, Houston	20.0	12.2	36.6	2748	20.1	140	14.8	[57]
11	Medium	Taipei Enterprise Center TNEC	35.0	19.7	40.0	1280	18.9	50	106.5	[53]
12		Lurie Medical Building, Chicago (East Wall)	16.5	10.0	68.0	20	19.0	36	43.2	[58]
13		Lurie Medical Building, Chicago (West Wall)	19.0	12.8	68.0	20	19.0	36	63.5	[58]
14		Taiwan, Taipei	31.0	18.4	35.0	1821	19.0	47.5	62.6	[52]
15		Tokyo Subway Excavation Project, Tokyo	32.0	17.0	30.0	2261	19.0	42	176.6	[30]
16		HDR - 4 Project for the Chicago Subway	19.2	12.2	12.2	420	19.0	30	172.6	[31]
17		Oslo Subway Excavation Project	16.0	11.0	11.0	901	17.0	30	223.6	[59–63]
18		Embarcadero BART Zone 1, San Francisco	30.5	21.3	21.3	2624	18.0	44	28.3	[64]
19		Metro Station South Xizan Road, Shanghai	38.0	20.6	22.8	510	19.0	35	48.1	[65]
20		Open Cut in Oslo	14.0	8.5	11.0	945	19.1	27.5	228.9	[66]

(continued)

Table 4.11 (continued)

Case No	Soil type	Location	H_w (m)	H_e (m)	B (m)	S	$\gamma_{s,avg}$ (kN/m^3)	s_u (kPa)	δ_{hm} (mm)	References
21	Soft	Chicago and State Street Excavation, Chicago	18.3	12.2	22.0	376	19.1	20	38.1	[67]
22		Mass Rapid Transit Line, Singapore	31.0	16.0	20.0	3344	17.6	10	38.6	[68]
23		Excavation adjacent to Shanghai Metro Tunnels	21.0	11.5	28.5	630	18.0	22	15.4	[69]
24		Excavation in Downtown Chicago	16.8	7.0	7.0	144	19.0	22.7	83.3	[70]
25		Peninsula Hotel Project, Bangkok	18.0	8.0	65.0	132	16.0	13.5	123.7	[71]
26		AT&T Corporate Center, Chicago	18.3	8.5	25.0	1697	19.0	21.5	37.4	[72]
27		Chicago Museum of Science & Industry Parking	13.7	10.3	85.0	547	19.0	45	3.6	[73]
28		One Market Plaza Building, San Francisco	30.5	11.0	11.0	1152	17.0	25	107.1	[64]
29		Sheet Pile Wall Field Test, Rotterdam	19.0	8.0	12.2	2	14.0	20	385.4	[74]
30		MUNI Metro Turnback Project, San Francisco	41.0	13.1	16.0	1491	16.5	27.5	48.1	[75]

Note The cases are generally distinguished by soil types based on the undrained shear strength (stiff clay, $s_u > 50$ kPa; medium clay, 25 kPa $\leq s_u \leq 50$ kPa; and soft clay, $s_u < 25$ kPa) found at the dredged level of the excavation. There may be some cases with the same value of s_u under different soil types. However, it would not affect the MARS modeling results since the whole data sets were used for the model development total of 30 case histories. For detailed information about the subsurface soil conditions, geometry characteristics, excavation support system details, raw inclinometer data, and maximum ground movements for each case history, Zapata-Medina [49] is referred.

Table 4.11 also summarizes the geometric (the wall length H_w, the final excavation depth H_e, and the excavation width B), soil (the averaged soil unit weight $\gamma_{s,avg}$ and the undrained shear strength s_u), and support system ($EI/\gamma_2 S_v^4$, in which EI is the wall stiffness, γ_w is the unit weight of water and S_v is the average vertical spacing of struts) parameters for the case histories on stiff, medium, and soft clay, respectively. In addition, the maximum horizontal wall movements recorded at the end of excavation are also included.

Figure 4.30 shows typically the subsurface soil conditions, geometry characteristics, excavation support system details, raw inclinometer data, and maximum ground movements for case history No. 1: Lion Yard Development in Cambridge, UK [50]. In Fig. 4.30, H is equal to the wall length H_w, I is the moment of inertia of the wall while E denotes the deformation modulus of the wall materials, S_H is the averaged horizontal strut spacing, $\delta_{H(max)}$ and $\delta_{V(max)}$ represent the maximum horizontal wall deflection and the maximum ground surface settlement, respectively.

Based on each case history listed in Table 4.11 and the corresponding wall deflection profiles against depth below the ground surface, the MARS model is developed. A total of 7 input variables comprising the geometric, soil, support system, and depth parameters as well as the output lateral wall deflections are listed in Table 4.12. Table 4.12 considers only the profiles in the final excavation stage since for most of the intermediate cases, the wall deflection profiles are unavailable. Of the wall deflection observations, only 75% of the samples were randomly selected as the training data and the remaining 25% of the samples were used for testing. For illustration of the data structures, Table 4.13 lists some sample testing data sets.

4.5.2 *MARS Modeling*

Various MARS models with different number of BFs were developed for comparison. The optimal MARS model is determined based on the smallest relative error (defined as the ratio of the difference between the MARS predicted and target values divided by the target value, in percentage). Figures 4.31 and 4.32 shows the relative error for both the location where maximum wall deflection occurs and the magnitude of the maximum wall deflection, respectively. From Figs. 4.31 and 4.32, it is obvious that the developed MARS model with 29 BFs is the most accurate.

The training and testing predictions of δ_h at a certain depth using the optimal MARS model are shown in Fig. 4.33. It is obvious that the MARS model has been able to learn the complicated relationship between the wall deflection and the geometric, soil, support system, and depth parameters. For the majority of the data points, especially those with significant wall deflections (i.e., greater than 100 mm), the relative errors fell within $\pm$ 20% of the target values. However, it should be noted

Wall Type = Diaphragm
Thickness (m) = 0.6
H (m) = 16.3
H_e (m) = 9.6
B (m) = 45
I (m^4/m) = 0.018
E (GPa) = 31
S_V (m) = 3.2
S_H (m) = 1.5
$\delta_{H(\max)}$ ($\delta_{V(max)}$) (mm) = 17.66 (10.13)
$EI/\gamma_w S_V^4$ = 543.01
$\gamma_{s,avg}$ (kN/m^3) = 20
s_u (kPa) = 120

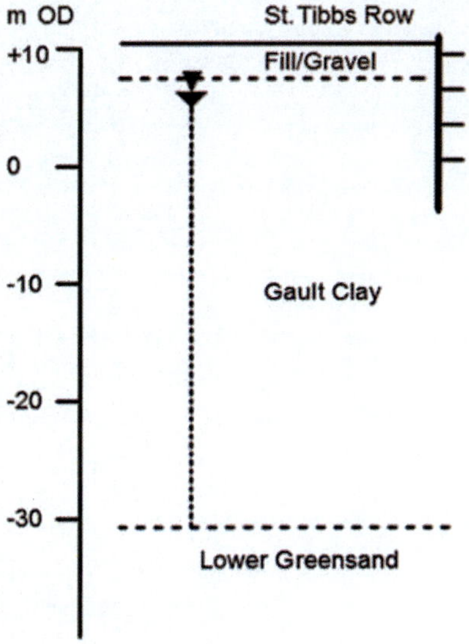

(a)

Depth (m)	δ_H (mm)
0.00	9.04
1.04	10.30
2.50	12.04
3.50	13.73
4.48	15.01
5.50	16.26
6.48	17.02
7.47	17.67
8.47	17.64
10.01	16.18
11.51	14.76
12.49	11.73
13.51	9.62
15.03	6.24
16.57	3.57

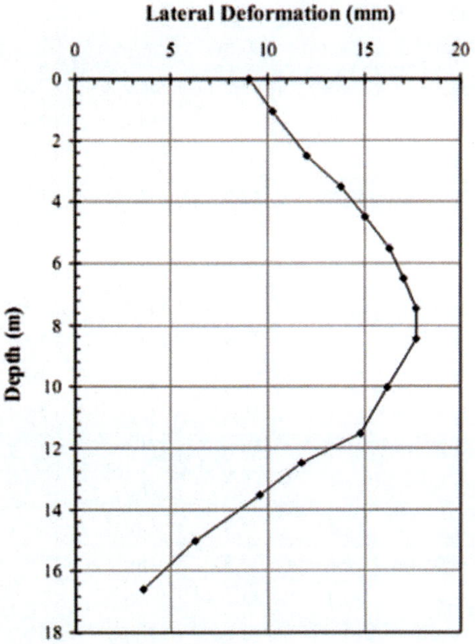

(b)

Fig. 4.30 A typical case history of case No. 1: **a** subsurface soil conditions, geometry characteristics, and excavation support system details, **b** raw inclinometer data and wall deflection profile (adapted from [49])

Table 4.12 Summary of input variables and outputs

Inputs	Parameters	Physical meaning	Values or ranges
1	H_w (m)	Wall length	13.7–41.0
2	H_e (m)	Excavation depth	8.0–20.6
3	B (m)	Excavation width	11–140
4	$EI/\gamma_2 S_v^4$	System stiffness	1.2–13,760
5	$\gamma_{s,avg}$ (kN/m^3)	Average unit weight of the soil	14.0–20.2
6	s_u (kPa)	Undrained shear strength	10–140
7	h (m)	Depth	0–H_w
Output δ_h (mm)		Lateral wall deflection profile	

Table 4.13 Sample testing data sets

H_w (m)	H_e (m)	B (m)	$EI/\gamma_2 S_v^4$	$\gamma_{s,avg}$ (kN/m^3)	s_u (kPa)	h (m)	δ_h (mm)
16.3	9.6	45.0	543	20.0	120	7.5	17.7
16.3	9.6	45.0	543	20.0	120	12.5	11.7
33.0	20.0	63.8	1443	19.0	77	5.7	77.6
33.0	20.0	63.8	1443	19.0	77	9.6	101.0
33.0	20.0	63.8	1443	19.0	77	14.0	122.3
33.0	20.0	63.8	1443	19.0	77	18.6	120.1
33.0	20.0	63.8	1443	19.0	77	22.8	85.6
33.0	20.0	63.8	1443	19.0	77	28.6	22.2
23.0	11.4	33.7	186	19.7	50	2.0	22.9
23.0	11.4	33.7	186	19.7	50	6.7	36.1
23.0	11.4	33.7	186	19.7	50	10.0	44.1
23.0	11.4	33.7	186	19.7	50	13.4	40.2
23.0	11.4	33.7	186	19.7	50	16.6	28.2
23.0	11.4	33.7	186	19.7	50	20.0	15.6
25.6	20.2	61.0	1760	20.2	91	3.6	33.8
25.6	20.2	61.0	1760	20.2	91	8.8	49.0

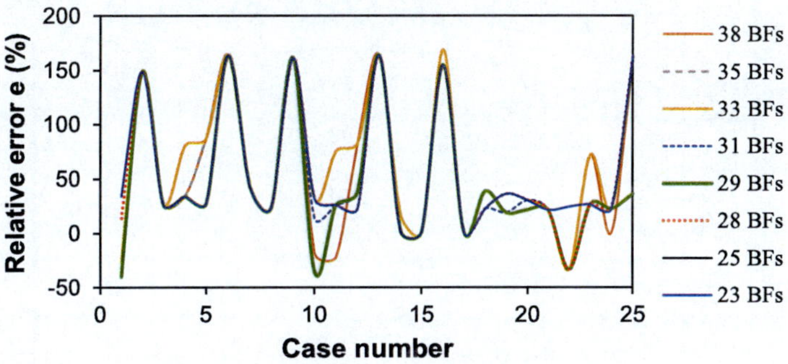

Fig. 4.31 Relative error for location where maximum wall deflection occurs

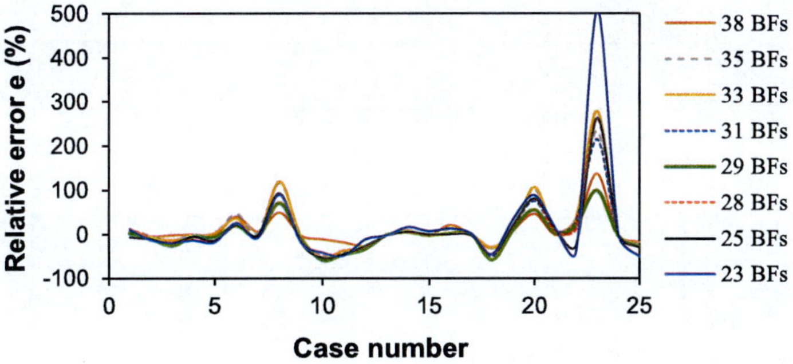

Fig. 4.32 Relative error for magnitude of the maximum wall deflection

that there are large scatters with regard to the estimations of the wall deflections in these intermediate excavation stages (smaller wall deflections within 50 mm). Figure 4.34 plots the comparison between the target and the MARS predictedδ_{hm}, together with the corresponding location where δ_{hm} occurs. It can be observed that the developed MARS model is also capable of providing accurate estimations of the maximum wall deflection values, as well as the depth at which this maximum wall deflection occurs.

For better illustration of the MARS modeling accuracy in wall deflection profile estimations, Fig. 4.35 shows the comparison of wall deflection profile between the observed and MARS predicted for some of the cases considered. It is obvious that the developed MARS modeling results fit the observed wall deflection profile well.

Table 4.14 displays the ANOVA decomposition of the developed MARS model. The first column lists the ANOVA function number. The second column gives an indication of the importance of the corresponding ANOVA function, by listing the GCV score for a model with all BFs corresponding to that particular ANOVA function removed. The third column provides the standard deviation of the function. It gives an indication of its relative importance to the overall model and can be interpreted in a manner similar to the standardized regression coefficient in a linear model.

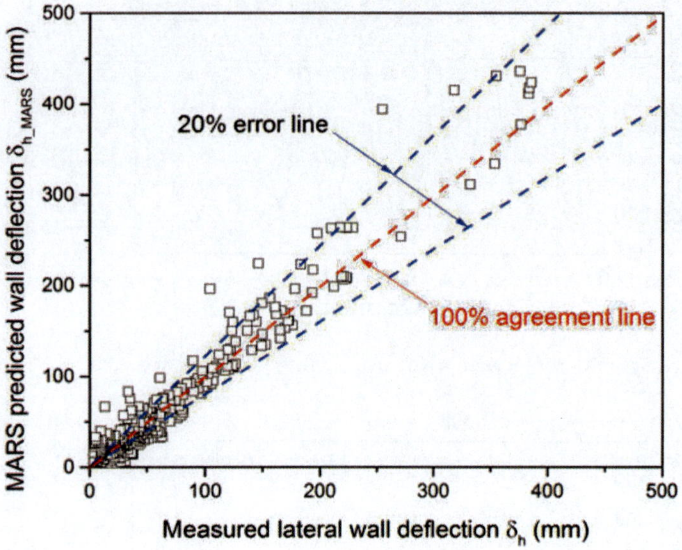

(a)

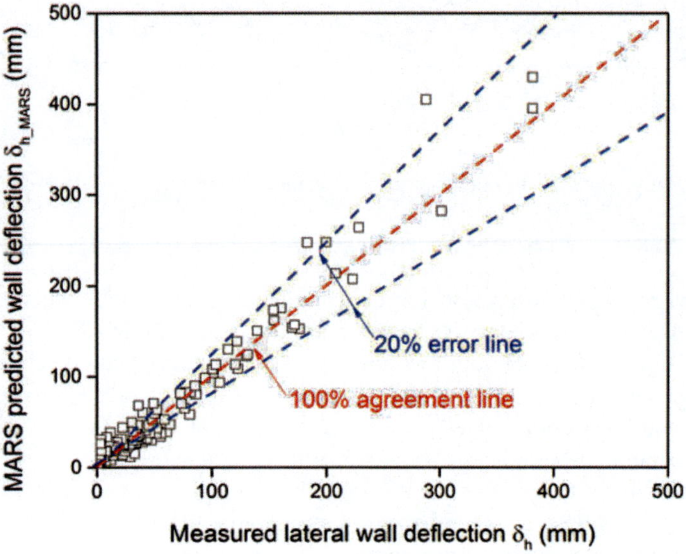

(b)

Fig. 4.33 Comparison between target and predicted δ_h: **a** training patterns, **b** testing patterns, **c** zoom of training patterns, and **d** zoom of testing patterns

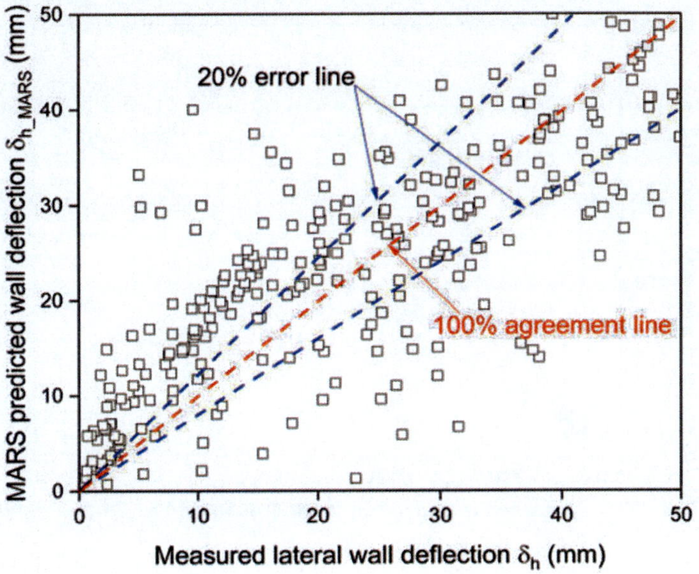

(c)

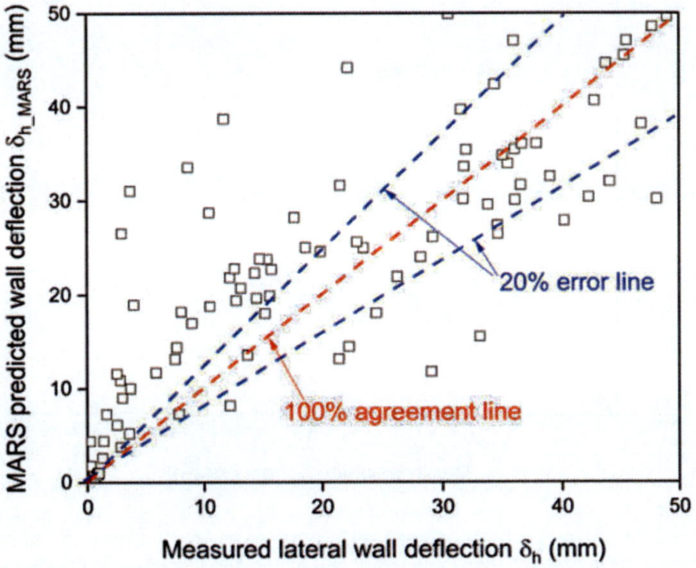

(d)

Fig. 4.33 (continued)

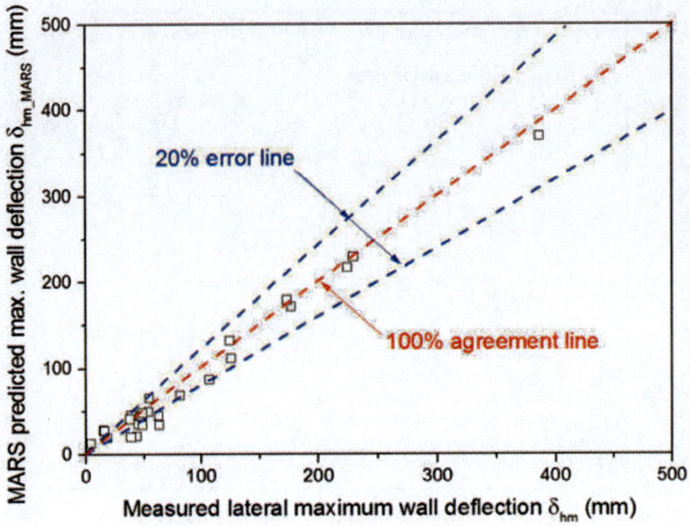

(a)

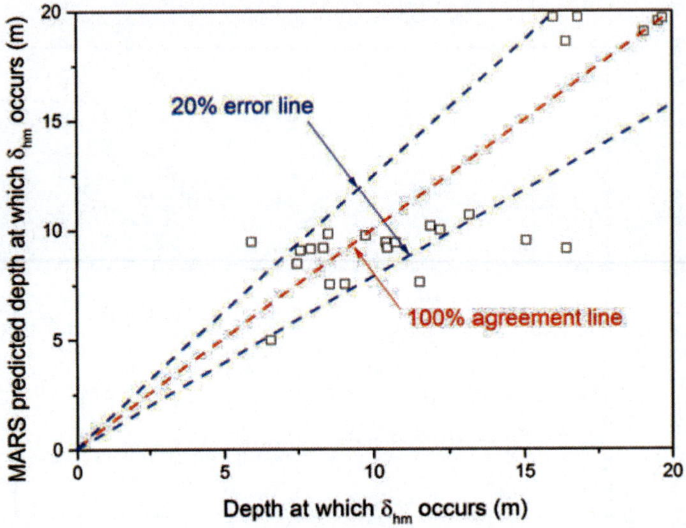

(b)

Fig. 4.34 Comparison between target and predicted: **a** δ_{hm}, and **b** depth at which δ_{hm} occurs

The fourth column gives the number of BFs comprising the ANOVA function. The last column gives the particular input variables associated with the ANOVA function. Figure 4.36 gives the plot of the relative importance of the input variables for the MARS model, which is evaluated by the increase in the GCV value caused by removing the considered variables from the developed MARS model. The results

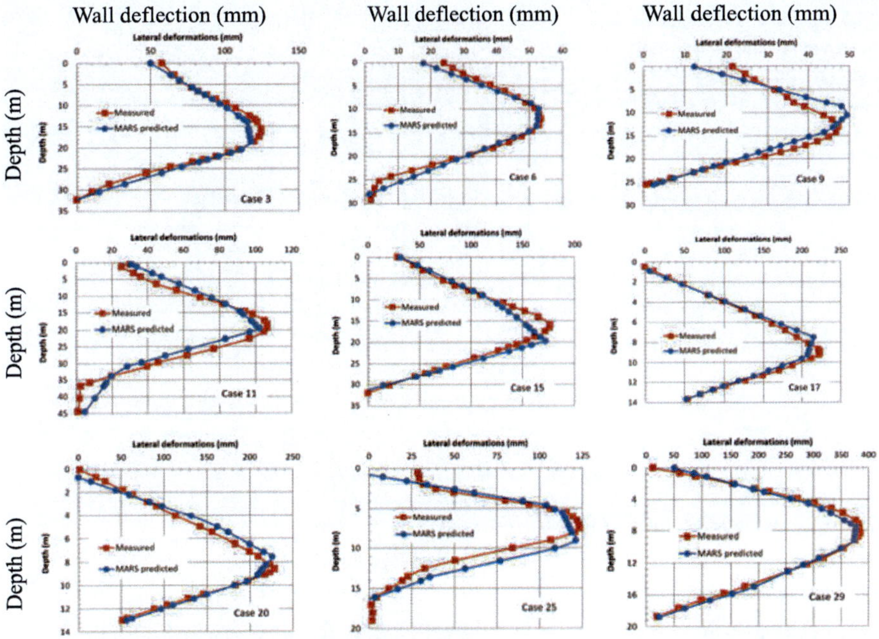

Fig. 4.35 Comparison of wall deflection profile between observed and MARS predicted

Table 4.14 ANOVA decomposition for MARS model

Function No	GCV	STD	#basis	Variable(s)
1	3533.72	60.85	4	H_w
2	1596.48	33.51	2	H_e
3	2987.19	53.96	2	B
4	1259.46	21.83	2	H_w, H_e
5	1248.76	21.81	6	$H_w,$
6	1902.33	25.51	5	H_e, h
7	1522.03	20.35	1	$B, EI/\gamma_2 S_v^4$
8	1647.64	29.37	5	B, h
9	881.58	15.89	2	$\gamma_{s,avg}, h$

indicate that the wall deflection at a certain depth is more sensitive to the excavation geometries and the depth compared with the soil and support system parameters. In addition, Fig. 4.36 indicates that the wall deflection profiles are not influenced by the soil's undrained shear strength s_u.

Table 4.15 lists the BFs and their corresponding equation for the developed MARS model. It is observed from Table 4.16 that interactions have occurred between BFs (21 of the 29 BFs are interaction terms). The presence of interactions suggests that

Fig. 4.36 Parameter relative importance of the developed MARS model

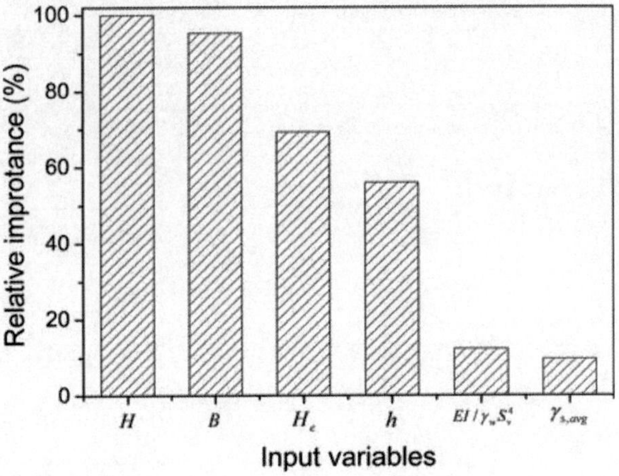

Table 4.15 Expressions of BFs for MARS model

BF	Equation	BF	Equation
1	max(0, B −16)	16	max(0, H_w −31) × max(0, h −19.71)
2	max(0, 16 – B)	17	max(0, H_w −31) × max(0, 19.71 – h)
3	BF2 × max(0, 7.51 – h)	18	BF12 × max(0, h −19.71)
4	BF2 × max(0, 420.57 −$EI/\gamma_2 S_v^4$)	19	BF12 × max(0, 19.71 – h)
5	max(0, 33 – H_w)	20	max(0, H_w −33) × max(0, h −19.71)
6	BF5 × max(0, H_e −8.5)	21	max(0, H_w −33) × max(0, 19.71 – h)
7	BF5 × max(0, 8.5 – H_e)	22	BF2 × max(0, h −14.25)
8	max(0, h −9.438) × max(0, 8.5 – H_e)	23	BF2 × max(0, 14.25 – h)
9	max(0, 9.438 – h) × max(0, H_e −8.5)	24	max(0, H_e −10.3)
10	max(0, 9.438 – h) × max(0, 8.5 – H_e)	25	max(0, 10.3 – H_e)
11	max(0, 31 – H_w)	26	BF25 × max(0, h −4.53)
12	max(0, H_w −32)	27	BF25 × max(0, 4.53 – h)
13	max(0, 32 – H_w)	28	max(0, h −9.438) × max(0, $\gamma_{s,avg}$−18.9)
14	max(0, h −9.438) × max(0, B −12.2)	29	max(0, h −9.438) × max(0, 18.9 −$\gamma_{s,avg}$)
15	max(0, h −9.438) × max(0, 12.2 – B)		

the built MARS model is not simply additive and that interactions play a significant role in building an accurate model for lateral wall deflection. This again indicates that MARS is capable of capturing the nonlinear and complex relationships between wall deflection profile and a multitude of influential parameters with interactions among each other without making any specific assumption about the underlying functional relationship between the input variables and the dependent response. The equation of the MARS lateral wall deflection model is given by

$$\begin{aligned}\delta_{\mathrm{h}}(\mathrm{mm}) &= 155.01 + 0.303 \times \mathrm{BF1} + 34.249 \times \mathrm{BF} - 27.487 \times \mathrm{BF3} + 0.064 \times \mathrm{BF4} + 39.826 \times \mathrm{BF5} \\ &+0.569 \times \mathrm{BF6} + 10.376 \times \mathrm{BF7} - 19.312 \times \mathrm{BF8} - 0.329 \times \mathrm{BF9} - 21.135 \times \mathrm{BF10} + 132.6 \times \mathrm{BF11} \\ &-8.454 \times \mathrm{BF12} - 177.65 \times \mathrm{BF13} - 0.012 \times \mathrm{BF14} - 25.352 \times \mathrm{BF15} - 14.802 \times \mathrm{BF16} - 5.977 \times \mathrm{BF17} \\ &+22.486 \times \mathrm{BF18} + 10.129 \times \mathrm{BF19} - 6.479 \times \mathrm{BF20} - 4.318 \times \mathrm{BF21} - 7.723 \times \mathrm{BF22} + 1.422 \times \mathrm{BF23} \\ &-4.003 \times \mathrm{BF24} + 29.485 \times \mathrm{BF25} - 1.832 \times \mathrm{BF26} - 6.107 \times \mathrm{BF27} - 1.181 \times \mathrm{BF28} - 1.54 \times \mathrm{BF29}\end{aligned} \tag{4.18}$$

4.5.3 Model Validations

To validate the proposed MARS deflection model, three well-documented excavation case histories as listed in Table 4.16 were analyzed. It should be noted that for simplicity, only the maximum wall deflection of the three case histories are listed and compared in Table 4.16 ($\delta_{\mathrm{hm\_m}}$ denotes the measured maximum wall deflection while $(\delta_{\mathrm{hm}}/H_{\mathrm{e}})_{MARS}$ represents the MARS model predicted maximum wall deflection). Figure 4.37 shows the predicted lateral wall deflection at different depth/profiles versus the measured ones. The plot indicates that the developed MARS wall deflection profile model is able to predict reasonably well both the excavation-induced wall deflection profiles and the maximum deflection, as well as the depth at which the maximum wall deflection occurs for case histories considered.

4.6 Predictive Models for Wall Deflection Envelope

For a symmetrical excavation, as shown in Fig. 4.38, the maximum horizontal deflection of the primary wall occurs at the middle section and there is a gradual reduction of the wall deflection away from the middle section of excavation. The wall deflection is close to zero at the corner due to the restraint of the secondary wall. Besides investigating the maximum wall deflection at the center line, it is also useful to determine the maximum horizontal wall deflection at a certain distance x, denoted as $\delta_{\mathrm{hm}}(x)$, in which x is defined as the distance from the center line along the excavation length L-direction. The profile of $\delta_{\mathrm{hm}}(x)$ is then denoted as the L-direction horizontal wall deflection envelope, which can give us a better understanding of the wall deflection behavior and will provide designers with useful information on the expected ground settlements along the length of the excavation [76].

4.6.1 Numerical Modeling Results

The data used to build the MARS model is extracted from the numerical analysis depicted in Sect. 4.4.2. In order to generalize the profiles, the $\delta_{\mathrm{hm}}(x)/\,\delta_{\mathrm{hm}}$ ratio is used. Due to the symmetry, half of the horizontal wall deflection envelopes are presented.

Table 4.16 Summary of two excavation case histories for model validation

Case ID	Case name	H_w (m)	H_e (m)	B (m)	$EI/\gamma_2 S_v^4$	$\gamma_{s,avg}$ (kN/m^3)	h (m)	δ_{hm_m} (mm)	δ_{hm_MARS} (mm)
1	Lavender MRT	28	15.7	23	1313	17.5	0–28	31.3	31.9
2	Shanghai CBD	35	20.0	65	1897	18.3	0–35	81	87
3	Kotoku Project	32	17.0	30	426	17.0	0–32	159	140.0

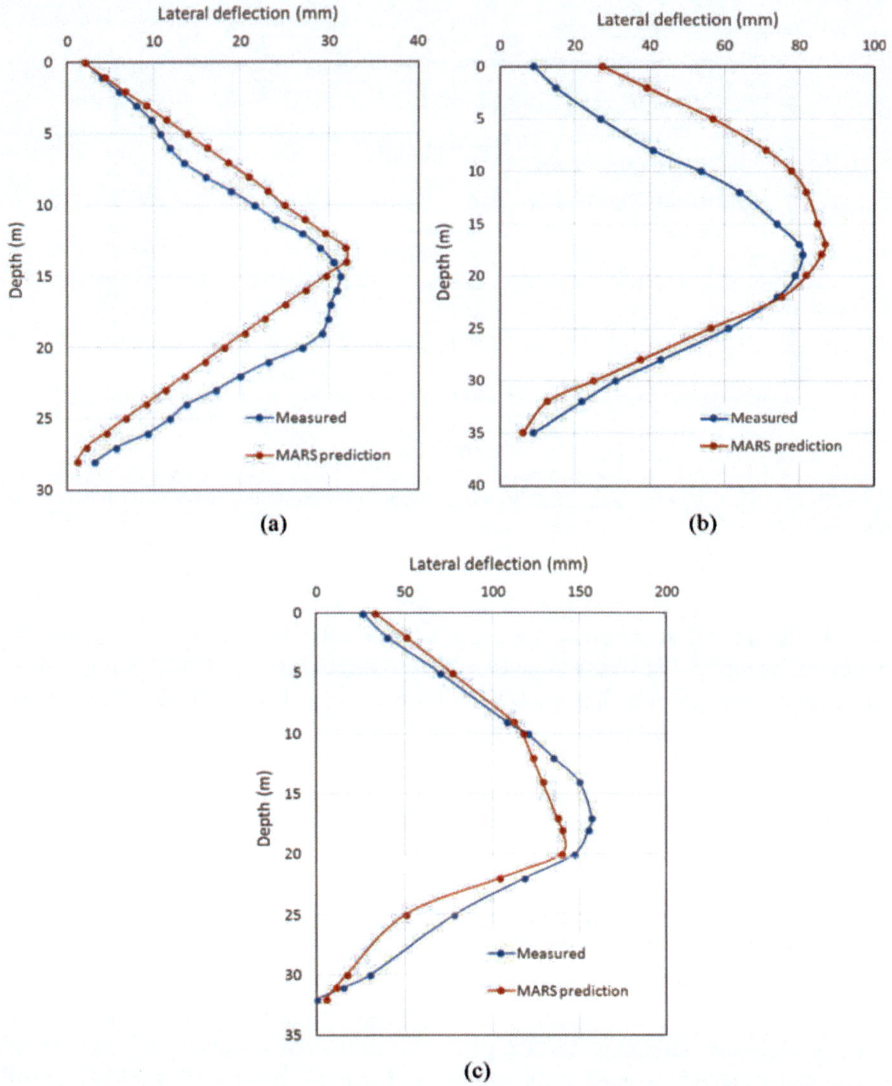

Fig. 4.37 Measured wall deflection profiles versus MARS predicted: **a** Case 1, **b** Case 2, and **c** Case 3

For the envelopes, when $x/L = 0$ (middle section), $\delta_{hm}(x)/\ \delta_{hm,3D} = 1.0$; when $x/L = 0.5$ (corner), $\delta_{hm}(x)/\ \delta_{hm,3D} = 0$.

The relationship between the $\delta_{hm}(x)/\ \delta_{hm}$ ratio and x/L for different clays (soft, medium, stiff clays), and various L/B ratios (2.2, 3.4, 4.2) with five wall stiffness values ($\alpha = 0.1, 0.2, 1.0, 2.0$, and 10) are plotted in Figs. 4.39, 4.40, and 4.41. It can be seen that the shape of the profiles is influenced by the retaining wall stiffness, soil types, H, and the L/B ratio. For $L/B = 2.2$ (Fig. 4.39), the $\delta_{hm}(x)/\delta_{hm}$ ratio is close to

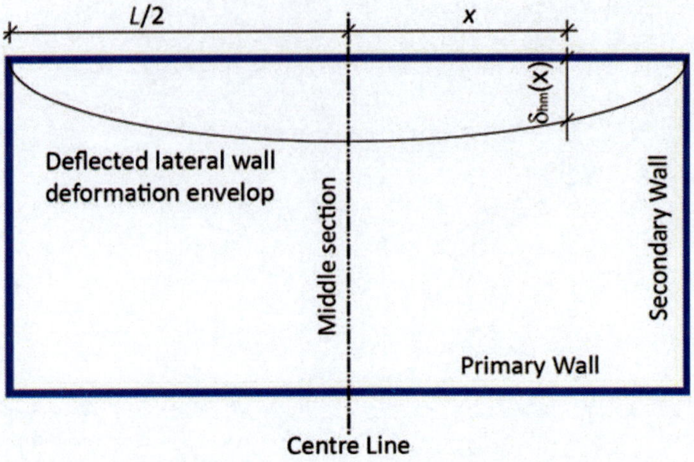

Fig. 4.38 Plan view of excavation and deflected envelope, middle section, and distance from center line

1.0 for each case when $x/L \leq 0.25$. For a given H and clay type, when $x/L > 0.25$, the ratio diminishes at different rates depending on the wall stiffness, with steeper reductions as the wall stiffness increases. As L/B increases (Fig. 4.40 and Fig. 4.41), a larger extent of the wall has $\delta_{hm}(x)/\delta_{hm}$ ratio close to 1.0 due to the diminishing influence of the restraint from the secondary wall.

4.6.2 MARS Surrogate Model

A summary of the input and output variables as well as their ranges are listed in Table 4.17.

Based on the numerical results, the database adopted in this study consisted of 617 data points. Of the 617 cases, 451 cases were randomly selected as the training patterns while the remaining 166 data samples were used for testing. The criterion of data pattern selection used in this study was based on ensuring that the statistical properties including the mean and standard deviations of the training and testing subsets were similar to each other. Table 4.18 lists some sample testing data sets.

The prediction of maximum wall deflection envelope using the MARS model with 2nd order interaction adopted 15 BFs of linear spline function. The predictions are shown in Fig. 4.42 along with the performance statistics (the coefficient of determination R^2, the coefficient of correlation R, the root mean squared error RMSE, and the mean average error MAE) for the training and testing patterns. It is obvious that the MARS model has been able to learn the complicated relationship between the maximum wall deflection envelope and the various interaction factors such as the soil

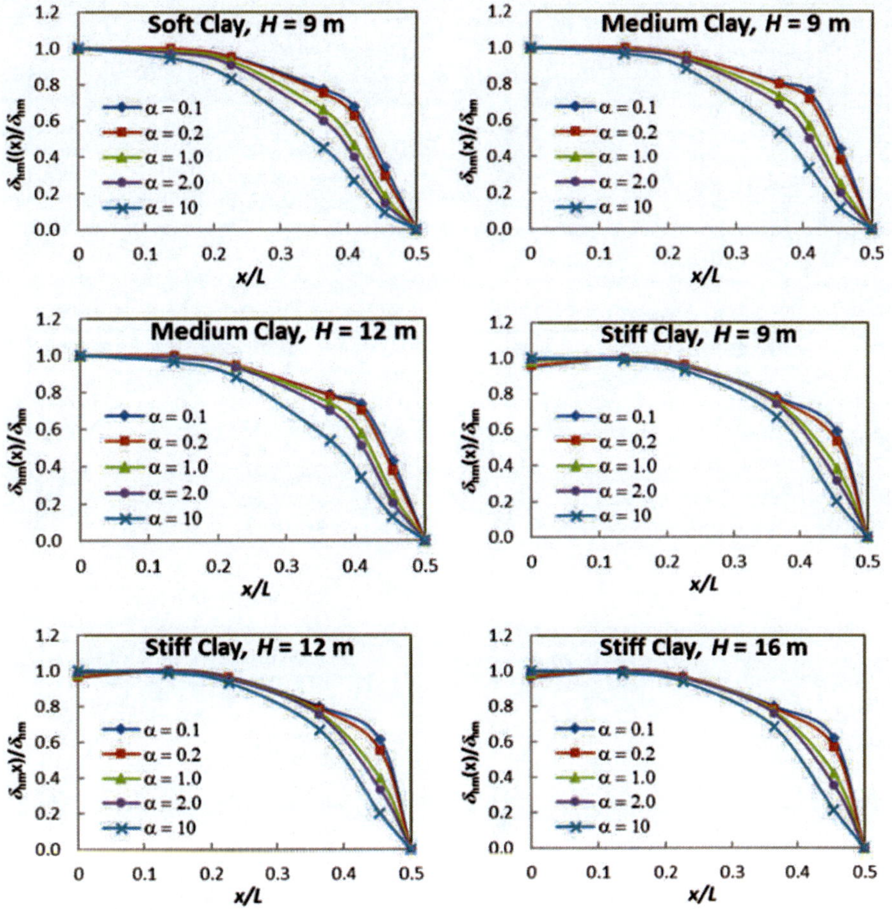

Fig. 4.39 *L*-direction horizontal wall deflection envelopes for $L/B = 2.2$

type, wall stiffness, ratio of excavation length to width, and ratio of point distance to excavation length. It is obvious that most of the MARS estimations of the data patterns fell within ± 20% of the target values, apart from some small $\delta_{hm}(x)/\delta_{hm}$ values, which may be due to the rounding errors when developing the MARS model. For the training patterns, all of the MARS estimations fell within ± 20% of the target values when $\delta_{hm}(x)/\delta_{hm}$ is greater than 0.4 while for the testing patterns, all MARS estimations fell within ± 20% of the target values when $\delta_{hm}(x)/\delta_{hm}$ is greater than 0.2.

Table 4.19 lists the BFs and their corresponding equations for the developed MARS model. It is observed from Table 4.19 that interactions have occurred between BFs (12 of the 15 BFs are interaction terms). The presence of interactions suggest that the built MARS model is not simply additive and that interactions play a significant role in building an accurate model for maximum wall deflection envelope estimation.

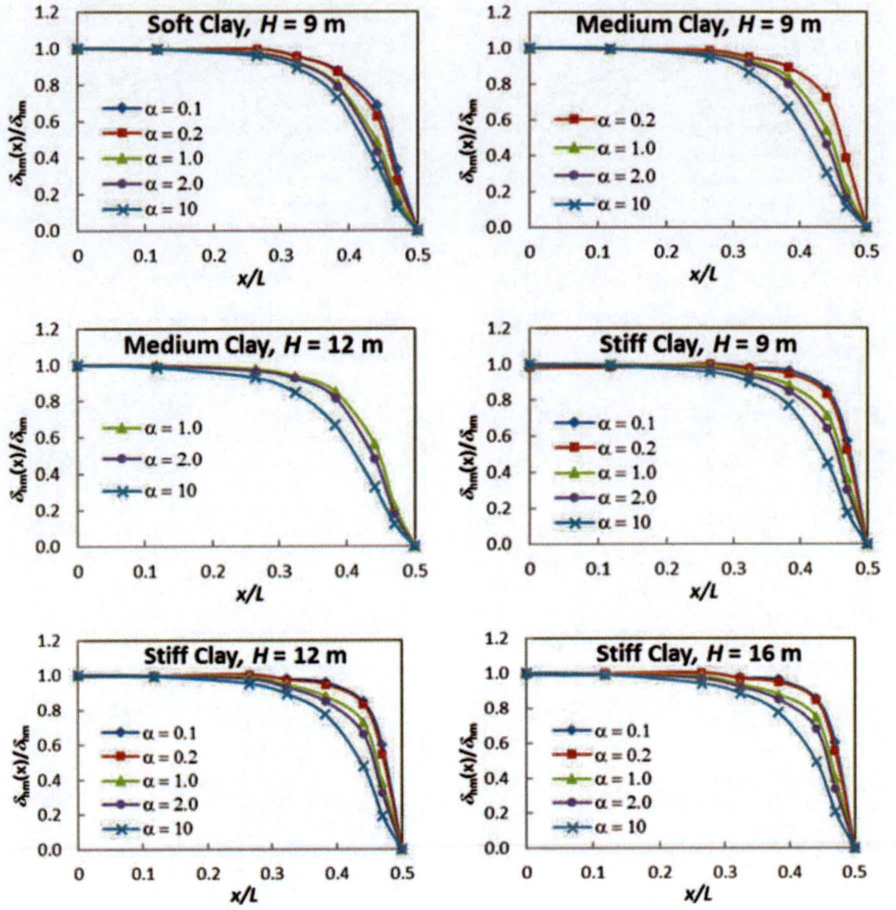

Fig. 4.40 *L*-direction horizontal wall deflection envelopes for *L/B* = 3.4

This again indicates that MARS is capable of capturing the nonlinear and complex relationships between envelope line $\delta_{\text{hm}}(x)/\delta_{\text{hm}}$ and a multitude of influential parameters with interactions among each other without making any specific assumption about the underlying functional relationship between the input variables and the dependent response. The equation of MARS maximum wall deflection envelope line $\delta_{\text{hm}}(x)/\delta_{\text{hm}}$ model is given by

$$
\begin{aligned}
\delta_{hm}(x)/\delta_{hm} &= 0.622 - 3.98 \times BF1 - 8.43 \times BF2 - 0.177 \times BF3 + 3.8 \times BF4 \\
&+ 0.43 \times BF5 - 2.1 \times BF6 + 0.5 \times BF7 - 0.26 \times BF8 - 0.03 \times BF9 + 0.63 \times BF10 \\
&- 10.6 \times BF11 + 3.49 \times BF12 + 14.38 \times BF13 - 10.4 \times BF14 + 2.23 \times BF15
\end{aligned} \tag{4.19}
$$

To further demonstrate the accuracy of the developed MARS model in estimating $\delta_{\text{hm}}(x)/\delta_{\text{hm}}$, the predicted maximum wall deflection envelope was plotted

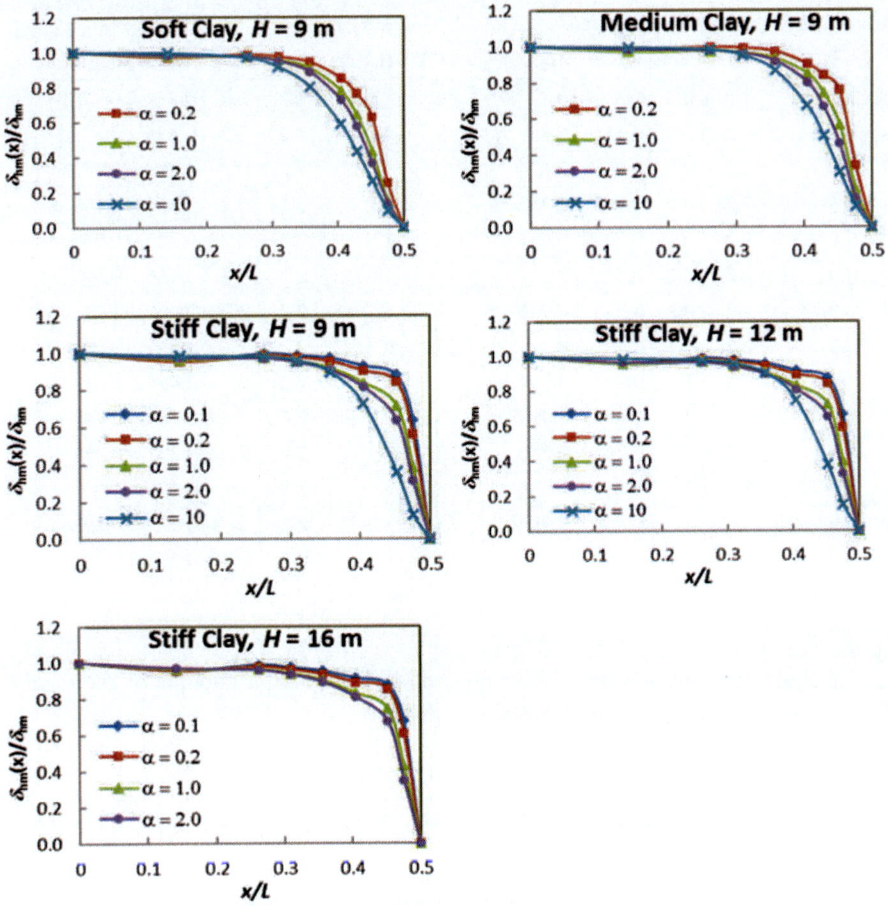

Fig. 4.41 *L*-direction horizontal wall deflection envelopes for $L/B = 4.2$

Table 4.17 Summary of input variables and outputs

Inputs	Parameters	Physical meaning	Values or ranges
1	CT	Clay type	1: soft, 2: medium, 3: stiff
2	H_e (m)	Excavation depth	9, 12, 16
3	L/B	Ratio of excavation length to width	2.2, 3.4, 4.2
4	α	Wall stiffness	0.1, 0.2, 1, 2, 10
5	*x*/L	Ratio of point distance to excavation length	0–0.5
Output $\delta_{hm}(x)/\delta_{hm}$		Normalized maximum wall deflection	0–1.0

Table 4.18 Sample testing data sets

CT	H_e (m)	L/B	α	x/L	$\delta_{hm}(x)/\delta_{hm}$	CT	H_e (m)	L/B	α	x/L	$\delta_{hm}(x)/\delta_{hm}$
1	9	2.2	0.2	0.5	0	2	12	2.2	0.2	0.5	0
1	9	2.2	0.2	0.455	0.289	2	12	2.2	0.2	0.455	0.377
1	9	2.2	0.2	0.409	0.624	2	12	2.2	0.2	0.409	0.702
1	9	2.2	0.2	0.364	0.752	2	12	2.2	0.2	0.364	0.779
1	9	2.2	0.2	0.227	0.951	2	12	2.2	0.2	0.227	0.947
1	9	2.2	0.2	0.136	1	2	12	2.2	0.2	0.136	0.997
1	9	2.2	0.2	0	0.996	2	12	2.2	0.2	0	1
2	9	2.2	2	0.5	0	3	9	2.2	2	0.5	0
2	9	2.2	2	0.455	0.197	3	9	2.2	2	0.455	0.318
2	9	2.2	2	0.409	0.497	3	9	2.2	2	0.364	0.743
2	9	2.2	2	0.364	0.686	3	9	2.2	2	0.227	0.953
2	9	2.2	2	0.227	0.934	3	9	2.2	2	0.136	1
2	9	2.2	2	0.136	0.984	3	9	2.2	2	0	0.992
2	9	2.2	2	0	1	2	12	2.2	0.2	0.5	0

against the numerically calculated results, as shown in Fig. 4.43, for some randomly chosen cases. It is obvious that the developed MARS model is capable of accurately determining the maximum wall deflection envelope.

4.6.3 Case Validations

Validations from the excavation case histories are provided in Table 4.20. It is obvious that the developed MARS model gives accurate estimations of the normalized maximum lateral wall deflections.

4.7 Estimation of Wall Deflections in Anisotropic Clays

4.7.1 Finite Element Modeling

The braced excavation system is analyzed by the finite element method using Plaxis2D. The numerical model comprises the diaphragm wall, four levels of struts, one soft clay layer, and one stiff clay layer. The soil bodies are simulated via 15-noded triangular elements. The structural elements of the diaphragm wall are assumed to be linearly elastic and modeled by 5-noded beam elements. The 3-noded bar elements represented the four levels of struts. The nodes on the left and right sides of the

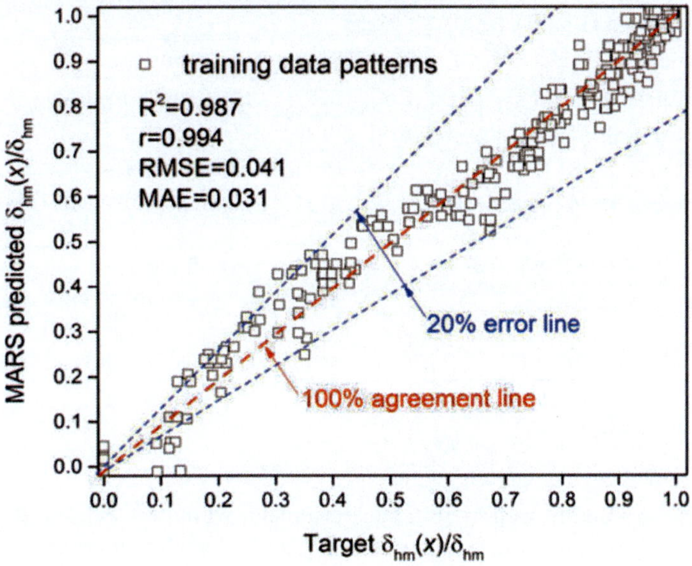

(a) training patterns

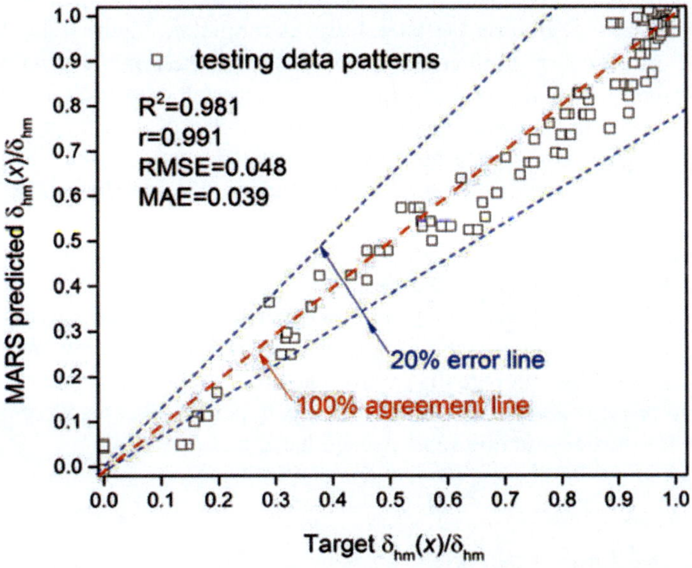

(b) testing patterns

Fig. 4.42 Comparison between target and MARS predicted $\delta_{hm}(x)/\delta_{hm}$

Table 4.19 Expressions of BFs for MARS model

BF	Equation	BF	Equation
BF1	max(0, 0.40476 – *x*/L)	BF9	BF6 × max(0, α –2)
BF2	max(0, 2 –α) × max(0, *x*/L –0.47619)	BF10	BF6 × max(0, 2 –α)
BF3	max(0, 3.4 – L/B)	BF11	max(0, 0.45238 – *x*/L)
BF4	BF3 × max(0, *x*/L –0.45455)	BF12	max(0, *x*/L –0.45238) × max(0, CT –2)
BF5	BF3 × max(0, 0.45455 – *x*/L)	BF13	max(0, 0.47059 – *x*/L)
BF6	max(0, *x*/L –0.22727)	BF14	max(0, *x*/L –0.47059) × max(0, CT –2)
BF7	BF6 × max(0, CT –2)	BF15	max(0, *x*/L –0.47059) × max(0, 2 – CT)
BF8	BF6 × max(0, 2 – CT)		

mesh are constrained from horizontal displacement, while the nodes at the bottom are constrained from moving both horizontally and vertically. In order to minimize the effects of the boundary restraints effect, the right vertical boundary is extended far enough from the excavation area.

A parametric study was carried out by adopting the NGI-ADP model for the soft clay with emphasizing on the lateral wall deformation. Figure 4.44 shows the schematical cross-section of the excavation system with the final excavation depth H_e. It is a simplified typical soil stratum in many coastal areas such as Singapore, which comprises a thick normally consolidated soft-clay deposit layer overlying a stiff clay layer. The Mohr–Coulomb constitutive model is used for the stiff clay. The struts are used at the depths of 1 m, 3 m, 5 m, and 7 m below the original ground surface, and the horizontal strut spacing ($L_{spacing}$) is 4 m. The strut stiffness per meter *EA* is assumed to be constant at 6.1×10^5 kN/m, and the elastic modulus of the diaphragm wall is $E_{conc} = 2.8 \times 10^7$ kPa.

The properties of soft and stiff clay are listed in Tables 4.21 and 4.22, respectively.

The parameters considered and ranges of properties for soft clay are shown in Table 4.23. The wall width is *b*, the excavation width is *B*, the penetration depth of the wall into the stiff layer is *D*. A total of 1778 hypothetical cases were analyzed. The details for the construction sequence are listed in Table 4.24.

4.7.2 Results and Analyses

Figure 4.45 shows the comparison of δ_{hmax} for the cases with s_u^P/s_u^A equals 0.5 and 1. The case with $s_u^P/s_u^A = 1$ implies that the soft clay is isotropic while the case with $s_u^P/s_u^A = 0.5$ means that the soft clay is anisotropic. It is shown that the influence of

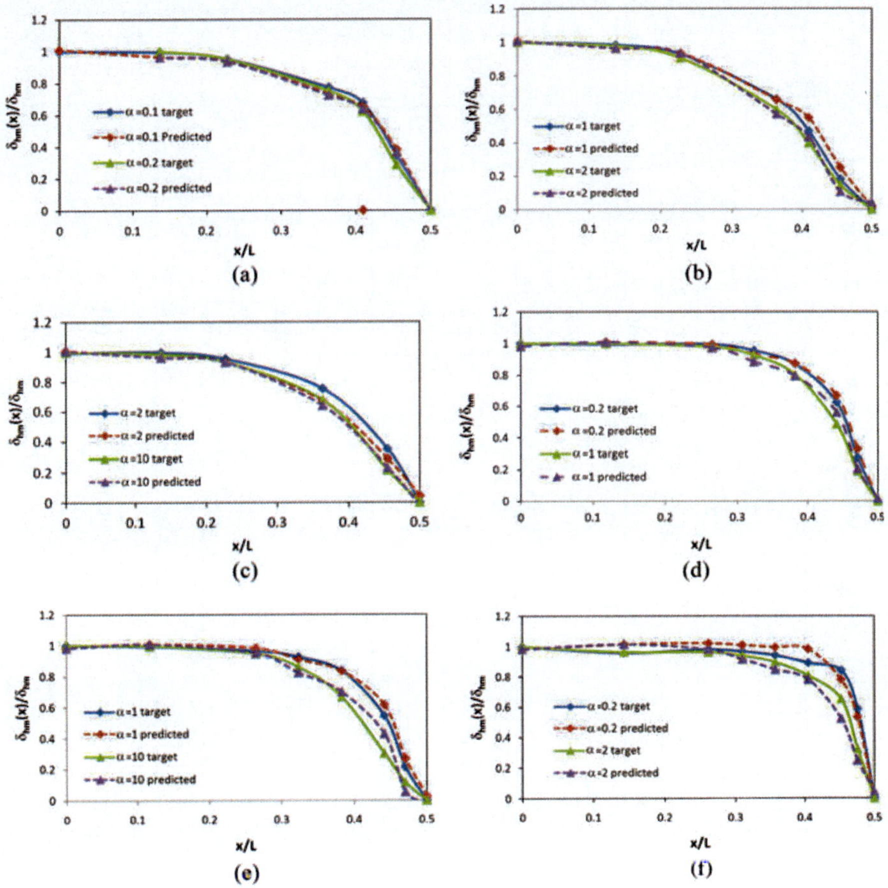

Fig. 4.43 Estimated maximum wall deflection envelope vs. calculated for cases: **a** CT = 1, H_e = 9 m, L/B = 2.2,α = 0.1 and 0.2, **b** CT = 1, H_e = 9 m, L/B = 2.2, α = 1 and 2, **c** CT = 3, H_e = 16 m, L/B = 2.2, α = 2 and 10, (d) CT = 1, H_e = 9 m, L/B = 3.4, α = 0.2 and 1, (e) CT = 2, H_e = 9 m, L/B = 3.4, α = 1 and 10, and (f) CT = 3, H_e = 12 m, L/B = 4.2, α = 0.2 and 2

s_u^P/s_u^A on the δ_{hmax} is significant. The δ_{hmax} of the case with $s_u^P/s_u^A = 1$ is less than half of δ_{hmax} for the case with $s_u^P/s_u^A = 0.5$.

Figure 4.46 shows the plastic failure points for $s_u^P/s_u^A = \{0.4, 0.5, 0.6, 0.8, 1.0\}$. It is shown that as the s_u^P/s_u^A decreases, not only the plastic failure point number increase but also the plastic failure area shape and coverage changed. Moreover, there is an evident sliding surface that develops behind the wall only for the cases with s_u^P/s_u^A smaller than 1.

As shown in Fig. 4.47, the excavation width, wall penetration, and wall width have little influence on the wall deformation, except for the case with the $s_u^P/s_u^A = 0.4$. Since the wall movement is rigid for the "stiff" walls, the soil of the passive side has not failed for the higher s_u^P/s_u^A. Thus, the influence on the wall deformation is less

Table 4.20 Validation of excavation case histories

No	Clay type (CT)	H_e (m)	L/B	α	x/L	*Measured $\delta_{hm}(x)/\delta_{hm}$	MARS predicted $\delta_{hm}(x)/\delta_{hm}$	**Relative error (%)	References
Xinye Bank									[75]
P3	1	8.55	2.6	4.0	0.25	0.84	0.92	9.5	
P4	1	12.4	2.6	4.0	0.0	0.94	0.98	4.2	
P4	1	14.4	2.6	4.0	0.0	1.00	1.00	0.0	
P5	1	8.55	2.6	4.0	0.25	0.87	0.92	5.7	
Cashew station									[75]
P1	2	20	6.9	3.23	0.36	0.41	0.58	29.3	
P2	2	20	6.9	3.23	0.38	0.36	0.55	34.5	
P3	2	20	6.9	3.23	0.0	1.00	1.00	0.0	
P4	2	20	6.9	3.23	0.0	1.00	1.00	0.0	
P5	2	20	6.9	3.23	0.15	0.68	0.78	12.8	
P6	2	20	6.9	3.23	0.15	0.76	0.78	2.6	
P7	2	20	6.9	3.23	0.44	0.28	0.43	34.9	
Hillview station									[75]
P1	2	24	4.0	3.23	0.24	0.58	0.65	10.8	
P2	2	24	4.0	3.23	0.00	1.00	1.00	0.0	
P3	2	24	4.0	3.23	0.24	0.61	0.65	6.5	
P4	2	24	4.0	3.23	0.42	0.31	0.53	41.5	
P5	2	24	4.0	3.23	0.24	0.65	0.70	7.6	
P6	2	24	4.0	3.23	0.00	1.00	1.00	0.0	
P7	2	24	4.0	3.23	0.24	0.53	0.70	24.3	

(continued)

Table 4.20 (continued)

No	Clay type (CT)	H_e (m)	L/B	α	x/L	*Measured $\delta_{hm}(x)/\delta_{hm}$	MARS predicted $\delta_{hm}(x)/\delta_{hm}$	**Relative error (%)	References
P8	2	24	4.0	3.23	0.42	0.29	0.51	43.1	
BKP station									[75]
P1	3	20.5	3.6	2.51	0.00	1.00	1.00	0.0	
P2	3	20.5	3.6	2.54	0.18	0.69	0.78	11.5	
P3	3	20.5	3.6	2.54	0.23	0.61	0.70	12.8	
BTW station									[75]
P1	3	21	7.1	2.54	0	1.00	1.00	0.0	
P2	3	21	7.1	2.54	0.33	0.49	0.57	17.5	

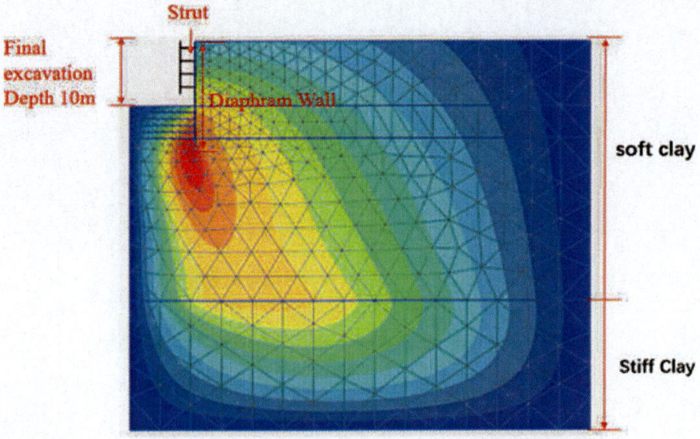

Fig. 4.44 Cross-sectional profile and typical mesh

Table 4.21 NGI-ADP soil model properties of soft clay (undrained C)

Parameters	Values	Units
γ	16	(kN/m^3)
$s_u^{C,TX}/s_u^A$	0.99	(–)
γ_f^C	0.75	(%)
γ_f^E	3.5	(%)
γ_f^{DSS}	1.735	(%)
R_{inter}	1.0	(–)
s_u^{DSS}/s_u^A	$(1 + s_u^P/s_u^A)/2$	(–)
τ_0/s_u^A	0.7	(–)
υ_u	0.495	(–)
y_{ref}	60	(m)
$s_{u,inc}$	0	(kPa/m)

Table 4.22 Stiff clay properties (undrained C)

Parameters and units	Values
Soil unit weight γ (kN/m^3)	16
Soil undrained shear strength c_u (kPa)	100
Soil modulus ratio E/c_u	300
0_u(°)	0
K_o = 1-sinϕ	1.0
Poisson's ratio υ_u	0.495
Interface R_{inter}	1.0

Table 4.23 Parameters and the ranges considered for soft clay

Parameter	Ranges	Units
Soil unit weight γ	15, 16, 18	(kN/m^3)
s_u^P/s_u^A	0.4, 0.5,0.6, 0.8, 1.0	(–)
$s_{u,ref}^A$	40, 50, 60	(kPa)
G_{ur}/s_u^A	300, 600, 900	(–)
Wall width b	0.6, 0.8, 1.2	(m)
Excavation width B	20, 30	(m)
Wall penetration D	3, 5, 10	(m)

Table 4.24 Construction sequences

Phase	Construction activities
1	Install wall (ground surface at y = 60 m)
2	Excavate to y = 58 m; reset displacement to zero
3	Install strut 1 at y = 59 m
4	Excavate to y = 56 m
5	Install strut 2 at y = 57 m
6	Excavate to y = 54 m
7	Install strut 3 at y = 55 m
8	Excavate to y = 52 m
9	Install strut 4 at y = 53 m
10	Excavate to y = 50 m
11	Set calculation type as safety

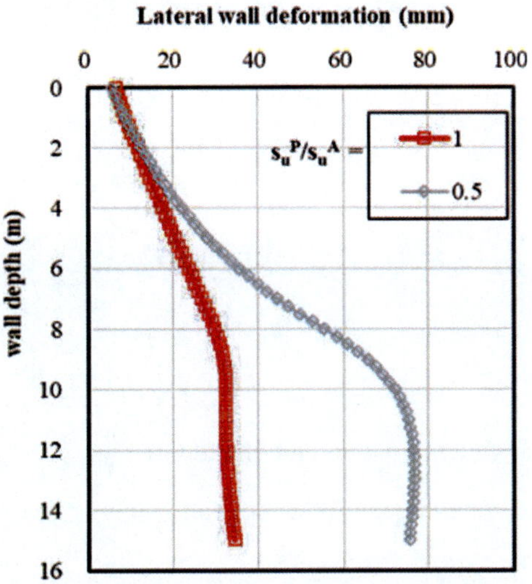

Fig. 4.45 Comparison of δ_{hmax} for isotropic and anisotropic clays

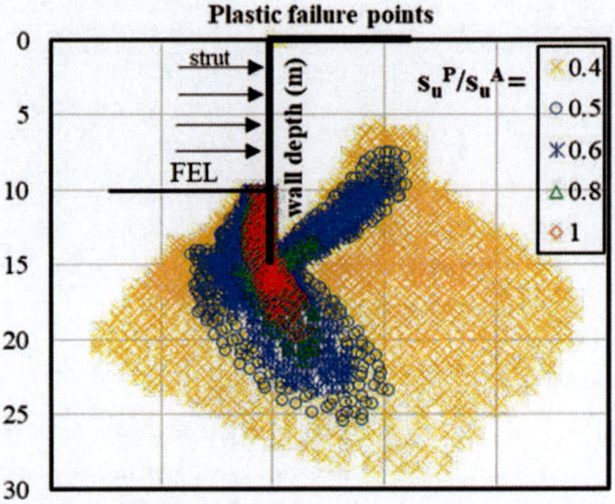

Fig. 4.46 Plastic failure points for $s_u^P/s_u^A = \{0.4,\ 0.5,\ 0.6,\ 0.8,\ 1.0\}$

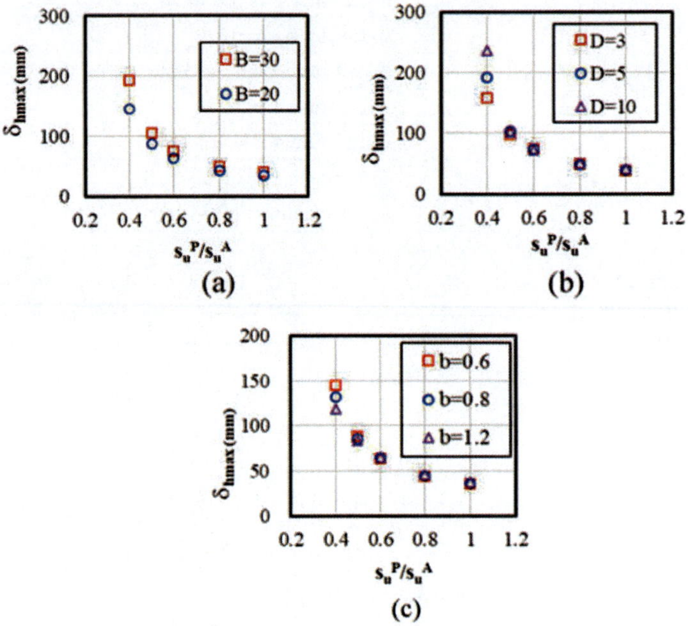

Fig. 4.47 Influence of *B*, *D*, and *b* on δ_{hmax} for $s_{u,ref^A} = \{0.4,\ 0.5,\ 0.6,\ 0.8,\ 1.0\}$

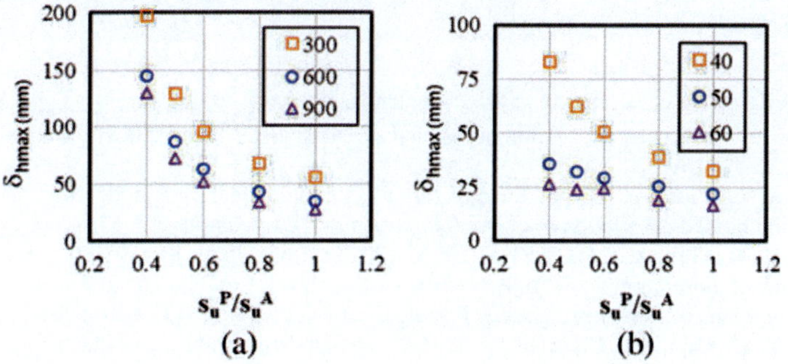

Fig. 4.48 Influence of $s^A_{u,ref}$ and G_{ur}/s^A_u on δ_{hmax} for $s^P_u/s^A_u = \{0.4, 0.5, 0.6, 0.8, 1.0\}$

significant. It requires further study in the future involving flexible walls, for which it is expected that the influence of *B*, *D*, and *b* will be more significant.

However, compared to the three parameters shown in Fig. 4.48, the soil strength parameters such as s^P_u/s^A_u, G_{ur}/s^A_u, and $s^A_{u,ref}$ have more significant influences on the δ_{hmax} than G_{ur}/s^A_u. Figure 4.8a shows that the influence of G_{ur}/s^A_u on the δ_{hmax} is less significant especially when s^P_u/s^A_u and G_{ur}/s^A_u are greater. Figure 4.48b indicates that with the s^P_u/s^A_u increases, the δ_{hmax} decreases more when the $s^A_{u,ref}$ is 40, compared with the case of $s^A_{u,ref} = 60$, indicating that s^P_u/s^A_u has more considerable influence on the wall deformation when $s^A_{u,ref}$ is smaller. It also can be concluded from (a) that the higher the s^P_u/s^A_u, the marginal the influence of $s^A_{u,ref}$ on the wall deformation.

4.7.3 Summary and Conclusions

This study carried out finite element simulations by adopting the NGI-ADP soil model to assess the lateral wall deflection for excavations in soft clays. The effects of seven key parameters on diaphragm wall deflections are evaluated, including excavation width, wall stiffness, wall penetration, soil parameters of s^P_u/s^A_u, G_{ur}/s^A_u, $s^A_{u,ref}$, and γ. The results show that the *b*, *B*, *D* have a marginal influence on the δ_{hmax} when the wall is stiff, except for the case with $s^P_u/s^A_u = 0.4$. The soil properties of the $s^A_{u,ref}$, G_{ur}/s^A_u, and s^P_u/s^A_u have significant influences on the δ_{hmax}. Moreover, the results are consistent with the sensitivity analysis results of the XGBoost. The anisotropic characteristic of soil parameters is essential in determining the lateral wall deflections induced by braced excavation supported by the diaphragm wall.

References

1. Zhang, W. G., Goh, A. T. C., Goh, K. H., Chew, O. Y. S., Zhou, D., & Zhang, R. (2018). Performance of braced excavation in residual soil with groundwater drawdown. *Underground Space, 3*, 150–165.
2. Zhang, W. G., Goh, A. T. C., & Xuan, F. (2015). A simple prediction model for wall deflection caused by braced excavation in clays. *Computers and Geotechnics, 63*, 67–72.
3. Zhang, W. G., Goh, A. T. C., & Zhang, Y. M. (2015). Probabilistic assessment of serviceability limit state of diaphragm walls for braced excavation in clays. *ASCE-ASME Journal of Risk and Uncertainty in Engineering Systems, Part A: Civil Engineering, 1*(3), 06015001.
4. Zhang, W. G., Goh, A. T. C., & Zhang, Y. M. (2015). Updating soil parameters using spreadsheet method for predicting wall deflections in braced excavations. *Geotechnical and Geological Engineering, 33*, 1489–1498.
5. Goh, A. T. C., Zhang, F., Zhang, W. G., Zhang, Y. M., & Liu, H. L. (2017). A simple estimation model for 3D braced excavation wall deflection. *Computers and Geotechnics, 83*, 106–113.
6. Zhang, W. G., Zhang, R. H., Wu, C. Z., Goh, A. T. C., Lacasse, S., Liu, Z. Q., & Liu, H. L. (2020). State-of-the-art review of soft computing applications in underground excavations. *Geoscience Frontiers, 11*, 1095–1106.
7. Zhang, R. H., Wu, C. Z., Goh, A. T. C., Böhlke, T., & Zhang, W. G. (2021). Estimation of diaphragm wall deflections for deep braced excavation in Anisotropic clays using ensemble learning. *Geoscience Frontiers, 12*, 365–373.
8. Symons, I. F., & Carder, D. R. (1992). Field measurement on embedded retaining walls. *Geotechnique, 42*, 117–126.
9. Powrie, W., & Kantartzi, C. (1996). Ground response during diaphragm wall installation in clay: Centrifuge model tests. *Geotechnique, 46*, 725–739.
10. Ou, C. Y., Hsien, P. G., & Chiou, D. C. (1993). Characteristics of ground surface settlement during excavation. *Canadian Geotechnical Journal, 30*, 758–767.
11. St., J. H. (1975). *Field and theoretical studies of the behaviour of ground around deep excavations in London clay*. University of Cambridge.
12. Simic, M., French, D.J., 1998. Three dimensional analysis of deep underground stations, In *Proceedings Seminar on the Value of Geotechnics in Construction*, pp. 93–100.
13. Richards, D. J., Homes, G., & Beadman, D. R. (1999). Measurement of temporary prop loads at Mayfair park. *Proceedings of the ICE-Geotechnical Engineering, 137*, 165–174.
14. Ou, C. Y., Chiou, D. C., & Wu, T. S. (1996). Three-dimensional finite element analysis of deep excavations. *Journal of Geotechnical Engineering-Asce, 122*, 337–345.
15. Finno, R. J., Blackburn, J. T., & Roboski, J. F. (2007). Three-dimensional effects for supported excavations in clay. *Journal of Geotechnical and Geoenvironmental Engineering, 133*, 30–36.
16. Rowe, P. W. (1952). Anchored sheet pile wall, Proceedings Civil Engineering, London: Thomas Telford, pp. 27–70.
17. Ou, C. Y. (2006). *Deep excavation: Theory and practice*. Taylor and Francis.
18. Gaba, A., Hard, S., & Doughty, L. (2017). Guidance on embedded retaining wall design. CIRIA, Griffin Court, 15 Long lane, London.
19. CGS. (2006). *Canadian foundafation engineering manual* (4th ed.). Canadian Geotechnical Society.
20. Clough, G. W., O'Rourke, T. D. (1990). Construction induced movements of in situ wall. Geotechnical Special Publication, pp. 439–470.
21. Winkler, E. (1867). Die Lehre Von Elasticitaet Und Festigkeit, Pray (H. Dominicus).
22. Brinkgreve, L., Swolfs, W., & Engin, E. (2017). *Plaxis manual*. PLAXIS BV.
23. Alpan, I. (1970). The geotechnical properties of soils. *Earth-Science Reviews, 6*, 5–49.
24. Kung, G. T. C., Ou, C. Y., & Juang, C. H. (2009). Modeling small-strain behavior of Taipei clays for finite element analysis of braced excavations. *Computers & Geotechnics, 36*, 304–319.
25. Kung, G. T. C. (2003). *Surface settlement induced by excavation with consideration of small-strain behavior of Taipei silty clay*. University of Science and Technology.

26. Kung, G. T. C., Hsiao, E. C. L., & Juang, C. H. (2007). Evaluation of a simplified small-strain soil model for analysis of excavation-induced movements. *Canadian Geotechnical Journal, 44*, 726–736.
27. Clayton, C. R. I. (2011). Stiffness at small strain: Research and practice. *Géotechnique, 61*, 5–37.
28. Poh, T. Y., Wong, I. H., & Chandrasekaran, B. (1997). Performance of two propped diaphragm walls in stiff residual soils. *Journal of Performance Constructed Facilities, 11*, 190–199.
29. Xuan, F. (2009). *Behavior of diaphragm walls in clays and reliability analysis*. Nanyang Technological University.
30. Miyoshi, M. (1977). Mechanical behavior of temporary braced wall, In: *Proceedings 9th Int Conference on Soil Mechanics and Foundation Engineering*, Tokyo, pp. 655–658.
31. Finno, R. J., Atmatzidis, D. K., & Perkins, S. B. (1989). Observed performance of a deep excavation in clay. *Journal of Geotechnical Engineering, 115*, 1045–1064.
32. Faheem, H., Cai, F., & Ugai, K. (2004). Three-dimensional base stability of rectangular excavations in soft soils using FEM. *Computers and Geotechnics, 31*, 67–74.
33. Bryson, L. S., & Zapata-Medina, D. G. (2012). Method for estimating system Stiffness for excavation support walls. *Journal of Geotechnical and Geoenvironmental Engineering, 138*, 1104–1115.
34. Terzaghi, K., & Peck, R. B. (1967). *Soil Mechanics in engineering practice*. John Wiley & Sons.
35. Chang, M.-F. (2000). Basal stability analysis of braced cuts in clay. *Journal of Geotechnical and Geoenvironmental Engineering, 126*, 276–279.
36. Friedman, J. H. (1991). Multivariate adaptive regression splines. *Annals of Statistics, 19*, 1–141.
37. Jekabsons, G. (2010). VariReg: A software tool for regression modeling using various modeling methods, Riga Technical University.
38. Hastie, T., Tibshirani, R., & Friedman, J. (2009). *The elements of statistical learning: Data mining, inference and prediction* (2nd ed.). Springer.
39. Goh, A. T. C., & Zhang, W. G. (2014). An improvement to MLR model for predicting liquefaction-induced lateral spread using multivariate adaptive regression splines. *Engineering Geology, 170*, 1–10.
40. Goh, A. T. C., Zhang, W. G., Zhang, Y. M., Xiao, Y., & Xiang, Y. Z. (2016). Determination of EPB tunnel-related maximum surface settlement: A Multivariate adaptive regression splines approach. Bulletin of Engineering Geology and the Environment.
41. Zhang, W. G., & Goh, A. T. C. (2013). Multivariate adaptive regression splines for analysis of geotechnical engineering systems. *Computers & Geotechnics, 48*, 82–95.
42. Zhang, W. G., Zhang, Y. M., & Goh, A. T. C. (2017). Multivariate adaptive regression splines for inverse analysis of soil and wall properties in braced excavation. *Tunneling and Underground Space Technology, 64*, 24–33.
43. Adoko, A. C., Jiao, Y. Y., Wu, L., Wang, H., & Wang, Z. H. (2013). Predicting tunnel convergence using multivariate adaptive regression splines and artificial neural network. *Tunnelling and Underground Space Technology, 38*, 368–376.
44. Xiang, Y. Z., Goh, A. T. C., Zhang, W. G., & Zhang, R. H. (2018). A multivariate adaptive regression splines model for estimation of maximum wall deflections induced by braced excavation in clays. *Geomechanics and Engineering, 14*(4), 315–324.
45. Fok, P., Neo, B. H., Veeresh, C., Wen, D., & Goh, K. H. (2012). Limiting values of retaining wall displacements and impact to the adjacent structures. *The IES Journal Part A: Civil & Structural Engineering, 5*, 134–139.
46. Alavi, A. H., Ameri, M., Gandomi, A. H., & Mirzahosseini, M. R. (2011). Formulation of flow number of asphalt mixes using a hybrid computational method. *Construction and Building Materials, 25*, 1338–1355.
47. Zhang, W. G., Zhang, R. H., & Goh, A. T. C. (2017). Multivariate adaptive regression splines approach to estimate lateral wall deflection profiles caused by braced excavations in clays. *Geotechnical and Geological Engineering, 36*(2), 1349–1363.

48. Zapata-Medina, D. G. (2007). *Semi-empirical method for designing excavation support systems based on deformation control*. University of Kentucky.
49. Ng, C. W. W. (1992). An evaluation of soil-structure interaction associated with a multi-propped excavation. Ph.D. Thesis University of Bristol 13, pp. 31–35.
50. Burland, J. B., & Hancock, R. J. R. (1977). Underground car park at the House of Commons: Geotechnical Aspects. *The Structural Engineer, 55*, 87–105.
51. Hsieh, P. G., & Ou, C. Y. (1998). Shape of ground surface settlement profiles caused by excavation. *Canadian Geotechnical Journal, 35*, 1004–1017.
52. Ou, C. Y., Liao, J. T., & Lin, H. D. (1998). Performance of diaphragm wall constructed using Top-Down Method. *Journal of Geotechnical and Geoenvironmental Engineering, 124*, 798–808.
53. Whittle, A. J., Hashash, Y. M. A., & Whitman, R. V. (1993). Analysis of deep excavation in Boston. *Journal of Geotechnical Engineering, 119*, 69–90.
54. Liao, H. J., & Hsieh, P. G. (2002). Tied-back excavations in alluvial soil of Taipei. *Journal of Geotechnical and Geoenvironmental Engineering, 128*, 435–441.
55. Becker, J. M., & Haley, M. X. (2011). Up/Down construction-decision making and performance. *Design & Performance of Earth Retaining Structures*, pp. 170–189.
56. Ulrich, E. J., Jr. (1989). Tieback supported cuts in over-consolidated soils. *Journal of Geotechnical Engineering, 115*, 521–545.
57. Finno, R. J., & Roboski, J. F. (2005). Three-dimensional responses of a tied-back excavation through clay. *Journal of Geotechnical and Geoenvironmental Engineering, 131*, 273–282.
58. NGI, N. G. I. (1962). Measurements at a strutted excavation, Oslo Subway, Enerhaugen South, km 1982, Technical Report No. 3. NGI, Oslo, Norway.
59. NGI, N. G. I. (1962). Measurements at a strutted excavation, Oslo Subway, Vaterland 1, km 1373 Technical Report No. 5. NGI, Oslo, Norway.
60. NGI, N. G. I. (1962). Measurements at a strutted excavation, Oslo Subway, Vaterland 2, km 1408., Technical Report No. 6. NGI, Oslo, Norwa.
61. NGI, N. G. I. (1962). Measurements at a strutted excavation, Oslo Subway, Vaterland 3, km 1450, Technical Report No. 5. NGI, Oslo, Norway.
62. NGI, N. G. I. (1962). Measurements at a strutted excavation, The new headquarters building for the Norwegian Telecommunications Administration, Oslo. , Technical Report No. 4. . NGI, Oslo, Norway.
63. Clough, G. W., & Buchignani, A. L. (1981). *Slurry walls in the San Francisco Bay Area* (pp. 81–142). ASCE.
64. Wang, J. H., Xu, Z. H., Chen, J. J., & Wang, W. D. (2005). Deformation properties of diaphragm wall due to deep excavation in Shanghai soft soil. *Chinese Journal of Underground Space and Engineering, 1*, 485–489.
65. Peck, R. B. (1969). Deep excavation and tunneling in soft ground. *Proceedings of the 7th International Conference on soil Mechanics and Foundation Engineering*, Mexico City, pp. 225–290.
66. Finno, R. J., Bryson, S., & Calvello, M. (2002). Performance of a stiff support system in soft clay. *Journal of Geotechnical and Geoenvironmental Engineering, 128*, 660–671.
67. Goh, A. T. C., Wong, K. S., Teh, C. I., & Wen, D. (2003). Pile response adjacent to braced excavation. *Journal of Geotechnical and Geoenvironmental Engineering, 129*, 383–386.
68. Hu, Z. F., Yue, Z. Q., Zhou, J., & Tham, L. G. (2003). Design and construction of a deep excavation in soft soils adjacent to the Shanghai Metro tunnels. *Canadian Geotechnical Journal, 40*(5), 933–948.
69. Gill, S. A., & Lucas, R. G. (1990). *Ground movement adjacent to braced cuts*. Geotechnical Special Publication, pp. 471–488.
70. Teparaksa, W. (1993). Behavior of deep excavations using sheet pile bracing system in soft Bangkok clay Third International Conference on Case Histories in Geotechnical Engineering, Louis, USA, pp. 745–750.
71. Baker, C. N. J., Bucher, S. A., & Baker, W. F. J. (1989). Complex high-rise foundation design and construction. *Proceedings Foundation Engineering: Current Principles and Practices*, Evanston, IL, pp. 1445–1458.

72. Konstantakos, D. C. (2000). *Measured performance of slurry Wall.* Massachusetts Institute of Technology.
73. Kort, D. A. (2002). *Steel sheet pile walls in soft soil.* Delft University of Technology.
74. Koutsoftas, D. C., Frobenius, P., Wu, C. L., Meyersohn, D., & Kulesza, R. (2000). Deformations during cut-and cover construction of MUNI Metro Turnback project. *Journal of Geotechnical and Geoenvironmental Engineering, 126*, 344–359.
75. Zhang, W. G., Zhang, R. H., Wei, W., Zhang, F., & Goh, A. T. C. (2019). A Multivariate Adaptive Regression Splines model for determining horizontal wall deflection envelope for braced excavations in clays. *Tunneling and Underground Space Technology, 84*, 461–471.
76. Xu, Z. H. (2007). *Deformation behaviour of deep excavations supported by permanent structure in Shanghai soft deposit*. Shanghai Jiao Tong University.

Chapter 5
Ground Settlements and Dewatering

5.1 Introduction

The excavation process causes unloading in the surrounding soil and ground settlement. Unfavorable ground movements result in detrimental effects on adjacent infrastructures, such as buildings, pipelines, roads, bridges, and tunnels.

This section will introduce the influential factors and some methods to predict ground settlement induced by excavation and dewatering.

5.2 Relation Between δ_{hm} and δ_{vm}

In general, the maximum ground surface settlement δ_{vh} can be estimated by referring to the value of the maximum wall deflection δ_{hm}. From the data of case histories reported by Mana and Clough [1], Ou et al. [2], and Hsieh and Ou [3], the relationship between maximum wall deflection and maximum ground surface settlement can be obtained as

$$\delta_{vh} \approx 0.5\delta_{hm} \text{ to } 1.0\delta_{hm} \tag{5.1}$$

Long [4] analyzed the effect of different support types on δ_{vm} and δ_{hm}. It was found that there is a variable response in the ratio of δ_{vm} and δ_{hm} between different support types and ground conditions. The findings of Long [4] are summarized in Table 5.1. This shows that, as the soil conditions deteriorate, the ratio of δ_{vm}/δ_{hm} changes from <1 to >1. Long [4] did not quantify the relationship between support stiffness and ground movements, despite identifying the possibility that existed. Long [4] suggested that there is a linear relationship between H and δ_{vm} and δ_{hm} once classification of different ground conditions has been made, although this was not

W. Zhang and H. Liu, *Design of Deep Braced Excavation and Earth Retaining Systems Under Complex Built Environment*,
https://doi.org/10.1007/978-981-16-5320-9_5

Table 5.1 Findings of Long [4] for magnitudes of ground movement based on the split database

Dataset	δ_{hm} /H_e (%)	δ_{vm} /H_e (%)	Description of data
1	0.18	0.14	$h < 0.6H_e$
2	0.38	0.48	$h > 0.6H_e$, FoS > 3
3	0.82	1.1	$h > 0.6H_e$, FoS < 3

proven in terms of statistical significance. He also identified that there is an increase in δ_{hm} with a decreasing FoS against basal heave, confirming the conclusions of the earlier study by Clough and O'Rourke [5].

Based on the cases by Long [4] using the same variables, in total, 389 case studies have been compiled, with 235 from the database constructed by Long [4], and additional 154 cases added by Harry et al. [6], dating from 2002 to 2016. The database is attached in Appendix I. It should be noted that only the propped or anchored excavation case studies from the database of Long [4] were included.

5.3 Analysis of Ground Settlements

FE method and beam on elastic foundation method can be also utilized to analyze ground settlements induced by deep braced excavation. Apart from it, there are also some empirical models.

5.3.1 Peck' Method

Peck [7] was the first to propose a method to predict excavation-induced ground surface settlement, based on field observations. He mainly employed the monitoring results of case histories in Chicago and Oslo and established the relation curves between the ground surface settlement (δ_v) and the distance from the wall (d) for different types of soil, as shown in Fig. 5.1. The method classifies soil into three types according to the characteristics of soil:

Type I: Sand and soft to stiff clay, average workmanship.
Type II: Very soft to soft clay.

1. 1. Limited depth of clay below the excavation bottom.
2. Significant depth of clay below the excavation bottom but $N_b < N_{cb}$.

Type III: Very soft to soft clay to a significant depth below the excavation bottom and $N_b \geq N_{cb}$

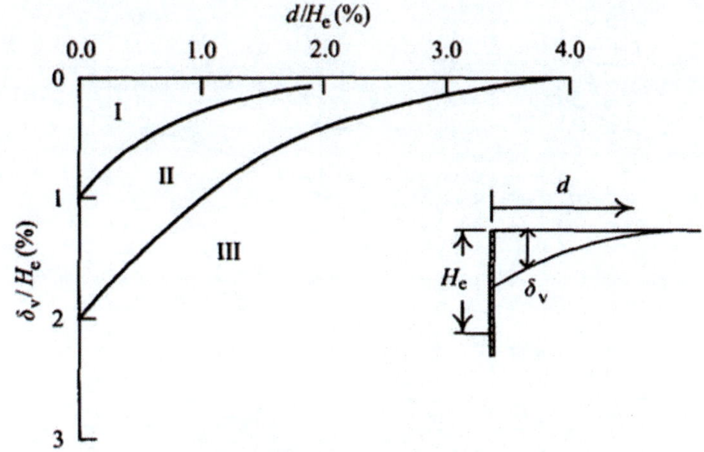

Fig. 5.1 Peck's method for estimating ground surface settlement [7]

where N_b, the stability number of soil, is defined as $FS\gamma H_e/s_u$, where FS is the factor of safety, γ is the unit weight of the soil, H_e is the excavation depth, and S_u is the undrained shear strength of soil, and N_{cb} is the critical stability number against basal heave.

Since Peck's method took the monitoring results of case histories before 1969, most of which employed steel sheet piles or soldier piles with laggings as the retaining wall, quite different from the more advanced design and construction methods (e.g., the diaphragm wall method which offers higher stiffness) employed in excavation projects in recent years, the relation curves proposed by Peck are not necessarily applicable to all excavations. Basically, the curves derived from Peck's method are envelopes. Since Peck's method is the first to derive an empirical formula to predict the ground surface settlement induced by excavation and is simple to apply, it is still used by some engineers.

5.3.2 Clough and O'Rourke's Method

Clough and O'Rourke [5] proposed various types of envelopes of excavation-induced ground surface settlements for different soils on the basis of case studies. According to their studies, excavation in sand or stiff clay will tend to produce a triangular ground surface settlement. The maximum settlement will be found near the retaining wall. The envelopes of the ground surface settlement are as shown in Fig. 5.2a and b, whose influence ranges are separately $2H_e$ and $3H_e$ where He is the final excavation depth. Excavation in soft to medium clay will produce a trapezoidal envelope of ground surface settlement, as shown in Fig. 5.2c. The maximum ground surface settlement occurs in the range of $0 < d/H_e < 0.75$ while $0.75 < d/H_e < 2.0$ is the transition zone where settlement decreases from the largest to almost none.

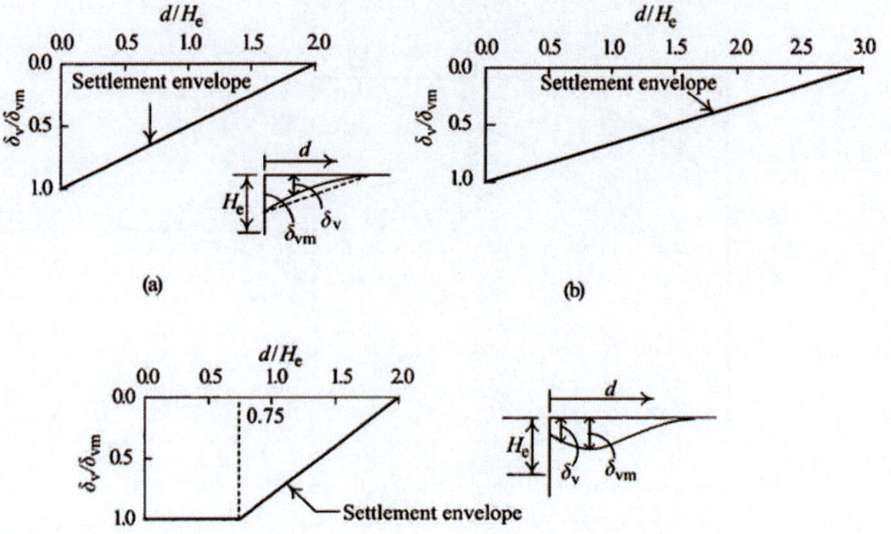

Fig. 5.2 Clough and O'Rourke's method [5] for estimating ground surface settlement: **a** sand, **b** stiff to very stiff clay, and **c** soft to medium soft clay

5.3.3 Bowles' Method

Bowles [8] suggested a procedure to estimate excavation-induced ground surface settlements, which can be described as follows (Fig. 5.3):

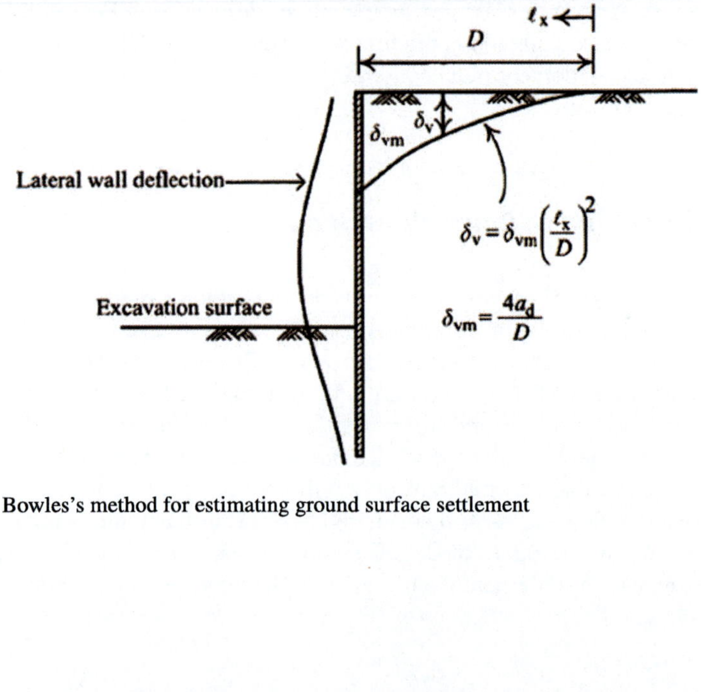

Fig. 5.3 Bowles's method for estimating ground surface settlement

1. Compute the lateral displacement of the wall using the finite-element method or the beam on elastic foundation method.
2. Compute the area of the lateral wall deflection.
3. Estimate the influence range of ground surface settlement (D) following Capse's method (1966):

$$D = (H_e + H_d)\tan\left(45^{\circ} - \varphi/2\right).$$

where H_e = the excavation depth, $H_d = B$ if $\varphi = 0$, and $H_d = 0.5B\tan(45° + \varphi/2)$ if $\varphi \geq 0$ B = the excavation width.
4. Maximum ground surface settlement $\delta_{vm} = 4a_d/D$.
5. The ground surface settlement δ_v is assumed to be parabolic $\delta_v = \delta_{vm}(I_x/D)^2$.

5.3.4 Ou and Hsieh's Method

Ou et al. [2] proposed a procedure to estimate excavation-induced ground settlement profile normal to the excavation support wall. Their work was based on the observation of 10 case histories in soft soils. From these data, they developed a trilinear settlement profile (Fig. 5.4) called the spandrel-type settlement, which presents the maximum settlement very near to the wall. The spandrel type of settlement profile occurs if a large amount of wall deflection occurs at the first phase of excavation

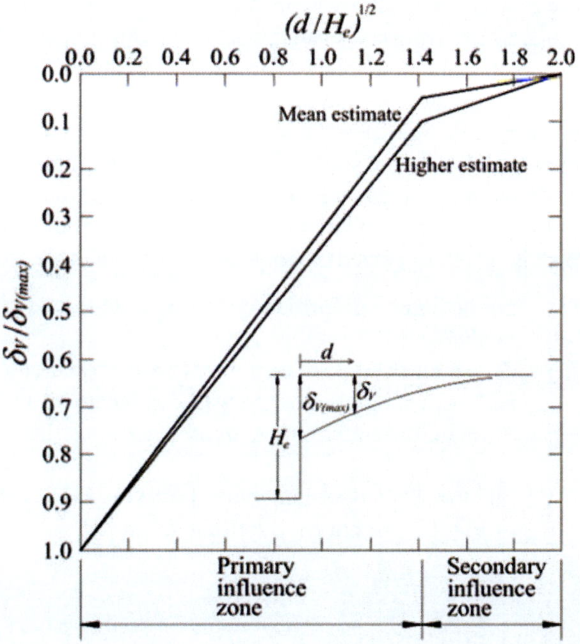

Fig. 5.4 Shape of "Spandrel" settlement profile (after Ou et al. [2])

Fig. 5.5 Proposed method for predicting concave settlement profile (after Hsieh and Ou [3])

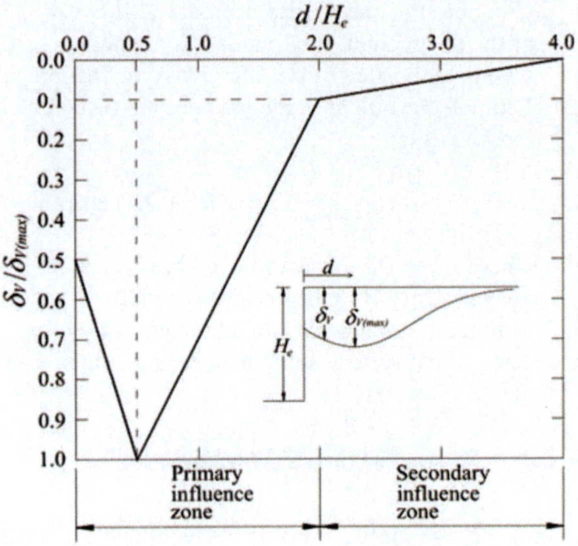

when cantilever conditions exist and the wall deflection is relatively small due to subsequent excavation. The data presented in Fig. 5.4 is the normalized settlement, δ_v/δ_{vh}, where δ_{vh} is the maximum ground surface settlement versus the square root of the distance from the edge of the excavation, d, divided by the excavation depth, H_e.

Hsieh and Ou [3], based on nine case histories worldwide, extended the work done by Ou et al. [2] by proposing the concave settlement profile (Fig. 5.5) induced by deep excavations. From Fig. 5.5, it can be seen that the maximum settlement occurs at a distance of $H_e/2$ from the wall and that the settlement at the wall can be approximated to $0.5\delta_{vh}$. The case history data also showed that the extent of the primary influence zone is approximately two excavation depths ($2H_e$), and after a distance of 4He, the settlement is basically negligible.

Hsieh and Ou [3] also established the relationship of cantilever area and deep inward area of wall deflection, as a first approximation to predict the type of settlement profile. They suggested the following procedures for predicting the settlement profile:

(1) predict lateral deformations using finite-element or beam on elastic foundation methods;
(2) determine the type of settlement profile by calculating the areas of the cantilever and inward bulging of the wall displacement profile;
(3) estimate the maximum ground surface settlement as $\delta_{vh} \approx 0.5\delta_{hm}$ to $1.0\delta_{hm}$; and
(4) plot the surface settlement profile using Fig. 5.4 for spandrel settlement profile or Fig. 5.5 for concave settlement profile.

5.4 Dewatering Effect

Analysis approaches in previous sections do not consider changes in groundwater conditions, thus they only work as if ground deformation is mainly caused by excavation and support process.

Generally, change of groundwater conditions is not a matter in excavations in clays because of the low permeabilities of soil layers. While in granular and residual soils, the dewatering process could result in significant groundwater level drawdown due to unsatisfied design or construction activities [9]. Section 10.1 reports a case in which significant groundwater level drawdown happened and the consequent surface settlement was marked.

This section presents studies on how dewatering impacts on excavation responds. The patterns of dewatering-induced wall displacement and field movement are investigated. And a logarithm regression (LR) model and an artificial neural network (ANN) model are developed to predict the ground settlements.

5.4.1 Logarithm Regression (LR) Model

5.4.1.1 Numerical Scheme and Results

The basic finite-element model is the same as that in Sect. 4.4.1; however, the ranges of parameters are different, listed in Tables 5.2 and 5.3.

The construction sequence comprised the following steps:

(1) the wall is installed (wished into place) without any disturbance in the surrounding soil; and
(2) the soil is excavated uniformly 1 m below each strut level prior to adding the strut support with struts at 3 m vertical spacing until the final depth H_e is reached. Details are listed in Table 5.4.

Table 5.2 Parameters considered and the values

Parameter	Ranges
Relative shear strength ratio c_u/σ'_v	0.25, 0.30, 0.35
Relative soil stiffness ratio E_{50}/c_u	100, 200, 300
Groundwater drawdown d_w (m)	0.3, 6.0, 12.0
*System stiffness S	7.309, 8.176, 8.846
Excavation width B (m)	30, 40
Soft clay thickness T (m)	25, 30
Excavation depth H_e(m)	14, 17, 20

Note: * System stiffness $S = \ln\left(EI/\gamma_w h_{avg}{}^4\right)$

Table 5.3 ϕ, K_0 and E_{50}^{ref} values for soft clay in HSS model

c_u/σ_v'		0.25	0.30	0.35
$\varphi(°)$		22.3	26.4	30.4
K_0		0.621	0.555	0.494
E_{50}^{ref} (kPa)	$E_{50}/c_u = 100$	4026	5405	7085
	$E_{50}/c_u = 200$	8052	10,811	14,170
	$E_{50}/c_u = 300$	12,077	16,216	21,255

Table 5.4 Construction procedures

Phases	Construction details
Initial phase	Generate the initial effective stress, pore pressure, and state parameters
Phase 1	Install the diaphragm wall
Phase 2	Reset displacement to zero, lowering water table inside the excavation to 2 m below ground surface, followed by the removal of soils of 2 m thickness, the water table behind the wall follows steady-state seepage pattern, into the excavated area
Phase 3	Install strut at 1 m below the ground surface
Phase 4	Dewatering of 5 m and excavate to 5 m below the ground surface inside the excavation
Phase 5	Install strut at 4 m below the ground surface
Phase 6	Dewatering of 8 m and Excavate to 8 m below the ground surface
Phase 7	Install strut at 7 m below the ground surface
Phase 8	Dewatering of 11 m and Excavate to 11 m below the ground surface
Phase 9	Install strut at 10 m below the ground surface
Phase 10	Dewatering of 14 m and Excavate to 14 m below the ground surface
Phase 11	Install strut at 13 m below the ground surface
Phase 12	Dewatering of 17 m and Excavate to 17 m below the ground surface
Phase 13	Install strut at 16 m below the ground surface
Phase 14	Dewatering of 20 m and Excavate to 20 m below the ground surface

The groundwater table was at the ground level when excavation started. As excavation proceeds, the water head inside the excavation drops to the excavation level with each stage. Meanwhile, the groundwater drawdown outside excavation may be caused by wall leakage, flow from beneath wall and perched water, flow along wall interface, or poor D-wall panel connections. The maximum drawdown is considered as 12 m, 6 m, and 0.3 m, respectively. The drawdown curve is determined through the steady-state seepage pattern, as shown in Fig. 5.6. It is obvious that due to the hydraulic head difference, the water behind the excavation flows into the excavated area through the zones right below the final elevation level.

Figure 5.7 shows the ground surface settlement profiles corresponding to the different excavation stages for case of $B = 30$ m, $H_e = 20$ m, $c_u/\sigma_v' = 0.35$, E_{50}/c_u

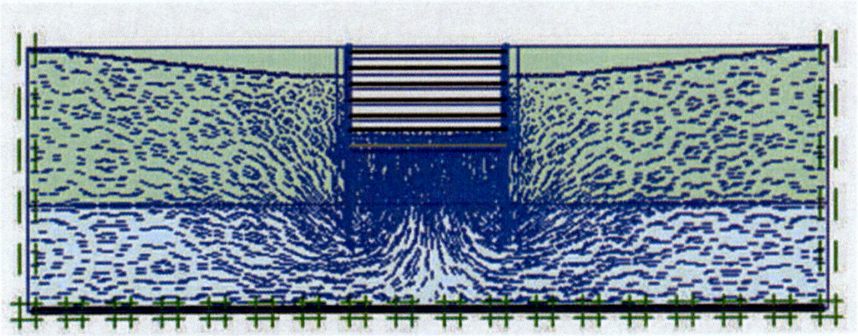

Fig. 5.6 Groundwater drawdown pattern ($d_w = 6$ m)

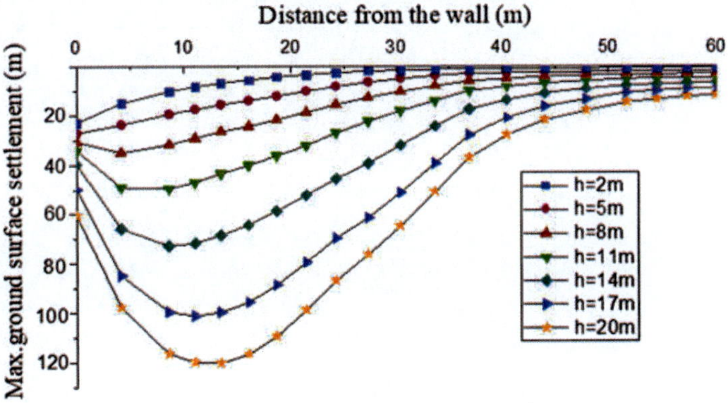

Fig. 5.7 Max. ground surface settlements δ_{vm} for different excavation stages

$= 200$, $\ln\left(EI/\gamma_w h_{avg}^4\right)$ $S = 8.176$, $d_w = 6.0$ m, and $T = 30$ m. It is obvious that the maximum ground surface settlement δ_{vm} increases as excavation proceeds. In addition, the ground surface settlement trough widens as excavation proceeds. For excavation depth h of 2 m, the width of the settlement trough is about 25 m. It increases to about 50 m when the excavation depth h is 11 m. It should also be noted that the maximum ground surface settlement at the final elevation level increases from 90.6 to 120.4 mm, compared with the case without consideration of groundwater drawdown, indicating that the consolidation settlement induced by the increased effective stress due to groundwater drawdown is considerable. Figure 5.8 shows the corresponding maximum wall deflections for each excavation stage. It is obvious that the wall deflects more as excavation proceeds.

Figure 5.9 shows the influence of the groundwater drawdown on the soil lateral movement behind the wall, for cases with $B = 30$ m, $T = 30$ m, $H_e = 20$ m, $S = 7.309$, $c_u/\sigma'_v = 0.30$, and $E_{50}/c_u = 200$, under $d_w = 0.3$ m, 6.0 m, and 12.0 m, respectively. It is obvious that for the same excavation depth, the more water drawdown, the less

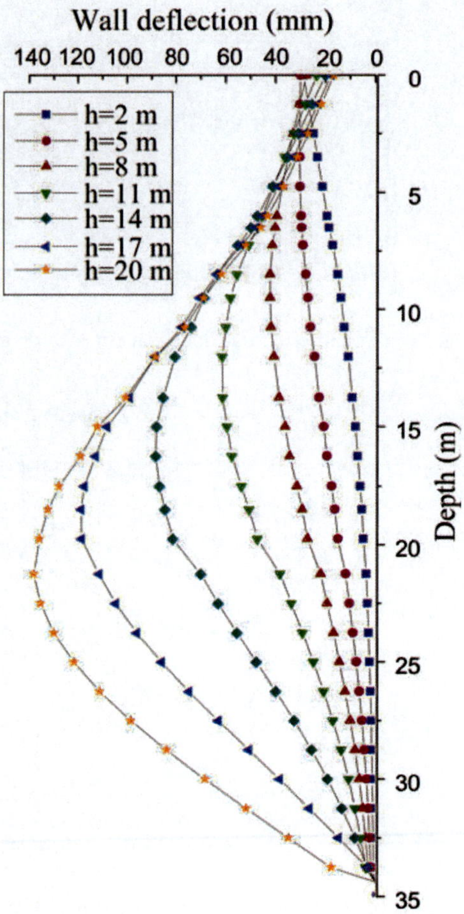

Fig. 5.8 Max. wall deflection δ_{hm_w} for different excavation stages

the maximum lateral soil movement (δ_{hm_s}) behind the wall due to a decreased pore water pressure acted on the wall. Accordingly, the diaphragm wall, as well as the soil behind, deflects less.

Figure 5.10 shows the influence of groundwater drawdown on the ground settlement profiles, the maximum ground surface settlement δ_{vm}, and the basal heaves δ_{bh}, for cases with $B = 30$ m, $T = 30$ m, $H_e = 17$ m, $S = 7.309$, $c_u/\sigma_v' = 0.30$, and $E_{50}/c_u = 200$, under $d_w = 0.3$ m, 6.0 m, and 12.0 m, respectively. It is obvious that δ_{vm} increases significantly as the water drawdowns more. Figure 5.10 also indicates that the settlement trough widens as the groundwater drawdown increases from 0.3 to 12.0 m, which confirms the finding in Fig. 5.6. In addition, the basal heaves δ_{bh} decrease considerably since the uplift pressures exerted on basal soils are reduced due to the groundwater drawdowns.

Figure 5.11 shows the influence of the soil shear strength ratio c_u/σ_v' and the groundwater drawdown d_w on the maximum ground surface settlement, for cases

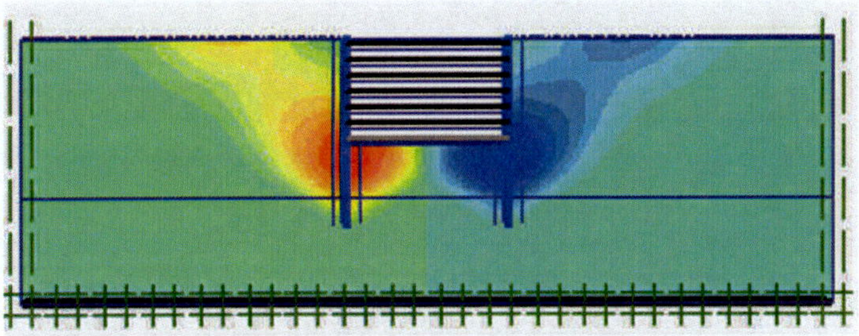

(a) d_w=0.3m, δ_{hm_s} = 252.8mm

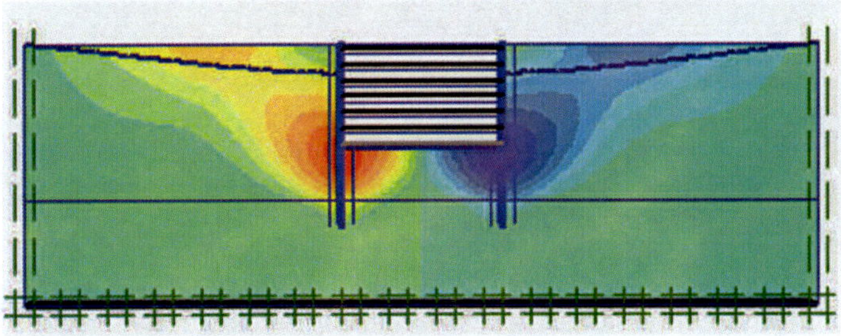

(b) d_w=6.0m , δ_{hm_s} = 203.2mm

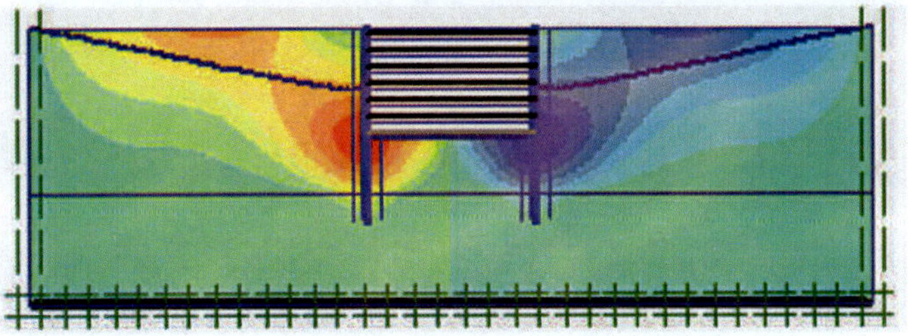

(c) d_w=12.0m, δ_{hm_s} = 151.0mm

Fig. 5.9 Ground horizontal movement for different groundwater drawdowns d_w

with $B = 30$ m, $T = 30$ m, $H_e = 14$ m, and $S = 8.846$ for $c_u/\sigma'_v = 0.25$, 0.3, 0.35 and $d_w = 0.3$ m, 6.0 m, and 12.0 m, respectively. It is obvious that the max. ground settlement increases as the soil shear strength ratio c_u/σ'_v is becoming smaller and the groundwater drawdown becomes more significant. In addition, for the same c_u/σ'_v and d_w, the max. ground settlement decreases as the relative soil stiffness ratio increases.

For brevity, only some general trends of the ground surface settlements are highlighted. The influence of the soil stiffness ratio E_{50}/c_u and the shear strength ratio c_u/σ'_v is shown in Fig. 5.11, for cases with $B = 30$ m, $T = 25$ m, $H_e = 17$ m, and $S = 8.176$ for $d_w = 0.3$ m, 6.0 m, and 12.0 m, respectively. It is obvious that

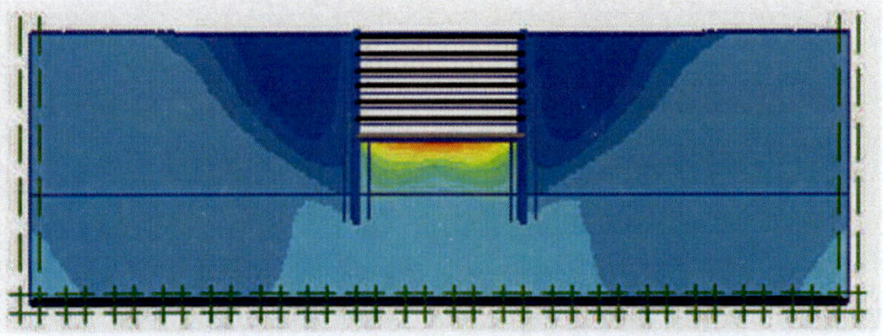

(a) d_w=0.3m δ_{vm}=217.6mm, δ_{bh}=417.1mm

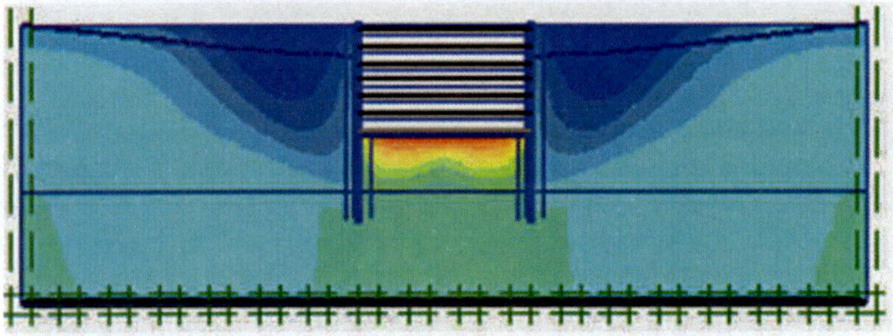

(b) d_w=6.0m δ_{vm}=237.8mm, δ_{bh}=315.1mm

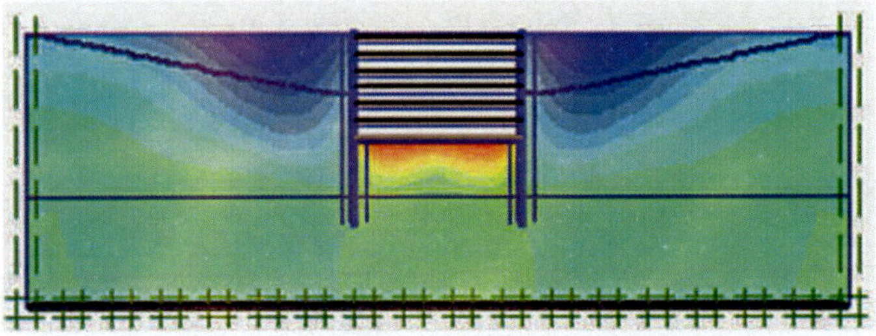

(c) d_w=12.0m, δ_{vm}=261.5mm, δ_{bh}=230.6mm

Fig. 5.10 Ground surface settlements for different groundwater drawdowns d_w

the max. ground surface settlement decreases with the increase of the relative soil stiffness ratio E_{50}/c_u and the relative soil shear strength ratio c_u/σ'_v.

Figure 5.12 shows the influence of the excavation depth H_e and the groundwater drawdown d_w on the maximum ground surface settlement, for cases with $B = 30$ m, $T = 30$ m, and $S = 8.176$ for $H_e = 14$ m, 17 m, and 20 m and $d_w = 0.3$ m, 6.0 m, and 12.0 m, respectively. It is obvious that the max. ground settlement increases as the excavation is becoming deeper and the groundwater drawdown becomes more significant.

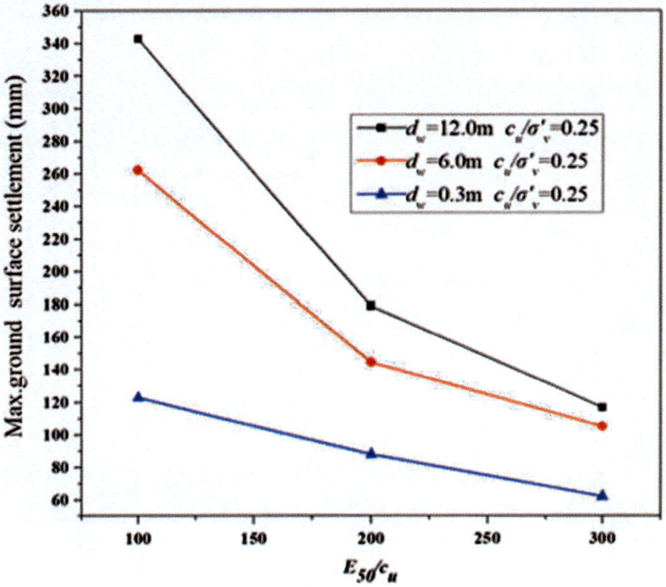

(a) Influence of E_{50}/c_u and d_w on δ_{vm} for c_u/σ'_v=0.25

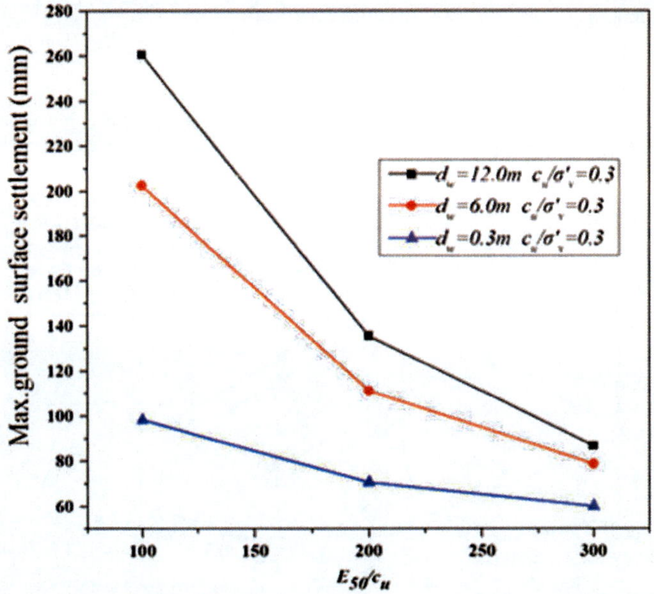

(b) Influence of E_{50}/c_u and d_w on δ_{vm} for c_u/σ'_v=0.3

Fig. 5.11 Effect of soil shear strength ratio and water drawdown on Max. ground surface settlement

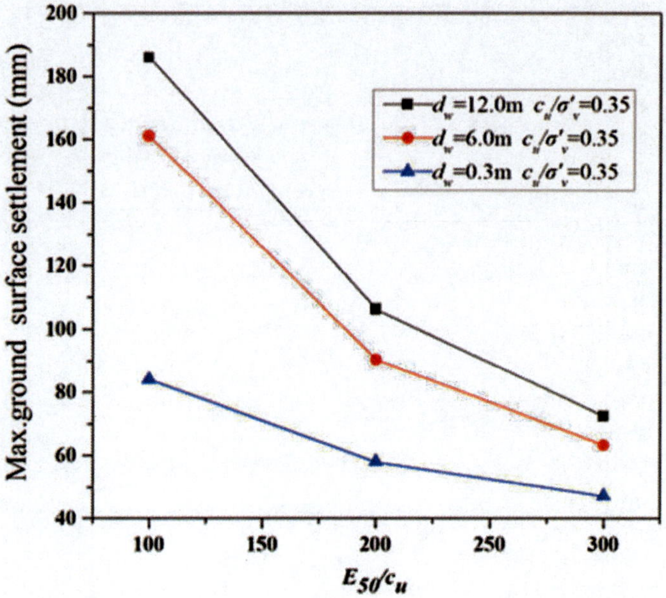

(c) Influence of E_{50}/c_u and d_w on δ_{vm} for c_u/σ'_v=0.35

Fig. 5.11 (continued)

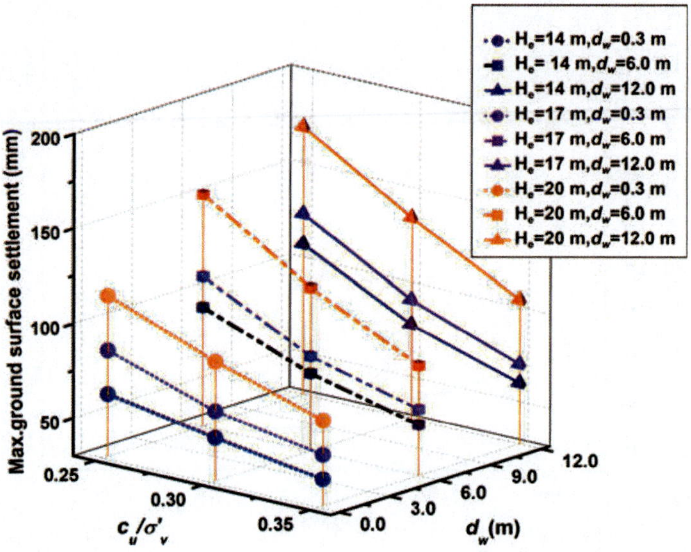

Fig. 5.12 Effect of excavation depth and water drawdown on Max. ground surface settlement

5.4.1.2 A Predictive Model

Based on the numerical results and the least-square method, a simple logarithm regression (LR) model has been developed for estimating the maximum ground surface settlement δ_{vm}, as a function of seven input parameters of excavation geometries B, T, H_e, soil parameters c_u/σ'_v, E_{50}/c_u, system stiffness S, and groundwater drawdown d_w. The optimal equation with the coefficient of determination R^2 = 0.9245 for δ_{vm} takes the following form:

$$\delta_{vm} = 24.26B^{0.3747}T^{0.7251}(H_e)^{1.2032}(c_u/\sigma'_v)^{-1.4687}(E_{50}/c_u)^{-0.5479}S^{-2.2223}(d_w)^{0.1013} \quad (5.2)$$

Based on the expression for δ_{vm}, the system stiffness S is the most influential parameter since its absolute power value is the greatest, followed by the relative shear strength ratio c_u/σ'_v and the excavation depth H_e. Figure 5.9 shows the plot of the estimation model of the maximum ground surface settlement, for cases with B = 30 m, T = 30 m, H_e = 17 m, and S = 8.176 assuming the d_w and E_{50}/c_u change within the assigned ranges. Also plotted in Fig. 5.13 are the 9 data points from FEM for d_w = 0.3 m, 6.0 m, and 12.0 m and E_{50}/c_u = 100, 200, and 300 respectively.

Figure 5.14 shows the plot of the maximum ground surface settlement estimations calculated using Eq. (5.2) versus the FEM values for the 746 hypothetical cases. The results indicates that Eq. (5.2) is reasonably accurate since most of the points are enveloped by the 20% error line.

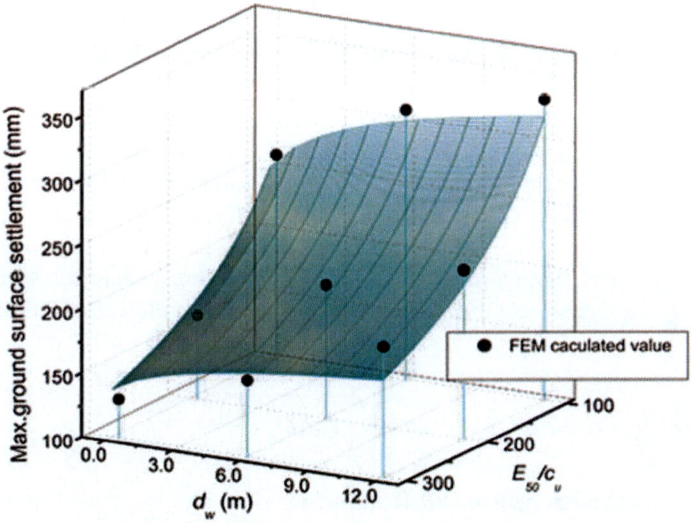

Fig. 5.13 Response surface for d_w and E_{50}/c_u and points of the FEM results

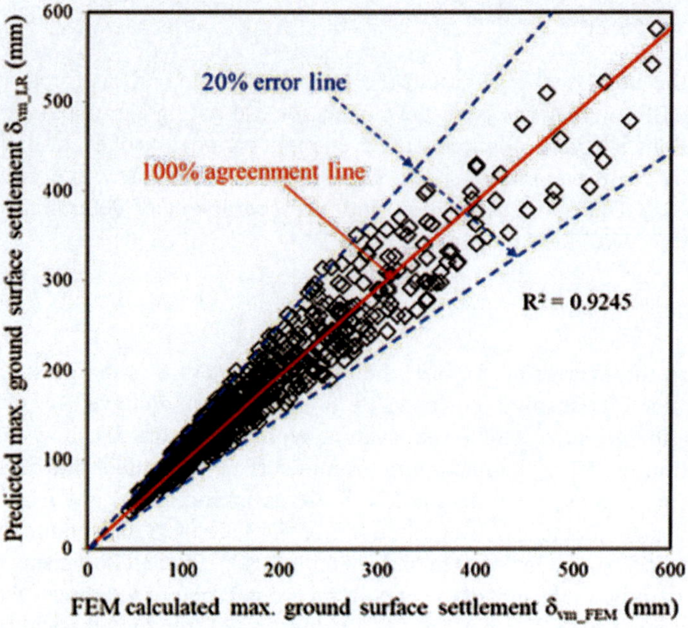

Fig. 5.14 Comparison between δ_{vm_LR} and δ_{vm_FEM}

5.4.1.3 Model Validation

To validate the proposed LR model, a total of 19 well-documented excavation case histories as listed in Table 5.5 were utilized. It should be noted that references with groundwater drawdown recordings are rarely reported. Figure 5.15 shows the predicted maximum ground surface settlement δ_{vm_LR} versus the measured δ_{vm_m} values for the 19 cases represented by the blue solid squares. The plot indicates that the developed LR model is able to predict reasonably well (within a relative error of 50%) the excavation-induced maximum ground surface settlements for the case histories considered, though many other influential factors such as the soil profiles, construction sequences, and the workmanship are not taken into consideration.

5.4.2 Artificial Neural Network Model

5.4.2.1 Numerical Schemes and Results

The plane strain finite-element program Plaxis 2D was used to perform a parametric study to assess the critical factors that influence the ground surface settlements. Figure 5.16 and Table 5.6 show the parameters that were varied. The parameter d_G is the depth of the GIII rock layer relative to the final excavation depth with a negative

Table 5.5 Summary of excavation case histories

No	Name	B (m)	T (m)	H_e (m)	c_u/σ_v'	E_{50}/c_u	S	d_w (m)	δ_{vm_m} (mm)	δ_{vm_LR} (mm)	References
1	Tunnel at race course road	14	12.0	17.4	0.25	200	8.11	11.3	51.0	63.0	[10]
2	Cashew station section 4	50	3.0	19.5	0.35	200	6.15	12.7	42.0	48.5	[11]
3	Cashew station section 8	30	6.9	19.5	0.35	200	6.15	13.5	87.0	73.8	
4	Cashew station section 9	30	12.5	19.5	0.35	200	6.15	13.6	118.3	113.6	
5	Cashew station section 13	30	8.7	19.5	0.35	200	6.15	13.1	71.7	87.0	
6	Hillview station section 7	36	7.0	24.5	0.35	200	8.83	12.8	25.9	46.8	
7	Hillview station section 9	40	7.5	24.5	0.35	200	8.83	13.1	28.0	51.3	
8	Hillview station section 12	40	13.0	24.5	0.35	200	8.83	14.1	69.5	77.1	
9	Hillview station section 13	40	3.0	24.5	0.35	200	8.83	13.0	14.9	26.4	
10	Hillview station section 14	40	25	24.5	0.35	200	8.83	16.1	171.2	125.5	
11	Hillview station section 16	40	7.5	24.5	0.35	200	8.83	12.8	31.7	51.2	
12	Hillview station section 19	18	8.5	24.5	0.35	200	8.83	7.8	48.2	39.5	
13	BKP section 1	54	6.2	20.0	0.35	200	8.17	4.7	48.2	42.0	
14	BKP section 5	30	5.8	20.0	0.35	200	8.17	4.4	33.0	31.9	
15	BTW Sect. 3	70	4.9	21.0	0.35	200	8.17	3.1	52.0	39.7	
16	BTW Sect. 4	70	8.5	21.0	0.35	200	8.17	6.5	54.0	63.7	
17	BTW section 5	45	5.6	21.0	0.35	200	8.17	4.2	25.0	38.2	
18	BTW section 7	45	4.2	21.0	0.35	200	8.17	2.0	37.0	28.8	
19	BTW section 8	45	5.1	21.0	0.35	200	8.17	6.3	41.0	37.2	

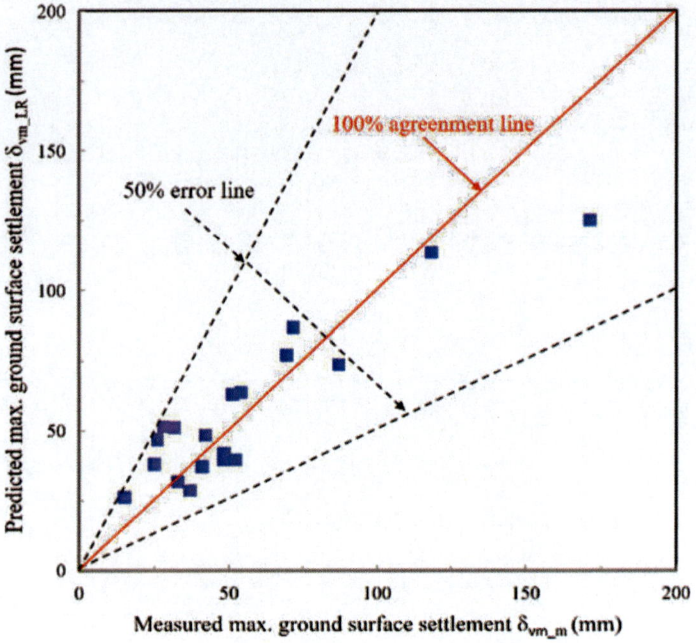

Fig. 5.15 Comparison between δ_{vm_LR} and δ_{vm}

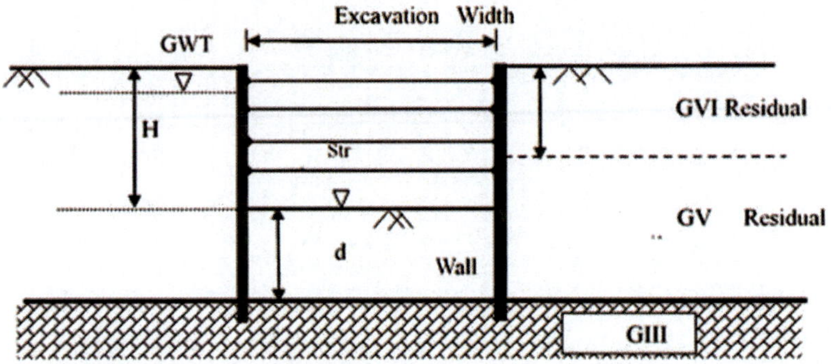

Fig. 5.16 Simplified cross-sectional profile

value of d_G denoting the GIII layer is below the final excavation level. In all the analyses, the wall is assumed to be embedded 1 m into the GIII layer.

A typical finite-element half-mesh is shown in Fig. 5.17. The structural elements that were assumed to be linear elastic with the wall represented by 5-noded beam elements and 3-noded bar elements were used for the struts. The soil was modeled by 15-noded triangular elements. The struts and wall members have not been shown, for ease of viewing. The nodes along the side boundaries of the mesh were constrained from displacing horizontally while the nodes along the bottom boundary were

Table 5.6 Summary of range of parameters

Parameter	Range of values
Depth of excavation H (m)	17–31
Excavation width B (m)	10–80
Thickness of GVI layer T (m)	2.6–35
Depth of groundwater drawdown d_w (m)	0–25
Depth of GIII layer d_G (m)	−18 to +14
SPT N of GVI soil	2–36
System stiffness S	181–2051

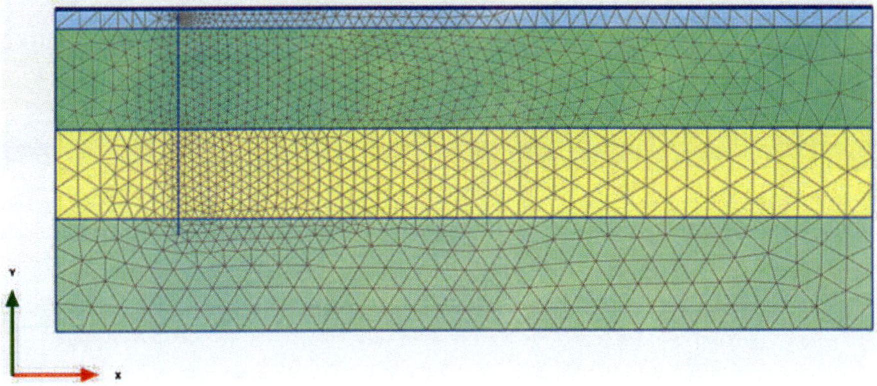

Fig. 5.17 Finite-element mesh

constrained from moving horizontally and vertically. The right vertical boundary extends far from the excavation to minimize the effects of the boundary restraints. In the analyses, groundwater drawdown was considered.

The properties of the soils as listed in Table 5.7 are based on extensive laboratory and in situ tests that were carried out at the four sites and were validated by finite-element back analyses.

For brevity, only one plot of typical back analysis is presented in this part. Figure 5.18 shows the measured and computed wall deflection and the ground surface settlement profiles at the final stage of excavation for one cross-section at C913. It is obvious that the computed lateral wall displacements and ground settlements are generally in good agreement with the measured results.

Subsequently, extensive finite-element parametric analyses were performed to assess the important factors that affect the maximum ground surface settlement. Figures 5.19, 5.20, 5.21, 5.22, 5.23 and 5.24 highlight the general trends that were observed. Figure 5.19 shows the influence of the SPT N of the GVI layer and the depth of groundwater drawdown d_w on the maximum ground surface settlement δ_{vm}. It is obvious that the maximum ground surface settlement increases with the increase in the depth of groundwater drawdown. The general trend was for the maximum ground settlement to decrease with the decrease of GVI thickness. In addition, it is obvious

Table 5.7 Summary of HS soil properties

Strata	γ_{unsat} (kN/m^3)	γ_{sat} (kN/m^3)	ν_{ur}	c' (kPa)	ϕ'(°)	Stress-dependent stiffness		
						M	E_{50}^{ref} (kPa)	E_{ur}^{ref} (kPa)
GVI	16.0	18.5	0.2	Related to SPT N		0.6	Related to SPT N	
GV	17.0	19.0	0.2	0.1	33	0.6	16,000	48,000
GIII	24.0	24.0	0.2	300	45	0	200,000	600,000

For GVI, $c_u = 5.627$ N for N < 15; otherwise, $c_u = 84.4 + 0.508$(N-15); $c' = 20$ kPa and $\varphi' = 28°$ for N < 15; otherwise, $c' = 17$ kPa and $\varphi' = 33°$; $E_u/c_u = 250$

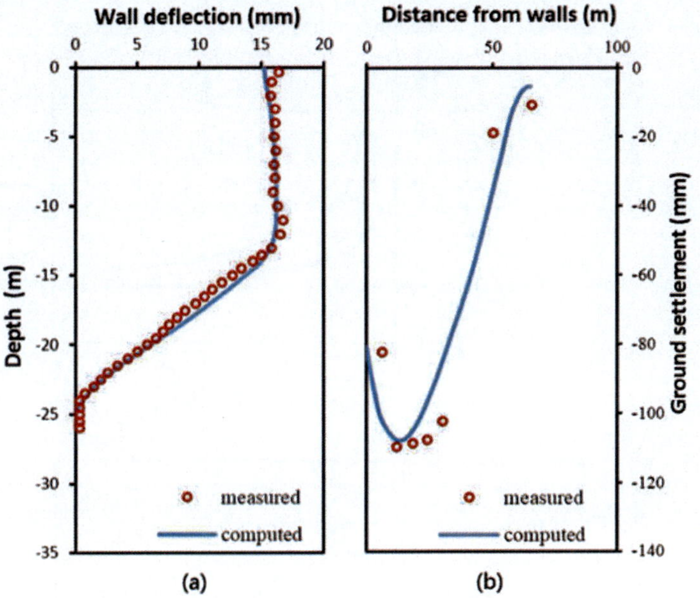

Fig. 5.18 Comparison of measured and computed excavation responses

Fig. 5.19 Plot of δ_{vm} versus d_w ($B = 20$ m, $H = 20$ m, $d_G = -4$ m, $T = 10$ m, and $S = 462$)

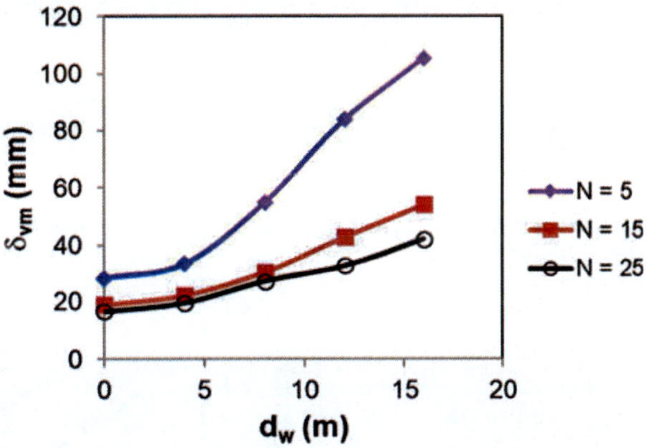

that the maximum ground surface settlement increases significantly as the SPT N value reduces from 15 to 5. The increase in the thickness of the GVI layer T was found to increase the maximum ground surface settlement significantly for SPT N values <10 as shown in Fig. 5.20. Negative values of d_G which denotes that the GIII layer is below the final excavation depth also resulted in increased δ_{vm} compared with the case of the GIII layer above the final excavation depth (positive d_G) as shown in Fig. 5.21. The width of the excavation B, the depth of the excavation H, and the wall system stiffness S were also found to influence δ_{vm} as shown in Figs. 5.22, 5.23 and

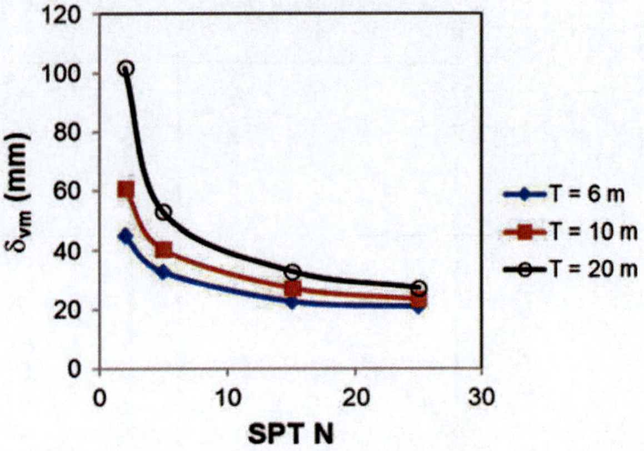

Fig. 5.20 Plot of δ_{vm} versus SPT N (B = 30 m, H = 20 m, $d_G = -4$ m, $d_w = 0$, and S = 462)

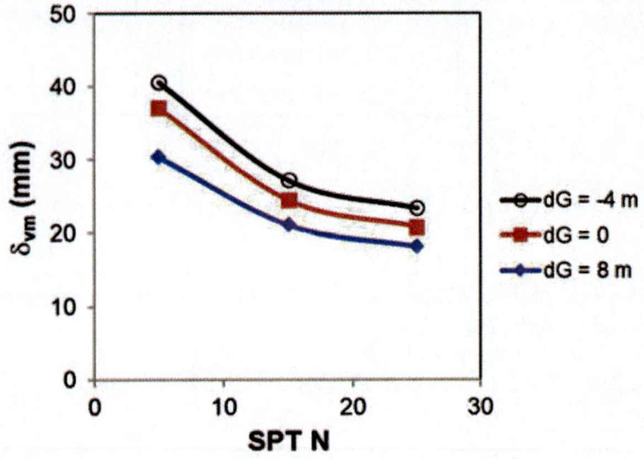

Fig. 5.21 Plot of δ_{vm} versus SPT N (B = 30 m, H = 20 m, $d_w = 0$, T = 10 m, and S = 462

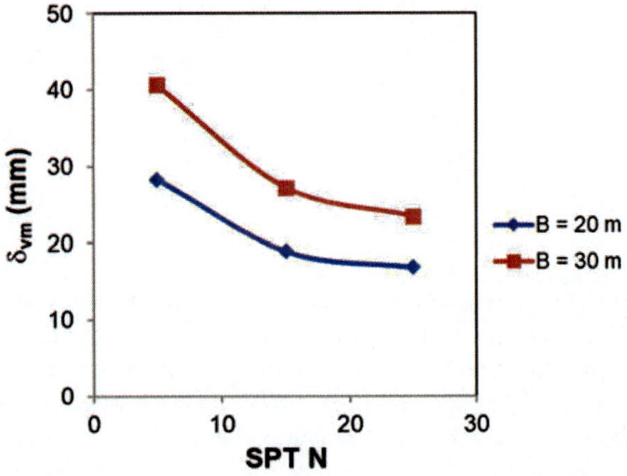

Fig. 5.22 Plot of δ_{vm} versus SPT N (H = 20 m, $d_G = -4$ m, $d_w = 0$, T = 10 m, and S = 462)

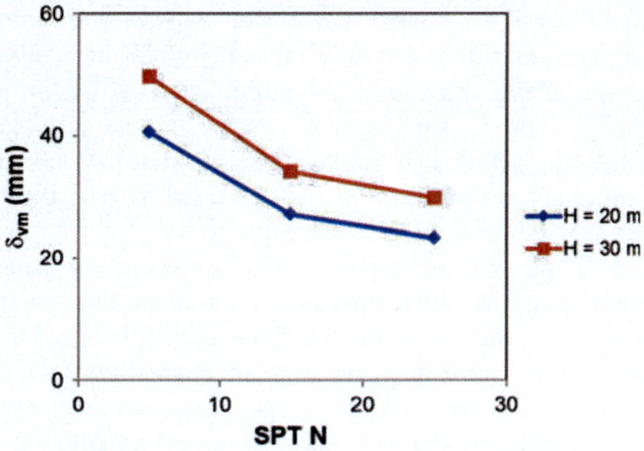

Fig. 5.23 Plot of δ_{vm} versus SPT N (B = 30 m, d_G = −4 m, d_w = 0, T = 10 m, and S = 181)

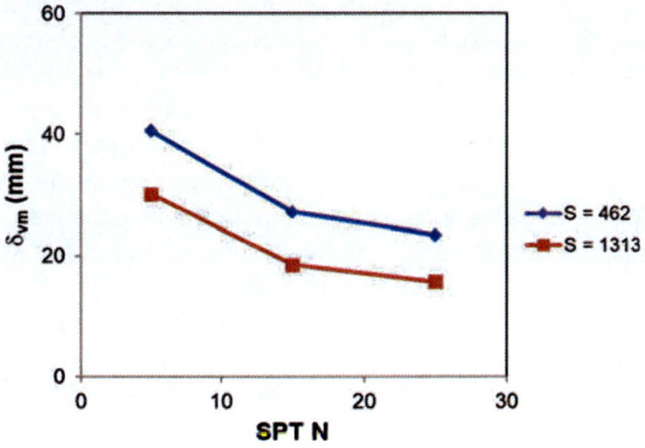

Fig. 5.24 Plot of δ_{vm} versus SPT N (B = 30 m, H = 20 m, d_G = −4 m, d_w = 0, and T = 10 m)

5.24. This is not surprising since the increase in B and H and the decrease in S will lead to an increase in the lateral wall deflection and correspondingly increase in the ground settlement as a result of undrained shear distortion.

5.4.2.2 Neural Network Analyses

The finite-element analyses in the previous section have highlighted the multiple factors that affect the ground surface settlement. Hence an artificial neural networks (ANNs) back-propagation algorithm was adopted to model the relationship between the various parameters. Artificial neural networks are computational methods that mimic very simplistically the structure and internal information processing operations of the human brain. In general, a neural network is made up of processing units (neurons). Each neuron receives signals, performs a simple computation, and then sends out signals through outgoing connections. The connection between neurons functions like a multiplicative filter with its connection strength or weight. Back-propagation neural networks [12] are multilayer feedforward networks consisting of an input layer of neurons, an output layer, and one or more hidden layers. Learning in a back-propagation network involves "supervised" learning in which the expected output is included in the training data that the network is supposed to learn. The "training" process involves an iterative process. In each training cycle, signals are propagated forward from the input to the output layer and then the error signals at the output layer are propagated backward, and subsequently, the network connection weights are modified through an algorithm that minimizes the sum-squared error using the gradient descent approach. Validation of the performance of the neural network to assess the generalization capability of the trained neural network model to produce the correct input–output mapping even when the input is different from the examples used to train the network is carried out by "testing" with a separate set of data that was never used in training the neural network. The learning capacity of a back-propagation network depends on the number of neurons in the hidden layer(s) and the connection strength (weights) between the neurons in the adjacent layers. Determining the optimal number of hidden neurons is commonly carried out by a trial-and-error approach through repeatedly increasing the number of hidden neurons until no further improvement in the network performance is obtained.

A total of seven input neurons representing the seven parameters as listed in Table 5.6 were used in the neural network analysis. The output consisted of the maximum ground settlement. A total of 865 different data patterns were obtained from the finite-element analyses described in Sect. 5.3. Of the 865 data patterns, 694 (i.e., 80%) were randomly selected as the training patterns while the remaining 171 (i.e., 20%) were used for testing purposes. Table 5.8 shows a summary of some of the training data patterns. The optimal back-propagation neural network model for predicting the maximum ground surface settlement consisted of 36 hidden neurons. The training and testing predictions are shown in Fig. 5.25 and Fig. 5.26, respectively. The performance statistics are summarized in Table 5.9. It is obvious that the neural network model has been able to learn the complicated relationship between the ground settlement and the various geotechnical and excavation parameters. Using the trained neural network model, further validation carried out using measured data from the four sites (Table 5.10) showed reasonable agreement between the neural network predicted values and the measured ground settlement as shown in Fig. 5.27.

The developed ANN model can also provide the parameter relative importance by partitioning the hidden-output neuron connection weights into components connected with each input neuron [13]. Table 5.11 shows the relative importance of the seven input variables for the neural network model. The results indicate that the three most important parameters the affect the ground settlement are the excavation depth H, the SPT N value of the residual soil, and the location of the GIII layer.

Table 5.8 Sample training data

S	B (m)	H (m)	d_G (m)	d_w (m)	T (m)	SPT N	δ_{vm} (mm)
462	30	20	−4	0	6	2	45
462	30	20	−4	0	10	2	60.7
462	30	20	−4	0	16	2	88.2
462	30	20	−4	0	20	2	101.8
462	20	20	−4	0	10	5	28.3
462	40	20	−4	0	10	5	45.3
1313	30	21	−4	0	10	5	33.6
918	30	17	−4	0	10	8	29.6
1313	30	17	−4	0	10	8	26.2
1313	40	17	−4	0	10	12	21.2
918	30	17	−4	0	10	12	23.4
1313	30	17	−4	0	10	12	20.5
181	30	31	−4	12	10	5	113.7
259	30	31	−4	12	10	5	111.5
405	30	31	−4	12	10	5	108.9
462	20	20	−4	16	10	5	105.4
462	40	20	−4	16	10	5	118.4
462	30	20	−4	16	10	5	112.6

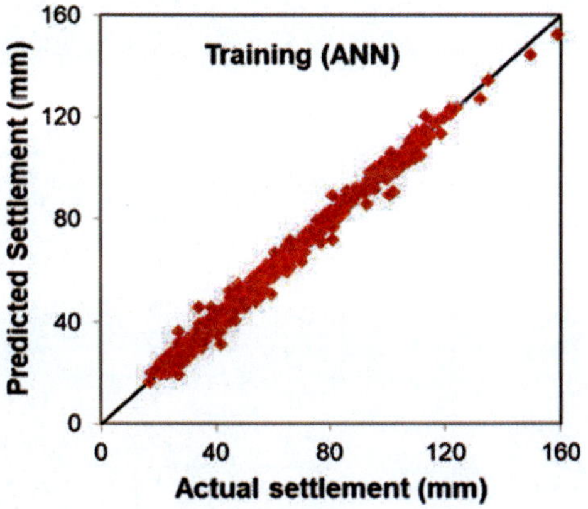

Fig. 5.25 Plot of predicted versus actual ground settlement (Training)

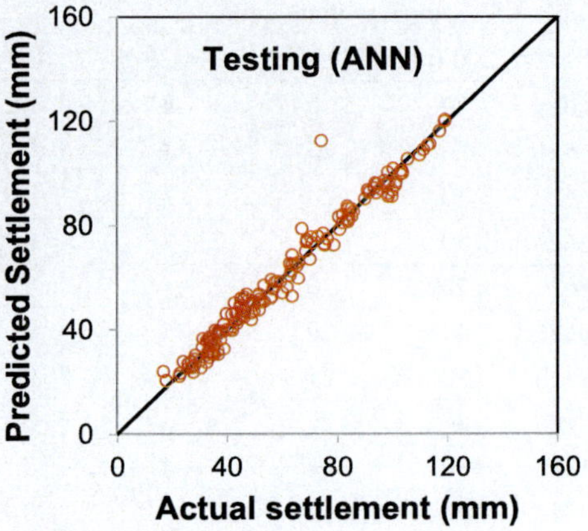

Fig. 5.26 Plot of predicted versus actual ground settlement (Testing)

Table 5.9 Summary of training and testing results

	Number of patterns	Mean squared error MSE	Coefficient of correlation r
Training data	694	6.200	0.9959
Testing data	171	20.062	0.9853

Table 5.10 Summary of measured data used in model validation

S	B (m)	H (m)	dG (m)	dw (m)	T (m)	SPT N	δ_{vm} (mm)
261	50	19.5	−16	15.4	8.5	12.3	40.5
261	50	19.5	−6.5	15.5	9	11	35.1
261	50	19.5	−4.8	15.7	3	12	42
261	40	19.5	−11.5	16.3	15	17.5	48.5
261	30	19.5	1.5	12.4	16	8.4	30.7
261	30	19.5	−6	14.9	14.2	16.2	70.8
261	30	19.5	1.9	14.9	12.2	14.2	28.6
261	30	19.5	−6.7	13.1	10	4.5	71.7
261	30	19.5	−0.5	15.7	14.5	16	72.2
1219	36	24.5	−11	23.4	23	20.6	28.6
705	45	24.5	9.5	15.1	7.5	20	28
1219	55	24.5	−12.3	13	18	15	26.6
1219	55	24.5	−5.8	16.1	13	10.9	69.5
705	12	24.5	6.7	5.8	8.5	5	31.7
1090	54	20	1.7	4.7	2.6	4	48.2
1090	30	20	−7.7	7.7	8.3	12.3	24.3
1090	30	20	0.2	4.4	9	8	33
1219	70	21	0.8	6.5	8.5	18.6	54
1219	45	21	−8.8	4.2	6	15	25

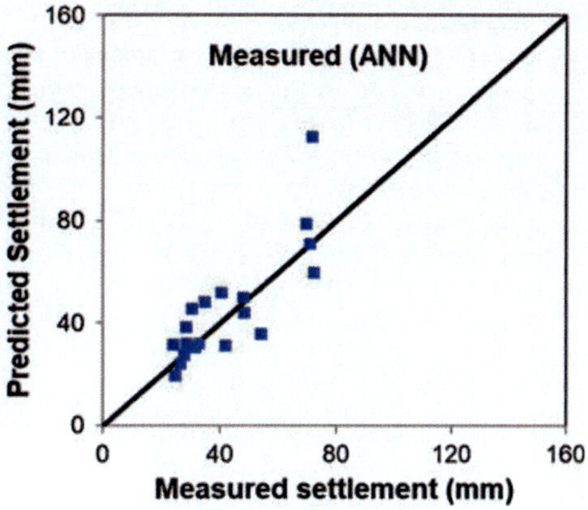

Fig. 5.27 Plot of predicted versus measured ground settlement

Table 5.11 Summary of relative importance of each parameter

Parameter	Relative importance (%)
H	22.2
N	21.3
d_G	17.8
d_w	12.7
T	12.3
B	7.9
S	5.8

References

1. Mana, A. I., & Clough, G. W. (1981). Prediction of movements for braced cuts in clay. *Journal of the Geotechnical Engineering Division, ASCE, 107*, 759–777.
2. Ou, C. Y., Hsien, P. G., & Chiou, D. C. (1993). Characteristics of ground surface settlement during excavation. *Canadian Geotechnical Journal, 30*, 758–767.
3. Hsieh, P. G., & Ou, C. Y. (1998). Shape of ground surface settlement profiles caused by excavation. *Canadian Geotechnical Journal, 35*, 1004–1017.
4. Long, M. (2001). Database for retaining wall and ground movements due to deep excavations. *Journal of Geotechnical and Geoenvironmental Engineering, 127*, 203–224.
5. Clough, G. W., & O'Rourke, T. D. (1990). *Construction induced movements of in situ wall* (pp. 439–470). Geotechnical Special Publication.
6. Harry, H., Emilio, G.-T., & Raul, F. (2019). Meta-analysis of ground movements associated with deep excavations using a data mining approach. *Journal of Rock Mechanics and Geotechnical Engineering, 11*, 409–416.
7. Peck, R. B. (1969). Deep excavation and tunneling in soft ground. In *Proceedings of the 7th International Conference on soil Mechanics and Foundation Engineering*, Mexico City, pp. 225–290.
8. Bowles, J.E., 1988. *Foundation analysis and design* (4th ed.). New York, USA: McGraw-Hill Book Company.
9. Zhang, W. G., & Goh, A. T. C. (2016). General behavior of braced excavation in Bukit Timah Granite residual soils: A case study. *International Journal of Geoengineering Case Histories, 3*(3), 190–202.

10. Wen, D. Z., & Lin, K. Q. (2002). The effect of deep excavation on pore water pressure changes in the Old Alluvium and under-drainage of marine clay in Singapore. In *Geotechnical aspects of underground construction in soft ground.*
11. Zhang, W. G., Goh, A. T. C., Goh, K. H., Chew, O. Y. S., Zhou, D., & Zhang, R. (2018). Performance of braced excavation in residual soil with groundwater drawdown. *Underground Space, 3*, 150–165.
12. Rumelhart, D. E., Hinton, G. E., & Williams, R. J. (1988). Learning internal representations by error-propagation. *Readings in Cognitive Science, 323*, 399–421.
13. Goh, A. T. C. (1994). Seismic liquefaction potential assessed by neural networks. *Journal of Geotechnical Engineering, 120*, 1467–1480.

Chapter 6
Probabilistic Analysis on Excavation Responses

6.1 Introduction

The soil parameters used for analysis by numerical methods are not necessarily the "true" values of these parameters, since such data obtained are estimated based on limited field tests, laboratory data, or local experience, generally involving considerable uncertainty. In addition, the wall and ground responses in a supported excavation may also be influenced by construction quality/workmanship and other environmental factors. Thus, deterministic modeling using numerical methods such as the finite-element or finite-difference method, even with advanced soil constitutive models, may not be able to take into account all the uncertainties and complexities mentioned above. To this end, the probabilistic framework combining a simplified wall deflection estimation model with reliability methods that are able to take into account the soil parameter uncertainties is desirable.

Based on the PR model, determined in Sect. 4.4.1 which is used to estimate the excavation-induced wall deflection, design charts for preliminary design and checking are given in this chapter. It demonstrates that with this PR model, the first-order reliability method (FORM) and Monte Carlo simulation (MCS) can be applied to determine the probability of serviceability limit state failure. Consequently, using this approach, the chapter presents a methodology that allows engineers to determine the required threshold limiting $(\delta_{hm}/H_e)_T$ ratios to meet the different target serviceability reliability indices.

W. Zhang and H. Liu, *Design of Deep Braced Excavation and Earth Retaining Systems Under Complex Built Environment*,
https://doi.org/10.1007/978-981-16-5320-9_6

6.2 Reliability Analysis on Serviceability Limit State

6.2.1 *Probability Density Determination*

Take Bugis MRT (Table 4.6) in Singapore as an example. Based on the PR model, the probability density of the wall deflection can be determined through MCS. The analysis involved 1,000,000 realizations assuming that parameters E_{50}/c_u and c_u/σ_v' are random variables while the other parameters are deterministic and that both E_{50}/c_u and c_u/σ_v' follow a lognormal distribution with a coefficient of variation (COV) of 0.15. The values for deterministic parameters are $\gamma = 16.5$ kN/m^3, B = 21 m, $H_e =$ 18 m, $\ln(S) = 8.18$, and T = 35 m. The mean values of c_u/σ_v' and E_{50}/c_u are 0.25 and 150, respectively. Figure 6.1 shows the plot of the probability density of δ_h^* based on the parameter statistics and the PR model. As can be seen from Fig. 6.1, the variation of δ_h^* best follows the normal distribution, ranging from 75 to 275 mm. The most probable δ_h^* value is 136 mm, very near to the actual measured value of 150 mm. This section shows that the combined use of the PR model and MCS can be used to determine the distribution of the wall deflections and to perform reliability analysis.

6.2.2 *Statistical Information of Input Variables*

Among the seven input variables for the PR model, the soil unit weight γ, the soft clay thickness T, the excavation width B and depth H_e, and the system stiffness $\ln(S)$ are simply treated as constants, as they have rather limited influence on the results of probabilistic analysis due to their low uncertainty [1]. The two key soil parameters E_{50}/c_u

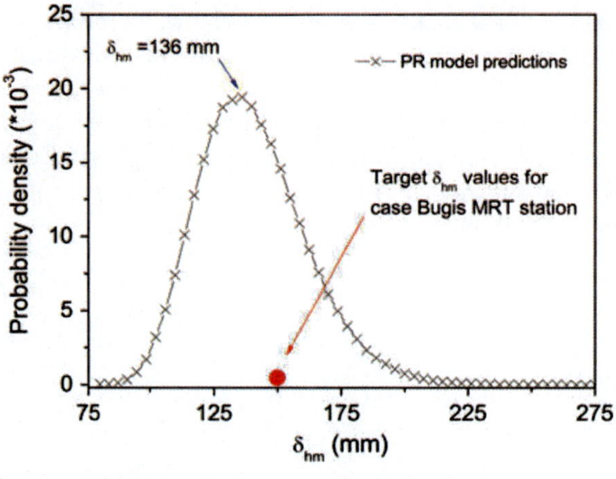

Fig. 6.1 Distribution of the predicted δ_h^* values for Bugis MRT

Table 6.1 Statistics of random variables and values used for deterministic variables

Parameter	Values considered
Random variables	
c_u/σ'_v	Mean = 200, 300; COV = 0.15
E_{50}/c_u	Mean = 0.25, 0.29; COV = 0.15
Deterministic	
γ (kN/m^3)	16.7, 17.5
B (m)	30, 40
T (m)	15 (for $B = 30$ m), 30 (for $B = 40$ m)
H_e (m)	18 (for $B = 30$ m), 30 (for $B = 40$ m)
ln(S)	7.30, 8.18

and c_u/σ'_v are treated as random variables and are assumed as lognormal (LN) distribution. The COV values for the two soil parameters are assumed to be 0.15. Useful guidelines on typical COVs of many common soil strength properties have been summarized by Phoon and Kulhawy [2] and Duncan [3]. The statistics of the random variables and values used for the deterministic variables are listed in Table 6.1.

6.2.3 Serviceability Criterion and Threshold $(\delta_{hm}/H_e)_T$

The ratio δ_{hm}/H_e is adopted as the serviceability limit state criterion, and in assessing the serviceability limit state of diaphragm walls for braced excavation, the limit state function based on the PR model can be expressed as

$$g(x) = R - S = (\delta_{hm}/H_e)_T - \delta_h^*/H_e \tag{6.1}$$

in which $(\delta_{hm}/H_e)_T$ is the limiting threshold value while δ_h^*/H_e is derived from the PR model. Serviceability limit state failure is deemed to occur if the predicted δ_h/H_e is greater than the limiting $(\delta_{hm}/H_e)_T$.

6.2.4 Reliability Assessment Methods: FORM and MCS

The probability that the serviceability limit state P_f is exceeded can be determined using Eq. (6.1). Such calculation of P_f involves the determination of the joint probability distribution of the resistance R and the loading S and the integration of the probability density function (PDF) over the failure domain $g(x) < 0$. Considering that the PDFs of the random variables are not known in engineering applications and the integration is computationally demanding when multiple variables are involved, an approximate method known as the first-order reliability method [4] is commonly used to assess P_f. The approach involves the transformation of the limit state surface into a space of standard normal uncorrelated variables, wherein the shortest distance from the transformed limit state surface to the origin of the reduced variables is the reliability index β [5]. For normal distributed random variables, $P_f \approx 1 - \Phi(\beta)$, in which Φ = cumulative probability of a standard normal distribution. Mathematically, Low and Tang [6] have shown that β can be computed using

$$\beta = \min_{x \in F} \sqrt{\left[\frac{x_i - \mu_i}{\sigma_i}\right]^T [\boldsymbol{R}]^{-1} \left[\frac{\mathrm{x_i} - \mu_i}{\sigma_i}\right]} \tag{6.2}$$

in which x_i = the set of n random variables; μ_i = the set of mean values; σ_i = the standard deviation; **R** = the correlation matrix; and F = the failure region. Low [7] has shown that a Microsoft Excel spreadsheet can be used to perform the minimization and determine β. As an alternative to the FORM method, MCS is adopted for comparison. In this study, the Microsoft Excel add-in program @RISK (http://www.palisade.com) was used.

6.2.5 Developed FORM_PR and MCS_PR Frameworks

For each MCS simulation, the number of iterations is 1,000,000, and Latin hypercube sampling is adopted. The MCS procedures and the calculation of probability failure P_f for the case with B = 30 m, γ = 16.7 kN/m3, H_e = 18 m, ln(T) = 7.30, T = 15 m, mean value of E_{50}/c_u = 200, and average value of $c_u/\sigma_v^{'}$ = 0.25 are illustrated in Fig. 6.2a. It can be observed that for a limiting threshold $(\delta_{hm}/H_e)_T = 0.5\%$, the probability that this value is exceeded is 4.3%, while for $(\delta_{hm}/H_e)_T = 0.2\%$, the P_f value can be as high as 56%. The typical FORM procedures and the calculation of P_f are illustrated in Fig. 6.2b. Cells D3:G4 are parameters that are set corresponding to the variable distribution types. For normal distributions, cells D3:D4 correspond to the mean values and cells E3:E4 correspond to the standard deviations. For non-normal distributions, the non-normal parameters are replaced by an equivalent normal ellipsoid, centered at the equivalent normal mean. The correlation matrix R cells K3: L4 are used to define the correlations between E_{50}/c_u and $c_u/\sigma_v^{'}$. The dimensionless $x_i^{'}$ vector in cells M3:M4 contains equations for $(x_i - \mu_i)/\sigma_i$. Cell I3 contains the estimations of the maximum wall deflection based on the predictive PR model δ_h. Cell N3 contains the serviceability limit state function $g(x) = (\delta_{hm}/H_e)_T - \delta_h^*/H_e$. The design point ($x_i^*$ values) was obtained by using the spreadsheet's built-in optimization routine SOLVER to minimize the cell, by changing the $x_i^{'}$ values, under the

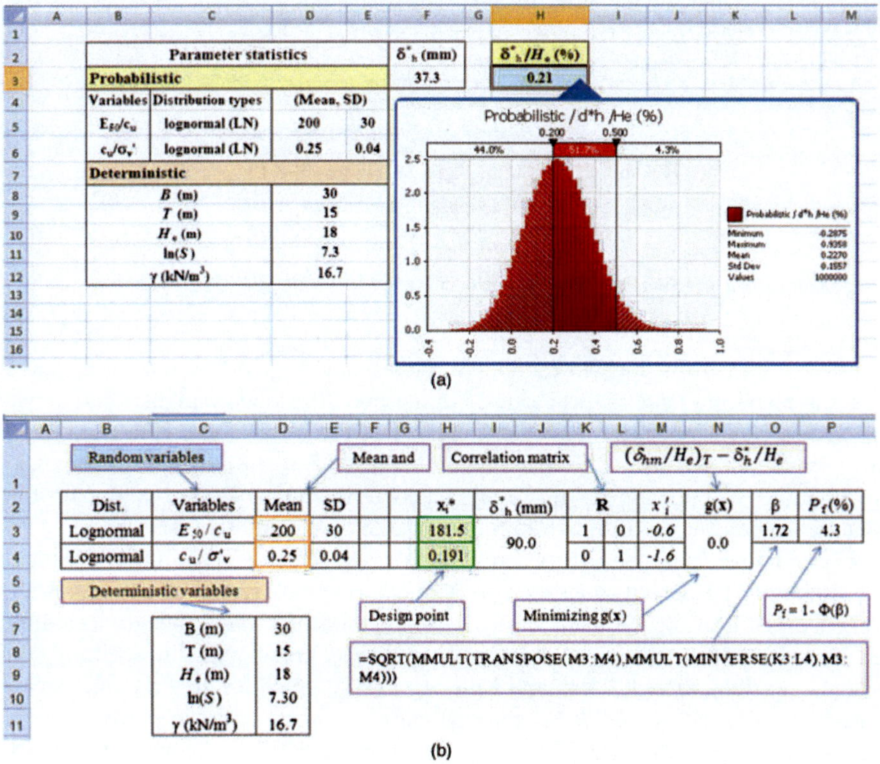

Fig. 6.2 The developed reliability assessment frameworks: **a** MCS_PR; **b** FORM_PR

constraint that $g(x_i^*) = 0$. Prior to invoking the SOLVER search algorithm, the $\times$ 0 i values in cells M3:M4 were set equal to the mean values (0, 0), corresponding to the mean values (200, 0.25) of the original two random variables. Iterative numerical derivatives and directional search for design point x_i^* were automatically carried out in the spreadsheet. It is obvious from Fig. 6.2a–b that both methods give the same P_f value of 4.3%.

6.2.6 Probabilistic Analyses and Target Reliability Indices

Based on the MCS_PR and FORM_PR frameworks illustrated in the section above, three hypothetical cases were adopted to show the influence of the chosen threshold $(\delta_{hm}/H_e)_T$ values on the P_f results and the determination of $(\delta_{hm}/H_e)_T$ according to the target serviceability reliability index βSLS. The assumed statistics of the seven input variables for the three cases are listed in Table 6.2. The lognormal distribu-

Table 6.2 Hypothetical cases and parameters considered

Case number	Parameters considered
1	$B = 30$ m; $T = 15$ m; $H_e = 18$ m; ln(S) = 7.30; $\gamma = 16.7$ kN/m^3 $E_{50}/c_u = LN(200, 30)$; $c_u/\sigma_v' = LN(0.25, 0.04)$
2	$B = 40$ m; $T = 30$ m; $H_e = 24$ m; ln(S) = 8.18; $\gamma = 17.5$ kN/m^3 $E_{50}/c_u = LN(300, 45)$; $c_u/\sigma_v' = LN(0.29, 0.045)$
3	$B = 30$ m; $T = 15$ m; $H_e = 12$ m; ln(S) = 6.10; $\gamma = 16.7$ kN/m^3 $E_{50}/c_u = LN(200, 30)$; $c_u/\sigma_v' = LN(0.25, 0.04)$

tion was assumed for the two random variables since it avoids allowing negative values in modeling material properties. Furthermore, the lognormal distribution has the advantage in modeling the right-hand tail of the distribution. The right-hand tail is important because it governs the probability of excessive deformations and the reliability of the serviceability limit state (SLS) design. Typical values for the reliability index for the ultimate limit state (ULS) β_{ULS} and the corresponding P_f have been proposed by the U.S.A.C.o. Engineers [8]. Wang and Kulhawy [9] stated that although ULS design codes apply reliability principles, serviceability limit state designs are still evaluated using conventional approaches where the serviceability reliability index β_{SLS} remains unknown. That is, there are no guidelines with regard to the target β_{SLS}, though Wang and Kulhawy [9] presented for the first time a relationship between β_{SLS} and β_{ULS} and applied it to augered cast-in-place piles. In this study, β_{SLS} values of 2.0, 2.3, 2.6, and 3.1 are selected as the target reliability indices since their corresponding Pf values are 2%, 1%, 0.5%, and 0.1%, respectively, which are commonly used for serviceability considerations. Figures 6.3, 6.4 and 6.5a show the significant influence of $(\delta_{hm}/H_e)_T$ on P_f. It is obvious that small limiting $(\delta_{hm}/H_e)_T$ will result in higher Pf since low thresholds are easily exceeded. The P_f decreases with the increase of the $(\delta_{hm}/H_e)_T$ values. Figures 6.3, 6.4 and 6.5b plot the curves of the threshold $(\delta_{hm}/H_e)_T$ values versus the target β_{SLS} of 2.0, 2.3, 2.6, and 3.1. It can be observed that $(\delta_{hm}/H_e)_T$ increases with the increase of the target βSLS, indicating that the choice of target βSLS has a significant effect on determination of the threshold $(\delta_{hm}/H_e)_T$ values.

6.3 Conclusions

This chapter presents a probabilistic framework combining a proposed polynomial regression model used for estimating wall deflections induced by excavation with reliability methods MCS and FORM to determine the probability of serviceability limit state failure. The main work and findings include the following.

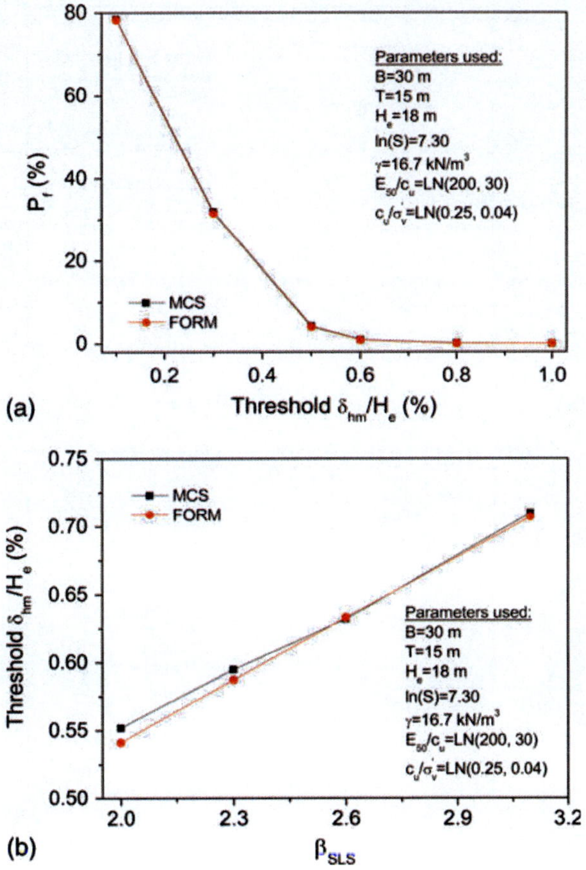

Fig. 6.3 Influence of $(\delta_{hm}/H_e)_T$ on P_f and determination of $(\delta_{hm}/H_e)_T$ according to the target β_{SLS}, Case 1

A series of design charts are proposed, which are potentially useful for preliminary estimation of the maximum wall deflections for given excavation geometries, soil parameters, and support conditions. In addition, the example case of Bugis MRT station shows that the combined use of the PR model and MCS can be used to determine the distribution of the wall deflections and to perform reliability analysis.

Probabilistic frameworks FORM_PR and MCS_PR are developed for reliability assessment of the probability of serviceability limit state failure.

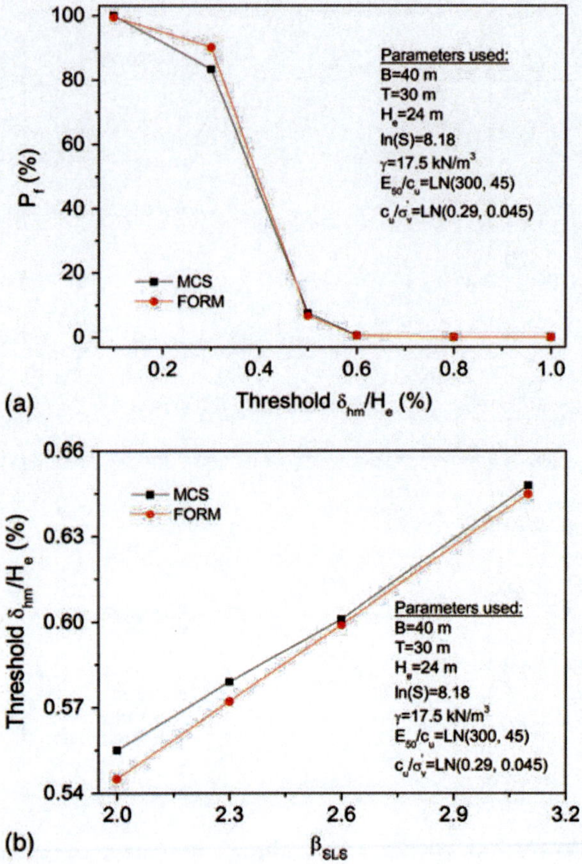

Fig. 6.4 Influence of $(\delta_{hm}/H_e)_T$ on P_f and determination of $(\delta_{hm}/H_e)_T$ according to the target β_{SLS}, Case 2

The influence of the threshold $(\delta_{hm}/H_e)_T$ values on the probability of serviceability limit state failure is significant. The choice of the threshold δ $(\delta_{hm}/H_e)_T$ values can be determined through the target serviceability reliability index β_{SLS}.

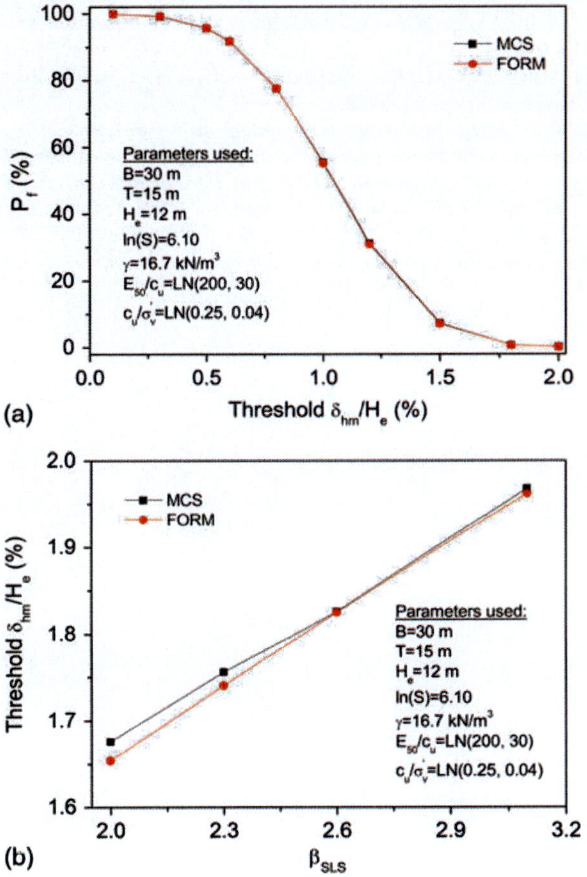

Fig. 6.5 Influence of $(\delta_{hm}/H_e)_T$ on P_f and determination of $(\delta_{hm}/H_e)_T$ according to the target βSLS, Case 3

References

1. Hsiao, E. C. L., Schuster, M., Juang, C. H., & Kung, T. C. (2008). Reliability analysis of excavation-induced ground settlement for building serviceability evaluation. *Journal of Geotechnical and Geoenvironmental Engineering, 134*(10), 1448–1458.
2. Phoon, K. K., & Kulhawy, F. H. (1999). Characterization of geotechnical variability. *Canadian Geotechnical Journal, 36*, 612–624.
3. Duncan, J. M. (2000). Factor of safety and reliability in geotechnical engineering. *Journal of Geotechnical and Geoenvironmental Engineering, 126*, 307–316.
4. Hasofer, A. M., & Lind, N. (1974). An exact and invariant first-order reliability format. *Journal of engineering mechanics-asce, 100*, 111–121.

5. Cornell, C. A. (1969). A probability-based structural code. *Journal of the American Concrete Institute, 66*, 974–985.
6. Low, B. K., & Tang, W. H. (2004). Reliability analysis using object-oriented constrained optimization. *Structural Safety, 26*, 69–89.
7. Low, B. K. (1996). *Practical probabilistic approach using spreadsheet. Uncertainty in the geologic environment—From theory to practice* (pp. 1284–1302), Madison, WI.
8. U.S.A.C.o. Engineers (1997). *Engineering and design: Introduction to probability and reliability methods for use in geotechnical engineering*. Department of the Army, Washington, DC.
9. Wang, Y., & Kulhawy, F. H. (2008). Reliability index for serviceability limit state of building foundations. *Journal of Geotechnical and Geoenvironmental Engineering, 134*, 1587–1594.

Chapter 7
One-Strut Failure Analysis

7.1 Introduction

The excavation sides are normally supported by concrete diaphragm walls or secant bored pile walls with two or more levels of struts. The purpose of the excavation support system is to provide rigid lateral support for soil surrounding excavation and to limit the movement of surrounding soils. One concern in the design of these earth retaining and stabilizing systems is the consequence of an isolated case of overloading or failure of any particular struts (due to accidental damage by machinery during construction or any other reason) in the bracing system and whether it would have sufficient redundancy to avoid any catastrophic collapse [1]. The 30 m deep excavation collapse adjacent to the Nicoll Highway in Singapore [1] and the excavation work collapse of Hangzhou Xianghu subway station [2] causing more deaths were both partly due to the poor connections between structural members or failure of the single strut, which further triggered the subsequent progressive failure. Therefore, part of the design requirement requires the braced retaining wall system to be structurally safe, robust, and to have sufficient redundancy to avoid catastrophic collapse due to one-strut failure (OSF).

One widely reported field study was carried out at four sites in Sweden and was described by Stille and Broms [3]. They reported that the maximum change of the anchor load expressed as a percentage of the initial load of the failed anchor was 35% for walls with two or more rows of anchors. The field study indicated that the increase in the loads of the adjacent anchors was of the same order of magnitude in the horizontal and vertical directions. As the associated wall movements were not significant, it was concluded that the load transferred to the adjacent anchors was not through the arching of the soil behind the wall, but through the wall and wale members. Based on a Singapore Mass Rapid Transit (MRT) case, Low et al. [4] presented the design approach and consideration for the temporary removable ground anchor. Pong et al. [5] proposed a simplified method to rationally idealize

W. Zhang and H. Liu, *Design of Deep Braced Excavation and Earth Retaining Systems Under Complex Built Environment*,
https://doi.org/10.1007/978-981-16-5320-9_7

the OSF problem from a 3D analysis to a 2D plane strain analysis. Unfortunately, in the case of braced strut systems, there is no reported case history detailing the load transfer mechanism due to one-strut failure.

The focus of this chapter is on OSF analyses, to investigate how loads from the failed strut are transferred to the adjacent struts and the whole support system. Since struts provide passive resistance to wall movement, while anchors rely on stresses in the ground being mobilized to retain the wall, it may not be realistic to compare or equalize the redistribution of forces for anchors and struts. In Singapore, part of the design requirement requires that the braced retaining wall system be structurally safe, robust, and has sufficient redundancy to avoid catastrophic collapse. In the conventional approach for one-strut failure using 2D analysis, the entire level of the failing strut is removed and thus the forces can only be distributed vertically. This generally leads to a more conservative design with heavier strut sections. Thus, 3D analysis of one-strut failure is essential to provide a more realistic understanding of the force/stress transfer behavior of the braced excavation system. This part describes the both use of 2D and 3D FE analyses to assess the impact of the failure of one strut on the remaining struts.

7.2 One-Strut Failure in Clay

7.2.1 Numerical Schemes

The basic FE test schemes are the same as those in Sect. 3.3.1.2, one-strut failure is done through the following deactivate of strut elements.

7.2.2 Analysis of Two Hypothetical Cases

In order to investigate the influence of the one-strut failure on the adjacent struts, the strut sforces before and after strut failure are examined. By tabulating the load transfer percentage, the influence of the strut failure can be demonstrated. Assume N_{pre} is the load on the strut before strut failure; N_{post} is the load on the strut after strut failure, and N_{fail} is the load on the failed strut before failure. Then the load transfer percentage is defined as

$$\text{Load Transfer}(\%) = \frac{N_{post} - N_{pe}}{N_{fail}} \times 100\% \tag{7.1}$$

7.2.2.1 One-Strut Failure for Soft Clay: $H_e = 12$ m with 3-Levels of Struts

The load transfer percentages from 2D analyses are tabulated in Table 7.1 and the 3D results are shown in Table 7.2. The 2D results indicate that failure of S3 leads to a considerable increase in the load for the 2nd level S2 strut. On the other hand, the 3D results indicate that the load of the failed strut is not only transferred vertically upward, but also to the adjacent horizontal and diagonal struts as shown in Fig. 7.1, where larger arrows denote the larger magnitude of the transferred loads and the red circles denote the struts with more than 10% load increase after the one-strut failure. It should be noted from Table 7.2 that the load transfer percentage values along the left and right sides of the failed strut (i.e., 9.7 and 9.2 for soft clay, $a = 0.1$, S2 in Table 7.2, 23.3, and 21.1 for soft clay, $a = 0.1$, S3 in Table 7.2) are not equal considering that the numerical model is not symmetrical about the failed ($x = 0$) strut. In addition, for the 3D analyses, the magnitude of the load transfer percentages is much smaller, up to approximately 20% for flexible walls and 50% for medium walls compared with the results from the 2D analyses. Therefore, compared to 3D results, the 2D analyses overestimate the possible consequence of the one-strut failure for soft clay, as it ignores the restraining effects of the adjacent horizontal struts.

Table 7.1 Load transfer percentages from 2D analyses for the one-strut failure of S3 (soft clay)

Struts	Load transfer (%)	
	$\alpha = 0.1$	$\alpha = 1.0$
S1	−37.7	−30.4
S2	122.2	120.3
S3	Failed strut	

Table 7.2 Load transfer percentage from 3D analyses for the one-strut failure of S3 (L/B = 2.2)

Struts	Load transfer (%)				
	$x = -8$	$x = -4$	$x = 0$	$x = 4$	$x = 8$
	Soft clay $\alpha = 0.1$				
S1	−0.1	−2.5	−9.2	−2.5	0.2
S2	−0.5	9.7	**15.4**	9.2	−2.2
S3	2.7	**23.6**	Failed strut	**21.1**	−0.1
	Soft clay $\alpha = 1.0$				
S1	−5.7	−4.4	−3.3	−3.7	−3.0
S2	1.5	**11.9**	**53.6**	**11.2**	−0.7
S3	4.2	**24.4**	Failed strut	**22.5**	3.3

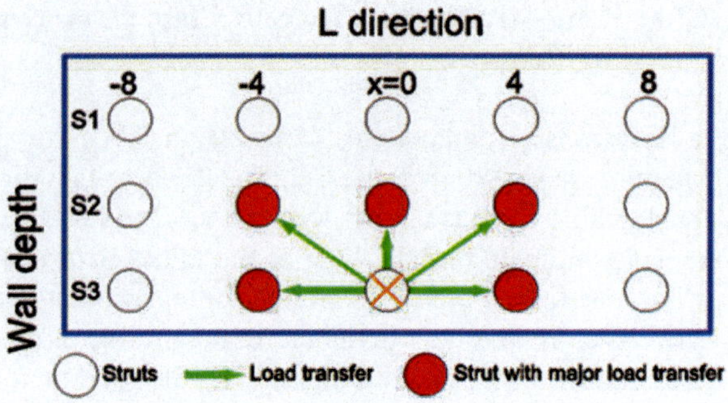

Fig. 7.1 Influence zone of the one-strut failure of S3 (soft clay)

7.2.2.2 One-Strut Failure for Medium and Stiff Clays: $H_e = 16$ M with 5-Levels of Struts

The load transfer percentages of the one-strut failure from 2D analyses for medium and stiff walls are shown in Table 7.3. For medium clay, most of the load of the failed strut is transferred to the struts immediately above and below the failed strut, with approximately 30–50% of the load transferred to the upper strut and the remainder carried by the lower strut. For stiff clay, the failure of strut S3 leads to the force redistribution for the S2, S4, and S5 struts. The S2 and S4 struts each carry approximately 30% of the load of the failed strut, and S5 carries approximately 10–20%. When the wall is also stiff, the top strut is able to carry approximately 10% of the load. For the 3D analyses, the percentage of the load transfer to adjacent struts of the relevant struts for medium clay with stiff walls are tabulated in

Table 7.3 Load transfer percentage from 2D analyses (medium and stiff clays)

Clay type	Struts	Load transfer (%)		
		$\alpha = 0.1$	$\alpha = 1.0$	$\alpha = 10$
Medium clay	S1	−0.9	−0.9	−0.4
	S2	−5.2	11.2	**15.5**
	S3	**46.0**	**41.7**	**31.1**
	S4	Strut failure		
	S5	**53.4**	**47.1**	**46.3**
Stiff clay	S1	−0.8	2.4	**15.4**
	S2	**27.6**	**30.2**	**23.1**
	S3	Failed strut		
	S4	**29.3**	**29.5**	**22.1**
	S5	**10.8**	**11.9**	**14.6**

Clay type	Struts	Load transfer (%)		
		$\alpha = 0.1$	$\alpha = 1.0$	$\alpha = 10$
Medium clay	S1	−0.9	−0.9	−0.4
	S2	−5.2	11.2	**15.5**
	S3	**46.0**	**41.7**	**31.1**
	S4	Strut failure		
	S5	**53.4**	**47.1**	**46.3**
Stiff clay	S1	−0.8	2.4	**15.4**
	S2	**27.6**	**30.2**	**23.1**
	S3	Failed strut		
	S4	**29.3**	**29.5**	**22.1**
	S5	**10.8**	**11.9**	**14.6**

Table 7.4 for L/B = 2.2 and Table 7.5 for L/B = 3.4, respectively. The tables show that the struts affected most are located directly above or below the failed strut S4 or diagonally across, as plotted in Fig. 7.2. The influence of the one-strut failure from 3D analyses for the stiff clay is shown in Fig. 7.2, The load transfer percentages for stiff clay are listed in Table 7.6 for L/B = 3.4. The load is mainly transferred to the struts directly above and below the failed strut S3, and the load transfer percentage is approximately 10–20%. The 3D results again highlight that the

Table 7.4 Load transfer percentage from 3D analyses for the one-strut failure of S4 for medium clay (L/B = 2.2)

Struts	Load transfer (%)				
	$x = -8$	$x = -4$	$x = 0$	$x = 4$	$x = 8$
S1	−1.1	0.2	1.6	0.2	−1.0
S2	−0.3	3.1	7.9	3.2	−0.3
S3	0.5	5.5	**15.3**	5.5	0.5
S4	1.2	7.2	Failed strut	7.2	1.3
S5	2.8	**9.0**	**19.5**	**9.0**	2.7

Table 7.5 Load transfer percentage from 3D analyses for the one-strut failure of S4 for medium clay (L/B = 3.4)

Struts	Load transfer (%)				
	$x = -8$	$x = -4$	$x = 0$	$x = 4$	$x = 8$
S1	−0.3	0.8	2.0	0.5	−0.7
S2	−0.1	3.4	8.0	3.2	−0.2
S3	−0.1	5.3	**15.1**	5.4	0.4
S4	0.2	6.5	Failed strut	6.5	0.7
S5	1.4	7.5	**17.9**	7.4	1.5

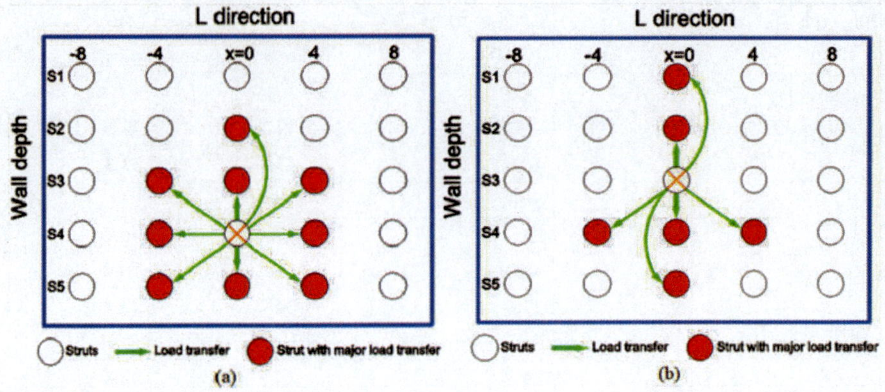

Fig. 7.2 Influence zone of the failure of strut: **a** failure of S4 (medium clay); **b** failure of S3 (stiff clay)

Table 7.6 Load transfer percentage from 3D analyses for the one-strut failure of S3 for stiff clay (L/B = 3.4)

Struts	Load transfer (%)				
	$x = -8$	$x = -4$	$x = 0$	$x = 4$	$x = 8$
	Stiff clay, $\alpha = 0.1$				
S1	−0.5	−3.7	−5.1	0.0	0.0
S2	0.1	4.8	**12.6**	5.8	1.6
S3	1.4	4.6	Failed strut	5.4	2.6
S4	3.4	6.8	**12.3**	6.6	7.2
S5	4.5	5.0	4.7	4.8	3.8
	Stiff clay, $\alpha = 1.0$				
S1	−0.2	0.6	0.6	0.8	0.1
S2	1.2	4.6	**17.8**	4.9	1.6
S3	1.3	4.1	Failed strut	4.4	1.8
S4	3.2	6.0	**17.3**	6.1	3.0
S5	4.6	5.9	6.6	5.8	4.2
	Stiff clay, $\alpha = 10$				
S1	−0.3	3.0	5.9	3.1	−0.1
S2	0.7	4.4	**12.1**	4.5	0.9
S3	1.6	5.0	Failed strut	5.1	1.7
S4	3.0	4.4	**11.1**	5.9	3.0
S5	4.7	6.6	9.3	6.5	4.4

2D analyses would result in fairly conservative (i.e., larger) estimates of the loads transferred to the adjacent struts from the failed strut as it ignores the restraining effects of the adjacent horizontal struts.

7.3 One-Strut Failure in Sand

7.3.1 Numerical Schemes

The FE software packages PLAXIS 2D and PLAXIS 3D were used to perform the excavation simulations. Figure 7.3 shows a typical cross-section and plan view for the cases considered (excavation depth $H_e = 20$ m). The parameters shown in the figure include: L = excavation length, B = excavation width, H_p = wall penetration depth, H_e = depth of final excavation, S_H = horizontal strut spacing, S_V = vertical strut spacing, all in meters, FEL is the final elevation level. In this study, the geometrical properties considered, as provided by Chowdhury et al. [6], include: excavation widths B fixed at 10 m, excavation lengths L of 16, 24, and 40 m, excavation depths He of 15 and 20 m, wall penetration depth H_p equaled to 40% of the excavation depth as suggested by Bahrami [7] to ensure that the structural bracing system was designed both safely and economically. Horizontal strut spacing S_H of 4 m. The symbol *x* represented the horizontal distance from the center of the excavation.

For both 2D and 3D analyses, D as marked in Fig. 7.4 was the distance from the primary wall to the boundary of the modeled zone and was taken to be four times

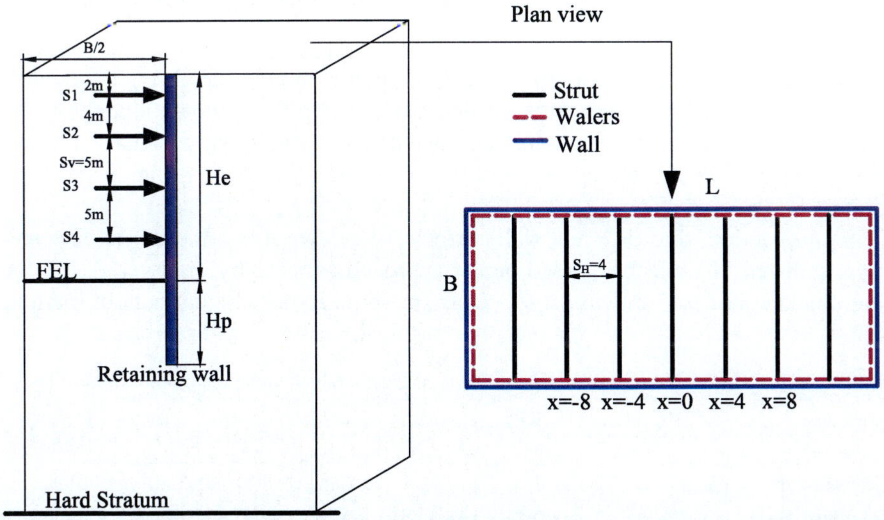

Fig. 7.3 Cross-section and plan view of the model for braced excavation ($H_e = 20$ m)

Fig. 7.4 3D mesh of the excavation from PLAXIS 3D

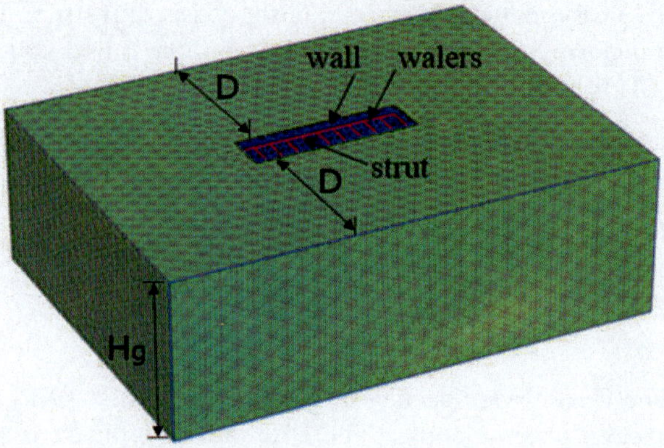

as large as the excavation depth to assure that the analysis results had an acceptable level of accuracy [7]. Besides, Hg represented the distance between ground level and hard stratum and was taken to be 50 m. The boundary conditions for 2D and 3D cases were: (1) rollers at side boundaries to allow vertical displacements and (2) pinned at the base to restrain any movements. For 2D analyses, a very fine mesh size was used to improve the accuracy of FE calculations. For 3D analyses, a medium mesh size was adopted to reach a balance between processing time and accuracy. For brevity, only a typical 3D mesh is shown in Fig. 7.4, comprising 95,665 nodes and 63,846 15-node wedge elements.

The research was conducted on three types of sandy soils, which were related to and adopted from the No. 06 station on the orange line in the Kaohsiung Rapid Transport System (KRTS) excavation [8]. The three soil types included the loose sand ($N < 10$), the medium dense sand ($10 < N < 30$), and the dense sand ($30 < N$), where N is the average standard penetration test (SPT) result. Table 7.7 summarizes the soil properties adopted in this study. E, G, and K denote the young's modulus, shear modulus, and bulk modulus, respectively. μ is the Poisson's ratio and φ is the internal friction angle of the soil. It is assumed that the groundwater table is far below the ground surface and the influence of the braced excavation on groundwater level change is marginal.

In this section, four different wall stiffness values were considered for each soil type, as listed in Table 7.8. Based on the approach adopted by Finno et al. [9], the wall thickness of 0.42 m was set to a constant value so that the moment of inertia I

Table 7.7 Properties of sandy soils used in the numerical analysis (adapted from Hsiung [8])

Soil type	SPT-N value	E (MPa)	G (MPa)	K (MPa)	μ	φ (deg)	Density (kg/m^3)
Loose sand	5–14	19.0	7.3	15.8	0.3	32.0	2008
Medium dense sand	19–29	48.0	18.5	40.0	0.3	33.0	2008
Dense sand	28–42	70.0	26.9	58.3	0.3	33.0	2028

Table 7.8 Wall properties for 2D and 3D analyses

Parameters		Wall types			
		Flexible			Medium
Plane strain (2D) FE parameters					
a		0.06	0.1	1.0	2.0
Wall stiffness *EI* (kNm^2/m)		3.224×10^4	5.407×10^4	5.407×10^5	1.081×10^6
Compressive stiffness *EA* (kN/m)		2.206×10^6	3.678×10^6	3.678×10^7	7.356×10^7
Poisson's ratio, ν		0.15	0.15	0.15	0.15
Three-dimensional (3D) FE parameters					
Young's modulus (kPa)	E_1	5.252×10^6	8.754×10^6	8.754×10^7	1.751×10^8
	E_2	2.626×10^5	4.377×10^5	4.377×10^6	8.754×10^6
Shear modulus (kPa)	G_{12}	2.627×10^5	4.378×10^5	4.379×10^6	8.757×10^6
	G_{13}	8.754×10^5	1.459×10^6	1.459×10^7	2.918×10^7
	G_{23}	2.626×10^5	4.377×10^5	4.377×10^6	8.754×10^6
Poisson's ratio	ν	0	0	0	0

and area A were kept constant, and only the wall elastic modulus E was varied. A coefficient α was introduced to represent walls with different rigidities. The baseline bending stiffness EI for the analysis is 5.407×10^5 kNm^2/m, which refers to a wall of medium stiffness based on the databases of Long [10]. Therefore, $\alpha = 1.0$ is used for cases with this medium wall stiffness, while a smaller $\alpha = 0.1$ and $\alpha = 0.06$ represent flexible walls and larger α represents stiff walls.

The shoring system of the braced excavation comprised of struts and walers, as described in Sect. 2.1.

7.3.2 Model Verification

The plot of the bending moment of an 18 m diaphragm wall with an excavation depth of 10 m as provided by Chowdhury et al. [11] was used for the validation. Material properties and geometry are described in detail in Bahrami [7]. Figure 7.5 shows the results of the present 2D and 3D analyses as compared to the plot presented by Chowdhury et al. [11]. It is obvious from the figure that, the bending moment of the wall changes suddenly at the location of jointed points of the struts and the wall. The 2D results in this study are close to the plane strain analysis by Chowdhury et al. [11], and the 3D results are generally smaller than plane strain analysis in the wall's embedded area, which is reasonable since, in that area, the surrounding soil steadily prevents the movement of the wall in 3D analysis. In other areas, the 3D effect is less significant. Overall, the 2D and 3D numerical analysis in this study is consistent with the plane strain analysis conducted by Chowdhury et al. [11] and the field instrumentations.

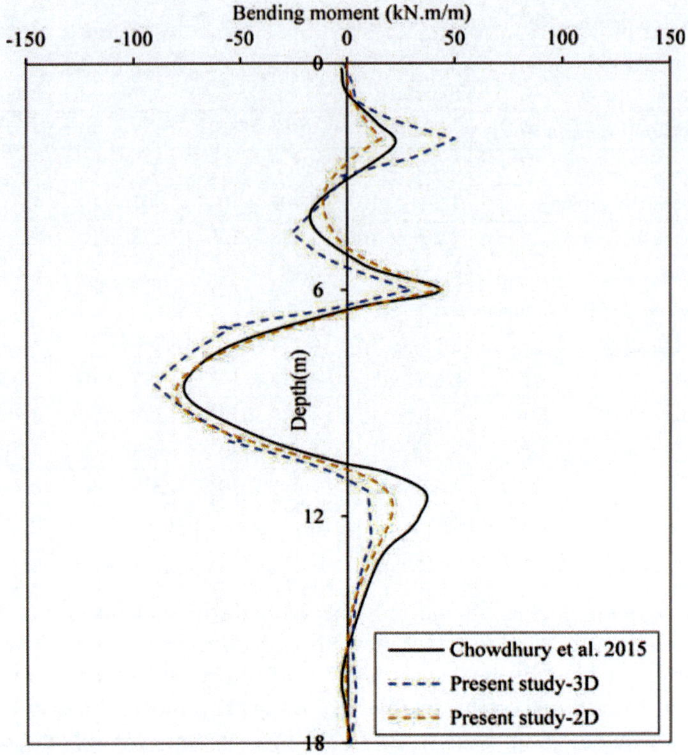

Fig. 7.5 Comparison of the bending moment of the wall between the current study and study by Chowdhury et al. [11]

7.3.3 *Analysis Results*

7.3.3.1 Load Transfer Mechanism of One-Strut Failure

Then the load transfer percentage is also defined as Eq. 7.1. To illustrate the load transfer mechanism of the OSF, cases with excavation depth $H_e = 20$ m, excavation length L of 40 m (L/B = 4.0) for loose sand are presented as an example. The load transfer percentages from 2 and 3D analyses are tabulated in Table 7.9 and Table 7.10, respectively. The 2D results indicate that failure of S3 leads to a major increase in the load for the 2nd level S2 strut and 4th level S4 strut, and the top struts bear the larger magnitude of the transferred loads, the maximum change of which is up to 66% of the initial load of the failed strut. While the failure of S4 leads to a considerable increase in the load for the 3rd level S3 strut and the maximum change of the struts S3 is approximately 90% of the initial load of the failed strut. On the other hand, the 3D results indicate that the load of the failed strut mainly transferred vertically and horizontally to the adjacent struts as shown in Table 7.10. It is found that the maximum change of the load is approximately 40% vertically and 10% horizontally of the initial

Table 7.9 Load transfer percentages from 2D analyses for one-strut failure of S3 and S4

Soil type	Struts	Load transfer percentage (%)							
		Failure of S3				Failure of S4			
		$a = 0.06$	$a = 0.1$	$a = 1.0$	$a = 2.0$	$a = 0.06$	$a = 0.1$	$a = 1.0$	$a = 2.0$
Loose sand	S1	−1.62	−6.19	−15.47	−16.52	11.75	9.18	3.43	0.96
	S2	48.80	52.31	64.13	66.16	−4.10	−11.59	−23.86	−23.09
	S3	Failed	Failed	Failed	Failed	72.39	78.19	85.27	89.29
	S4	46.01	48.03	53.76	54.35	Failed	Failed	Failed	Failed

Note Generally the struts are in a compressive stress state; negative values mean the decrease of the compressive stress due to the load transfer induced by OFS while positive values represent the increase brought by the failed strut

Table 7.10 Load transfer percentages from 3D analyses for one-strut failure of S3 and S4 (L/B = 4.0)

Wall stiffness	Struts	Load transfer percentage (%)									
		Failure of S3					Failure of S4				
		$x=-8$	$x=-4$	$x=0$	$x=4$	$x=8$	$x=-8$	$x=-4$	$x=0$	$x=4$	$x=8$
$a=0.06$	S1	0.64	1.07	0.56	1.05	0.62	0.37	0.63	0.82	0.63	0.35
	S2	1.35	5.33	**11.93**	5.34	1.36	0.51	0.41	−0.24	0.42	0.52
	S3	1.56	**10.94**	Failed	**10.95**	1.54	0.90	4.65	**11.83**	4.61	0.89
	S4	1.19	4.82	**13.06**	4.79	1.18	2.22	**21.97**	Failed	**21.97**	2.18
$a=0.1$	S1	0.53	0.67	−1.00	0.67	0.53	0.22	0.48	0.76	0.49	0.22
	S2	1.13	5.24	**16.27**	5.25	1.13	0.46	0.05	−1.68	0.05	0.46
	S3	1.26	**10.49**	Failed	**10.53**	1.26	0.70	4.38	**16.49**	4.35	0.68
	S4	0.95	4.61	**17.66**	4.59	0.94	1.36	**21.06**	Failed	**21.08**	1.38
$a=1.0$	S1	0.27	−1.24	**−8.77**	−1.24	0.27	0.15	0.04	0.36	0.04	0.15
	S2	−1.14	4.85	**38.01**	4.85	−0.31	−0.61	−1.06	**−8.37**	−1.07	0.22
	S3	−0.67	**9.79**	Failed	**9.79**	−0.67	−0.31	3.22	**39.38**	3.23	−0.34
	S4	−0.26	3.23	**36.36**	3.21	−0.25	−2.23	**20.68**	Failed	**20.66**	−2.10
$a=2.0$	S1	0.15	−1.76	**−8.91**	−1.75	0.15	0.20	−0.17	−0.58	−0.17	0.21
	S2	−0.73	5.32	**40.93**	5.31	−0.73	0.05	−1.30	−7.61	−1.30	0.06
	S3	−1.27	**10.09**	Failed	**10.09**	−1.27	−0.83	3.50	**41.79**	3.51	−0.85
	S4	−0.76	3.32	**38.30**	3.29	−0.74	−3.19	**22.72**	Failed	**22.63**	−3.06

load of the failed strut when subjected to OSF of S3, while the maximum change of the load is approximately 40% vertically and 20% horizontally when subjected to OSF of S4. For 3D analyses, the magnitude of the load transfer is much smaller, up to approximately 20% for flexible walls and 50% for medium walls compared with the results from the 2D analyses when subjected OSF of the lowest strut S4, which comes to the same conclusions drawn by Goh [12] for excavation in clay. Therefore, the 2D analyses overestimate the possible consequence of the one-strut failure for loose sand as it ignores the restraining effects of the adjacent horizontal struts. It is also observed from Tables 7.9 and 7.10 that load transfer percentages to adjacent horizontal struts change slightly when wall stiffness α varies, while the load transfer percentages to adjacent vertical struts have a tendency to grow with wall stiffness α, such as the load transfer percentage for the 2nd level S2 strut arises from 48.8 to 66.16% when S3 failed in 2D analysis and from 11.93 to 40.93% in 3D analysis when wall stiffness α increases.

In Table 7.10, struts with more than 8% load transfer percentage are outlined by the framework as well as boldface, indicating the load mainly transferred vertically and horizontally to the adjacent struts after the one-strut failure as depicted in Fig. 7.6, which shows the influential zone of the one-strut failure of S3 with wall stiffness $\alpha = 1.0$ for L/B = 4.0 in loose sand, where larger arrows denote the larger magnitude of the transferred loads and the red circles denote the struts with more than 8% load transfer after the one-strut failure.

For medium dense sand and dense sand, most of the load of the failed strut S3 is transferred to the struts immediately above and below the failed strut, as indicated in Table 7.11, with approximately 35–55% of the load transferred to the upper strut and the remainder carried by the lower strut. For the one-strut failure of S4, the load transfer to the 3rd level S3 strut is modulated near 75%, and S2 carries approximately 10–20% in medium dense sand and 15–30% in dense sand. Generally, it is found that the more flexible the wall is, the less forces are distributed to the adjacent struts when subjected to one-strut failure of S3 in medium dense sand, which is similar to the trend indicated in loose sand.

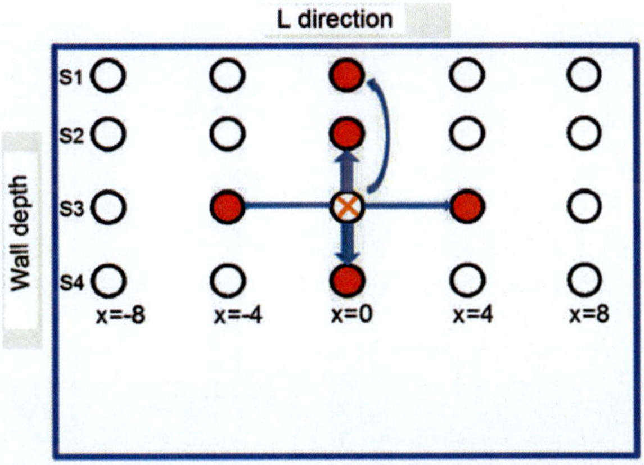

Fig. 7.6 Influential zone of the one-strut failure of S3 ($\alpha = 1.0$)

Table 7.11 Load transfer percentages from 2D analyses for one-strut failure of S3 and S4 ($H_e = 20$ m)

Soil type	Strut	Load transfer percentage (%)							
		Failure of S3				Failure of S4			
		$a = 0.06$	$a = 0.1$	$a = 1.0$	$a = 2.0$	$a = 0.06$	$a = 0.1$	$a = 1.0$	$a = 2.0$
Medium dense sand	S1	3.62	3.08	**−9.44**	**−10.57**	5.81	**9.83**	2.15	0.48
	S2	**36.67**	**43.27**	**55.13**	**57.33**	**19.85**	7.27	**−15.74**	**−13.77**
	S3	Failed	Failed	Failed	Failed	**72.48**	**74.10**	**75.52**	**74.63**
	S4	**50.16**	**47.31**	**50.49**	**51.52**	Failed	Failed	Failed	Failed
Dense sand	S1	−3.07	0.39	−7.58	**−8.23**	1.82	7.77	3.39	0.18
	S2	**33.97**	**44.71**	**52.58**	**53.83**	**29.52**	**14.83**	**−15.47**	**−11.21**
	S3	Failed	Failed	Failed	Failed	**76.13**	**72.28**	**76.23**	**71.47**
	S4	**55.63**	**49.03**	**49.54**	**49.90**	Failed	Failed	Failed	Failed

To illustrate the effect of various L/B on load transfer percentages of the OSF, cases with wall stiffness $\alpha = 1.0$ for dense sand with excavation depth $H_e = 20$ m for various L/B were studied, as indicated in Table 7.12. For dense sand with medium rigid walls, the load is mainly transferred to the struts directly above and below the failed strut, for L/B = 1.6, the forces transferred to adjacent struts are greater than those in another two situations. Generally, the load transfer percentage to adjacent struts decreases gradually with increasing L/B ratio.

7.3.3.2 Effect of One-Strut Failure on Axial Stress of Struts

Based on the evaluations of Bahrami [7], Peck's soil pressure distribution method is acceptable for lateral struts in excavations up to 20 m in the sand. For the strut force F_i, the apparent earth pressure σ_h can be obtained by

$$\sigma_h = \frac{F_i}{S_H h_{si}} \tag{7.2}$$

The axial stress of struts is presented in Figs. 7.7 and 7.8, in these figures, the nomenclature used for the diagrams is represented by a triple system (such as LS-BF-S3). The first component indicates the soil type, where LS indicates loose sand, MD indicates medium dense sand and DS indicates dense sand; The second component indicates the condition of the excavation, where BF denotes conditions before failure, FS2 denotes one-strut failure of 2nd level S2 strut; the third component represents the specific strut in a certain level influenced most by OSF, where S3 in Fig. 7.7 denotes the strut influenced most in the 3rd level when subjecting to OSF of S2.

In Figs. 7.7 and 7.8, f_a represents the existing axial stress in the steel strut, while F_a specifies the allowable stress calculated by Eq. 7.2. Increasing the ratio of f_a/F_a to values greater than one may put the lateral steel strut at risk of failure, and thereby challenge the safety of the excavation.

As can be inferred from the diagrams illustrated in Fig. 7.7, for 15 m excavations in all soil types, the ratio of f_a/F_a of strut S2 and strut S3 before strut failure was around 0.3–0.5 for flexible walls and 0.3–0.7 for rigid walls. When subjected to OSF of S2 or S3, the ratio of f_a/F_a increases up to around 0.4–0.6 and 0.4–1.0 for flexible walls and rigid walls, respectively. It can also be found that the 2nd level S2 strut installed at a depth of 0.4He bears stresses beyond the allowable value when subjected to OSF of S3 for rigid walls in loose sand (in Fig. 7.7b green solid line), which means the second strut in such situations falls in an unsafe zone and requires a larger cross-section than those required for the first and the third struts, which is consistent with the conclusion drawn by Bahrami [7]. Other cases in Fig. 7.7 show the failure of one strut due to an accident would not result in the detrimental failure of the entire excavation system.

Similarly, Fig. 7.8 shows the states of stress in the strut for an excavation depth of 20 m in different soil types. According to the plots presented in Fig. 7.8, the ratio of f_a/F_a of strut S3 and S4 before strut failure was around 0.3–0.5 for flexible walls and

Table 7.12 Load transfer percentages from 3D analyses for the one-strut failure of S3 and S4

L/B	Struts	Load transfer percentage (%)									
		Failure of S3					Failure of S4				
		$x = -8$	$x = -4$	$x = 0$	$x = 4$	$x = 8$	$x = -8$	$x = -4$	$x = 0$	$x = 4$	$x = 8$
L/B = 1.6	S1	–	0.16	−4.32	0.15	–	–	0.19	0.54	0.18	–
	S2	–	4.90	**27.84**	4.90	–	–	−0.41	−4.79	−0.42	–
	S3	–	**9.22**	Failed	**9.23**	–	–	4.54	**28.91**	4.54	–
	S4	–	4.51	**27.61**	4.46	–	–	**17.83**	Failed	**18.23**	–
L/B = 2.4	S1	0.37	0.23	−4.21	0.23	0.36	0.13	0.24	0.53	0.23	0.12
	S2	0.80	4.79	**25.69**	4.78	0.81	0.32	−0.36	−4.57	−0.34	0.33
	S3	0.89	**8.75**	Failed	**8.72**	0.91	0.44	4.41	**27.80**	4.37	0.46
	S4	0.65	4.48	**26.17**	4.46	0.69	0.78	**18.12**	Failed	**17.64**	0.81
L/B = 4.0	S1	0.40	0.13	−4.05	0.12	0.40	0.07	0.07	0.35	0.07	0.07
	S2	0.83	4.77	**24.86**	4.77	0.84	0.39	−0.24	−4.31	−0.25	0.39
	S3	0.93	**8.53**	Failed	**8.55**	0.93	0.61	4.23	**26.89**	4.22	0.60
	S4	0.63	4.19	**25.83**	4.16	0.62	0.34	**17.14**	Failed	**17.18**	0.39

Note "–" represents the absence of the struts due to the limited excavation length

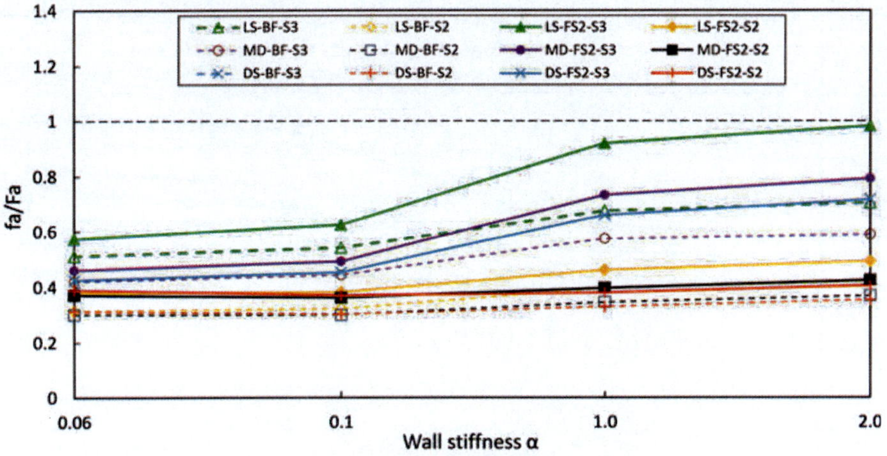

(a). One-strut failure of S2

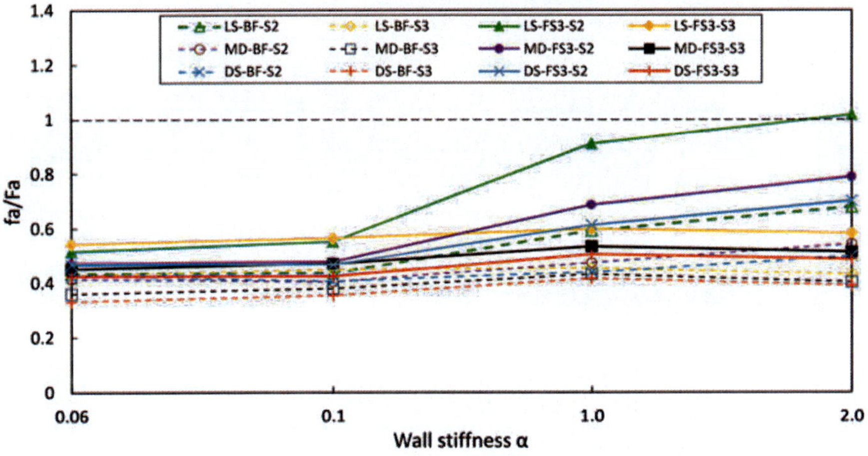

(b). One-strut failure of S2

Fig. 7.7 Variations of axial stress in struts with various wall stiffness α when $H_e = 15$ m

0.4–0.7 for rigid walls. When subjected to OSF of S3 or S4, the ratio of f_a/F_a increases up to around 0.4–0.6 and 0.4–1.0 for flexible walls and rigid walls, respectively. It can also be found that the 4th level S4 strut installed at depth of 0.8He bears stresses beyond the allowable value when subjected to OSF of S3 for rigid walls in loose sand (in Fig. 7.8a green solid line), which means the 4th level S4 strut falls in an unsafe zone.

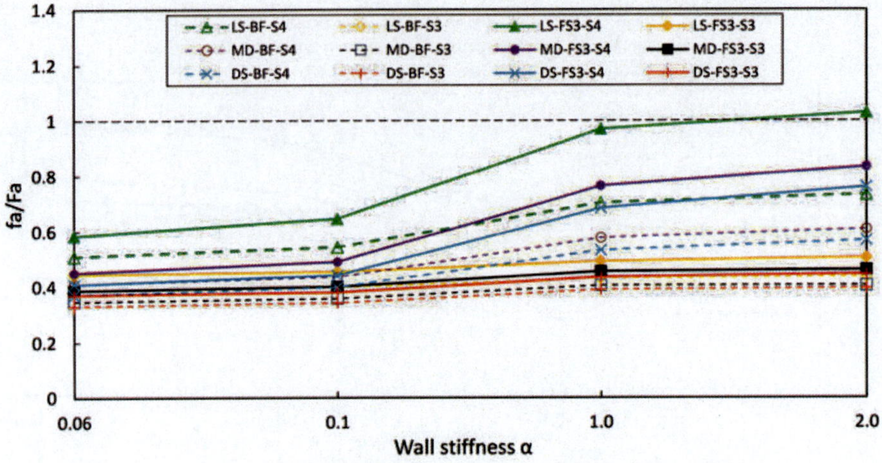

(a). One-strut failure of S3

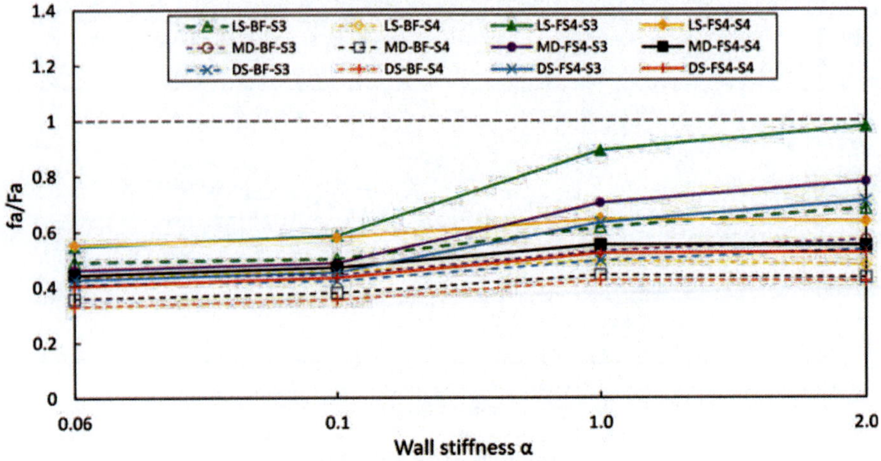

(b). One-strut failure of S4

Fig. 7.8 Variations of axial stress in struts with various wall stiffness α when $H_e = 20$ m

7.3.3.3 Effect of One-Strut Failure on Wall Deflection

For brevity, only 3D numerical results are presented and discussed here, and the lateral wall deflections before and after strut failure for excavation depth $H_e = 15$ m are plotted in Fig. 7.9, in which BF represents horizontal wall deflection before failure, and OSF-S1 denotes the horizontal wall deflection when subjected to one-strut failure of S1. Generally, the lateral wall deflection decreases as the soil strength increases, and the maximum lateral wall deflection occurs at a location of 13 m below

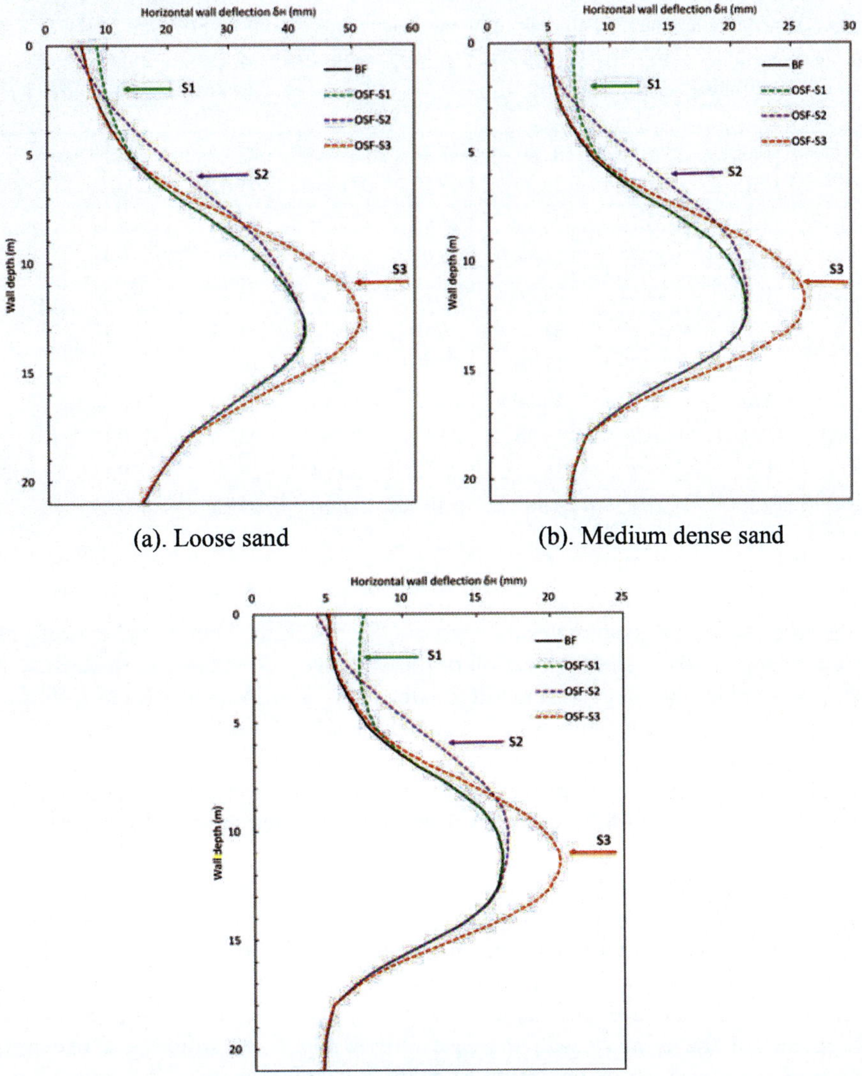

(a). Loose sand

(b). Medium dense sand

(c). Dense sand

Fig. 7.9 Effect of One-strut failure on horizontal wall deflection ($H_e = 15$ m)

the ground surface, approximately 60% of the total length of the wall. Besides, with a certain type of sand, the lateral wall deflection increases in varying degrees when subjected to different failures of struts. However, the influenced zones for lateral wall deflection in different types of sandy soils are nearly the same, i.e., the influenced zone of OSF-S1 ranges from ground level to the 2nd level of the strut, and the influenced zone of OSF-S2 begins with the ground level and ends in the 3rd level

Table 7.13 Maximum lateral wall deflection: before and after strut failure ($H_e = 15$ m)

Soil type	Wall stiffness α	δ_{max}-BF	OSF-S1	OSF-S2	OSF-S3	Maximum increase ratio % (%)	Maximum δ_{max}/H_e (%)
Loose sand	0.06	47.56	47.6	47.48	57.95	21.84	0.39
	0.1	42.38	42.42	42.2	51.29	21.02	0.34
	1.0	25.19	25.19	24.94	27.93	10.88	0.19
	2.0	22.1	22.11	21.91	23.93	8.28	0.16
Medium dense sand	0.06	23.62	23.66	24.23	29.39	24.42	0.20
	0.1	21.22	21.25	21.35	26.13	23.14	0.17
	1.0	12.25	12.26	12.2	14.22	16.08	0.09
	2.0	10.67	10.69	10.62	11.95	12.00	0.08
Dense sand	0.06	19.2	19.25	20.31	23.57	22.76	0.16
	0.1	16.7	16.74	17.12	20.64	23.59	0.14
	1.0	9.55	9.55	9.55	11.2	17.28	0.07
	2.0	8.23	8.23	8.23	9.39	14.09	0.06

of the strut. When the system is subjected to OSF of S3, the lateral wall deflections start to change at the installed location of the 2nd level of strut and keep consistent with those before failure at wall depth greater than 18 m. Actually, when suffering from the one-strut failure of a certain strut, the load acted on the failure strut will distribute to adjacent struts as discussed previously, and the wall deflection will increase undoubtedly. However, the horizontal wall deflection will be restrained due to the buttressing effect of the shoring system (including retaining walls, walers, and struts), which makes the displacement increment of the horizontal wall deflection small at the location of installed struts and walers. When subjected to OSF of S2, the restraining effect of the top struts is not enough to confine its displacement as the short cantilever wall provides a weak constraint of the deformation.

Tables 7.13 and 7.14 below presents the comparison of maximum lateral wall deflection before and after strut failure, as indicated in these two tables, the lateral movements of the primary wall increase remarkably when suffering a one-strut failure of the lowest strut (i.e., strut S3 with excavation depth of 15 m and strut S4 with excavation depth of 20 m, respectively.), such as the maximum lateral wall deflection in loose sand with maximum increasing ratio reaching to 21.84%. The maximum δ_{max}/H_e in excavation depth of 15 m rises up to 0.39%, 0.20% and 0.16% for loose sand, medium dense sand, and dense sand, respectively. For $H_e =$ 20 m, it reaches to 0.47%, 0.23%, and 0.18%, respectively. As Zhang et al. [14, 15] summarized 63 cases of braced excavations in the sand and obtained that the average maximum lateral wall deflection was less than 0.35% of excavation depth, some cases of the 3D numerical results of OSF go beyond this bound, i.e., cases for OSF of S3 in loose sand with wall stiffness $\alpha = 0.06$ and cases for OSF of S4 in loose sand with wall stiffness $\alpha = 0.06$ or $\alpha = 0.1$, which should be paid more attention

Table 7.14 Maximum lateral wall deflection: before and after strut failure ($H_e = 20$ m)

Soil type	Wall stiffness α	δ_{max}-BF	OSF-S1	OSF-S2	OSF-S3	OSF-S4	Maximum increase ratio % (%)	Maximum δ_{max}/H_e (%)
Loose sand	0.06	79.14	79.15	79.19	79.59	93.91	18.65	0.47
	0.1	70.72	70.72	70.75	70.72	83.48	18.04	0.42
	1.0	41.34	41.34	41.31	41.06	45.85	10.91	0.23
	2.0	35.78	35.78	35.75	35.53	38.65	8.02	0.19
Medium dense sand	0.06	39.3	39.32	39.37	40.79	45.66	16.12	0.23
	0.1	35.33	35.34	35.36	36.28	41.28	16.80	0.21
	1.0	20.37	20.38	20.36	20.41	23.23	13.98	0.12
	2.0	17.66	17.66	17.64	17.67	19.64	11.21	0.10
Dense sand	0.06	30.84	30.85	30.92	32.64	35.23	14.20	0.18
	0.1	27.54	27.55	27.99	28.6	31.98	16.08	0.16
	1.0	15.92	15.92	15.91	16.09	18.31	15.01	0.09
	2.0	13.65	13.66	13.64	13.74	15.38	12.595	0.08

when designing the shoring system considering the probable ground movement associated with the excessive wall deflections. On the other hand, Tables 7.13 and 7.14 also indicated that one-strut failure of the upper struts, representing the struts except the lowest one (for example, S1, S2 in 15 m excavation and S1, S2, S3 in 20 m excavation), generally have little influence on maximum lateral wall deflection.

7.4 One Column Failure of Structure Above Excavation Pit

Aiming at the failure of a single force transfer column at different locations caused by construction accident or excessive loading, alos revealing the load transfer mechanism to surrounding beams and columns, a series of FE analysis is carried out based on project in Changsha.

7.4.1 Case Details and Numerical Model

7.4.1.1 Project Description

The project is a commercial reconstruction project of Xiangjiang happy city in a deep disused mine. The site is a deep mine leftover from mining for more than 50 years. The overall length is about 440 m and the width is 350 m. The main building structure was constructed above the north deep pit and supported on all levels of rock walls of the pit through the column structure as shown in Fig. 7.10 Contact between pit wall and superstructure. The columns are arranged in rectangle and connected by beams [16].

For the specific project, there is no obvious poor geological condition in the site, and the stable groundwater level is between 5.00 and 6.10 m beneath the ground. Site soil is mainly comprised of mix backfill soil, silty clay, and two layers of limestone with slightly different properties, the specific situation is shown in Fig. 7.11.

The groundwater of the site mainly consists of upper phreatic water and fissure water in bedrock. The upper phreatic water is distributed in the artificial fill of the

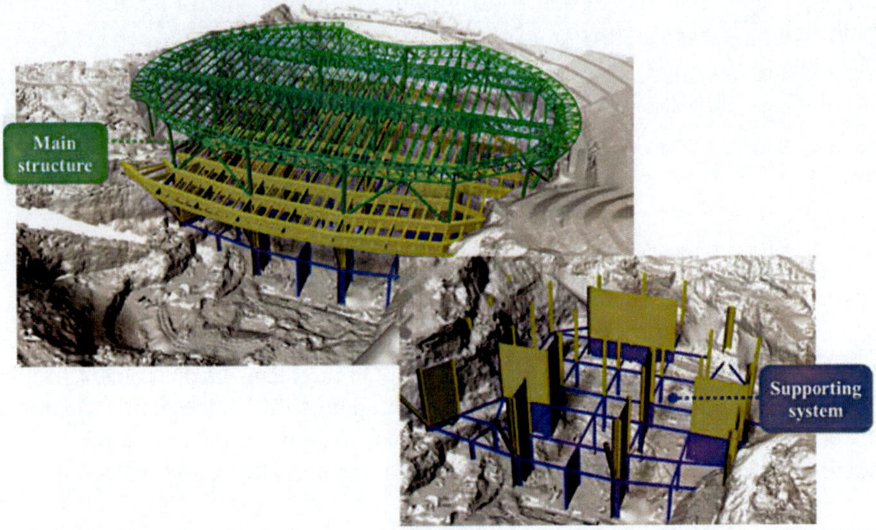

Fig. 7.10 Contact between pit wall and superstructure

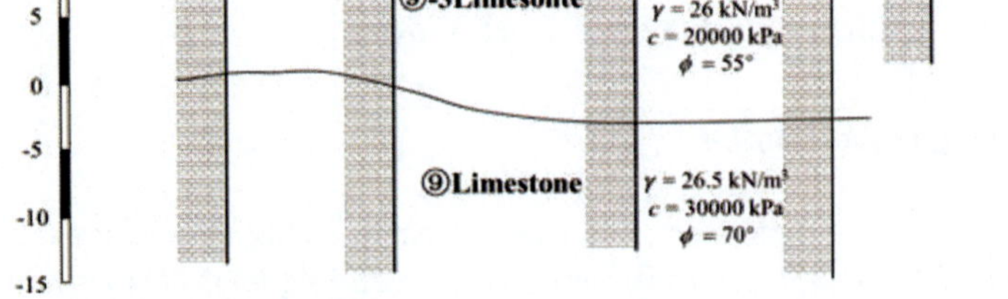

Fig. 7.11 Site soil distribution and properties

site, and the stable water level buried depth is between 5.00 and 6.10 m as Fig. 7.11 indicates; the fissure water exists in the bedrock limestone. The hydraulic connections between these two are positive.

7.4.1.2 Numerical Model

The plane geometry of the underlying the main area is approximate to the ellipse with 240 min long 160 m in short. To simplify the modeling, the irregular plane shape of the pit is approximated to multiple break lines to carry out numerical simulation; the underlying area of the main building is assumed to be the same slope shape with four levels of grading, as shown in Fig. 7.12a. The model size is 500 × 500 × 80 m (*X*–*Y*–*Z* axis). The overall model is represented in Fig. 7.12b. Tetrahedral element is used to simulate rock and soil. This type of mesh element can adapt to and simulate complex geometries.

Due to the limitation of computational efficiency and the lack of joint information, the discontinuities in the rock mass are not considered for modeling limestone, so

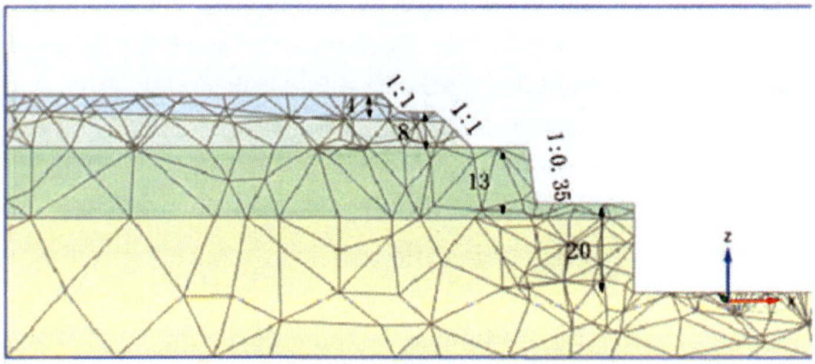

(a) Simlified pit wall geometry

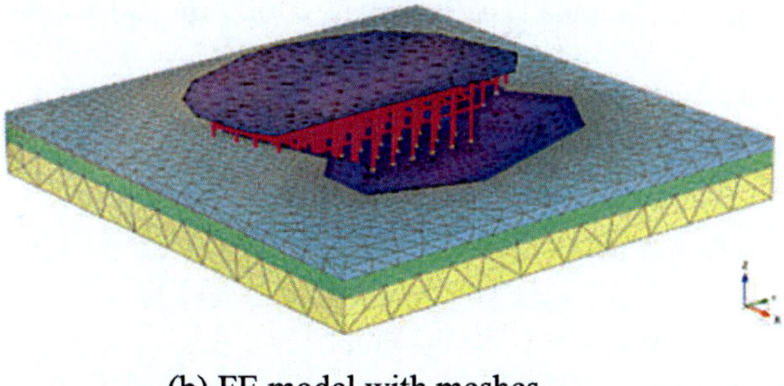

(b) FE model with meshes

Fig. 7.12 Established 3D model

Table 7.15 Structural properties

Structural members	Unit weight (kN/m^3)	Size (m)	Spacing (m)	Young's modulus (MPa)	Poisson ratio	Element type
Column	24	2.2 × 2.2	24	28,000	–	Beam
Connecting beam	24	1.2 × 1.0	24	28,000	–	Beam
Foundation	25	3.6 × 3	24	31,500	0.35	Soil
Pit wall treatment	26	Thickness 0.5	–	40,000	0.2	Plate
Building base slab	26	Thickness 1.0	–	31,500	0.35	Plate

Table 7.16 Construction stages for numerical model

Stage no.	Construction activities
1	Consolidation of site rock and soil
2	De-activate soil and rock mass inside the mining pit, the consolidation calculation type and long time interval are used to simulate the long-term deformation process
3–6	Reset displacement; from bottom to top, simulate pit wall treatment in four steps
7	Column, foundation and beam construction, cast building base slab
8	Activate the vertical load on the base slab to simulate the construction and working load of the superstructure
9	De-activate columns supported on different locations to simulate the failure of single columns

the rock mass is assumed to be isotropic in the numerical simulation. The structural components mainly include: the rock wall treatment for the slope and the bottom of the mining pit, the construction panel, the base slab of the upper building (the upper building structure is simplified as the building floor bearing the vertical uniform load), the columns and the foundation to transfer building servicing load, and the beam connecting the columns. The information of structure elements is listed in Table 7.15.

For the construction steps, nine stages are established, as shown in Table 7.16.

7.4.2 Analysis Results

According to the calculation results, the failure of a single column causes the local suspension of its connecting beam structure, and the settlement at this position increases sharply from 30.92 to 178.4 mm, as shown in Fig. 7.13a. The significant deformation also causes the change of beam internal forces. Figure 7.13b

Fig. 7.13 Deformation and force of the connecting beam before and after the single pile failure

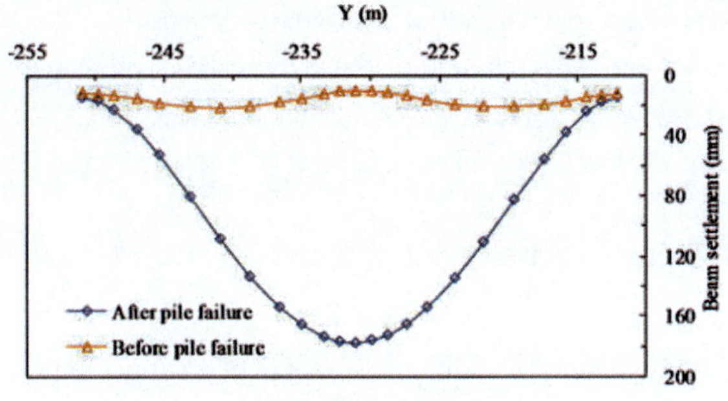

(a) Deformation

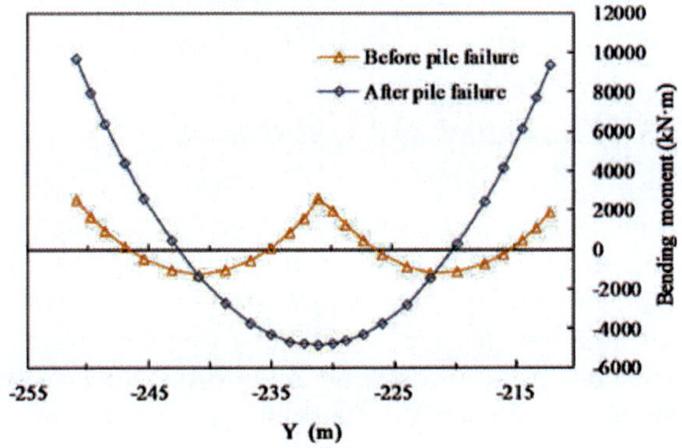

(b) bending moment

is the bending moment diagram of the beam before and after the column failure. The bending moment distribution has changed significantly, and its value has also increased to five times comparing to the normal servicing state. The large increase of beam deformation and stress may cause concrete cracking, therefore, affect the bearing capacity of the beam.

After the failure of a single column, the original vertical load on the column is transferred to other columns through the connecting beam and the base slab of the upper structure. Through the analysis, it is concluded that the load on the column supported in miscellaneous fill and silty clay layer is far less than that on the support and rock mass, so the load increment of the surrounding column caused by its failure can be ignored. As for the columns supported on the limestone, after its sudden failure, the service load of the surrounding columns increases in various degrees. Table 7.17 lists the axial force changes of the surrounding columns before and after the failure of a single column supported on the second step wall. The situation caused by the failure of columns on the lower rock wall is similar, therefore, has not been displayed.

It can be seen from the results in Table 7.17 that the failure of a single column or foundation transferring the load from the superstructure mainly produce large internal force and deformation increment to the transverse and longitudinal beams connected

Table 7.17 Axial forces of surrounding columns

Locations	Axial force ($\times 10^3$ kN) (before single column failure/after)		
First level	21/22	28.48/43.5	23/31
Second level	126/146	140/Failed	168/198
Third level	85/116	100/131	114/128
Pit base	90/102	95/97	101/107

with it. After the failure of a single pile supported on the rock wall, the original load is transferred to the surrounding columns, resulting in the minimal increment of the internal force, which will not cause progressive and continuous column failure.

7.5 Summary and Conclusions

In this chapter, the FE simulation of one-strut failure of multi-level braced excavations was considered. The results were validated against the semi-empirical Clough design chart before performing the one-strut failure analysis. The 3D one-strut failure analysis indicates that the 2D analyses would result in fairly conservative (i.e., larger) estimates of the loads transferred to the adjacent struts from the failed strut as it ignores the restraining effects of the adjacent horizontal struts. Therefore, the 2D analysis would result in a more conservative design. This may result in a larger strut size and larger waler size which would increase costs. Actually, the one-strut failure analysis involves an interaction process between neighboring struts, between struts and wall, and is influenced by the strut location, system stiffness, soil types, etc. It should be noted that the load transfer paths and the percentages for the one-strut failure were purely from 3D analyses and the use of them for the design of strutting system should be done with caution.

Generally, the load transfer percentage to adjacent struts decreases gradually with increasing L/B ratio. 3D results indicate that the load of the failed strut mainly transfers vertically and horizontally to the adjacent struts, load transfer to adjacent horizontal struts change slightly when wall stiffness α varies, while the load transfer percentages to adjacent vertical struts have a tendency to grow with wall stiffness α.

The lateral movements of the primary wall increase remarkably when suffering from a one-strut failure of the lowest strut, while upper struts failure have little influence on the maximum lateral wall deflection.

While for the deep mining pit under annular heavy point loading from uplying large space structure, the failure of a single foundation or load transfer column mainly causes great increment to the displacement and internal force of the connecting beam, which may lead to local cracking of the beam structure, and even gradually develop into fracture failure. However, the load transferred to the surrounding columns is limited and will not cause progressive continuous column failure.

References

1. Endicott, J. (2013). Case histories of deep excavation. Examination of where things went wrong: Nicoll Highway Collapse, Singapore. In: *International Conference on Case Histories 357 in Geotechnical Engineering*, Paper 7.
2. Gong, X. N., & Zhang, X. C. (2012). Excavation collapse of Hangzhou subway station in soft clay and numerical investigation based on orthogonal experiment method. *Journal of Zhejiang University SCIENCE, 13*(10), 760–767.
3. Stille, H., Broms, B.B. (1976). Load redistribution caused by anchor failures in sheet pile walls. Sechste Europaeische Konferenz Fuer Bodenmechanik und Grundbau 197–200.
4. Low, Y. H. S., David Ng, C. C., Chin, Y. P. Y., & Ting, S. K. E. (2012). A Singapore case history of temporary removable ground anchor design to TR26: 2010. *The IES Journal Part A: Civil & Structural Engineering, 5*, 181–194.
5. Pong, K. F., Foo, S. L., Chinnaswamy, C. G., Ng, C. C. D., & Chow, W. L. (2012). Design considerations for one-strut failure according to TR26—a practical approach for practising engineers. *The IES Journal Part A: Civil & Structural Engineering, 5*, 166–180.
6. Chowdhury, S. S., Deb, K., & Sengupta, A. (2013). Estimation of design parameters for braced excavation: Numerical study. *International Journal of Geomechanics, 13*, 234–247.
7. Bahrami, M., Khodakarami, M.I., Haddad, A. (2018). 3D numerical investigation of the effect of wall penetration depth on excavations behavior in sand. *Computers and Geotechnics, 98*, 82–92.
8. Hsiung, B.-C.B. (2009). A case study on the behaviour of a deep excavation in sand. *Computers and Geotechnics, 36*, 665–675.
9. Finno, R. J., Blackburn, J. T., & Roboski, J. F. (2007). Three-dimensional effects for supported excavations in clay. *Journal of Geotechnical and Geoenvironmental Engineering, 133*, 30–36.
10. Long, M. (2001). Database for retaining wall and ground movements due to deep excavations. *Journal of Geotechnical and Geoenvironmental Engineering, 127*, 203–224.
11. Chowdhury, S. S., Deb, K., & Sengupta, A. (2015). Behavior of underground strutted retaining structure under seismic condition. *Earthquakes and Structures, 8*, 1147–1170.
12. Goh, A.T.C., Zhang, F., Liu, H.L., Zhang, W.G., Zhou, D. (2018). Numerical analysis on strut responses due to one-strut failure for braced excavation in clays. *Proceedings of the 2nd International Symposium on Asia Urban GeoEngineering.*
13. Peck, R. B. (1969). Deep excavation and tunneling in soft ground. In *Proceedings of the 7th International Conference on soil Mechanics and Foundation Engineering*, Mexico City (pp. 225–290).
14. Zhang, R. H., Goh, A. T. C., Li, Y. Q., Hong, L., & Zhang, W. G. (2021). Assessment of apparent earth pressure for braced excavations in anisotropic clay. *Acta Geotechnica.* https://doi.org/10.1007/s11440-020-01129-x
15. Zhang, W. G., Zhang, R. H., Fu, Y. R., Goh, A. T. C., & Zhang, F. (2018). 2D and 3D numerical analysis on strut responses due to one-strut failure. *Geomechanics and Engineering, 15*(4), 965–972.
16. Zhang, W. G., Wu, J. H., Han, F. C., He, C. J., Li, J. X., & Li, Y. Q. (2021). Numerical investigation on stability of disused deep mining pit under annular heavy point loading from large space structure. *Journal of Civil and Environmental Engineering.* https://doi.org/10.11835/j.issn.2096-6717.2021.039.

Chapter 8
Jet Grouting Slabs and Jet Grouting Piles

8.1 Inroduction

Excessive lateral deformation of supporting wall system can result in excessive ground settlement and subsequent damage to adjacent infrastructures. Above the final excavation level, the lateral struts/props can be used to control and restrain the wall deflection and ground movement. However, for deep excavations in soft clay, due to the lack of restraint of the wall below the final excavation level, the maximum wall deflection generally occurs below the final excavation level. Under such circumstances, ground improvement methods such as jet grouting or deep mixing are commonly adopted to strengthen the soil beneath the excavation area and consequently to restrain the wall deformation.

Ground improvement through jet grouting or deep mixing cement has become a common method to control soil movement induced by excavation in some districts such as Chinese Taiwan and Singapore. The effectiveness of these methods in ensuring the safety of the excavation system and adjacent structures has been widely demonstrated via various successful projects [1].

In deep excavation projects involving soft clay deposits, the jet grouting piles are commonly constructed directly below the jet grout slab, which consists of a series of short overlapping grouting columns below the final excavation level [2]. The slab serves as part of the earth retaining stabilizing system to resist the lateral movement of supporting wall [3]. In this section, braced excavations with ground improved by jet grouting slabs and piles are computed and analyzed, based on comparison with the measured excavation responses. Subsequently, 2D FE analyses are presented to evaluate several influential factors including the depth-width ratio of the excavation, the thickness of jet grouting slabs, the depth of diaphragm wall, the thickness and elastic modulus of marine clay. Their influences on jet grouting slab behavior are explored through parametric analysis. Later, extensive 3D FE analyses are carried out, focusing on the performance of the jet grouting pile in controlling the excavation response. The effects of various geometric and stiffness properties of JGP on the

W. Zhang and H. Liu, *Design of Deep Braced Excavation and Earth Retaining Systems Under Complex Built Environment*,
https://doi.org/10.1007/978-981-16-5320-9_8

excavation response are considered and discussed. FEM software Plaxis 2D and Plaxis 3D are employed in the computation in this chapter.

8.2 Effect of Jet Grouting Slabs (2D) on Braced Excavations

8.2.1 Case Study

8.2.1.1 Excavatio Details

The site for this case is located in Race Course Road, Singapore. Soil profiles and the arrangement of retaining structure system are shown in Fig. 8.1. The first layer located from surface to −1.5 m is fill layer. The second layer from −5 to −10 m is fluvial sand. The third layer from −10 to −27.5 m is marine clay. The layer below −27.5 m is old alluvium. The standard penetration test (SPT) of the layer below −33 m is greater than 100. The final excavation depth is 17.5 m. A jet grouting slab of 1.5 m thickness is adopted to constrain the wall deflection and control seepage. The groundwater level was about 5 m below the ground surface. The supporting system consists of a 0.8 m thick and 35 m deep diaphragm wall and six axial struts (the average vertical distance is 3 m). There is a 3-storey frame structure building which is 12 m away from the excavation. The construction of diaphragm walls and jet grouting slabs is prior to excavation.

Due to the lack of geological survey reports, the value of soil parameters in this part are from other cases in Singapore [4]. Considering that the soil properties of

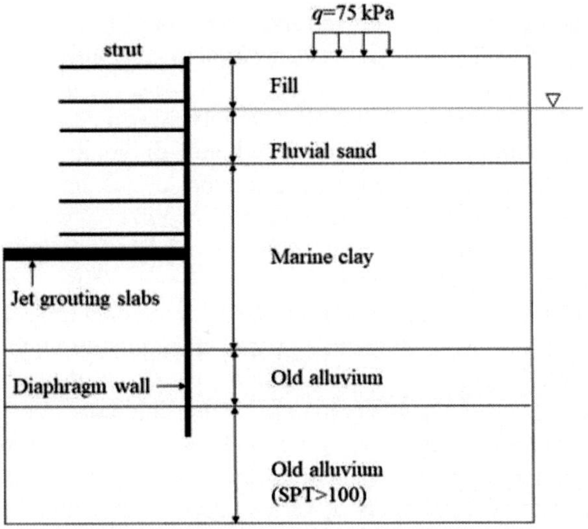

Fig. 8.1 Schematic plot of soil profiles and support systems

different regions and depths vary greatly, this chapter selects engineering cases with similar soil distribution in adjacent areas.

8.2.1.2 FEM Modeling

Hardening Soil with small strain effect (HSS) model is adopted to represent the stress–strain relationship of all the soil layers. The input parameters of soils are shown in Table 8.1, in which fill and fluvial sand are simulated by the drainage mode, marine clay and old alluvium are simulated by undrained B mode. In Table 8.1, the soil weight γ, friction angle φ, undrained shear strength c_u, and permeability coefficient k, etc., are listed. The range of small strain hardening is 0–0.0002%. The parameters are derived from the case histories from Xuan [4].

Linear elastic model is adopted to simulate the diaphragm and axial struts (Table 8.2). The jet grouting slab is simulated by Mohr–Coulomb model. The performance of jet grouting slab is greatly affected by the quality of construction. Therefore, conservative values are adopted for the equivalent modulus of elasticity and equivalent undrained shear strength. In this part, elastic modulus of $E = 120$ MPa and undrained shear strength of $c_u = 300$ kPa (undrained B mode in PLAXIS 2D) are used (Table 8.2).

Considering that the quality of diaphragm wall is not easy to control during construction, the reduction factor of elastic modulus is set at 0.7.

For adjacent buildings, each floor is simplified to the uniform loading of 25 kPa, and the maximum ground settlement is approximated to replace the settlement of the adjacent building.

The total simulation steps follow: (1) initial consolidation process (K_0 process); (2) activation of uniform load to simulate building construction; (3) reset displacements to zero; activate diaphragm wall and jet grouting slab at the base of the excavation; (4) excavation to 3 m below ground surface (BGS), while setting up the first strut; (5) similarly, gradually excavation to 6 m, 10 m, 12.5 m, 15 m, 17.5 m BGS, respectively, and setting up the corresponding strut at the same time; and (6) excavation proceeds till the final elevation level.

The FE model is shown in Fig. 8.2.

8.2.1.3 Model Validation

After carrying out FE analysis, numerical simulation result demonstrates that the maximum deflection of the diaphragm wall without jet grouting slab is 82.08 mm, and 49.82 mm with jet grouting slab. The test data indicated that the deflection of diaphragm wall was 80 mm (without jet grouting slabs) and 55 mm (with jet grouting slabs), respectively. The difference is marginal, indicating that the parameters of soil and jet grouting slab are basically reasonable. The differences may lie in that the mechanical property of jet grouting slab is non-uniform.

Table 8.1 Soil properties at race course road

Soil type	Model	γ (kN/m^3)	c_u (kPa)	E_{50} (MPa)	E_{50}^{ref} (MPa)	E_{ur}^{ref} (MPa)	m	G_0^{ref} (MPa)	$\gamma_{0.7}$	φ (°)	k (m/day)
Fill	HSS	19	–	5.25	33	99	0.8	130	0.0002	30	1×10^{-4}
Fluvial sand	HSS	19	–	7.5	23	69	0.8	67	0.0002	35	1×10^{-5}
Marine clay	HSS	16	44	7.95	7.95	23.85	1	30	0.0002	23	1×10^{-8}
Old alluvium	HSS	20	150	90	107	321	0.8	380	0.0002	40	1×10^{-8}
Old alluvium (SPT > 100)	HSS	20	500	300	270	810	0.8	820	0.0002	45	1×10^{-8}

Table 8.2 Structural properties at race course road

Structural members	E (kN/m^2)	A (m^2/m)	I (m^4/m)
Strut (at −1 m)	2.07×10^8	0.019	–
Strut (at −4 m)	2.07×10^8	0.019	–
Strut (at −7 m)	2.07×10^8	0.019	–
Strut (at −10 m)	2.07×10^8	0.038	–
Strut (at −13 m)	2.07×10^8	0.038	–
Strut (at −16 m)	2.07×10^8	0.038	–
Diaphragm wall of 800 mm thickness	2.8×10^7	0.8	0.04267

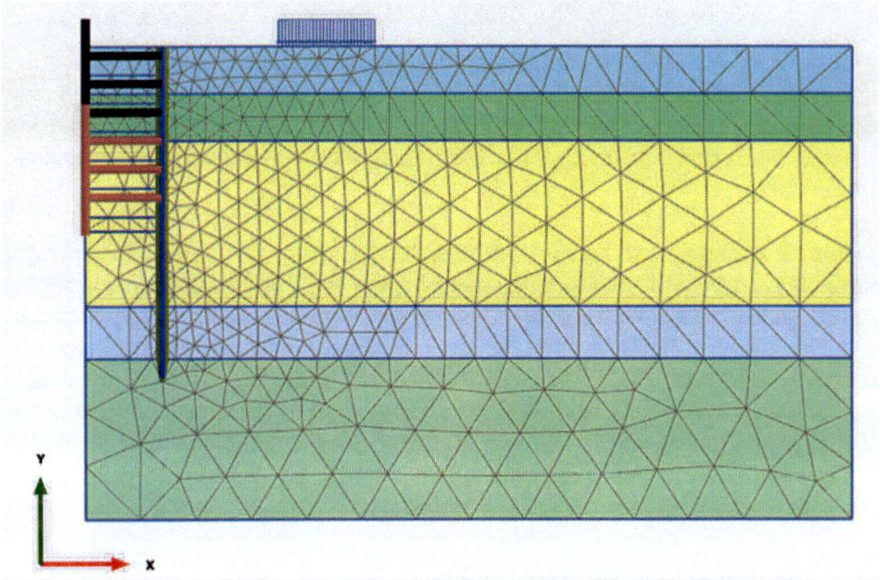

Fig. 8.2 Finite-element numerical model

8.2.2 *Parametric Study*

8.2.2.1 Influence of Slabs Thickness

To evaluate the influences of slabs thickness on excavation responses, 4 cases with different slabs thickness of 1.5 m, 2.0 m, 2.5 m, and 3.0 m were simulated, respectively. The deformation of diaphragm wall and the ground settlement are shown in Table 8.3. Figure 8.3a, b shows the deformation curve and bending moment of diaphragm wall, respectively. It is obvious that the slabs with thickness of 1.5–3.0 m can effectively reduce the deflection and bending moment of the diaphragm wall, the basal heave and the additional settlement of the adjacent building. The slabs

Table 8.3 excavation responses with different slab thicknesses

	δ_{hm} (mm)	h_{max} (mm)	δ_{vm} (mm)	M_{max} (kN/m)
Without jet grouting slabs	82.08	128.8	61.20	2028
Slabs of 1.5 m thickness	49.82	88.21	39.13	1474
Slabs of 2.0 m thickness	41.14	76.32	33.27	1326
Slabs of 2.5 m thickness	34.54	65.48	28.90	1183
Slabs of 3.0 m thickness	29.17	54.63	25.37	1034

Notes δ_{hm}—the max deflection of diaphragm wall
h_{max}—the max value of basal heave
δ_{vm}—the max ground settlement
M_{max}—the max bending moment of diaphragm wall

with thickness of 2.0–2.5 m can reduce the deformation and bending moment by about 50%. If the deformation should be controlled strictly, the jet grouting slab with 3.0 m thickness can be considered. Generally, the slabs with thickness more than 3 m would not be carried out for economic considerations. Consequently, the reasonable thickness of jet grouting is set at 1.5~3.0 m, which is about 10~20% of the excavation depth.

Figure 8.3a shows that there is no remarkable difference in the shape between the deformation curve with jet grouting slabs and without. The maximum wall deflection δ_{hm} occurs at the depth of 22~23 m because the main deformation occurs in marine clay layer. The location at which the maximum negative bending moment of diaphragm wall develops is approximately the same as the location of maximum deflection (Fig. 8.3b). The location of max negative bending moment is deeper with the thicker jet grouting slabs. The maximum positive bending moment is located at the interface between marine clay and old alluvium (at about 28 m).

8.2.2.2 Influence of Excavation Width

The ratio of excavation width to excavation depth will affect the excavation responses. Therefore, it is necessary to analyze the influence of the geometric size of excavation with jet grouting slab. Four cases with different slab thickness were simulated (width B = 16 m, 20 m, 32 m, 48 m, respectively; depth = 17.5 m). With the increase of excavation width, the maximum deflection and ground settlement increase slightly, while the deformation of basal heave decreases (Fig. 8.4a–c). However, for the jet grouting slab, the performance of reducing basal heave is poor when the excavation width increases (From Fig. 8.4c, the gradient of curve is smaller with the excavation width increasing). Therefore, the jet grouting slab will have a better performance in excavation in which depth is greater than width. In addition, there is a same conclusion that the reasonable thickness of slab is 1.5–3.0 m (Table 8.4).

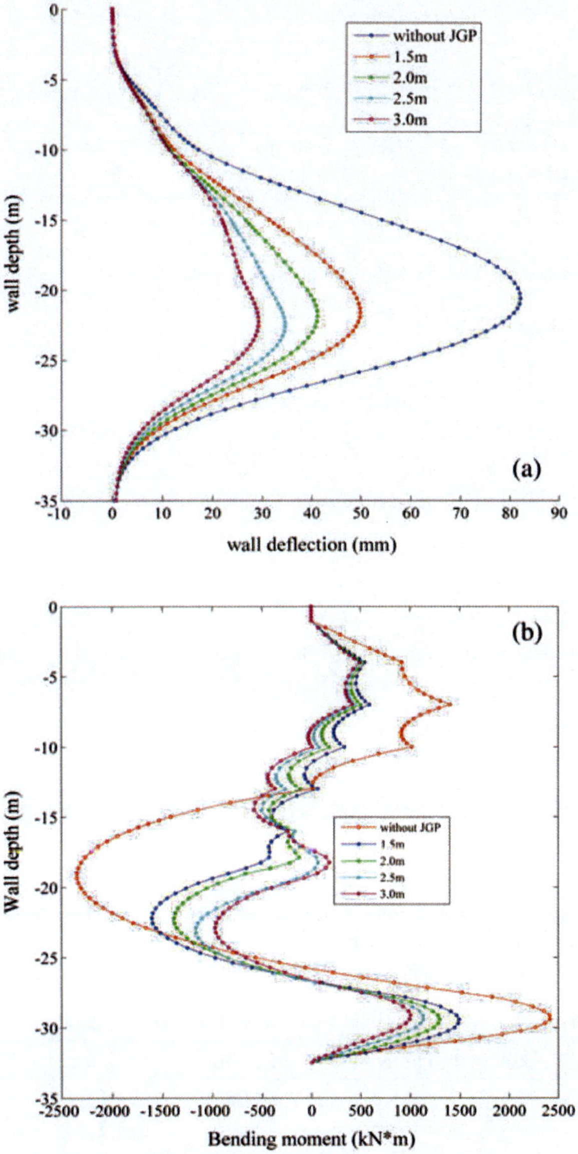

Fig. 8.3 Influence of jet grouting slab thickness on diaphragm wall **a** deflection, and **b** bending moment

8.2.2.3 Influence of Diaphragm Wall Stiffness

The support system consists of the diaphragm wall and the jet grouting slab. To evaluate the influence of wall depth and stiffness on excavation responses, 4 cases with different wall depths (30 m, 35 m, 40 m, 45 m) and 5 cases with different wall stiffness values are simulated. According to pre-determination, when the depth of the

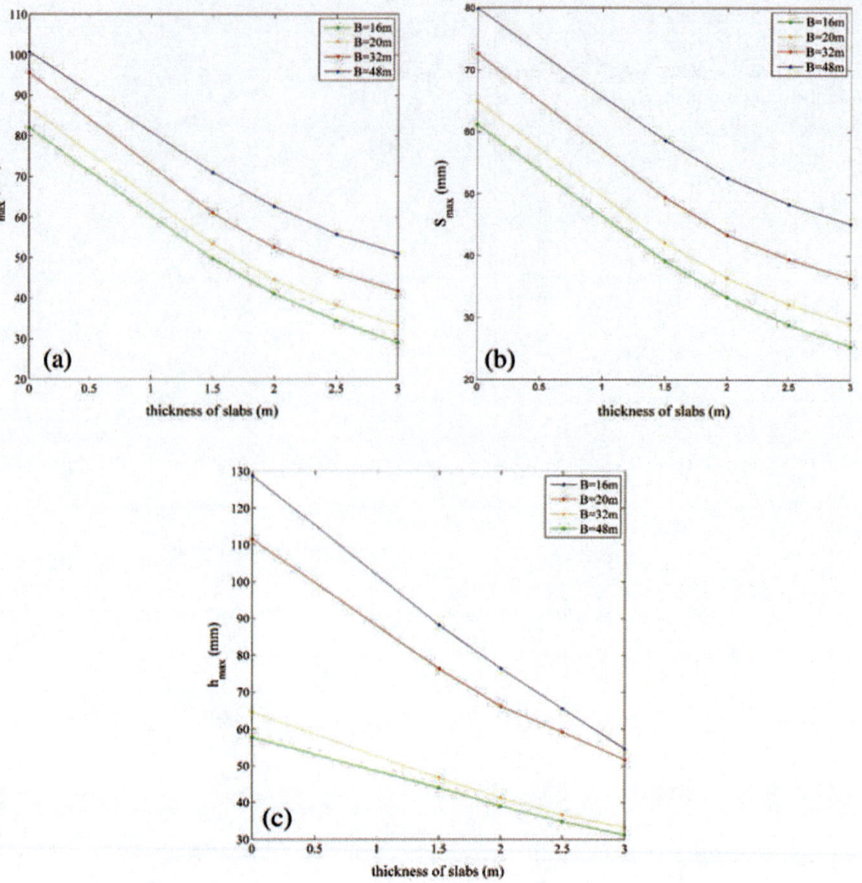

Fig. 8.4 Influence of excavation width on **a** diaphragm wall deflection, **b** max ground surface settlement, and **c** basal heave

Table 8.4 Excavation responses with different wall depths

Diaphragm wall length (m)	δ_{hm} (mm)	h_{max} (mm)	δ_{vm} (mm)	M_{max} (kN/m)
30	54.66	94.12	41.50	1678
35	49.82	88.21	39.13	1474
40	49.44	87.83	38.70	1467
45	49.44	87.82	38.70	1467

retaining wall is 35, 40, and 45 m, respectively, the deflection of the diaphragm wall is almost the same, considering that the diaphragm wall has been embedded in the hard stratum (old alluvium, SPT >100). On the contrary, reducing the embedding depth of the diaphragm wall, the maximum deflection increases slightly since the diaphragm

Table 8.5 Excavation responses with different wall stiffnesses

Wall thickness (mm)	Average vertical distance (m)	ln (*S*)	δ_{hm} (mm)	h_{max} (mm)	δ_{vm} (mm)
800	6	4.1671	59.57	102.0	47.60
400	3	4.8602	98.11	153.4	66.86
600	3	6.0766	67.65	113.1	50.23
800	3	6.9397	49.82	88.21	39.13
1000	3	7.6091	38.76	72.51	31.68

wall is still embedded in the hard stratum at 30 m (old alluvium, SPT <100). In short, if the support system which consists of jet grouting slab and diaphragm is adopted, the diaphragm wall can be embedded in the hard layer of 2–3 m. Additional depth will not decrease the wall deflection and ground settlement considerably.

Besides the embedded depth of diaphragm wall, the stiffness of diaphragm wall is also a key factor. By introducing system stiffness *S*, the support effect of the axial struts on diaphragm wall can be well illustrated.

By changing the vertical spacing of struts and the thickness of diaphragm wall, five cases were simulated. In case 1, the vertical distance of struts is 6 m (ln (*S*) = 4.1671). The case 2–5, the thicknesses of diaphragm wall are 400 mm (ln (*S*) = 4.8602), 600 mm (ln (*S*) = 6.0766), 800 mm (ln (*S*) = 6.9397), 1000 mm (ln (*S*) = 7.6091), respectively.

The results of analysis are shown in Table 8.5 and Fig. 8.5. It is obvious that the stiffness of diaphragm wall plays an important role in support system. If the thickness of diaphragm wall decreases, the deformation of excavation increases rapidly. Nevertheless, the vertical spacing of struts has marginal effect on the deformation of the whole excavation than wall width (Fig. 8.5a). The equivalent stiffness in case 1 is lower than in case 2, but the max wall deflection and ground settlement in case 1 is smaller than in case 2 (Fig. 8.5b). Therefore, in engineering practice, even if the jet grouting slab is used, the thickness of diaphragm wall should be guaranteed first. The number of axial struts can decrease appropriately, if the deformation control is not stringent.

8.2.2.4 Influence of Marine Clay Stiffness and Thickness

In coastal or marine areas such as Singapore, the distribution of marine clay is not uniform. In adjacent area for this case, the total depth to the base of the marine clay was between 18 and 35 m [5]. To explore the influence of marine clay layer thickness on the reinforced effect of jet grouting slab, 5 cases of marine clay layer thickness (12.5 m, 15 m, 17.5 m, 20 m, and 22.5 m, respectively) are chosen.

In addition, empirical formulas are generally used to predict the elastic modulus of marine clay in FE simulation. According to the undrained shear strength (c_u) measured by experiments, it is commonly assumed that $E = 100 \sim 400c_u$. By

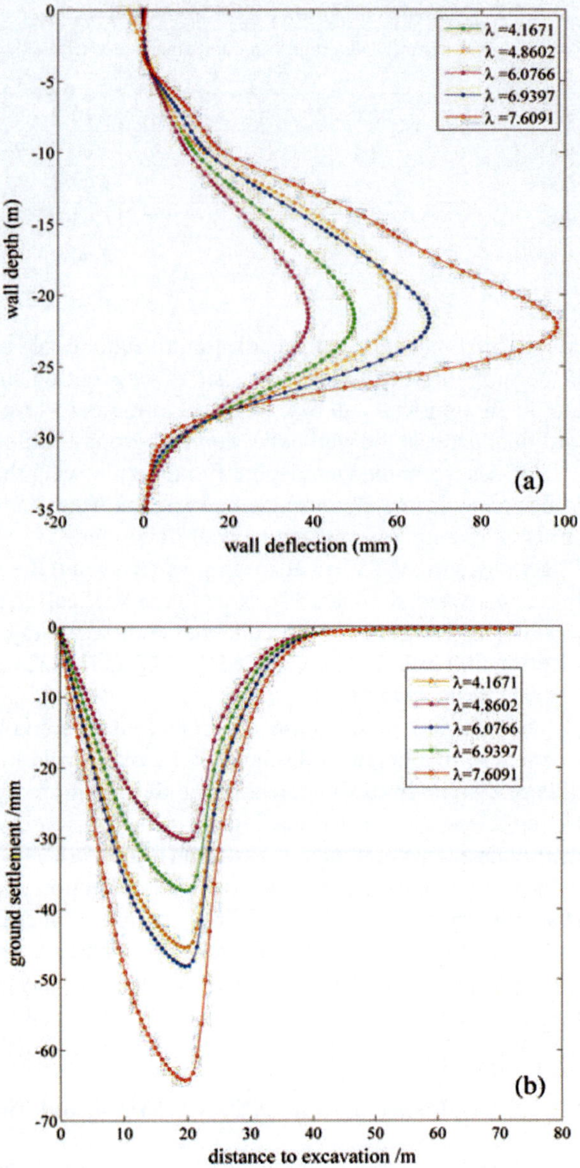

Fig. 8.5 Influence of wall stiffness on **a** diaphragm wall deflection and **b** ground settlement

comparing the testing results with those from FE model analysis, an appropriate value is adopted. Therefore, it is necessary to study the stability of jet grouting slab with different soil stiffness values. Four cases (assume $E = 150c_u$, $200c_u$, $250c_u$, and $300c_u$, respectively) are carried out.

Table 8.6 and Fig. 8.6 imply that the thickness of soft soil layer is one of the important factors affecting the deformation of excavation. With increase of the marine clay

Table 8.6 Excavation responses with different marine clay thicknesses

Marine clay thickness (m)	δ_{hm} (mm)	h_{max} (mm)	δ_{vm} (mm)	M_{max} (kN/m)
12.5	24.11	20.50	19.03	721.5
15	39.75	52.47	28.62	1074
17.5	49.82	88.21	39.13	1474
20	75.80	153.1	56.28	1895
22.5	114.5	248.3	80.02	2395

thickness, the wall deflection, basal heave, and ground settlement increase significantly, especially basal heave. With the increase of the thickness of marine clay layer, the deflection curve of the diaphragm wall changes dramatically, and the maximum deflection of the diaphragm wall appears in deeper layer. The reason is that the diaphragm wall is not embedded in the hard soil layer. On the contrary, with the thickness of marine clay decreases, the deformation of excavation decreases rapidly.

In a word, in Singapore, if the project is located in marine clay area, the main deformation would occur in marine clay. For deep excavation with soft clay layer, the deformation of diaphragm wall, basal heave, and ground settlement can be effectively controlled by the support system of jet grouting slab combined with diaphragm wall; if the soft clay is thick, the diaphragm wall should be embedded deeper so that the support system can have a better performance.

In general, Table 8.7 shows that the wall deflection, ground settlement, and basal heave tend to decrease when the undrained shear strength of marine clay is not changed, while the elastic modulus is increased. In all four cases, there is a common phenomenon that the max deflection, basal heave and ground settlement both decrease by about 30% compared without the jet grouting slab. It shows that the jet grouting slab can effectively control deformation and has good stability in general marine clay areas.

8.2.3 Summary

By a series of numerical simulations, the main conclusions are as follows:

(1) This part presents the reasonable thickness of jet grouting slabs. According to results of parameter studies, the thickness of marine clay and the stiffness of diaphragm wall play an important role indetermination of the deep excavation responses.

(2) The diaphragm wall may be embedded in the hard stratum for 2–3 m, too deep is not so economical.

(3) For most of the soft clay ($E_u = 100 \sim 400c_u$), the jet grouting slab can achieve good reinforcement effect. In this case, the slab with 1.5 m thickness can reduce basal heave, deflection, and ground settlement by about 30%.

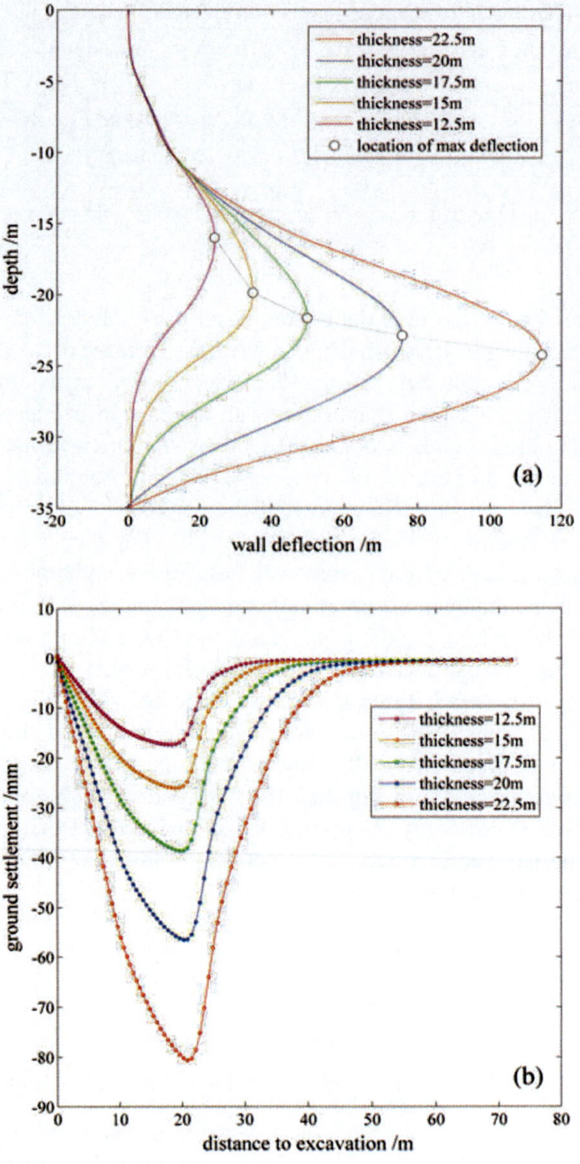

Fig. 8.6 Influence of marine clay thickness on **a** diaphragm wall deflection and **b** ground settlement

The results of calculation, analysis, and discussion in this part can provide reference for the design of support schemes for similar projects.

Table 8.7 Excavation responses with different marine clay stiffnesses

	δ_{hm} (mm)	h_{max} (mm)	δ_{vm} (mm)	M_{max} (kN/m)
Without jet grouting slabs				
$E_u = 150C_u$	82.08	128.8	61.20	2028
$E_u = 200C_u$	71.33	112.3	53.74	1831
$E_u = 250C_u$	64.19	101.1	48.71	1667
$E_u = 300C_u$	58.86	92.94	44.90	1542
With jet grouting slabs				
$E_u = 150C_u$	49.82	88.21	39.13	1474
$E_u = 200C_u$	43.02	76.09	34.24	1305
$E_u = 250C_u$	38.50	68.29	30.92	1188
$E_u = 300C_u$	35.30	62.65	28.54	1102

8.3 Effect of Jet Grouting Piles (3D) on Braced Excavations

3D FE back analysis of the case history reported by Ou et al. [6] is firstly carried out. In this excavation project in Taipei, the ground was improved by jet grout piles, with the treated slab serving as a structural restraint below the formation level to resist the lateral compressive pressure transferred from the retaining walls. Subsequently, extensive 3D FE analyses are carried out, focusing on the performance of the JGP in controlling the excavation response. The effects of various geometric and stiffness properties of the JGP, jet grout slab thickness, wall stiffness, and soil properties on the excavation responses are considered and discussed.

8.3.1 Numerical Back Analysis

8.3.1.1 Case History

The Song-san excavation project is located at K1 area of the Taipei basin. The Taipei basin is composed of a thick alluvium formation, which has several alternating sand and clay layers. Underlying this formation is the gravel formation known as Chingmei. The groundwater level for the excavation site is at GL—3.3 m. The ground consists of a 1.5 m thick fill layer and is underlain by multiple layers of silt and clay as shown in Fig. 8.7.

The plan view of the site was shown in Fig. 8.8. At the time of this investigation, the site was surrounded by several old buildings founded on spread footings. The final excavation depth for this project was 9.31 m. Diaphragm wall of 60 cm thickness and 21 m depth, together with three levels of struts were used to support the excavation as shown in Figs. 8.7 and 8.8. The lateral wall movements were monitored using

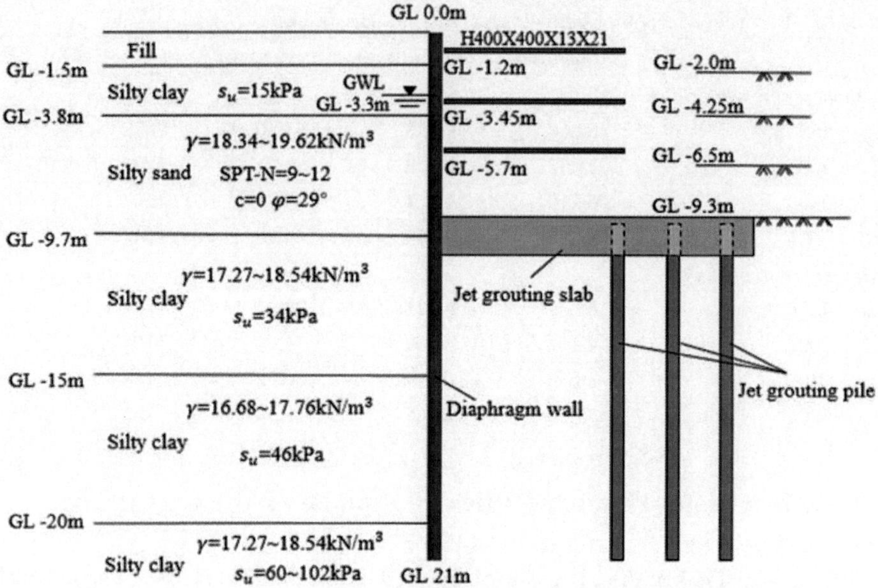

Fig. 8.7 Soil profile along with the excavation process of the Song-san excavation (adapted from ou et al. [6])

Fig. 8.8 Plan view of Song-san excavation with strut layout

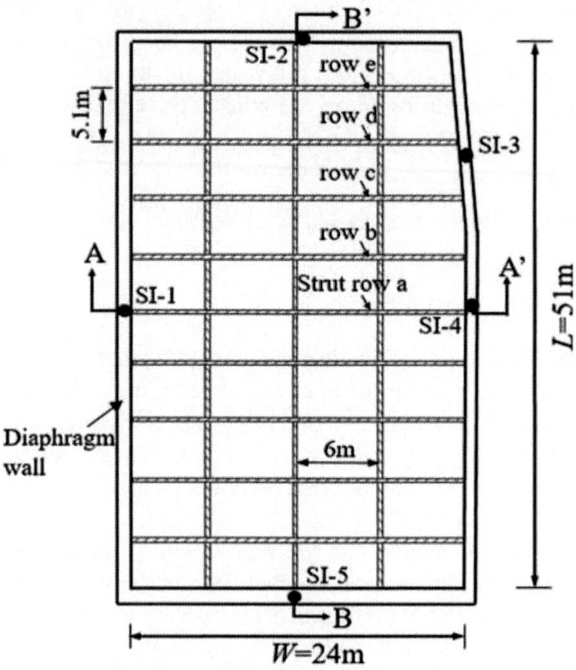

Table 8.8 Excavation and dewatering procedures of Song-san excavation

Stages	Constructions
1	Wall penetration
2	Jet grouting pile construction
3	Jet grouting slab cast
4	Excavate to GL-1.2 m, install the first strut level, lower the groundwater level to GL-5.25 m
5	Excavate to GL-4.25 m, install the second strut level, then lower the groundwater level to GL-7.5 m
6	Excavate to GL-6.5 m, install the third strut level, lower the groundwater level to GL-9.7 m
7	Excavate to GL-9.3 m

five inclinometers (SI-1 to SI-5). Unfortunately, inclinometers SI-3 and SI-5 were damaged during construction. The excavation was completed in 7 stages, as shown in Table 8.8.

8.3.1.2 Numerical Model

The site geometry was approximated as a rectangular shape (51 $\times$ 24 m in plan), and for symmetry a quarter of the site was modeled. The model size of 50 $\times$ 100 $\times$ 40 m (x-y-z axes) was used in the analyses. The horizontal movement along the vertical boundaries was restrained while both the horizontal and vertical movements along the base were restrained. The vertical boundaries extended far from the excavation to minimize the boundary restraints. Ten-node tetrahedral elements were used to model the subsoil and jet grouting materials.

Hardening soil (HS) model is adopted to represent the stress–strain responses of all the soil layers. Table 8.9 lists the main soil parameters used in the HS model.

Diaphragm wall of 0.6 m thickness was modeled using linear elastic beam elements, with Young's modulus calculated using $E = 15{,}000(f_c')^{0.5}$, where f_c' is the compressive strength of C30 concrete. The steel struts, which was modeled by linear elastic bar elements with cross-sectional area $A = 218.69\ \text{cm}^2$. The details of the supporting structural properties are shown in Table 8.10.

The Mohr–Coulomb model was used to simulate the jet grouting slab and piles. According to Wen [7] and Shirlaw et al. [1], the properties of the jet grouting composite material are strongly dependent upon the construction factors, as a series of unconfined compressive strength tests were conducted, the actual values of s_u varied significantly from 600 to 2500 kPa. Therefore, for safety consideration, the determined undrained shear strength for JGPs in numerical simulation is conservatively set around $s_u = 300$ kPa, while Young's modulus is typically around $400s_u$. According to Ou et al. [6], the size and conservative values of Young's modulus of $E = 67$ MPa, undrained shear strength of $s_u = 115$ kPa were used for JGP properties

Table 8.9 Soil parameters used for numerical simulation

Depth (m)	Soil type	Type	γ (kN/m^3)	s_u (kPa)	E_{50}^{ref} (kPa)	c (kPa)	φ (°)	ν_{ur}	R_f
0–1.5	Fill	Drained	18	–	5000	0	28	0.2	0.9
1.5–3.8	Silty clay	Undrained	17.76	15	7500	–	–	0.2	0.9
3.8–9.7	Silty sand	Drained	18.34	–	7500	0	29	0.2	0.9
9.7–15	Silty clay	Undrained	18.54	34	20,400	–	–	0.2	0.9
15–20	Silty clay	Undrained	17.76	46	27,600	–	–	0.2	0.9
20–39.4	Silty clay	Undrained	18.05	87	54,000	–	–	0.2	0.9

Table 8.10 Retaining system parameters

Structural members	EA (kN/m)	EI (kN m^2/m)
Strut	4.5×10^6	–
Diaphragm wall	8.06×10^7	2.4×10^7

Table 8.11 Parameters of jet grouting composite material

Members	γ(kN/m^3)	s_u (kPa)	E (MPa)	Size (m)
Base slab	18.5	170	100	Thickness 0.8 m
Jet grouting pile	18.5	115	67	Length 11.4 m, radius 1.6 m

in the case study. Properties of jet grouting material are listed in Table 8.11. The JGP improvement is displayed in Fig. 8.9. It can be seen that the reinforced area covers the whole length along the short side of excavation in the central area, while the percentage of improved length of the long side equals 72%, which defines the area improvement ratio, the Z type of layout was adopted with the pile spacing of 3.05 m × 3.23 m. The treated depth was from GL-9.4 m to GL-21 m.

8.3.1.3 Model Validation

The significant effect of constructing jet grout piles can be reached by comparing the calculated soil displacements with another hypothesis case, in which no JGP was considered. Figure 8.10 shows the contour plots of soil displacement in the most critical section, which is the central cross-section A-A'. With JGP conducted at the bottom of the excavation, the soil displacement behind the diaphragm wall and underneath excavation level was obviously controlled.

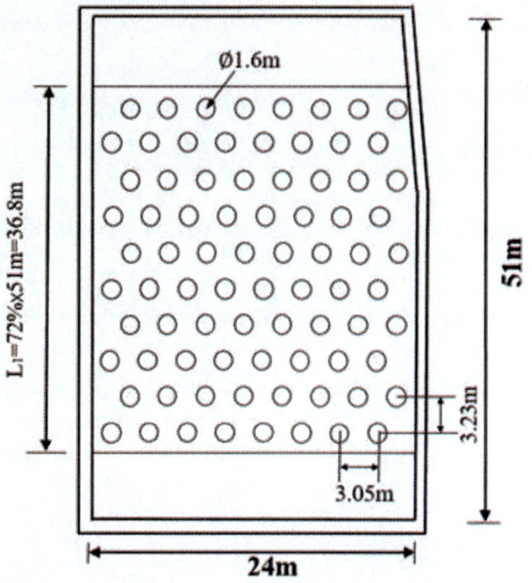

Fig. 8.9 Plan view of JGP layout

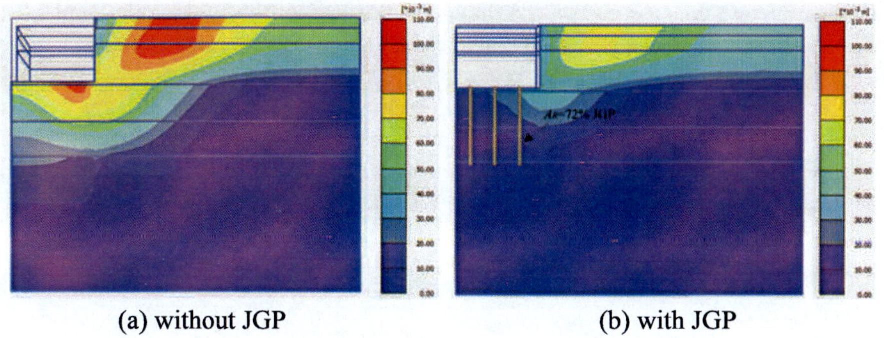

Fig. 8.10 JGP effect on controlling soil displacement

Based on the measured inclinometer results, the observed maximum wall deflection was approximately 35 mm at the center section of the primary wall (i.e., Section B-B' which is the short side of the excavation). This observation agreed well with the FE predicted maximum wall deflection of 38.5 mm. Figure 8.11 shows the comparison of observed maximum wall deflection of the secondary wall (i.e., Section A-A') with FE results from both Ou et al. [6] and this study. The wall deformations computed from the above analysis at four excavation stages were in reasonably good agreement with the observed values. The minor differences could be caused by the conservative estimates of Young's modulus and strength of the jet grout material assumed in the FE analysis. Based on the satisfactory performance of the numerical model, parametric analysis was performed.

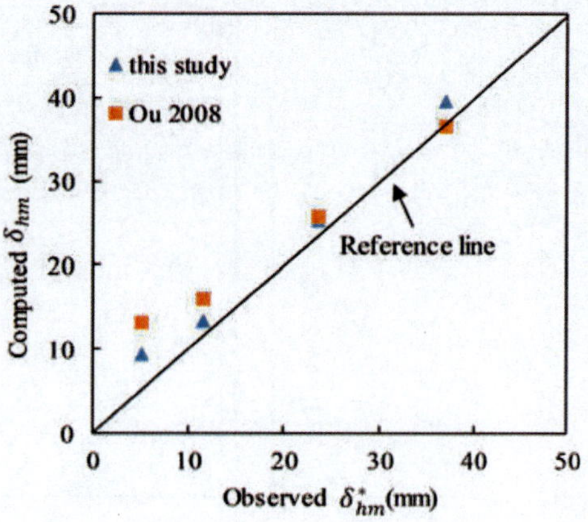

Fig. 8.11 Comparison of observed δ^*_{hm} with computed δ_{hm} at various excavation stages

8.3.2 Parametric Analysis

Subsequent parametric studies were carried out to determine the impact of the geometric arrangements and JGP material property on the excavation response. Five JGP parameters considered include the area improvement ratio A_R, pile length L_p, pile size *d*, pile stiffness *E,* and pile spacing *S.* The range of values considered are listed in Table 8.12. As the pile spacing S in the original case is 3.05 × 3.23 m, different pile spacings were considered. Since the original secant pile stiffness E_0 is relatively conservative, the larger values are considered in the parametric study. In addition, the properties of the wall, jet grout slab, and site soil were also considered to be variables to investigate the applicable situation of jet grout pile. A total of 48 cases were considered.

Table 8.12 Cases for parametric sensitivity study

Parameters	Values
Area improvement ratio A_R	0, 22%, 34%, 50%, 72%, 100%
Jet grout pile spacing S (m)	2 × 2.12, 3.05 × 3.23, 5 × 5.30, 7 × 7.41
Jet grout pile diameter d (m)	1.2, 1.4, 1.6, 2.0
Jet grout pile length L_p (m)	5.6, 11.6, 18.6, 25.6
Jet grout pile stiffness E (E_0 = 67 MPa)	E_0, $2E_0$, $3E_0$, $4E_0$
Wall stiffness E_wI_w (kN m²/m)	4.12×10^5, 3.37×10^4
Jet grout slab thickness T (m)	0.8, 2, 3
Soil strength ratio	0.7, 1, 1.3

8.3.2.1 Effect of Area Improvement Ratio

To evaluate the effectiveness of the JGP ground improvement in reducing structural forces and soil displacement, six improvement ranges were considered: $L_1 = 0$ m, 11.2 m, 17.6 m, 24.0 m, 36.8 m, and 51.0 m. They correspond to 0%, 22%, 34%, 50%, 72%, and 100% of the entire excavation length, respectively. The JGP length, stiffness, and spacing as well as the diaphragm wall and jet grout slab properties were consistent with the Song-san case study.

For structural members, Fig. 8.12 shows the wall deflection profiles after the final excavation for the two central cross-sections A-A' and B-B'. Besides, for section A-A', the maximum lateral deformation δ_{hm} decreased from 68 to 41 mm as the partial JGP improvement ratio increased from 0 to 34%. However, the decrease in wall deflection is not as significant for the secondary wall (Section B-B').

Table 8.13 summarizes the strut loads for the no JGP improvement case. Row *a* corresponds to the row of struts supporting the excavation center, while Rows b, c, d, and e are located at y = 5, y = 10, y = 15, y = 20 m planes, respectively, as shown in Fig. 8.8. Since the results indicate that the 3rd level struts in A-A' section which support the long side of diaphragm wall were subjected to the highest compression force, the effects of A_R on the forces acting on the 3rd level of struts for Rows a to e were studied in detail in Fig. 8.13. The results indicate a significant decrease

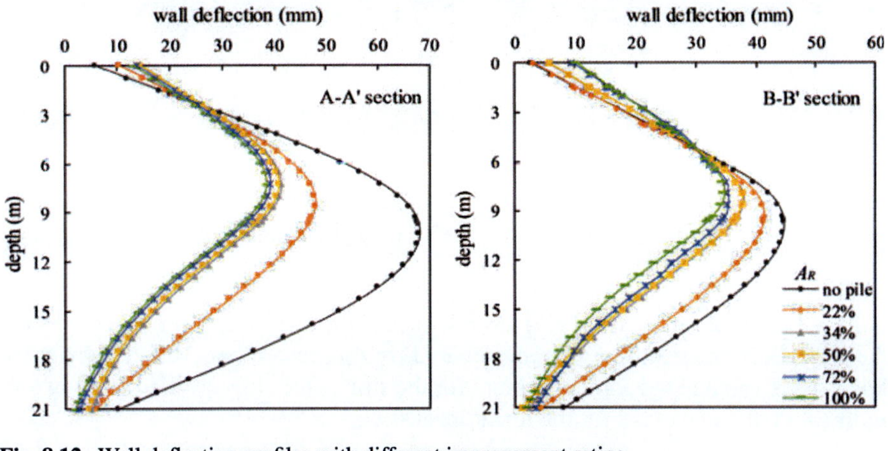

Fig. 8.12 Wall deflection profiles with different improvement ratios

Table 8.13 Strut force for no JGP case

	Strut force (kN)				
	Row a	Row b	Row c	Row d	Row e
1st level	392.43	174.26	176.18	107.26	76.26
2nd level	944.96	697.29	589.59	398.26	308.26
3rd level	1389.56	1129.22	931.74	623.80	493.80

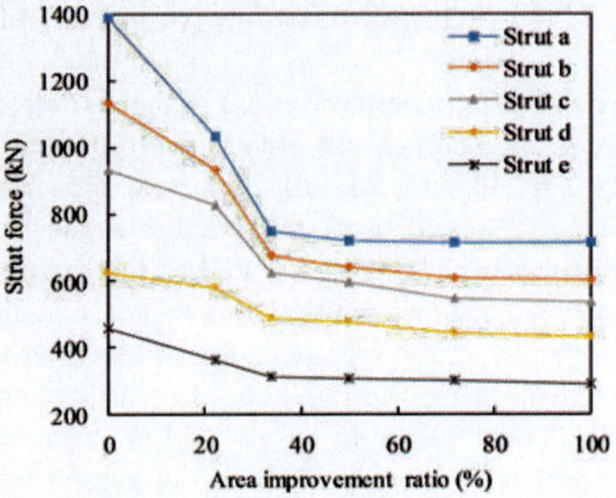

Fig. 8.13 Lowest level strut forces for different A_R

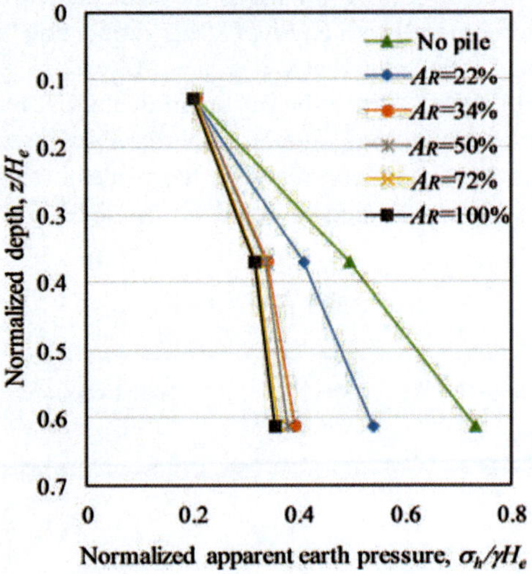

Fig. 8.14 Normalized apparent earth pressure for different A_R

in the strut forces as A_R increased from 0 (i.e., no ground improvement) to 34%. The changes in the strut forces were minimal for A_R greater than 34%. Figure 8.14 shows the normalized apparent pressure for the strut forces in Row a, indicating the significant influence of A_R for the lower level struts.

8.3.2.2 Effect of Pile Spacing

Table 8.14 shows the parameter setup for the four different jet grout pile spacings for the case with $A_R = 50\%$, $E = 67$ MPa, $L_p = 11.6$ m, and 0.8 m thick jet grout slab. For simplicity, these are denoted as S = 2 m, 3 m, 5 m, and 7 m, while large spacing suggests less pile constructed within the same area.

Table 8.14 Jet grout pile spacing properties

Improvement area (m)	Pile size	Spacing (m)	Quantity of jet grout pile
12 × 12.5	Length 11.6 m, Diameter 1.6 m	2 × 2.12	36
		3.05 × 3.23	16
		5 × 5.30	9
		7 × 7.41	4

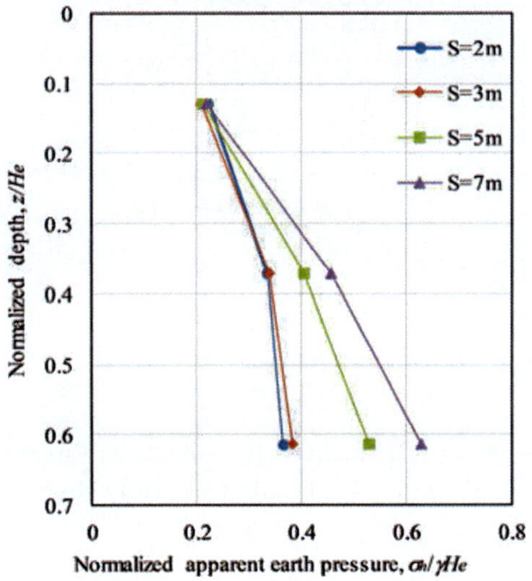

Fig. 8.15 Normalized apparent earth pressure for different jet grout pile spacing

According to the results, the spacing S has a significant influence on the wall deflections. For the longer side of the diaphragm wall (section A-A'), the maximum wall deflection decreased by 37.5% when S was reduced from 7 × 7.41 m to 2 × 2.12 m.

The effects of S on the maximum force for the 3rd level struts in section A-A' are shown in Fig. 8.15. The results indicate minimal differences in the maximum strut forces for S = 2 and 3 m. These differences became more significant as S increased to 5 and 7 m. Figure 8.15 shows normalized apparent earth pressure for different jet grout pile spacings and the normalized apparent pressure for the strut forces in Row a, indicating the significant influence of S for the lower level struts.

8.3.2.3 Effect of Pile Size

Based on the results above, large pile spacing, implying less treated volume, results in a reduction in JGP reinforcement effect. In this section, the area improvement ratio and the jet grout volume are set to be constant. The pile size was enlarged while at the same time the spacing was widened, as demonstrated in Fig. 8.16.

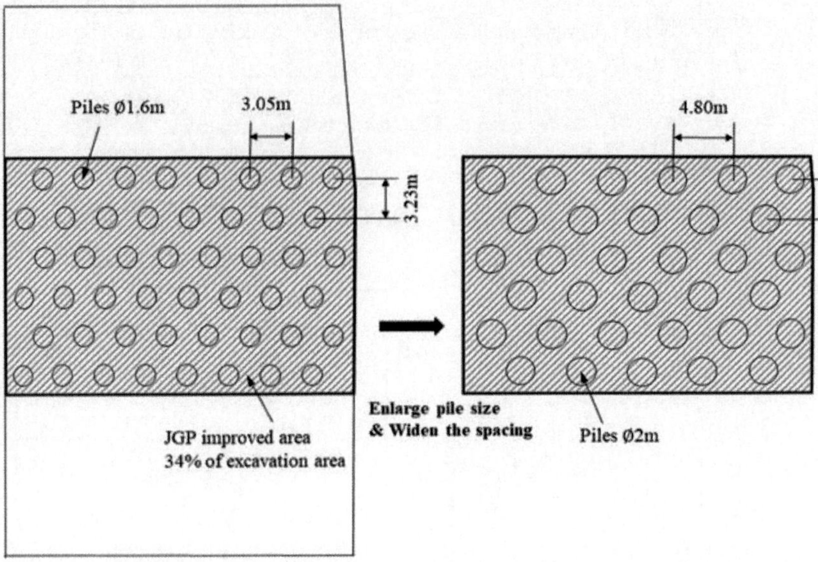

Fig. 8.16 Illustration of pile size variation

Results indicate that despite the constant injection volume, smaller pile size which means more piles installed can only slightly minimize the diaphragm wall deformation and strut force. Hence, the pile diameter of 1.6 m were simulated in following analysis.

8.3.2.4 Effect of Pile Stiffness

The influence of the jet grout pile stiffness on the strut forces and wall deflections is investigated through the case with A_R = 50%, L_p = 11.6 m, S = 3 m, 0.8 m thick slab, and 0.6 m thick diaphragm wall. The results indicate minimal differences even for the case when the jet grout pile stiffness was increased from $E = E_0 = 67$ MPa to $E = 4E_0 = 268$ MPa. As the variations in the maximum wall deformations and strut forces were neglectable, it can be inferred that the jet grout pile stiffness produces minor effects on the excavation responses.

8.3.2.5 Effect of Pile Length

The jet grout piles penetrating into the GL-15 m, GL-21 m, GL-28 m, and GL-35 m cases were analyzed to investigate the effects of JGP length. Other conditions of A_R = 22%, E = 67 MPa, and S = 3.05 m × 3.03 m were considered. The computed results show that the differences in strut forces and wall deflections were minimal when L_p was increased from 5.6 to 11.6 m and even negligible when L_p becomes

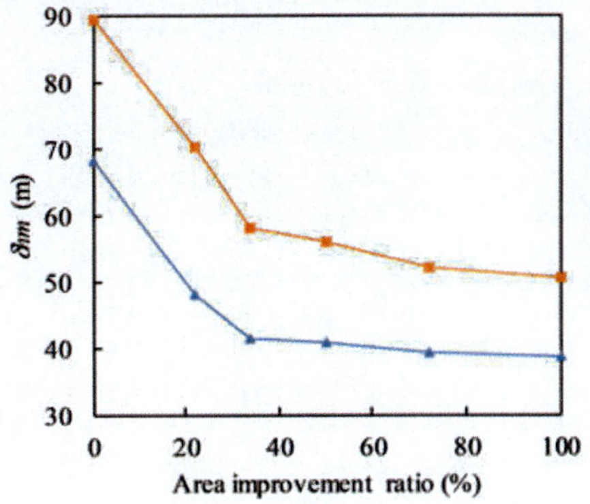

(a) Wall deformation

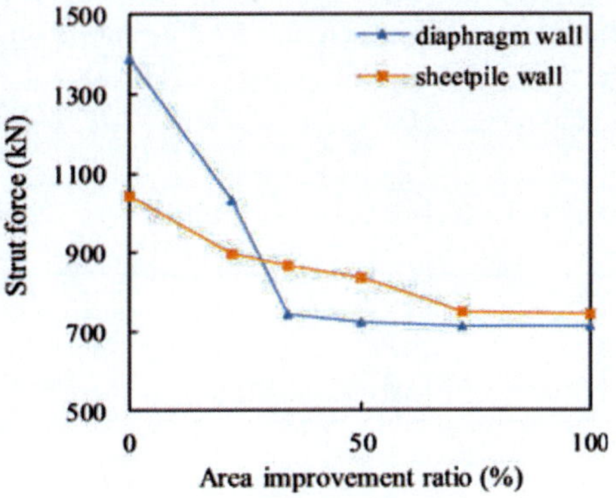

(b) Strut force

Fig. 8.17 Excavation responses for two types of walls

greater, indicating that further lengthening of the JGP had no considerable influence on the excavation response.

8.3.2.6 Effect of Wall Rigidity

In order to study the influence of wall stiffness on the excavation system performance, the ground improvement for the Song-san case was reanalyzed assuming a flexible sheet pile wall with $E_w I_w = 3.37 \times 104$ kNm^2/m. The results as shown in Fig. 8.17 generally indicate the same trends for both the diaphragm wall and the flexible sheet pile wall with the maximum wall deflection decreasing significantly with increase of A_R up to 34%. Subsequently, the decrease in the maximum wall deflection was marginal for A_R greater than 34% for both wall types. Similarly, the results indicate a significant decrease in the strut forces as A_R increased from 0 (i.e., no ground improvement) to 34% for the diaphragm wall. The changes in the strut forces were minimal for A_R greater than 34%. However, the reduction in the maximum strut force is not as significant for the sheet pile wall, compared with the more rigid diaphragm wall.

8.3.2.7 Effect of Jet Grout Slab Thickness

A combinative study was carried out to investigate the influence of the jet grout slab thickness on the performance of the excavation system. Cases with jet grout slab thickness of 0.8, 2, and 3 m were analyzed, both with and without the JGP piles. As seen in Table 8.15 and Fig. 8.18, as expected, increasing the jet grout slab thickness resulted in a decreased maximum wall deflection and strut force. The results also indicate that there is a fairly significant reduction in both the wall deflections and

Table 8.15 Computed results for different slab thicknesses

Slab thickness T (m)	Maximum strut force (kN)		Maximum wall deflectionδ_{hm} (mm)			
			A-A' section		B-B' section	
	No JGP	With 72% JGP	No JGP	With 72% JGP	No JGP	With 72% JGP
0.8	1386	716	70	46	45	40
2.0	979	623	56	39	42	37
3.0	780	586	44	36	41	33

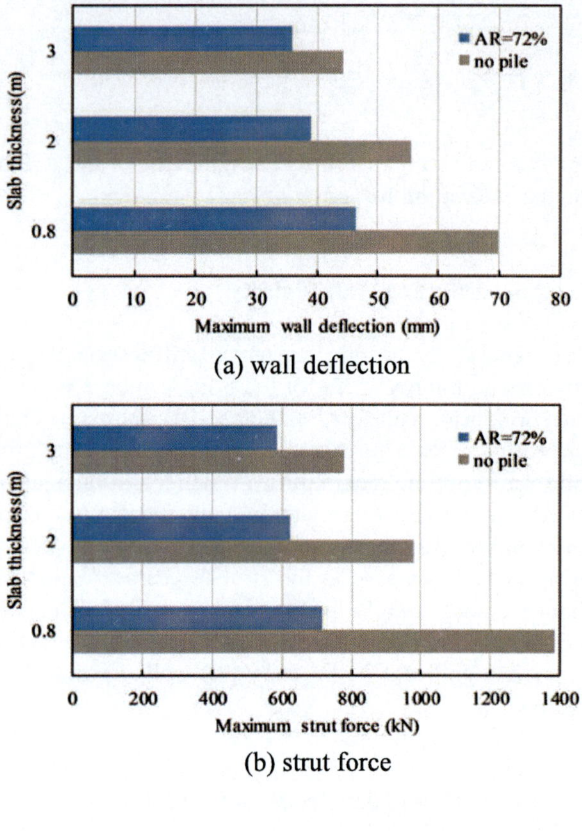

Fig. 8.18 Maximum values of excavation responses for various slab thickness

strut forces for the cases with jet grout slab and piles compared with the cases with jet grout slab of 0.8 m thickness.

As for the influence on the surrounding soil displacement, Fig. 8.19 shows the different soil deformation modes with jet grout slab or jet grout pile reinforcements. By installing piles into deeper soil, the potential sliding surface is prevented, the risk of basal heave failure is greatly reduced.

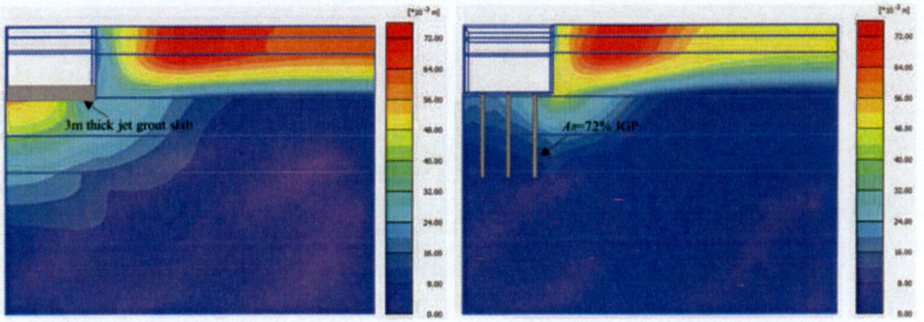

(a) with 3m thick jet grout slab (b) with 72% JGP

Fig. 8.19 Soil displacement fields with different jet grout members

8.3.2.8 Effect of Jet Grout Soil Strength

To investigate the effect of JGP on the excavation responses in softer and stiffer clay, six cases considering three groups of soil strength values with or without jet grout piles underneath the final excavation level were analyzed. Soil strength variation ratios of 0.7, 1, 1.3 were considered, as shown in Fig. 8.20. Soil strength variation ratios of 0.7 and 1.3 mean that the soil strength was taken as 0.7 and 1.3 times those in the prototype case study. It was observed that when constructing 34% jet grout piles underneath the excavation, one-third reduction in maximum wall deflection and strut force was achieved, compared with the no ground improvement case for weaker soils. For stiffer soil, this reduction is less significant. Therefore, for excavation in weaker soils, ground improvement can greatly reduce deformation and internal forces acting on the supporting members.

8.3.3 Summary

Extensive analyses of wall deflections and strut forces for excavation in soft clay stabilized with jet grout piles (JGP) were performed. The FE analysis procedures were firstly calibrated through back analyses of the Song-san excavation case history. Subsequently, parametric analyses were carried out to investigate the influence of the area improvement ratio, jet grout pile length, jet grout pile stiffness and spacing on the maximum wall deflection and strut forces. In addition, the effects of the slab thickness, wall stiffness, and soil strength were also investigated. The findings include the following:

(1) Increase of the improvement area or number of piles results in significant reduction in wall deflections and strut forces. However, the jet grout pile stiffness and strength has only marginal effects on the structural responses induced by excavation.

Fig. 8.20 Maximum excavation responses for various soil strength (with and without jet grout piles)

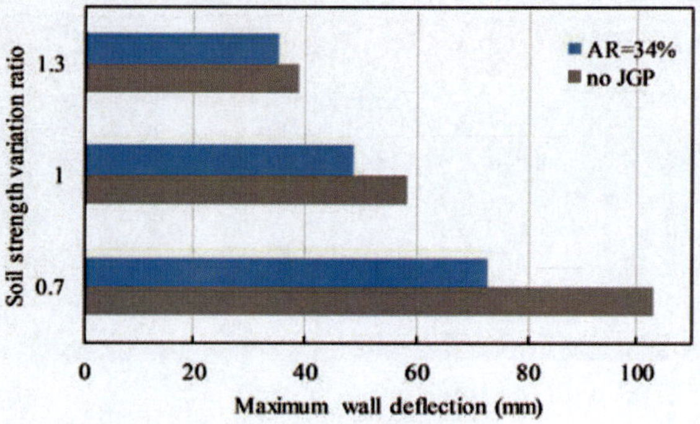

(a) wall deflection

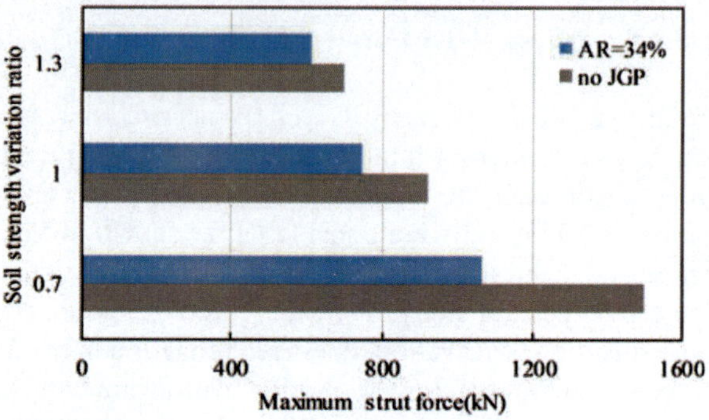

(b) strut force

(2) JGP was more effective in reducing the strut forces for the stiffer diaphragm wall compared with the flexible sheet pile wall. The impact of JGP is more significant for weaker soils.

(3) The jet grout slab acts as a compression member to reduce wall deflections and strut forces underneath the final excavation level. Under certain circumstances, it may be feasible to omit the jet grout piles or one level of struts by constructing a thicker jet grout slab to simplify the construction process and save time.

References

1. Shirlaw, J. N., Wen, D., Nadarajah, P., Yoon, S. I., Sugawara, S. (2000). Construction issues related to jet grouted slabs at the base of excavations. In *International Conference on Tunnels and Underground Structures* (pp. 639–643). Singapore

2. Zhang, W. G., Hong, L., Li, Y. Q., Zhang, R. H., & Goh, A. T. C. (2020). Assessment of effect of jet grouting slabs on responses for deep braced excavations. *Underground Space*. https://doi.org/10.1016/j.undsp.2020.02.002
3. Zhang, W. G., Li, Y. Q., Goh, A. T. C. & Zhang, R. H. (2020). Numerical study of the performance of jet grout piles for braced excavations in soft clay. *Computers and Geotechnics*. https://doi.org/10.1016/j.compgeo.2020.103631.
4. Xuan, F. (2009). *Behavior of diaphragm walls in clays and reliability analysis*. Nanyang Technological University.
5. Shirlaw, J. N., Tan, T. S., Wong, K. S. (2005). Deep excavations in Singapore marine clay. In *Fifth International Conference on Geotechnical Aspects of Underground Construction in Soft Ground*. Amsterdam, Netherlands.
6. Ou, C. Y., Teng, F. C., Wang, I. W. (2008). Analysis and design of partial ground improvement in deep excavations. *Computers and Geotechnics* , *35*(4), 576–584.
7. Wen, D. Z. (2015). Use of Jet Grouting in Deep Excavations. In B. Indraratna, J. Chu, & C. Rujikiatkamjorn (Eds.), Ground improvement case histories (pp. 109–122). E-Publishing Inc.

Chapter 9
Protection of Adjacent Infrastructures

9.1 Introduction

In the past, with shallower excavations, the influence of excavation on the surrounding environment was not so significant. Recently, with increase of excavation depth and scale in urban areas, the magnitude and extent of ground settlement increases, which frequently damages the adjacent buildings and infrastructures. This chapter contains studies on interactions between braced excavations and urban infrastructure systems including the slope, piles, buildings, and tunnels.

9.2 Influence of Deep Braced Excavation on Adjacent Upper Slope

For deep excavations in cities with mountainous terrain such as Hong Kong, Chongqing, and Guiyang, one major concern is that the excavated soils and resulted stress relief may have an influence on the stability of nearby existing slopes or the slopes subjected to excessive deformation [1].

In this section, the global factor of safety FS obtained via the shear strength reduction (SSR) technique (also called c-φ reduction method) is used as the criterion to determine the ultimate limit state and the calculated maximum lateral wall deflection is adopted as the serviceability limit state criterion. A detailed study was carried out to evaluate the influence of the excavation geometry, the supporting system stiffness, and the distance between the braced excavation and the existing slope on the excavation-slope system response using the finite-element program PLAXIS2D. Based on these results, response surface models were developed for the ultimate and serviceability limit states. Subsequently, the first-order reliability method (FORM) is adapted to perform probabilistic analysis on the global factor of safety through setting the threshold maximum wall deflection as an optimization

W. Zhang and H. Liu, *Design of Deep Braced Excavation and Earth Retaining Systems Under Complex Built Environment*,
https://doi.org/10.1007/978-981-16-5320-9_9

constraint to enable a rational design approach for deep braced excavations adjacent to slopes in mountainous terrain.

9.2.1 Numerical Scheme

The Plaxis2D software was utilized for the numerical simulations. The Mohr–Coulomb constitutive model was selected for the soil. A typical cross-section of the excavation and slope system, as well as the properties of the soil and the supporting elements, are shown in Fig. 9.1.

The phreatic level is below the bottom boundary, so the pore water pressure and water percolation are not considered. It should be noted that the adopted soil properties are typical of the ground conditions in Chongqing and that the soil and wall profiles, the excavation geometries as well as the boundary conditions are simplified from a deep braced project in Chongqing.

The analyses considered a plane strain excavation near an unreinforced slope supported by a retaining wall system and a total of 6 levels of struts located below the original ground surface at depths of −1 m, −4 m, −7 m, −10 m, −13 m, and −16 m, respectively. The soil was modeled by 15-noded triangular elements, following the Mohr–Coulomb failure criterion. The structural elements were assumed to be linear elastic in Plaxis2D, the struts and the wall is represented by 3-noded bar elements and 5-noded beam elements, respectively.

The soil properties are regarded as deterministic and presented in Fig. 9.1. This study mainly examines the influences of the retaining wall system stiffness and the excavation geometries on the responses including the global factor of safety of the braced excavation-upper slope system, as well as the retaining wall deflection. The ranges of these key design parameters are listed in Table 9.1. It should be explained

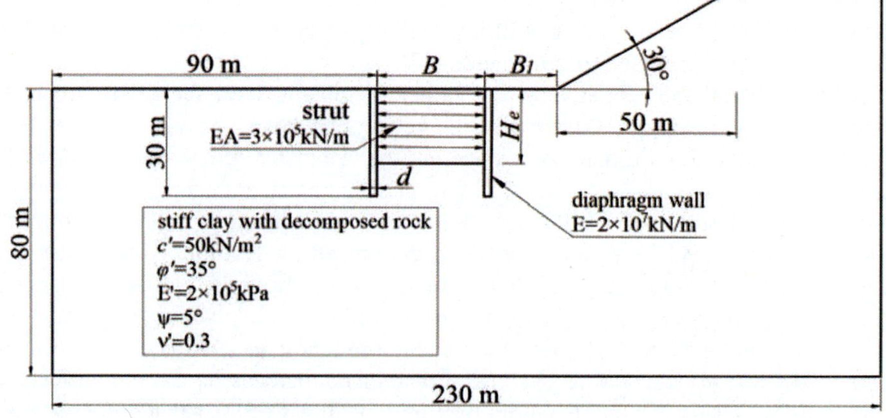

Fig. 9.1 Cross-sectional soil and wall profile

Table 9.1 Parameters considered and the ranges to be examined

Parameters	Ranges
System stiffness in logarithmic scale ln(*S*)	3.794, 4.605, 5.187
Excavation width *B* (m)	20, 30, 40
Excavation depth H_e (m)	14, 17, 20
Wall thickness *d* (m)	0.6, 0.9, 1.2
Distance between wall and toe of slope B_1 (m)	5, 10, 15, 20, 30, 40
Wall penetration ratio D/H_e	0.50, 0.76, 1.14

that ln(S) represents the natural logarithm of the system stiffness, e.g., $\ln(EI/\gamma_w h_{avg})$. The three values of 3.794, 4.605, 5.187 are realized by changing the corresponding wall thickness of 0.6, 0.9, and 1.2 m with average vertical strut spacing $h_{avg} = 3$ m, $\gamma_w = 10$ kN/m^3 and keeping Young's modulus of the wall constant ($E = 2.1 \times 10^7$ kN/m2).

According to Clayton [1], the strut stiffness does not have a very significant influence on wall deflection when the strut is stiff. A constant strut stiffness EA of 3.0×10^5kN/m is assumed. Based on Table 9.1, more than 160 hypothetical cases were analyzed.

The construction assumption and sequence are as follows:

(1) The wall is installed without any disturbance (wished into place);
(2) The soil is excavated prior to installing the strut, uniformly 1 m below each strut level until the final depth He is reached;
(3) Each phase of strut installation is followed by a subsequent phase in which the global safety factor is determined by the SSR method proposed by Zienkiewicz et al. [2], and improved by Brinkgreve and Bakker [3]. The procedure of the SSR method involves systematically reducing the soil shear strength until failure occurs. It has been verified by Lian et al. [4] that the SSR FE method can be widely applied in the engineering practice. In addition, Cheng et al. [5] and Schneider-Muntau et al. [6] have also recently demonstrated that the SSR technique performs well in the many slope cases analyzed.The detailed construction sequence is listed in Table 9.2.

9.2.2 Result and Analysis

9.2.2.1 Factor of Safety

Figure 9.2 plots the variation of incremental displacement contours, as excavation proceeds, for cases with B = 30 m, $B_1 = 5$ m, ln(S) = 4.605. The corresponding FS values for excavation depths He of 0, 14, 17, 21 m are also shown. It can be

Table 9.2 Detailed construction procedures

Phases	Construction procedures
Initial phase	Generate the soil initial effective stress and pore water pressure
Phase 1	Determine the initial global safety factor by SSR method
Phase 2	Install the diaphragm wall
Phase 3	Excavate to 2 m below the ground surface inside the excavation, install the first level strut
Phase 4	Excavate to 5 m below the ground surface
Phase 5	Install the second level strut
Phase 6	Calculated the global safety factor by SSR method
Phase 7	Excavate to 8 m below the ground surface
Phase 8	Install the third level strut
Phase 9	Calculated the global safety factor by SSR method
Phase 10	Excavate to 11 m below the ground surface
Phase 11	Install the fourth level strut
Phase 12	Calculated the global safety factor by SSR method
Phase 13	Excavate to 14 m below the ground surface
Phase 14	Install the fifth level strut at
Phase 15	Calculated the global safety factor by SSR method
Phase 16	Excavate to 17 m below the ground surface
Phase 17	Install the sixth level strut
Phase 18	Calculated the global safety factor by SSR method
Phase 19	Excavate to 20 m below the ground surface
Phase 20	Calculated the global safety factor by SSR method

observed that as the excavation depth H_e increases, the zone showing the incremental displacement (in red) enlarges and the corresponding FS values decrease. A comparison shows that after the final stage of excavation, the FS is about 1.705, which is a decrease of 0.636 from the original FS of 2.341 (prior to any excavation). The results indicate that excavations close to an existing slope may result in a significant reduction of the FS of the slope, which may cause slope instability or even failure.

Figure 9.3 presents some typical plots of the decrease of the factor of safety values ΔFS with increasing distance between the wall and the toe of the slope B_1 for three different excavation widths B with $H_e = 20$ m and S = 4.605. As expected, ΔFS decreases with increasing distance between the wall and the existing slope B1. The ΔFS was also influenced by the width of the excavation B. The ΔFS was minimal for $B_1 = 20$ m when B = 20 m. For the larger B values of 30 m and 40 m, ΔFS was minimal at a greater distance of B1 = 40 m.

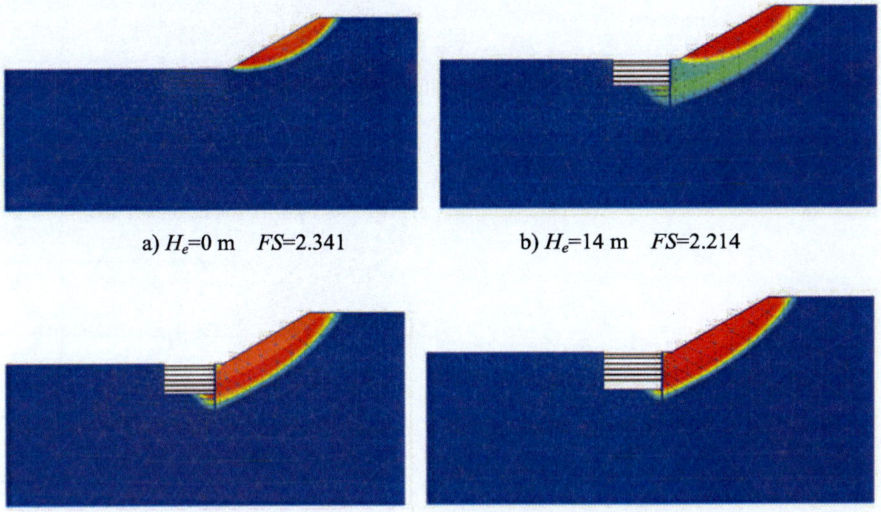

a) H_e=0 m FS=2.341 b) H_e=14 m FS=2.214

c) H_e=17 m FS=2.095 d) H_e=20 m FS=1.705

Fig. 9.2 Contour of slip surface and FS for different excavation depths He for B = 30 m, B1 = 5 m, S = 4.605

Fig. 9.3 Decrease of the factor of safety ΔFS for different B_1 with H_e = 20 m, S = 4.605

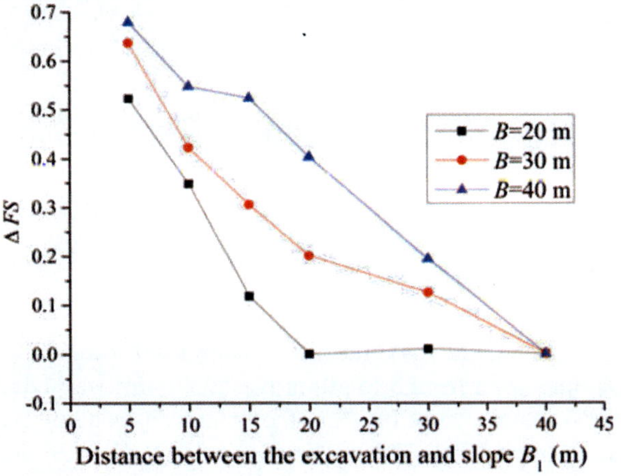

9.2.2.2 Wall Responses

Figures 9.4 and 9.5 show the variation of the wall lateral deformation profiles and the bending moments of the right wall (near the slope) for cases with different B_1,, respectively. The final excavation depth (H_e) is 20 m and supported by 1.2 m-thickness and 30 m-deep diaphragm wall with 6 levels of struts for B_1 ranging from 5 to 40 m, respectively. All the plots show a concave wall deflection shape. As shown

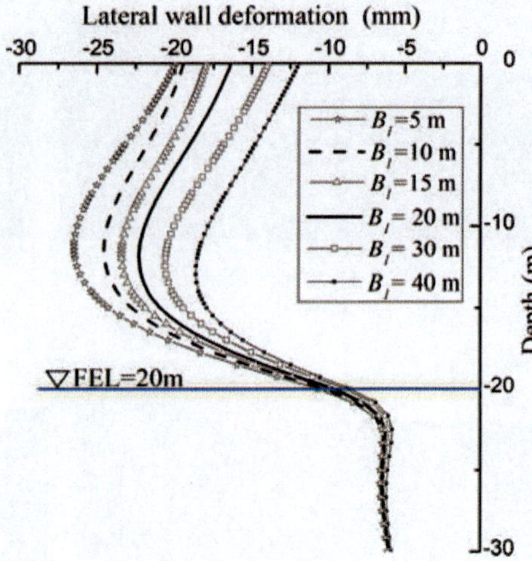

Fig. 9.4 Lateral deformation of the wall for different B_1

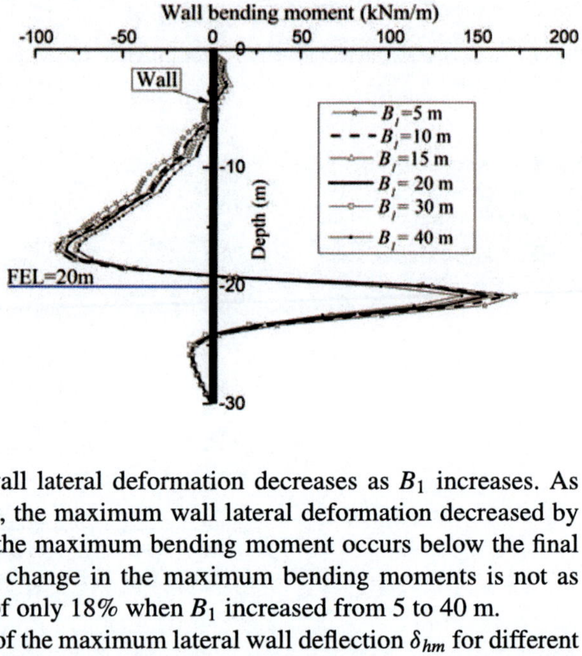

Fig. 9.5 Wall Bending moments for different B_1

in Fig. 9.4, the maximum wall lateral deformation decreases as B_1 increases. As B_1 increased from 5 to 40 m, the maximum wall lateral deformation decreased by 30%. As shown in Fig. 9.5, the maximum bending moment occurs below the final excavation level (FEL). The change in the maximum bending moments is not as significant with a reduction of only 18% when B_1 increased from 5 to 40 m.

Figure 9.6 shows the plot of the maximum lateral wall deflection δ_{hm} for different B and B_1 with $H_e = 20$ m, $S = 4.605$. The trends are similar to that observed for ΔFS in Fig. 9.3, i.e., δ_{hm} decreases as the excavation width B decreases and the distance between the wall and the toe of the slope B_1 increases. The influence of excavation width B on δ_{hm} is more significant when B_1 is smaller, i.e., the slope toe is closer to excavation. For example, the difference $\Delta\delta_{hm}$ between $B = 40$ and $B = 20$ is 7.5 mm at $B_1 = 5$ m. It decreases to 3.2 mm at $B_1 = 10$ m.

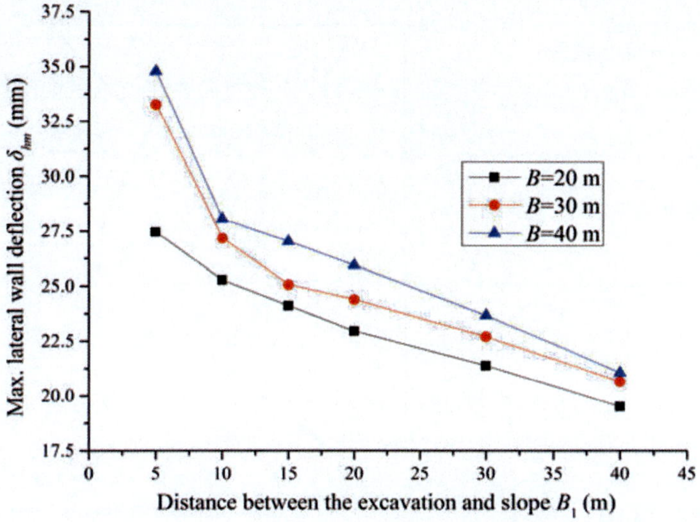

Fig. 9.6 Max. lateral wall deflection δ_{hm} for different B_1 with $H_e = 20$ m, $S = 4.605$

9.2.3 *Surrogate Models for Limit State Functions*

In order to assess the performance in deep braced excavations adjacent to an existing slope, both the ultimate limit state (ULS) and the serviceability limit state (SLS) of the excavation-slope system should be satisfied. In the following subsections, the limit state functions for ULS and SLS are developed respectively, based on the numerical results presented in the previous section.

9.2.3.1 Ultimate Limit State Surrogate Model

Based on the calculated *FS* results, a polynomial regression (PR) model has been developed for estimating the global factor of safety *FS* as a function of four input parameters B, B_1, H_e, and S, with a coefficient of determination R^2 of 0.881. The expression is shown in Eq. 9.1:

$$\begin{aligned}
\mathrm{FS}_{_PR} = {} & 7.35 \times 10^{-2}B - 1.57 \times 10^{-1}B_1 + 3.51 \times 10^{-2}H_e + 1.02\mathrm{S} \\
& - 2.5 \times 10^{-6}B^2 - 2.29 \times 10^{-4}B_1^2 - 8.07 \times 10^{-4}H_e^2 \\
& - 2.22 \times 10^{-2}S^2 + 2.4 \times 10^{-4}BB_1 - 4.08 \times 10^{-3}BH_e + 1.65 \times 10^{-4}BS \\
& + 8.01 \times 10^{-3}H_eB_1 - 1.56 \times 10^{-3}SB_1 - 2.8 \times 10^{-2}H_eS - 1.84 \times 10^{-2}B(D/H_e) \\
& + 4.83 \times 10^{-2}B_1(D/H_e) + 9.4 \times 10^{-2}H_e(D/H_e) \\
& - 3.55 \times 10^{-1}\mathrm{S}(D/H_e)
\end{aligned} \tag{9.1}$$

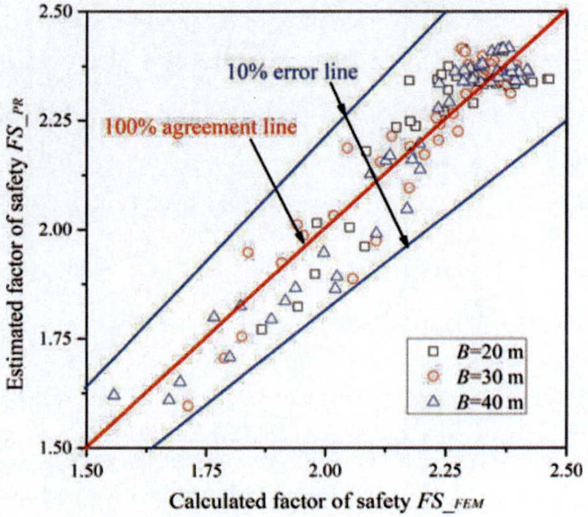

Fig. 9.7 Comparison between FS_{FEM} and FS_{PR}

Figure 9.7 plots the factor of safety values determined from the finite-element analyses FS_{FEM} against the calculated FS_{PR} results based on Eq. 9.1. Also shown in the plot are the 100% agreement line and the 10% error lines. The results indicate that Eq. 9.1 is fairly accurate in predicting the global factor of safety for deep braced excavations adjacent to slopes.

9.2.3.2 Serviceability Limit State Surrogate Model

Similarly, a logarithmic regression (LR) model was developed for predicting the maximum lateral wall deflection δ_{hm}, with a fairly high coefficient of determination $R^2 = 0.946$. The expression is shown in Eq. 9.2:

$$\delta_{hm\_LR} = 0.1133B^{0.1086}{B_1}^{-0.223}(H_e)^{2.1247}(D/H_e)^{0.0568}S^{-0.4448} \tag{9.2}$$

Figure 9.8 plots the comparison between δ_{hm_FEM} and δ_{hm_LR}, the estimated maximum lateral wall deflections δ_{hm_LR} values against the finite element calculated results δ_{hm_FEM}. Also shown in the plot are the 100% agreement line and the 20% error lines, indicating that the proposed Eq. 9.2 is fairly accurate in predicting the maximum wall deflections for deep braced excavations adjacent to slopes.

9.2.4 *Probabilistic Assessment of the Limit State Functions*

The first-order reliability method (FORM) is adopted in this study for probabilistic assessment of the two limit state functions. It can deal with the unknown PDFs of the

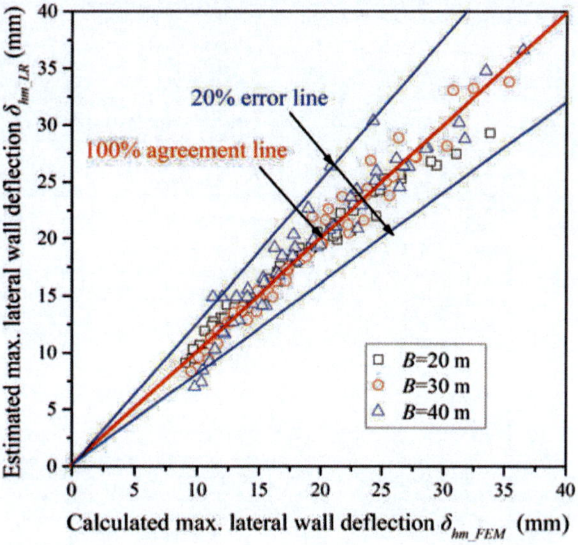

Fig. 9.8 Comparison between δ_{hm_FEM} and δ_{hm_LR}

random variables in most geotechnical applications and the demanding of computation of integration when multi-variables are involved [7]. FORM has been proved that the Microsoft EXCEL spreadsheet can be used to perform the minimization and determine reliability index β.

The reliability index β and the probability of failure P_f for both the ultimate and the serviceability limit states can be performed using FORM based on the built PR and LR models presented in the previous section. Based on the approach by Low and Tang [8], the ULS model expressed as Eq. 9.2 has been incorporated into an EXCEL spreadsheet environment, from which the reliability index β can be determined. A sample spreadsheet for computing the factor of safety *FS* is shown in Fig. 9.9a, in which the values of the input random variables are the same as those used in the simulation model. According to Low and Tang [8], the spreadsheet cells B3: B5 allows the random variables distribution that has the selection of various types including lognormal, normal, triangular, etc. Cells D3: E5 are average values and standard deviation which are set corresponding to the random variables in Cells B3: B5, three random variables are of the normal distribution. Cells G3:I5 represent the correlation matrix **R** used to define the correlations between B, H_e and S. Cells J3: J5 denote the n_i vector which contains equations for $(x_i–u_i^N)/\sigma_i^N$. The $\mathbf{x_i}^*$ values were firstly set equal to the mean values (30, 17, 4.5) of the original random variables (B, H_e, S). Then by invoking the spreadsheet's built-in optimization routine SOLVER to search algorithm by minimize the cell L3, through changing the $\mathbf{n_i}$ values subject to G(x) = 0, the design point ($\mathbf{x_i}^*$ values) can be obtained. It should be stressed that in the spreadsheet environment, iterative numerical derivatives and directional search for the design point $\mathbf{x_i}^*$ were automatically carried out via SOLVER search. The probabilistic assessment of SLS in Fig. 9.9b is almost identical to Fig. 9.9a except

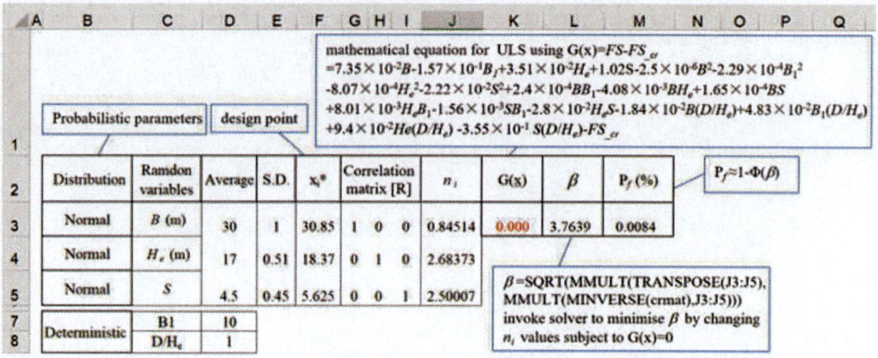

(a)

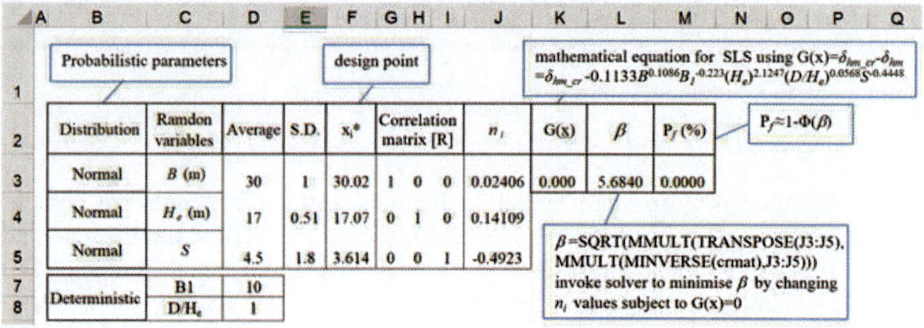

(b)

Fig. 9.9 Calculation on β and P_f using FORM spreadsheet: **a** for the ultimate limit state; **b** for serviceability limit

for the G(x) formulations. For the detailed procedures in performing the FORM spreadsheet framework to derive β and the corresponding P_f, please refer to Zhang and Goh [7].

9.2.4.1 Probabilistic Assessment of the Ultimate Limit State

For either the braced excavation or the slope, design codes or guidelines specify the minimum factor of safety. For example, in Chongqing, the building authorities specify a minimum factor of safety of 1.30 for slopes in mountainous terrains. However, for the braced excavation-upper slope system, there are no guidelines or codes with regard to the critical safety factor values. Thus, the influence of the critical factor of safety FS_cr on the reliability index β and the probability of failure P_f of the ultimate limit state ULS of the braced excavation-upper slope system are examined in this study, which can justify the choice of certain threshold factor of safety.

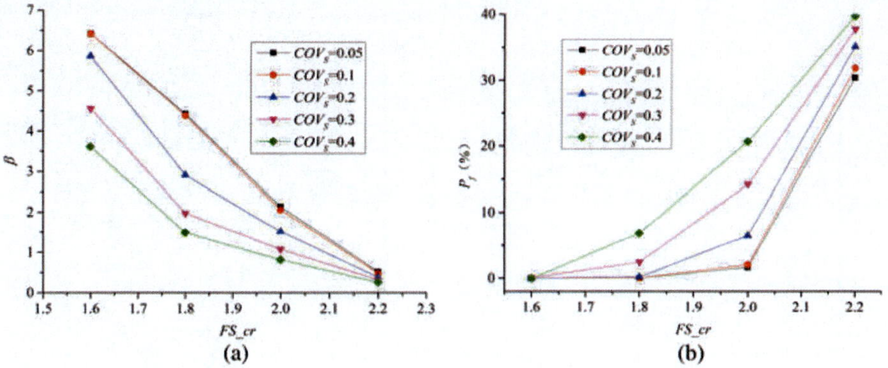

Fig. 9.10 Influence of COV_S and FS_cr on **a** β and **b** P_f for $B_1 = 5$ m, $B = 30$ m, $H_e = 17$ m, $S = 4.5$

Figure 9.10 plots the influence of the various design parameters on the reliability index β and the probability of failure P_f of the ULS. Both the coefficient of variation of the system stiffness COV_S and the critical factor of safety FS_cr have significantly influence on the β and P_f. A larger COV results in a smaller β. However, there is less significant difference on β and P_f between $COV_S = 0.05$ and $COV_S = 0.1$. In addition, the reliability index β decreases with FS_cr and converged at $FS\_cr = 2.2$, similarly P_f increases with FS_cr and converged at $FS\_cr = 1.6$.

Figure 9.11 compares the influence of both the COV_S and B_1 on P_f for $B = 20$, 30, 40 m, respectively, for $H_e = 17$ m, $S = 4.5$, and assuming the critical factor of safety $FS\_cr = 2.0$. It is obvious that P_f decreases with the increase of B_1 while P_f increases with the increase of excavation width B. A larger COV_S results in a larger P_f. However, there is no difference on P_f between $COV_S = 0.05$ and $COV_S = 0.1$. The P_f decreases with B_1 and converged at $B_1 = 15$ when B = 20 m, $B_1 = 20$ when B = 30 m, $B_1 = 30$ when B = 40 m. It suggests that COV_S has little influence on P_f when $B_1 \geq 15$ when B = 20 m, $B_1 \geq 20$ when B = 30 m, $B_1 \geq 30$ when B = 40 m.

9.2.4.2 Probabilistic Assessment of the Serviceability Limit State

In this section, the serviceability limit state of the braced excavation-upper slope system is considered. Figure 9.12 plots the influence of COV_S and the threshold max. wall deflection δ_{hm_cr} on β and P_f for $B = 30$ m, $H_e = 17$ m, $S = 4.5$, and $B_1 = 20$, 20, 40 m, respectively. The results indicate that both COV_S and δ_{hm_cr} significantly influence the β and P_f values. However, the influence of COV_S on β and P_f is not as significant as that of δ_{hm_cr}, especially when COV_S is greater than 0.20.

Figure 9.13 shows the influence of COV_S on β for $H_e = 17$ m; $S = 4.5$, $\delta_{hm\_cr} = 23$ mm; $B = 20, 30, 40$ m; and $B_1 = 10, 15$ m, respectively. It is clear that β decreases

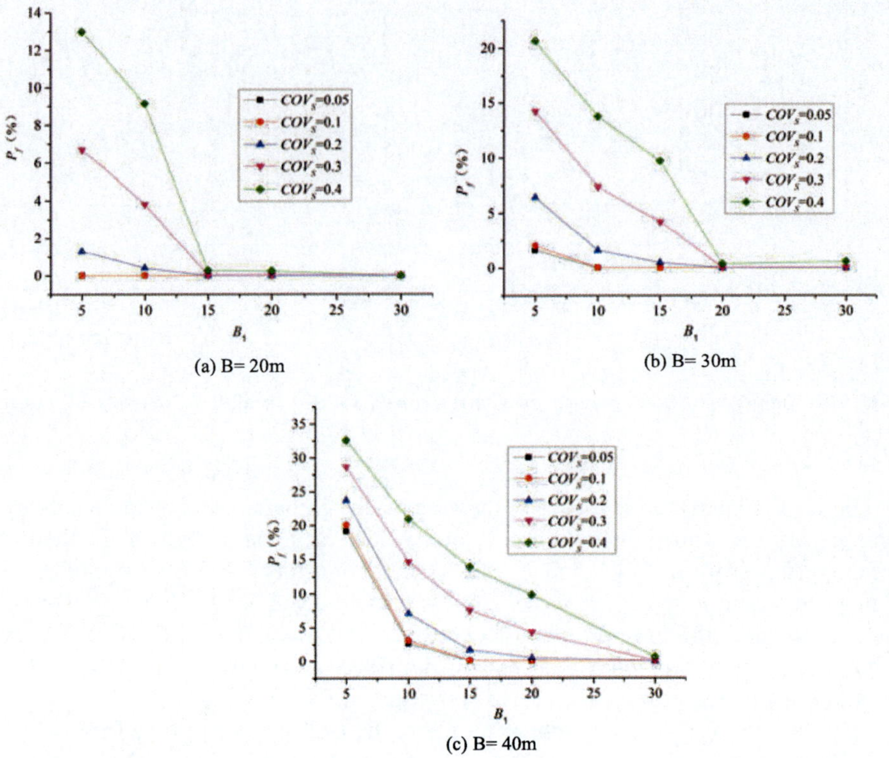

Fig. 9.11 Influence of COV_S and B_1 on β and P_f for $B = 20, 30, 40$ m, $H_e = 17$ m, $S = 4.5$, $FS\_cr = 2.0$

exponentially as the COV_S become greater. It is logical that β increases when the excavation is further away from the slope.

9.2.4.3 System Reliability Analysis

In the assessment of deep braced excavations adjacent to slopes, it is logical to consider the interaction between the global ULS factor of safety and the SLS of the wall deflections together rather than independently. Therefore, it requires a system reliability analysis approach that can consider each limit state and the interaction between the various limit states [7]. The proposed system reliability is ensured in this study by performing probabilistic analysis on the global factor of safety and at the same time satisfying the criterion that the threshold maximum wall deflection should not be exceeded. As shown in Fig. 9.14, the regression model predicting the maximum lateral wall deflection δ_{hm} is implemented into Cell K8 representing the constraint from SLS, in which mathematical expression of $\delta_{hm-cr} - \delta_{hm} \geq 0$ must be satisfied for each combination of parameters B, H_e, and S. Though this way,

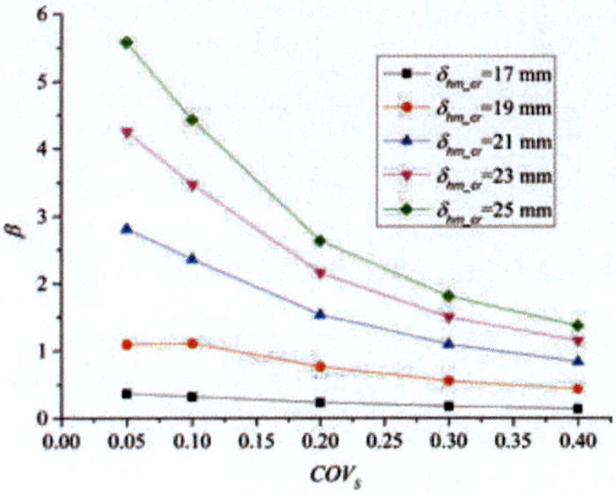

(a) B_1=20 m

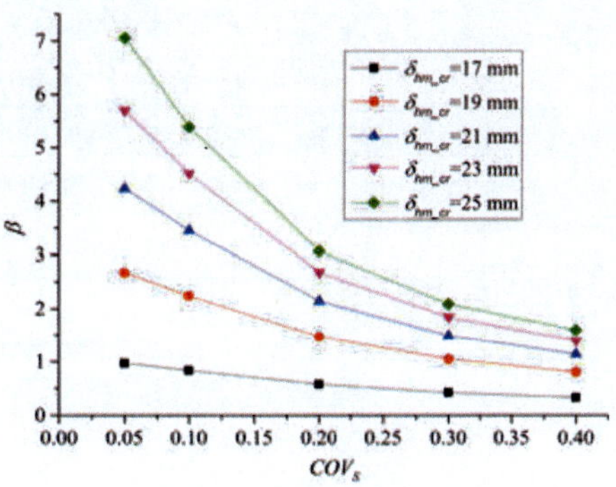

(b) B_1=30 m

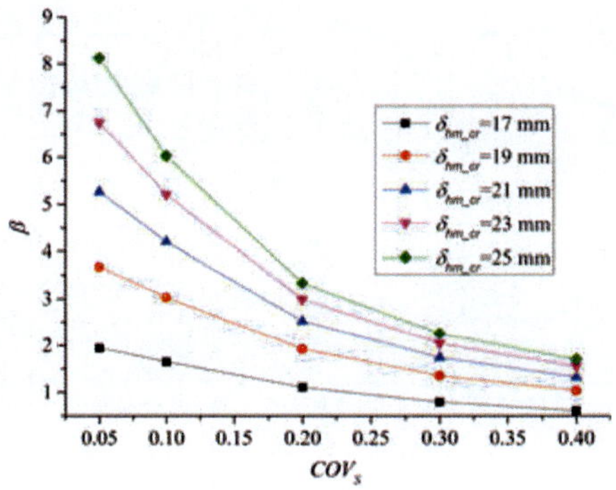

(c) B_1=40 m

Fig. 9.12 Influcnce of COV_S and δ_{hm_cr} on β for $B = 30$ m, $H_e = 17$ m, $S = 4.5$, $B_1 = 20, 20, 40$ m

Fig. 9.13 Influence of COV_S on β for $H_e = 17$ m; $S = 4.5$; $\delta_{hm\_cr} = 23$ mm; $B = 20, 30, 40$ m; and $B_1 = 10, 15$ m

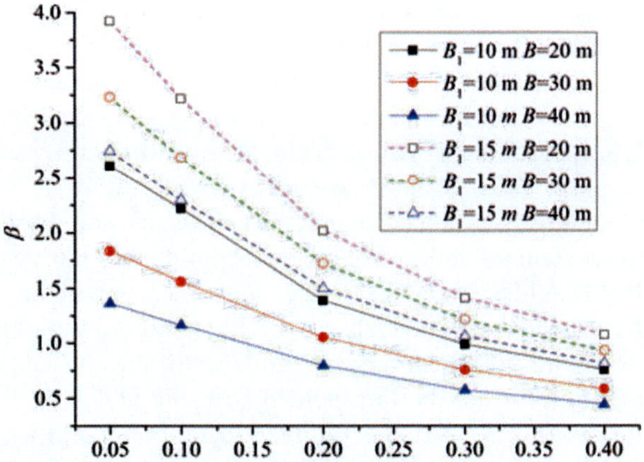

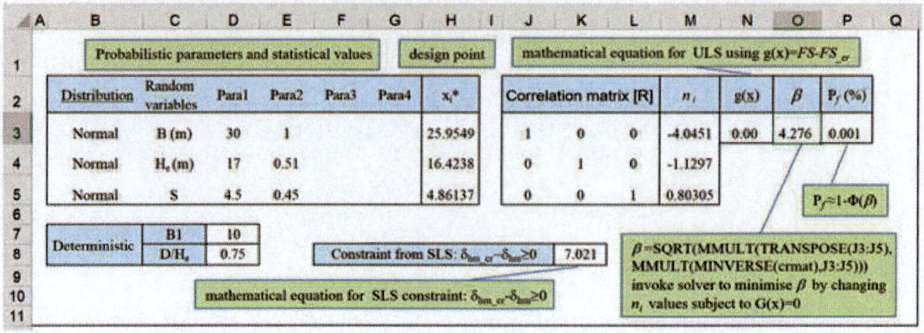

Fig. 9.14 Calculation on β and P_f for system reliability using FORM spreadsheet

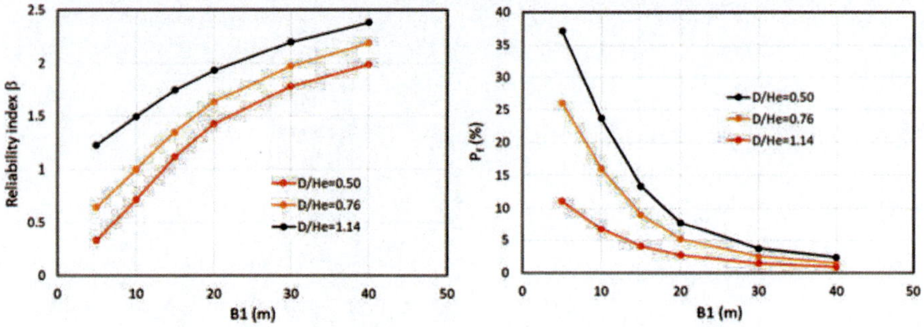

Fig. 9.15 System reliability analysis results for different excavation configurations

the parameter combinations for derived design point satisfy both the ULS and SLS simultaneously and the system reliability is ensured.

Figure 9.15 plots the system reliability analysis results for configurations with different B_1 values and penetration depths for the case with B = N(30, 3), H_e= N(20, 0.6), ln(S) = N(4.5, 0.45), FS_{cr} = 1.5, and δ_{hm} = 30 mm, where N(a,b) represents the normal distribution with mean value of a and standard deviation of b. It is obvious that as B_1 increases, the β value increases. The influences of penetration depth D/He on β and P_f are not as significant as B_1.

Figure 9.16 shows the influence of the chosen threshold factor of safety FS_{cr} on the system reliability when the threshold maximum wall deflection is set at 30 mm. It is should be noted that β decreases linearly with the FS_{cr} with different B_1 while P_f increases exponentially as the FS_{cr} becomes greater.

Figure 9.17 illustrates the influence of the chosen threshold δ_{hm_cr} on the system reliability when the threshold FS_{cr} is set at 1.5. It is obvious that the greater the threshold δ_{hm_cr} value is set, the greater the reliability index β derived, which indicating a much smaller possibility that the actual δ_{hm_cr} is exceeded. It also can be observed that the P_f decreases with δ_{hm_cr} and converged at δ_{hm_cr} = 40 mm.

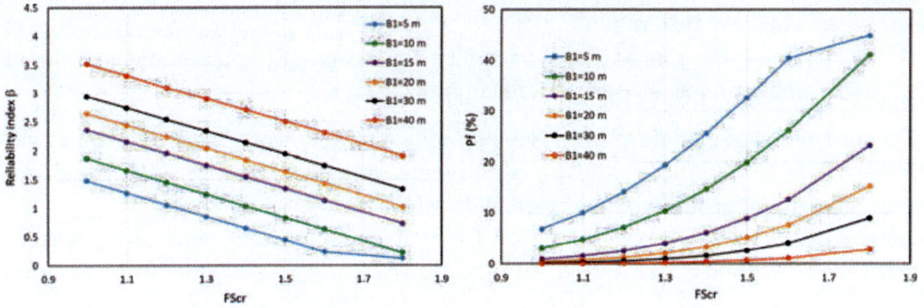

Fig. 9.16 Influence of threshold *FS* on system reliability

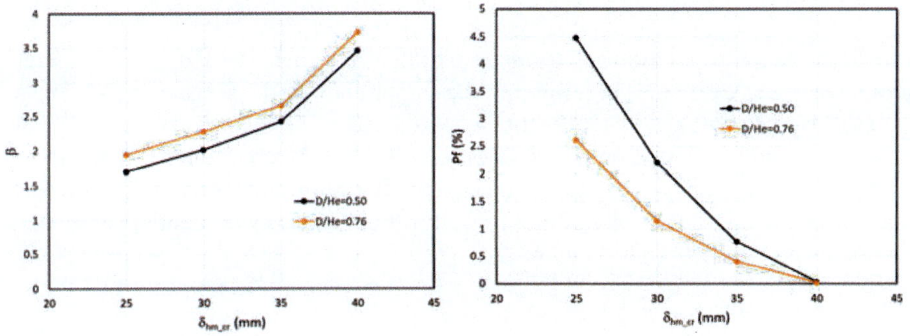

Fig. 9.17 Influence of threshold δ_{hm} on system reliability

9.2.5 Summary

This section presents numerical investigations on the performance of a braced excavation adjacent to an existing slope, from the perspectives of the global factor of safety and the maximum lateral wall deflections. It also proposed a probabilistic framework for the quantitative assessment of both the ultimate and the serviceability limit states in view of some design and construction uncertainties. The main conclusions arrived at include the following:

(1) Based on the case study, excavations close to an existing slope may result in a significant reduction of the global FS from 2.341 to 1.705 for the braced excavation-upper slope system. The amount of reduction is closely associated with the excavation depths and widths, the penetration depth and the distance between the excavation and the slope toe.

(2) The wall deflections are also significantly influenced by the key factors listed in conclusion point (1).

(3) Surrogate models for ULS and SLS are developed and implemented into FORM for reliability analysis.

(4) The proposed system reliability can be assessed by performing a probabilistic analysis on the global factor of safety while setting the threshold maximum wall deflection as a constraint for optimization.

It should be noted that the proposed probabilistic framework is based on an actual case study in Chongqing as illustrated in Fig. 9.2 in which the slope angle and soil properties are deterministic and fixed. The effects of the inherent variability of the soil properties and the slope angle on the braced excavation-upper slope system performance will be considered in a future study.

9.3 Influence of Deep Braced Excavation on Adjacent Piles

Due to the dense population and limited space in urban areas, development of underground space is desirable, highlighting the importance of deep excavation. Consequently, it is inevitable that deep braced excavations being constructed adjacent to buildings which footed on pile foundations. Table 9.3 complied the summary of studies on foundation pile responses adjacent to excavation and the detailed studies on pile responses adjacent to excavations as well as the chosen key influential parameters on pile responses.

9.3.1 Numerical Modeling

The prototype for numerical simulation is from Goh et al. [21] in which the results of an actual full-scale instrumented case history carried out to examine the behavior of an existing pile due to nearby excavation activities resulting from construction of a deep cut-and-cover tunnel were presented. The pile was located 3.0 m behind a 0.8 m thick diaphragm wall. Excavation to the formation level that was 16.0 m below the ground surface resulted in a maximum lateral pile movement of 28.0 mm.

The site consisted of sand fill over soft marine clay underlain by the geological formation is known as Old Alluvium. The soil at this section of the cut-and-cover tunnel is highly variable with a number of buried valleys infilled with marine clay and fluvial sand. The groundwater level (GWL) was about 4 m below the ground surface. The soil and wall profile and soil information inferred from the nearby boreholes and used for numerical simulation via Plaxis are shown in Fig. 9.1. The soft marine clay layer has a liquid limit of about 75% and a plasticity index of about 40%. The Old Alluvium can be divided into two zones. Old Alluvium I is made up of a weathered zone of residual soil consisting of sandy clay and silty fine to medium sand. Old Alluvium II is a cemented zone of hard and very dense clayey sand and silty sand. It was found that the standard penetration test increased rapidly to more than 100 blows within a short distance.

Table 9.3 Summary of studies on foundation pile responses adjacent to excavation

#	References	Key design parameters						Soil type	Soil model	Method
1	[99]	Pile diameter	Soil undrained shear strength	Depth of excavation	Soil tension capacity	Pile head condition	–	Clay	Tresca	FEM
2	[10]	Pile stiffness	Nonhomogeneous soil strength conditions	Pile head fixity	–	–	–	Clay and sand	Elastic soil spring	FEM
3	[11]	Excavation depth	3D effect of excavation	Diaphragm wall stiffness	Distance between diaphragm wall and pile	Pile head constraint condition	Pile stiffness	Soft clay and sand	MC	FEM
4	[12]	Pile flexibility	Magnitude of soil movement	Pile head boundary conditions	Shape of the soil movement profile	Thickness of the moving soil mass	–	Soft clay	MC	FEM
5	[13]	Pile rigidity	Distances between walls and piles	Vertical load on pile head	Constraint condition of pile head	–	–	Multi-layered soils	MC	FEM
6	[14]	Constraint conditions of pile head	Stiffness of diaphragm wall	Stiffness of pile	Distance between excavation and pile	Excavation depth	Length of pile	Soft clay, medium sand	MC	FEM
7	[15]	Pile head conditions	Distance from the excavation	The retaining wall stiffness	The pile stiffness and length	Excavation depth	Axial load	Modified 5th layer soil of Tianjin	MCC	FDM
8	[16]	Depth of excavation	Soil properties	Wall support system	Pile fixity conditions	Pile location	–	Malaysian Kaoline clay	MCC	FEM

(continued)

Table 9.3 (continued)

#	References	Key design parameters						Soil type	Soil model	Method
9	[17]	Pile diameter	Stiffness of supporting structure	Pile stiffness and length	Distance between pile and pit	Pile head constraint	Cap thickness	Multi-layered soils	MCC	FEM
10	[18]	Pile head constraint	Lateral deflection of wall	Distance from the excavation	Pile stiffness	Axial load on the pile	Pile length	Multi-layered soils	MCC	FDM
11	[19]	Strap stiffness	Pile diameter	Pile length	–	–	–	Sand and clay	HS	FEM
12	[20]	Excavation depth	Permeability and soil state	Pile group location from excavation	Supporting system stiffness	Pile length	Working load	Kaolin clay	HSS	FEM

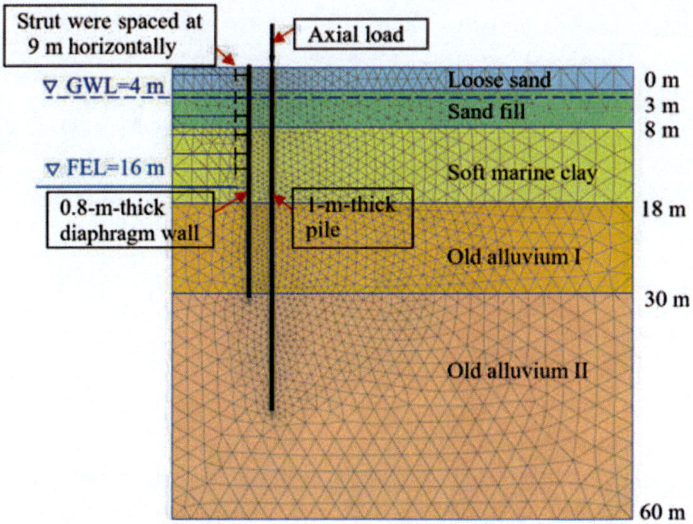

Fig. 9.18 Finite element half-mesh

For brevity and considering asymmetry, half-mesh model is developed. A typical two-dimensional (2D) numerical model setup is shown in Fig. 9.18. The struts and wall members are also shown, for better understanding. The structural elements were assumed to be linear elastic with the wall represented by 5-noded beam elements and 3-noded bar elements were used for the struts. To transfer the 3D stiffness of piles into a 2D case for numerical simulation use, the equivalent stiffness concept is adopted, i.e., pile stiffness value for plain strain is equal to the 3D stiffness divided by the pile spacing. The soil was modeled by 15-noded triangular elements. The nodes along the side boundaries of the mesh were constrained from displacing horizontally while the nodes along the bottom boundary were constrained from moving both horizontally and vertically. The right vertical boundary extends far from the excavation to minimize the effects of the boundary restraints effect. In the analyses, groundwater drawdown was not considered. The drainage condition of sand layers is set as drained, and clay layers as undrained. The HSS parameters of soil layers derived from the commonly used empirical values are shown in Table 9.4.

The numerically calculated and measured pile lateral deformation profiles and bending moments are compared in Figs. 9.19 and 9.20, respectively. The measured pile bending moments were obtained by differentiating the pile deflection profile twice. It is obvious that the computed displacements and bending moments are generally in good agreement with the measured results, in both the magnitude and the profiles.

Table 9.4 HSS parameters for soils

Parameters	E_{50}^{ref} kPa × 10^3	E_{oed}^{ref} kPa × 10^3	E_{ur}^{ref} kPa × 10^3	G_0^{ref} kPa × 10^3	ckPa	φ °	m	γ kN/m³
Loose sand	15	15	45	56	0	30	0.8	17
Sand fill	33	33	100	130	0	30	0.8	17
Soft marine clay	8	8	24	15	0.05	25	1	16
Old alluvium I	107	107	320	190	0.05	40	0.8	19
Old alluvium II	270	260	900	410	0.05	45	0.8	19

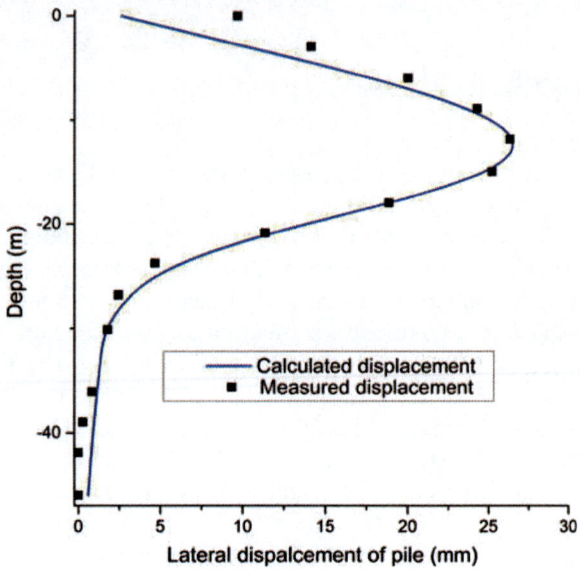

Fig. 9.19 Pile lateral displacement at the end of excavation

9.3.2 *Parameteric Analysis*

Numerical analyses were performed to investigate the response of an axially loaded pile adjacent to deep braced excavation. The final excavation depth (H_e) is 16.0 m and supported by 0.8 m-thickness and 31 m-depth diaphragm wall with 6 levels of struts. The struts were set at the bottom of each stage of excavation vertically and 9.0 m horizontally. The elastic behavior is defined by the following parameters: EA: normal stiffness, EI: bending stiffness, m: Poisson's ratio. The struts stiffness per meter is 3.87×10^5 kN, the elastic modulus of the wall is 2.1×10^7 kN/m².

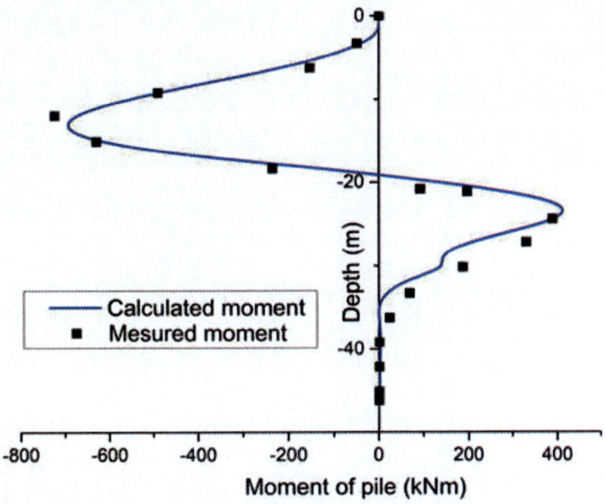

Fig. 9.20 Bending moment of pile at the end of excavation

The stiffness of wall is applied with a reduction factor of 75%. Bottom up excavation supported by struts was adopted in this study, which is commonly used. For consistency consideration with the field conditions, single pile was selected for this study. There are 7 modification factors considered in this study which are regarded as key design variables, the excavation depth He (m), unsupported depth of excavation du(m), distance between the wall and the pile D (m), pile length Lp (m), pile diameter d(m), pile head fixity of free/pinned/ translatable/fixed, axial load P (kPa). It is reported that the supporting system stiffness does not have a significant influence on pile behavior. Consequently, this aspect is not discussed in this study. A total of 42 numerical simulations were conducted, and the numerical simulation schemes are listed in Table 9.5 and shown in Fig. 9.21.

9.3.2.1 Excavation Depth

Figures 9.22 and 9.23 show the excavation induced pile lateral movements and bending moments at different stages of excavation. For this case, the pile is located 3.0 m away from the wall. According to Fig. 9.22, the maximum induced lateral deformation increases almost linearly with the excavation depth and the maximum pile lateral deformation is about (1.65–2.06%)H_e. At the excavation depth of 16.0 m, the point of maximum pile lateral deformation is located slightly above the elevation level and during excavation, there is no displacement at the pile toe since the pile is long and inserted deeply into the stiff soil layer. Even though the first strut is installed at the 1 m below the surface ground after excavating the first layer, the pile has a slight lateral deformation at the pile head. Figure 9.23 shows the increase of the bending moment along the pile shaft during excavation. The maximum bending moment is increasing approximately linearly with excavation depth. The induced maximum bending moment is about (43–62)H_e kN·m. When excavation goes deeper,

Table 9.5 Numerical simulation schemes

#	Description	Excavation depth, H_e(m)	Pile length, L_p (m)	Pile location from excavation, D(m)	Pile diameter, d (m)	Pile head fixity	Unsupported depth of excavation d_u (m)	Axial load, P (kN)
1	Excavation depth	4, 6.5, 9, 11.5, 13.5, 16	46	3	1	Free	1	1500
2	Pile length	16	31, 34, 37, 40, 43, 46, 49	3	1	Free	1	1500
3	Pile location from excavation	16	46	1, 3, 5, 7.5, 10, 20, 30	1	Free	1	1500
4	Pile diameter	16	46	3	0.4, 0.6, 0.8, 1, 1.2, 1.5, 2	Free	1	1500
5	Pile head fixity	16	46	3, 7.5, 20	1	Free/ pinned/ translatable /fixed	1	1500
6	Unsupported depth of excavation	16	46	3	1	Free	1, 2, 4	1500
7	Axial load	16	46	3	1	Free	1	0,1500, 3250, 5000,

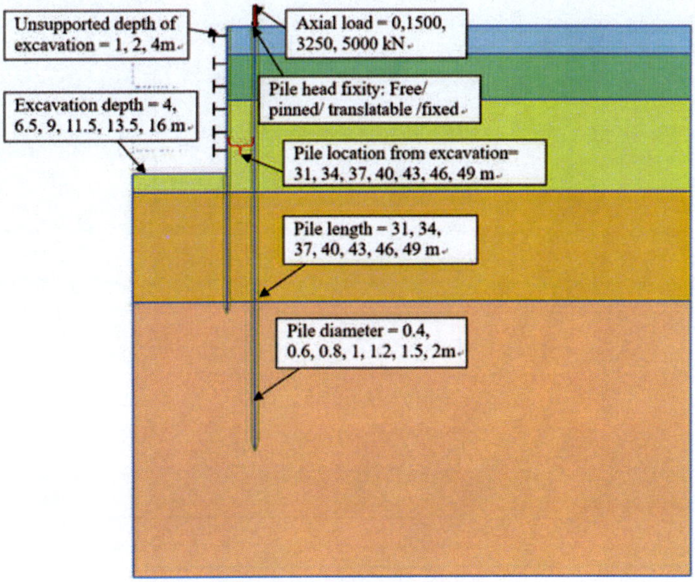

Fig. 9.21 Scheme of the design variables

Fig. 9.22 Pile lateral displacement at different excavation depths

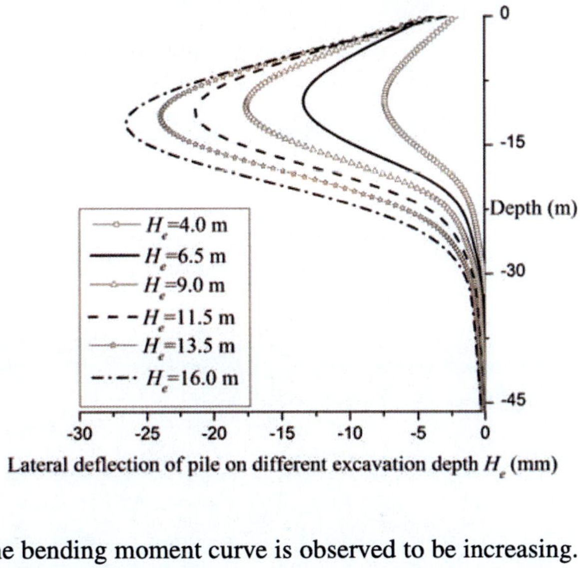

the maximum curvature of the bending moment curve is observed to be increasing. As excavation proceeds, though the bending moment exerted on pile is becoming greater, the moment curve of the top part tends to be concave at the opposite direction due to the support provided by the struts, and the maximum bending moment occurs where the pile is subjected to the maximum lateral deformation.

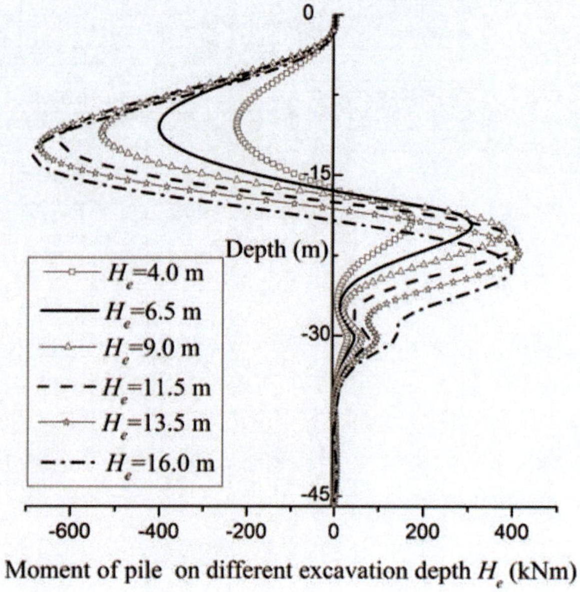

Fig. 9.23 Bending moment of pile at different excavation depths

9.3.2.2 Pile Length

Figure 9.24 shows the pile lateral displacements at different pile lengths. It is obvious that when the pile toe located in the Old alluvium I, i.e., the pile length is shorter

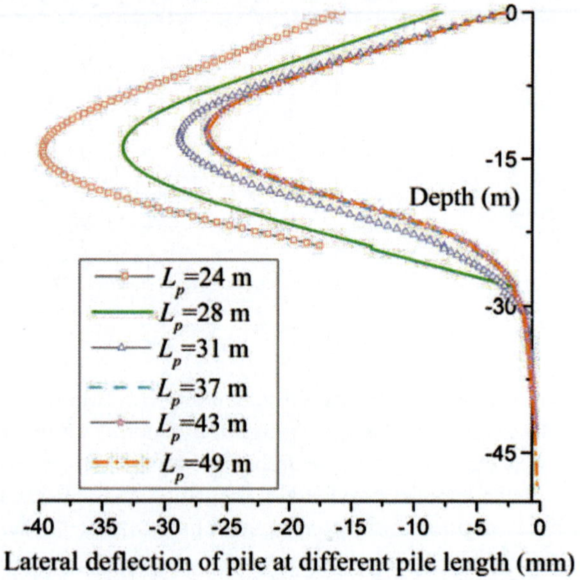

Fig. 9.24 Pile lateral displacement at different pile lengths

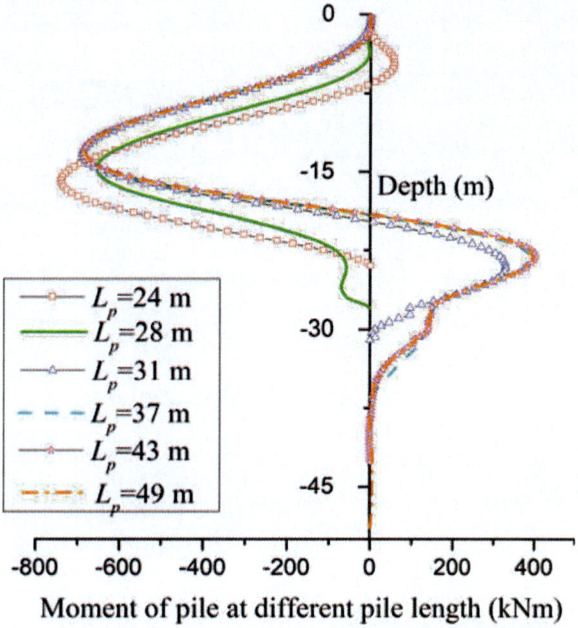

Fig. 9.25 Bending moment of pile at different pile lengths

than 30.0 m, there is a considerable displacement at the pile toe, about 44.1% of the maximum pile lateral deformation when L_p is 24.0 m and it decreases to 7.4% of the maximum pile lateral deformation when L_p is lengthened to 28.0 m. Besides, the pile head displacement is also significant. When the pile toe is located in the Old alluvium II, the displacement at the pile toe is negligible. As L_p is greater than 37.0 m, the difference of pile lateral deformation is marginal. Figure 9.25 shows the bending moment of pile at different pile lengths. When the pile toe is in Old alluvium I and L_p is 24.0 m or 28.0 m, the pile moment is similar to truss beam. When the pile toe located in the Old alluvium II, the pile moment is similar to cantilever beam. When L_p is greater than 37.0 m, there is little change of the bending moment, indicating a conservative design of the pile.

9.3.2.3 Pile Distance From Excavation

Figures 9.26 and 9.27 show the variation of pile lateral deformation and bending moment at various pile locations, respectively. It is apparent that the lateral displacements and bending moments for the cases of D = 1.0, 3.0, 5.0, 7.5, 10.0 m are dominated by concave movement, with a maximum negative moment occurring at strut level 5. For the case with greater distances of D = 20.0, 30.0 m, it shows a domination of cantilever wall characteristics. As shown in Fig. 9.26, the maximum value decreases as the distance increases. For cases of D ≥ 20.0 m, the maximum pile lateral deformation is at the pile head. As shown in Fig. 9.27, for cases of D ≥

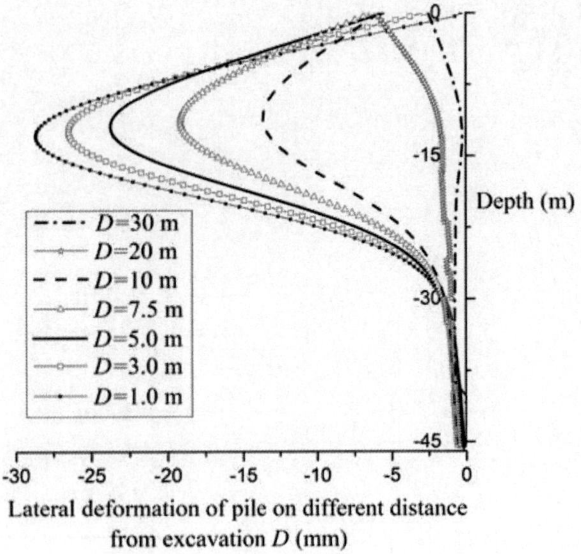

Fig. 9.26 Pile lateral displacement at different pile distances from excavation

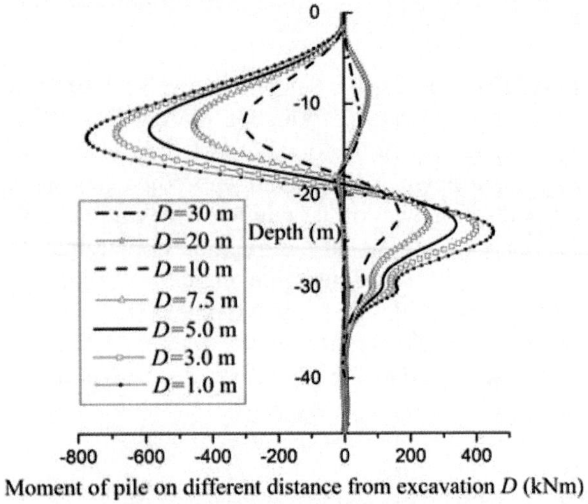

Fig. 9.27 Bending moment of pile at different pile distances from excavation

20.0 m, the curvatures of the bending moment curve is on the opposite side of the excavation, which are different from the other five cases, and the values are very small, which indicates that the excavation has minimal influence on the pile bending moment.

9.3.2.4 Pile Diameter

Figures 9.28 and 9.29 present the calculated pile lateral displacement and bending moment for the final stage of excavation for different pile diameters *d* of 0.4, 0.6, 0.8, 1.0, 1.2, 1.5, 2.0 m, respectively, indicating that different pile diameters have

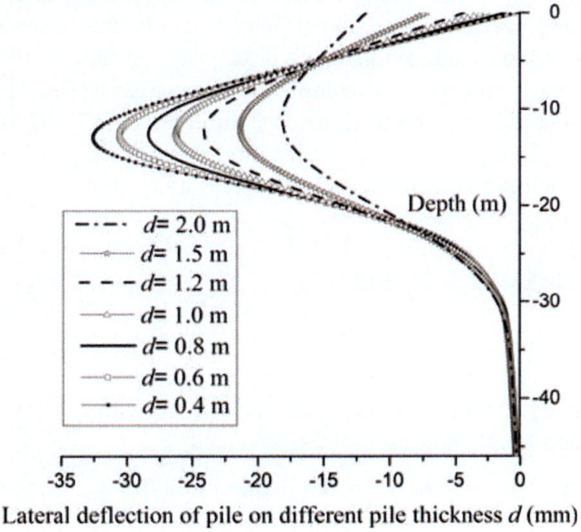

Fig. 9.28 Pile lateral displacement at different pile diameters

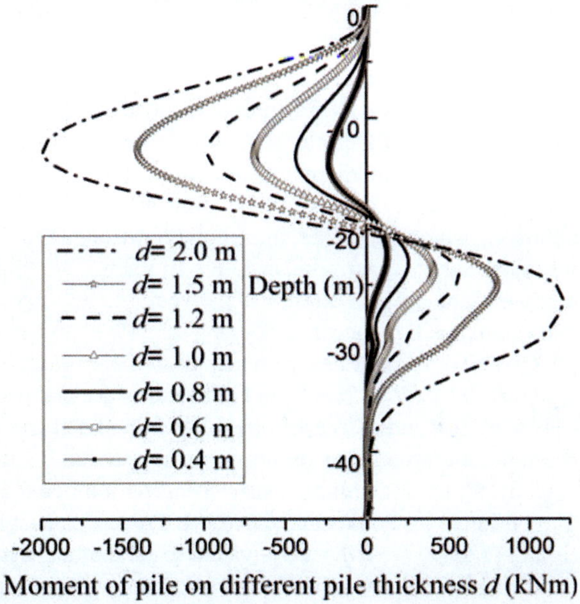

Fig. 9.29 Bending moment of pile at different pile diameters

significant influence on both the pile lateral displacement and bending moment. It is apparent that the lateral deflections and the bending moments for the cases of $d = 0.4, 0.6, 0.8, 1.0$ m are dominated by concave movement, with a maximum negative moment occurring at strut level 5. According to Fig. 9.28, maximum pile lateral displacement shows almost linear growth with a decreasing pile diameter. In contrast, the pile head lateral displacement increases as d increases. Nevertheless, the pile lateral displacement of all 7 cases are almost equal at pile depth below 20.0 m from the ground surface, possibly due to the great pile length and the fact that it was inserted into a very stiff clay. As shown in Fig. 9.29, the bending moment shows a uniform reduction with an increasing pile diameter d, and the zero value of bending moments are located at the depth of (17.520.5) m, and with the pile diameter increasing, the zero-point location move downwards. For these cases the maximum bending moment occurs at a level slightly above the bottom of the excavation.

9.3.2.5 Pile Head Fixity

Several studies have shown that the pile head fixity condition has a significant impact on the pile adjacent to excavation, e.g., Poulos and Chen [9]. The effect of pile head condition on pile response is investigated through four cases of four different constraint conditions at the pile head:

(a) Both translation and rotation free (free),
(b) Translation fixed and rotation free (pinned),
(c) Translation free and rotation fixed (translatable),
(d) Both translation and rotation fixed (fixed).

Figures 9.30 and 9.31 show the pile response for four different head fixity conditions for the case with unsupported excavation depth of 1 m, excavation depths of 16 m, and the pile is 3 m from the excavation. It is clear that greater the degree of freedom (DOF) of the pile head, smaller are the maximum lateral pile deformation developed in the middle part of the pile.

For Case d), high positive bending moments of pile head of 1245 kN·m were obtained, which is higher than maximum negative moment of 748 kNm. While for other cases, maximum negative moments are bigger than maximum positive moments, and these maximum negative moments vary from about 1.1 to 1.7 times the maximum positive moments varying the pile head conditions. Normally, in practice, piles are tied using either a pile cap or tie beams and finally connected to the superstructure. Therefore, pile head connections are normally partially fixed, in between Cases (b), (c), and (d), depending on the pile head and cap connection condition, for example, the amount of reinforcements extending into the cap. For the four cases considered, the maximum positive bending moments does not have much difference as mentioned by Poulos and Chen [9]. The maximum lateral deformation for pile of case (b) and (c) are almost equivalent to each other, while the pile head displacement is different. Figure 9.31 also shows that below the depth of about 10 m, the bending moment curves of pile in case (a), (b), (c), (d) are almost consistent with each other.

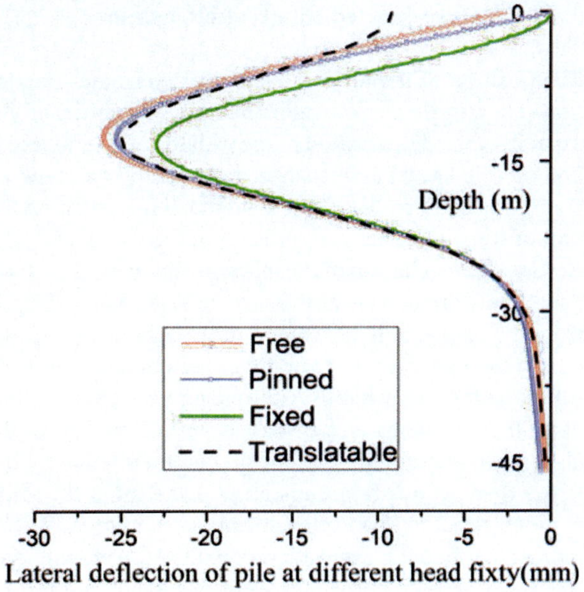

Fig. 9.30 Pile lateral displacement at different pile head fixities

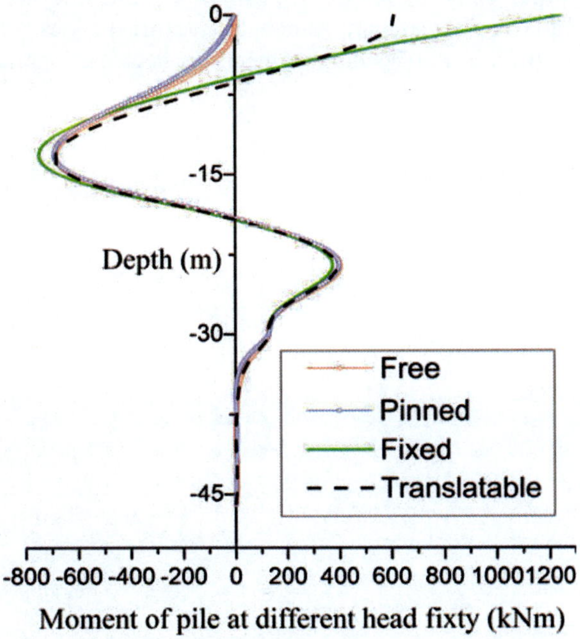

Fig. 9.31 Bending moment of pile at different pile head fixities

9.3.2.6 Unsupported Depth of Excavation

The position of the top strut actually determines the depth of the cantilever excavation. The cantilever excavation depth will significantly affect the initial deformation, especially the deformation of the wall top. The effect of unsupported depth of excavation on pile lateral deformation and bending moment was investigated. Even though the unbraced depth does not have much influence on the maximum lateral deformation of the pile, it has a significant influence on the pile head as shown in Fig. 9.32. Before the installation of the initial strut, the wall and soil behind it experience higher lateral deformations near the surface level. Since the pile head is located 3.0 m away from the excavation and free to move, cantilever deformation occurs at the installation of the first strut. Further increase in pile curvature occurs with the installation of subsequent struts which synchronized with excavation procedure. A similar pattern of wall movements was observed by O'Rouke [22] during deep excavations. It is also worth mentioning that the maximum pile lateral displacement on $d_u = 1.0$ m is larger than $d_u = 2.0$ m, indicating that strut at the midpoint of the excavated depth is more effective than other locations. Figure 9.33 shows the variation of bending moment. When the excavation depth is 16.0 m, with an unsupported depth of 4.0 m, less bending moment values were observed than those obtained with lower unsupported depths ($d_u = 1.0, 2.0$ m). This occurs due to the increased flexibility of the wall from less support. Although pile stiffness does not change with support spacing or stiffness, pile bending moments are decreased with increasing support spacing in

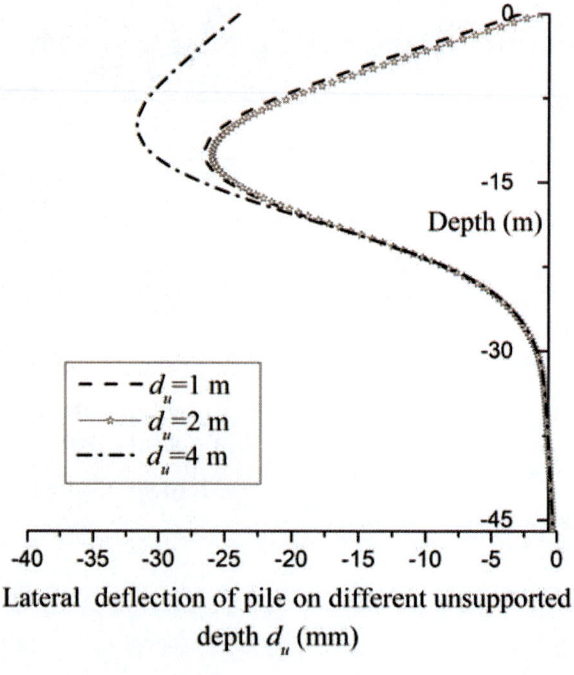

Fig. 9.32 Pile lateral displacement at different unsupported depths

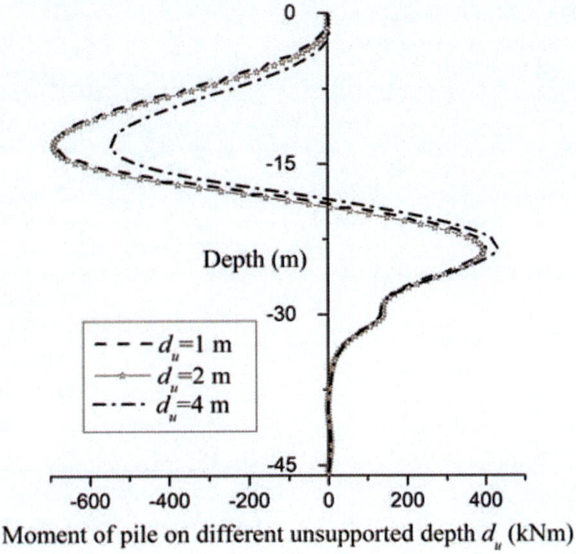

Fig. 9.33 Bending moment of pile at different unsupported depths

agreement with wall behavior. For the case which $H_e = 16.0$ m and the unsupported depth of the excavation was increased from 1.0 to 4.0 m, the maximum bending moment has a 20% reduction.

9.3.2.7 Axial Load

Figures 9.34 and 9.35 show the pile lateral displacement and bending moment at different axial load of 0, 1500, 3250, 5000 kPa, indicating that different axial loads exerted on the pile have marginal influence on both the pile lateral displacement and bending moment. It is observed that the displacement of pile toe is very small and absolutely meet regulatory criterion requirements. As a result, the axial load applied on the pile head has negligible influence on the pile lateral deformation and bending moment, once the load exceeded the serviceability limit state, damage occurs without any sign. Consequently, the collapse caused by over loading is suddenly and unpredictable. The same regular pattern was observed by Zhang et al. [23] and Liyanapathirana and Nishanthan [16].

9.3.3 *Modification Factors and the Proposed Design Charts*

The induced maximum lateral deformation and bending moment of piles for cases considered can be approximated through Eqs. 9.3 and 9.4, respectively. It should be

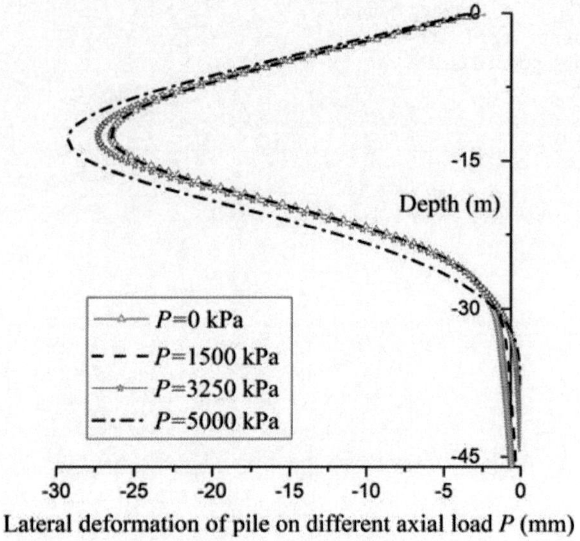

Fig. 9.34 Pile lateral displacement at different axial loadings

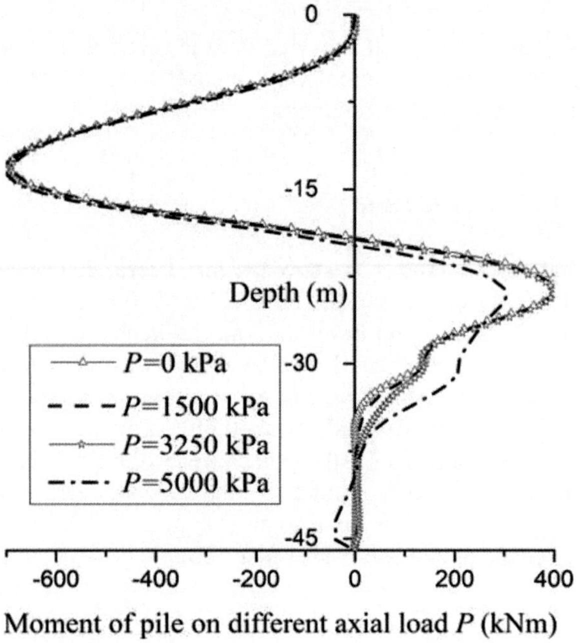

Fig. 9.35 Bending moment of pile at different axial loadings

noted that these equations are based on the base case response values and a set of modification factors, similar to the method proposed by Poulos and Chen [9].

$$\delta_{h\_max} = \delta_b \times \mu_l \times \mu_D \times \mu_{He} \times \mu_d \times \mu_{Lp} \times \mu_f \tag{9.3}$$

$$M_{h\_max} = M_b \times \mu_l' \times \mu_D' \times \mu_{He}' \times \mu_d' \times \mu_{Lp}' \times \mu_f' \tag{9.4}$$

where δ_{h_max} is the maximum lateral deformation after modification from those modification factors, δ_b is the maximum lateral deformation for the base case, M_{h_max} is the maximum bending moment after modification while M_b is the maximum bending moment for the base case. μ_l and μ_l' are the modification factors from the axial load exerted on the pile head for lateral pile deflection and bending moment, respectively. μ_D and μ_D' are the modification factors from the pile distance from excavation D; μ_{He} and μ_{He}' are the modification factors from excavation depth H_e. μ_d and μ_d' are the modification factors from the pile diameter d. μ_{Lp} and μ_{Lp}' are the modification factors from the pile length L_p. μ_f and μ_f' are the modification factors from the pile head fixity. δ_b and M_b values can be obtained from results for the base case, of which the excavation depth $H_e = 16.0$ m, $D = 3.0$ m, $L_p = 46$ m, $d = 1.0$ m, free pile head fixity, $d_u = 1.0$ m, and $P = 1500$ kPa. δ_b of the base case is 26.51 mm while M_b of the base case is 685.5 kNm.

Figures 9.36, 9.37, 9.38, 9.39, 3.40 and 9.41 show modification factors for each parameter variation, which is obtained by dividing the pile lateral deformation and bending moment for each case by the corresponding base case values.

Figure 9.36 shows the modification factors from the axial load exerted on the pile head μ_l and μ_l'. $\mu_l$increases with the axial load exerted on the pile head. However, the increase is not so significant since μ_l increases from 1.00 to 1.10 when the axial load increases from 0 to 5000 kN. The influence of axial load on wall bending moment is even less significant, in view of the slight increases of μ_l' from 0.99 to 1.03. The slight influence is understandable since the axial load is vertical while the lateral deflection and bending moment responses are mainly horizontal.

Figure 9.37 shows the modification factors from the pile distance from excavation μ_D and μ_D'. μ_D decreases from 1.08 to 0.10 as the pile distance from excavation changes from 1.0 to 30.0 m, indicating great influences of distance from excavation

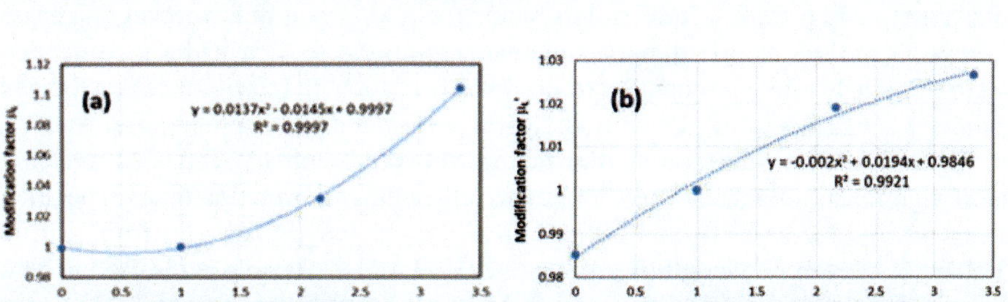

Fig. 9.36 Modification factors for the axial load exerted on the pile head: **a** deformation and **b** bending moment

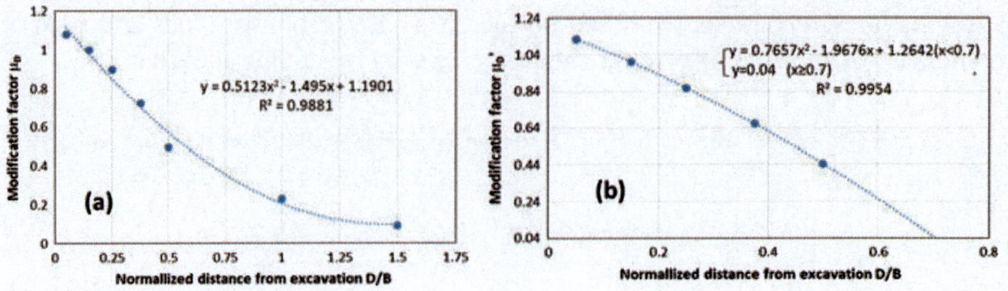

Fig. 9.37 Modification factors for the pile location from excavation: **a** lateral deformation and **b** bending moment

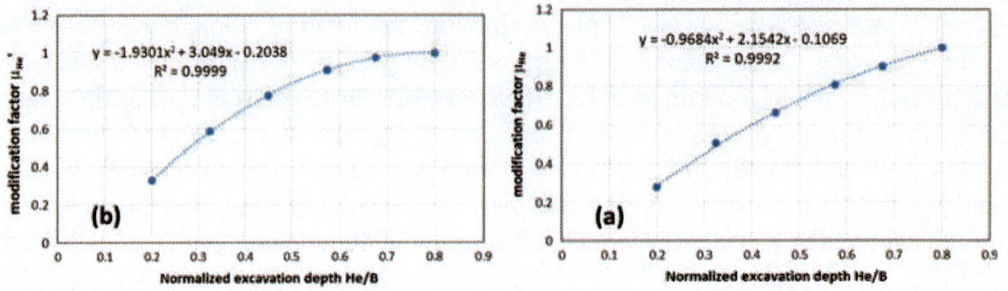

Fig. 9.38 Modification factors for excavation depth: **a** lateral deformation and **b** bending moment

on the pile lateral deflection. μ_D' decreases from 1.13 to 0.04 as the distance between pile and excavation rises from 1.0 to 18 m, i.e., D/B = 0.7. However when it comes to D/B > 0.7, μ_D' keeps as a constant 0.04, which indicates that the influence zone is about 18.0 m, when the pile is out of this excavation influence zone, the induced pile deformation and bending moment are negligible.

Figure 9.38 shows the modification factors from the excavation depth μ_{He} and μ_{He}'. It can be observed that μ_{He} increases from 0.28 to 1.00 as the excavation deepens from 4.0 to 16.0 m. At the same time, μ_{He}' increases from 0.33 to 1.00, indicating the significant influences of excavation depth on pile lateral deformation and the bending moments.

The influence of pile diameter on lateral deformation and bending moment is illustrated in Fig. 9.39. Modification factor for pile lateral deformation decreases from 1.23 to 0.68 as pile diameter increases from 0.4 to 2 m while modification factor for pile bending moment increases from 0.13 to 2.90. It is obvious that the pile diameter is of great influence on both lateral deformation and bending moment.

Figure 9.40 shows the modification factors from pile length μ_{Lp} and μ_{Lp}', revealing more significant influence of pile length on pile lateral deformation than on bending moment. μ_{Lp} decreases from 1.50 to 1.01 as the pile length increases from 24.0 to 37.0 m. When the pile length is greater than 37.0 m, i.e., the pile is inserted in hard stratum, μ_{Lp} keeps as a constant of 1.0. Meanwhile, μ_{He}' drops from 1.08 to 1.00 and keeps as a constant value of 1.0 when pile length is greater than 37.0 m.

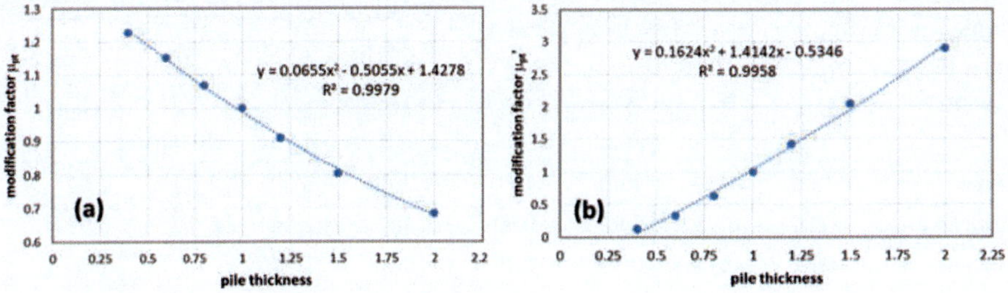

Fig. 9.39 Modification factors for pile diameter: **a** lateral deformation and **b** bending moment

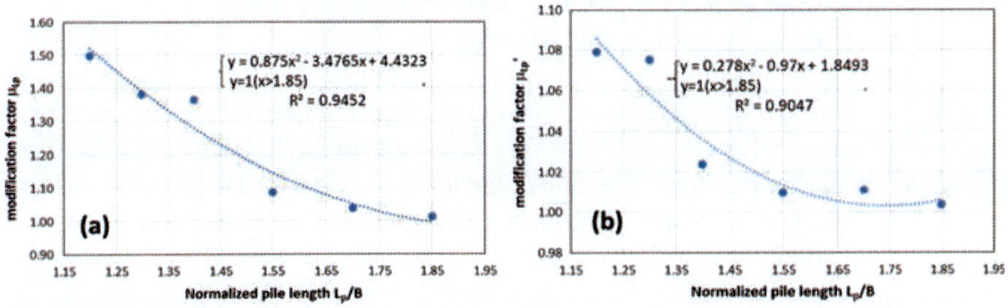

Fig. 9.40 Modification factors for pile length: **a** lateral deformation and **b** bending moment

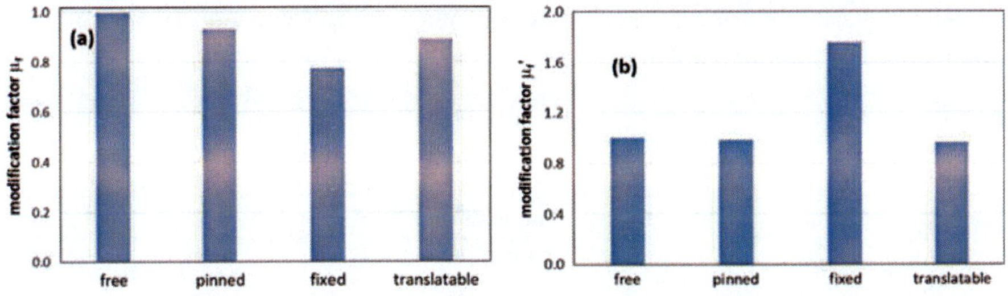

Fig. 9.41 Modification factors for pile head fixity: **a** lateral deformation and **b** bending moment

The influence of pile head fixity on pile lateral deformation and bending moment μ_f and μ_f' is shown in Fig. 9.41. In view that pile head fixity is categorical instead of continuous numbers, to ensure accuracy, the pile distance from excavation $D =$ 3.0, 7.5, 20.0 m are considered for four types of pile head fixity, respectively, as listed in Table 9.6, the average of modification factors of three different pile location is adopted. The bar charts values for each fixity condition are referred, as shown in Fig. 9.41. It is obvious that the fixed condition results in the smallest pile lateral deformations, yet the most bending moments. The other three conditions are more or less the same, slightly less than 1.0.

Table 9.6 Modification factors for pile head fixity

Modification factor	μ_f				μ_f'			
	$D =$ 3.0 m	$D =$ 7.5 m	$D =$ 20.0 m	Average	$D =$ 3.0 m	$D =$ 7.5 m	$D =$ 20.0 m	Average
Free	1.00	1.00	1.00	**1.00**	1.00	1.00	1.00	**1.00**
Pinned	0.97	0.93	0.90	**0.93**	1.01	1.04	0.90	**0.98**
Fixed	0.88	0.74	0.70	**0.77**	1.82	1.77	1.70	**1.76**
Translatable	0.96	0.97	0.75	**0.89**	0.89	0.95	1.04	**0.96**

9.3.3.1 Validation of the Proposed Design Procedure Using the Finite-Element Model Results

More FE analysis are carried out to compare the predictions with the design charts. The design variables for the synthetic cases were considered for the validation of the proposed sets of design charts. The pile diameter is of 1.5 m and located 7.5 m away from the proposed 16 m excavation with pile length of 34.0 m embedded in soft clay layer. The unsupported depth of excavation is 1 m, and the axial load exerted on the fixed pile head is 0 kN/m. The maximum pile lateral deformation and bending moment are calculated from the design charts for excavation depths of 4.0, 6.5, 9.0, 11.5, 13.5, and 16.0 m.

To obtain the maximum pile lateral deformation and bending moment based on the design charts, the following calculation steps were followed, and a summary of modification factors are shown in Table 9.7.

Table 9.7 Summary of modification factors obtained from the design charts based on synthetic case information

Factors	normalize	x	μ	μ'
Axial load	P/P_b ①	0.00	1.00	0.98
Distance between pile and excavation	D/B ②	0.375	0.70	0.63
Excavation depth	H_e/B	0.2	0.28	0.33
		0.325	0.49	0.58
		0.45	0.67	0.87
		0.575	0.81	0.91
		0.675	0.90	0.97
		0.8	1.0	1.0
Pile diameter	d/d_b ③	1.50	0.82	1.95
Pile length	L_p/B	1.70	1.05	1.05
Fixity	–	fixed	0.77	1.76

① P_b is the axial load exerted on the pile head of base case, $P_b = 1500$kN; ② B is the excavation width, $B = 20$ m; ③ d_b is the pile diameter of base case, $d_b = 1.0$ m

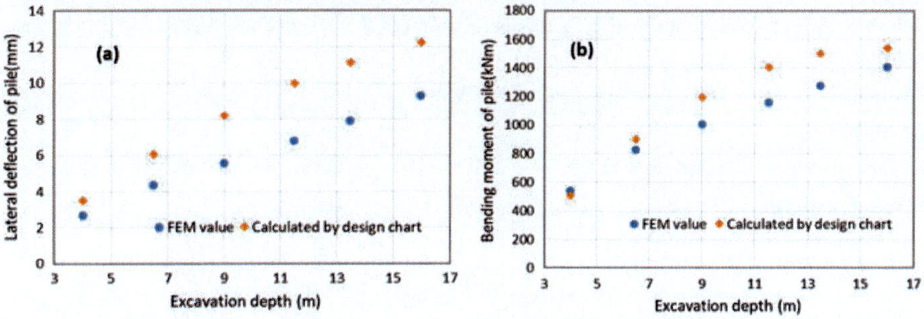

Fig. 9.42 Comparisons of **a** lateral deformation and **b** bending moment from FEM and the proposed design charts

Step 1: Base case maximum lateral deformation and bending moment: $\delta_b = 26.51$ mm and $M_b = 685.5$ kNm;

Step 2: Axial load on the fixed pile head is 0 kN/m, modification factors read from Fig. 9.36 are $\mu_l = 1.00$, $\mu_l' = 0.98$;

Step 3: Pile located 7.5 m away from excavation, modification factors read from Fig. 9.37 are $\mu_D = 0.70$, $\mu_D' = 0.63$;

Step 4: For excavation depths of 4, 6.5, 9, 11.5, 13.5, and 16m, modification factors read from Fig. 9.38 are $\mu_{He} = 0.28, 0.49, 0.67, 0.81, 0.90, 1.0$, $\mu_{He}' = 0.33, 0.58, 0.87, 0.91, 0.97, 1.0$;

Step 5: Pile diameter = 1.5 m, modification factors read from Fig. 9.39 are $\mu_d = 0.82$, $\mu_d' = 1.95$.

Step 6: Pile length of 34 m, modification factors read from Fig. 9.40 are $\mu_{Lp} = 1.05$, $\mu_{Lp}' = 1.05$

Step 7: Fixed pile head condition, modification factors read from Fig. 9.41 are $\mu_f = 0.77$, $\mu_f' = 1.76$

Figure 9.42 shows the maximum pile lateral deformations and the bending moments, for FE calculations and estimations based on design charts. The plots show that all the FEM results are enveloped by the value calculated from the design charts. In addition, the prediction of pile bending moment is more accurate while the prediction of deformation is not satisfactory but acceptable. Nevertheless, it indicates that the proposed design charts are useful for practicing geotechnical engineers to assess the influence of excavations on existing adjacent foundation pile responses.

9.3.3.2 Comparison of Design Chart Estimations with Measured Data

In this section, the results from the design charts will be compared with the case study reported by [21]. The specific case information is presented in Sect. 2.2. The maximum pile lateral deformations at excavation depths of 6.50 m and 16.0 m reported by [21] are 15.0 mm and 27.9 mm, respectively. The maximum pile bending moments are 439.0 kNm and 722.63 kNm, respectively.

Table 9.8 Summary of modification factors obtained from the design charts based on specific case information

Factors	normalize	x	μ	μ'
Axial load	P/P_b ①	0.00	1.00	0.98
Distance between pile and excavation	D/B ②	0.15	0.98	0.98
Excavation depth	H_e/B	0.33	0.49	0.58
		0.8	1.0	1.0
Pile diameter	d/d_b ③	1.00	1.00	1.00
Pile length	L_p/B	x > 37	1.00	1.00
Fixity	–	free	1.00	1.00

① P_b is the axial load exerted on the pile head of base case, P_b = 1500kN; ② B is the excavation width, B = 20 m; ③ d_b is the pile diameter of base case, d_b = 1.0 m

To obtain the maximum pile lateral deformation and bending moment based on the design charts, the following calculation steps were followed, and a summary of modification factors are shown in Table 9.8.

Step 1: Base case maximum lateral deformation and bending moment: δ_b = 26.51 mm and M_b = 685.5 kNm;

Step 2: Axial load on the fixed pile head is 0 kN/m, modification factors read from Fig. 9.36 are $\mu_l = 1.00$, $\mu_l' = 0.98$;

Step 3: Pile located 3.0 m away from excavation, modification factors read from Fig. 9.37 are $\mu_D = 0.98$, $\mu_D' = 0.98$;

Step 4: Excavation depth is 6.5 and 16 m, modification factors read from Fig. 9.38 are μ_{He} = 0.49, 1.0, μ_{He}' = 0.58, 1.0;

Step 5: Pile diameter = 1.0 m, modification factors read from Fig. 9.39 are μ_d = 1.0, μ_d' = 1. 0;

Step 6: Pile length is 28 m, modification factors read from Fig. 9.40 are μ_{Lp} = 1.0, μ_{Lp}' = 1.0;

Step 7: Free pile head condition, modification factors read from Fig. 9.41 are μ_f = 1.0, μ_f' = 1.0.

The estimations of pile lateral deformation based on proposed design charts for excavation depth of 6.5 m and 16.0 m are 12.7 mm and 25.8 m, respectively. The estimations of pile bending moments based on proposed design charts are 386.8 kNm and 663.3 kNm, respectively. Figure 9.43 plots the comparison of the pile lateral deformation and the bending moments between the measured values and the estimations from the design charts, respectively, indicating the estimations from the design charts are consistent with the instrumentations, as shown in Fig. 9.43, thus it indicates that the proposed design charts are useful for practicing geotechnical engineers to assess the influence of excavations on existing adjacent foundation pile responses.

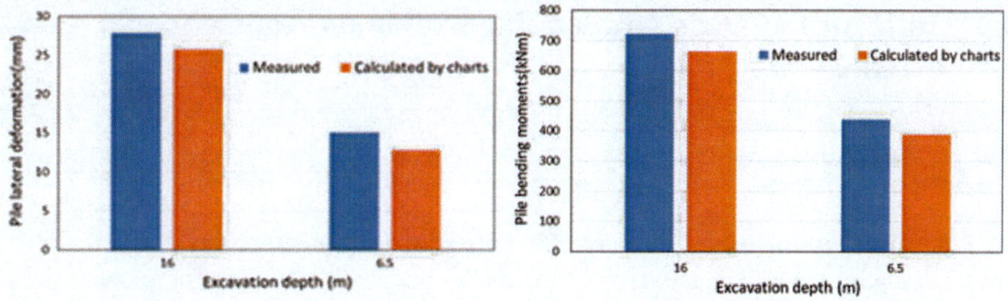

Fig. 9.43 Comparison between the measured and estimated values from design charts

9.3.4 Summary

FE analyses considering soil of HSS constitutive model were carried out to investigate the excavation induced pile behaviors influenced by various key design parameters.

(1) Among the chosen key factors, excavation depth, pile length, and distance from excavation have the most significant influences on the pile lateral deformations and bending moments. The maximum pile lateral displacement and the induced bending moment show a significant reduction with pile diameter. Pile head fixity has an obvious impact on the bending moments along the pile, as well as the pile lateral deformation. The axial load applied on the pile head has a negligible impact.
(2) Design charts with regard to the determinations of the modification factors from the key influential factors on pile lateral deformation and the bending moment are proposed.
(3) Synthetic cases are numerically formulated to validate the proposed design charts and the FE results are consistent with the estimations based on values of the base cases and modification factors from the charts.
(4) Real case history data confirmed the feasibility and accuracy of the proposed design charts, based on comparisons between instrumentations and estimations from design charts and base case, providing design guidelines for preliminary estimations of pile lateral deformation and bending moment capacity evaluations.

It should be emphasized that the applicability of the proposed design charts is limited to certain soil types and support schemes. It should also be noted that the measured response is from a single passive pile. Therefore, the instrumented performances of the pile groups are more desirable.

9.4 Influence of Deep Braced Excavation on Adjacent Tunnel

9.4.1 Introduction

With development of urban subway and underground infrastructures, the impact of deep excavation on adjacent existing tunnels has become a hot topic for geotechnical design engineers. Table 9.9 lists some cases with existing tunnels affected by adjacent excavations.

In this section, FEM is adopted to carry out numerical calculation. The influence of excavation on additional bending moment and displacement of tunnel at the base or side of excavation is analyzed. The influential factors including the relative spatial position relationship, and stiffness ratio of tunnel excavation on the interaction among tunnel, soil, and diaphragm wall is discussed.

9.4.2 Numerical Model and Validation

Zheng et al. [24] reported a case study in which the displacement and deformation of an existing shield tunnel occur near a braced excavation in Tianjin, and proposed a simplified prediction model. The project is for building of a three-storey basement, located at the intersection of two main roads. The shape of the excavation is of irregular polygon. The surrounding conditions of the project are complex. There are three roads in the east, west, and north of the excavation, respectively. The existing shield tunnel is located in the south of the excavation. The nearest distance between the outer lining of the tunnel and the edge of the retaining structure of the excavation is only 16.5 m.

Zheng et al. [24] compared the field instrumentation results with the FEM numerical calculations, which shows that the model can well reflect the deformation of the retaining structure of the excavation and the existing tunnel outside the excavation.

However, the influence of excavation on existing tunnels at the bottom of the excavation is not considered in Zheng et al. [24]. Based on that, the deformation of tunnels in different locations relative to the braced excavation can be further revealed. The plan of excavation is shown in Fig. 9.44.

Based on the simplified numerical model, the calculations and monitoring data are compared as shown in Figs. 9.45 and 9.46. The parameters of the simplified model are detailed in Sect. 9.4.2. Comparisons between the test results and the FEM simulations indicate that the numerical model adopted in this section can well reflect the deformation of retaining wall and the adjacent existing tunnels.

Table 9.9 Summary of case histories of excavations adjacent to existing tunnels (modified after [24])

No.	References	Excavation depth /m	Maximum deflection of diaphragm wall /mm	Data about subway tunnel			Maximum deformation of tunnel①	
				Tunnel type	Top depth /m	Distance to excavation /m	Maximum deflection /mm	Maximum vertical displacement /mm
1	[25]	About 16	–	Shield	14.4	2.8–5	13	−5
2	[26]	21	53	Shield	14.5	6.9	27	33
3	[27]	15	–	Shield	17	18	6.0/4.0③	3.8/2.1
4	[28]	14.4	–	Shield	9.3	7–11	10	−10
5	[28]	12.5	14.2	Shield	7.4	3	9	−5
6	[29]	6–6.7	20.7	Shield	11.8	5.45	4	7.1
7	[30]	–	–	Shield	–	–	3.5	5
8	[31]	12	–	Shield	7	6	8	−12.3
9	[32]	23.35	–	Shield	8.5	15	3.2	About −1.3
10	[33]	18.02	–	Shield	8.5	5.4	–	−16.67
11	[34]	15.9	54	Shield	16.1	7.3	50(21.5)②	(−21.4)
12	[34]	Groove of diaphragm wall	–	Shield	12.9	3	−17	(−2)
13	[35]	About 14	22	Shield	8.5	10.4–13.5	–	5.5/−3.5
14	[35]	About 10	–	Shield	7	7.2	4.8	−3.3

(continued)

Table 9.9 (continued)

No.	References	Excavation depth /m	Maximum deflection of diaphragm wall /mm	Data about subway tunnel			Maximum deformation of tunnel①	
				Tunnel type	Top depth /m	Distance to excavation /m	Maximum deflection /mm	Maximum vertical displacement /mm
15	[36]	19.9	–	Shield	11	25	–	−5.54
16	[37]	12.2	–	Shield	11.4–12.7	9.5–14	1.05	6.5
17	[38]	About 13	19.5	Shield	7.3	8.1	1.52	−4.11
18	[39]	8.2	–	Shield	15	3.6	3.2/1.9	3.1/1.6
19	[40]	7	–	Shield	10	4	1.4	6.8
20	[41]	21	–	Mining	14.5	5.7–6.6	6.9	–
21	[42]	16.9	–	Shield	9.9	8.4	8.5	–
22	[43]	18	–	Box	11	14.8	5.8	−6.8
23	[44]	13.85	6	Shield	8.37	8	–	5
24	[24]	15	50	Shield	15	16.6	14.56	9.09
25	[45]	14.2–16.4	–	Shield	[0.5]④	11.4	28.18	24.18
26	[46]	14.2–16.4	–	Shield	[0.5]	22.8	11.23	7.3
27	[47]	17.1	–	Shield	13	–	(9.2)	5.98
28	[47]	11.4	–	Shield	9–15	7.2–13	33.5 (21.9)	25.3 (16)
29	[48]	11.4–18.6	108.5	Shield	14.6–21.5	9.7	9.0	–
30	[49]	4.5	–	Shield	[9.1–15.0]	3.2–6.7	9.7	31.4
31	[50]	16.25–17.55	50	Shield	15	16.5	14.56	9.09

Note ①In the horizontal direction, positive value represents the displacement of tunnel toward excavation. In the vertical direction, positive value represents basal heave. ②Numerical values in parentheses indicate horizontal and vertical deformation of tunnel, positive value represents tension, negative value represents compression; ③Deformation value of double-track tunnel, the former is the adjacent line, and the correlation distance refers to the distance between the adjacent line and excavation. ④The numerical value in brackets indicates the distance between the top of tunnel at the base and final excavation surface

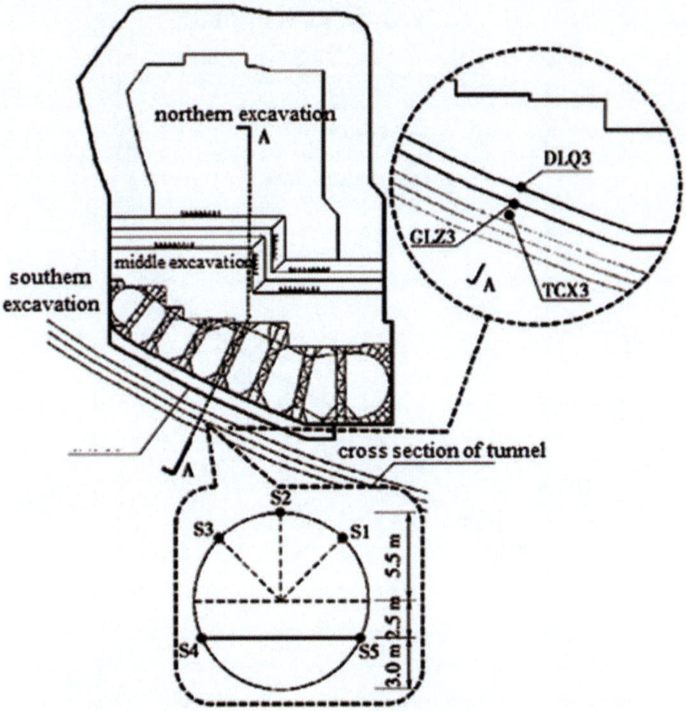

Fig. 9.44 Layout of excavations

9.4.3 Simplified Model

9.4.3.1 Assumptions

In this section, the assumptions are as follows:

(1) The mechanical deformations of soil obey the hardening soil with small strain considerations (HSS), while the behaviors of the diaphragm walls and tunnels obey the linear elastic assumption.
(2) The soil is of normal consolidation state before excavation.
(3) The initial groundwater level is 2 m below the ground surface, and the influence of rainfall on excavation and tunnel is marginal.
(4) The soil layer is homogeneous, with only silty clay considered.

9.4.3.2 Parameters

As shown in Fig. 9.47, FE analysis is carried out via PLAXIS. Considering symmetry, only the right half of model is adopted to conduct calculation. The width and height of model are 130 m and 60 m, respectively. The excavation depth (H_e) considered

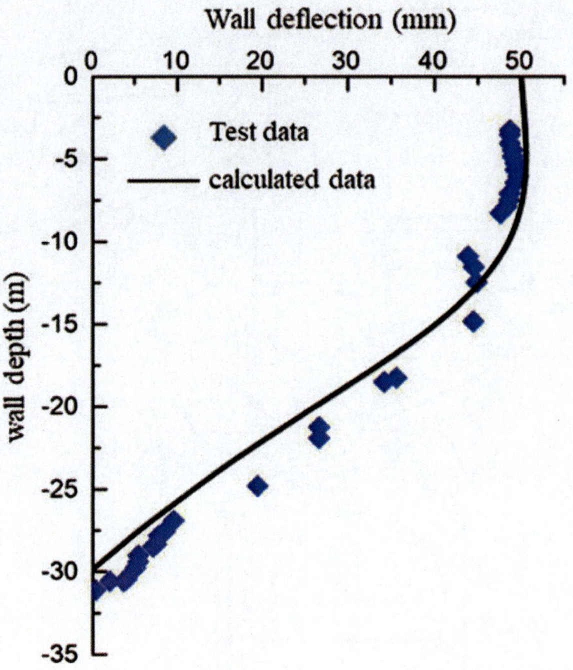

Fig. 9.45 Comparison between numerical and measured lateral wall displacements

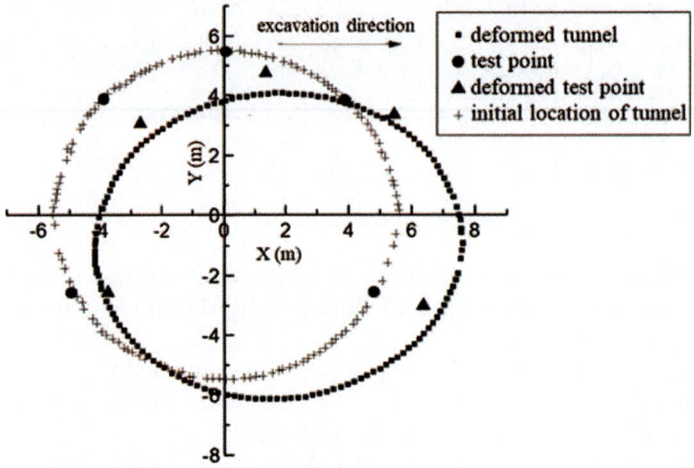

Fig. 9.46 Comparison between numerical and measured tunnel deformation after excavation

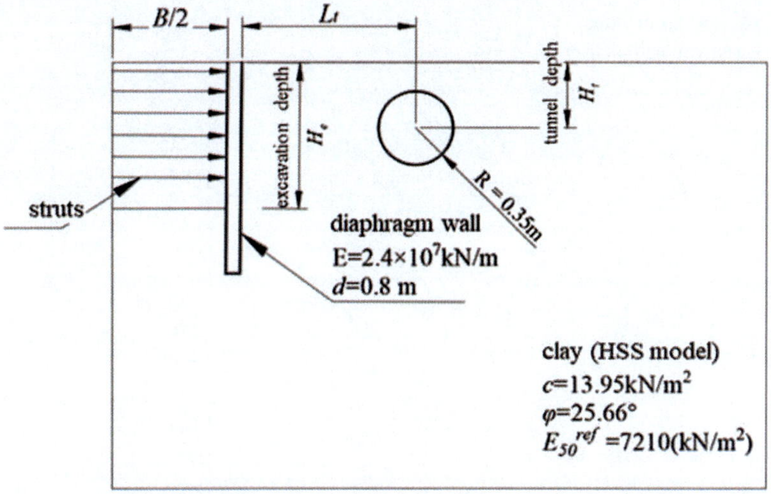

Fig. 9.47 A2D finite-element numerical model

includered is 11, 15, 20 m, respectively. The thickness and length of diaphragm wall are 0.8 m and 30 m, respectively. The inner diameter of tunnel is 3 m.

Fifteen-node triangular element is adopted to simulate soil while five-node plate element and three-node bar element are adopted to simulate diaphragm wall and strut, respectively. Interface element is set between enclosure structure and soil to simulate the interaction between soil and structure [51].

The boundary conditions of the model include 4 parts:

(1) vertical boundary conditions which restrict horizontal displacement on the vertical boundary;
(2) bottom boundary conditions which restrict displacement in three directions at the bottom;
(3) the upper surface is free to any movement;
(4) the hydraulic condition is set as the global water level.

In Fig. 9.47, L_t is defined as the horizontal distance between the center of the tunnel and the retaining wall (the origin is the point of diaphragm wall on the ground surface; the left is negative and the right is positive). H_t is defined as the vertical distance between the center of the near tunnel and the final surface (the lower part is positive), and H_e is defined as the final depth of the excavation.

The detailed soil strength parameters adopted in the model are summarized in Table 9.10 and the support structure strength parameters are listed in Table 9.11. The existing tunnel has operated normally during the process of excavation. The specific construction procedure is shown in Table 9.12.

Table 9.10 Soil parameters used for numerical calculation

Parameters	Input
γ (kN/m^3)	19.78
e	0.699
$E_{50}{}^{ref}$ (kN/m^2)	7210
$E_{oed}{}^{ref}$ (kN/m^2)	5055
$E_{ur}{}^{ref}$ (kN/m^2)	36,770
m	0.8
$K_0{}^{nc}$	0.567
c (kN/m^2)	13.95
φ (°)	25.66
$\gamma_{0.7}$	0.0002
G_0 (kN/m^2)	147,100
R_f	0.9
p^{ref} (kN/m^2)	100
R_{inter}	0.65

Table 9.11 Parameters used for structural elements

Structure element	Parameters and unit	Value
Diaphragm wall	EA (kN/m)	2.4×10^7
	d_w (m)	0.8
tunnel	EA (kN/m)	9.06×10^6
	d_t (m)	0.35
First strut	EA (kN/m)	2.0×10^5
	preload (kN)	0
Second strut	EA (kN/m)	1.4×10^6
	preload (kN)	0
Third struts	EA (kN/m)	1.4×10^6
	preload (kN)	0
Fourth struts	EA (kN/m)	1.4×10^6
	preload (kN)	0

9.4.4 *Numerical Results and Analysis*

9.4.4.1 Displacement with Different Locations in the Tunnel Cross-Section

Figure 9.48 shows the deflection of four key points on the cross-section of tunnel as the excavation depth increases. The four points are left (B), upper (C), right (E), and lower (D), respectively. With increase of excavation depth, some differences with regard to the deformation rate occur. The displacement rate of right (E) and lower

Table 9.12 Construction procedures

Step	Introduction
Initial phase	Initial effective stress, pore pressure, and state parameters are generated to balance crustal stress
Phase 1	Reset initial displacement to zero, and the interaction between soil and tunnel is activated
Phase 2	Reset initial displacement to zero, and the interaction between soil and diaphragm wall is activated
Phase 3	Set the underwater level to −2 m by drainage
Phase 4	Excavate to depth of −2 m
Phase 5	Install first struts
Phase 6	Set the underwater level to −7 m by drainage
Phase 7	Excavate to depth of −7 m
Phase 8	Install second struts
Phase 9	Set the underwater level to −11 m by drainage
Phase 10	Excavate to depth of −11 m
Phase 11	Install third struts
Phase 12	Set the underwater level to −15 m by drainage
Phase 13	Excavate to depth of −15 m
Phase 14	Install fourth struts
Phase 15	Set the underwater level to −20 m by drainage
Phase 16	Excavate to depth of −20 m

(D) vertices increases, while that of left (B) and upper (C) vertices increases first and then decreases. Figure 9.49 shows the deformation of tunnels that the horizontal distance from retaining wall is 6 m, tunnel depth is 15 m, and lining thickness is 0.45 m at different excavation depths. When the excavation depth is less than 7 m, the deformation of the tunnel is less obvious. However, when the excavation depth reaches 15 m, the horizontal deformation is greater than the vertical deformation. When the excavation depth reaches 20 m, the tunnel obviously moves toward the bottom of the excavation and was subjected to large deformations.

9.4.4.2 Comparison of Deflection Between Tunnel and Diaphragm Wall at Different Depths

The deflection of diaphragm wall and tunnel displacement at different tunnel depths is shown in Fig. 9.50, at the final excavation depth (H_e) of 20 m. According to Fig. 9.50, with increase of excavation depth, the deflection of diaphragm wall increases. At the final excavation depth, the deflection reaches the maximum value, and then gradually decreases due to the passive earth pressure in front of diaphragm wall.

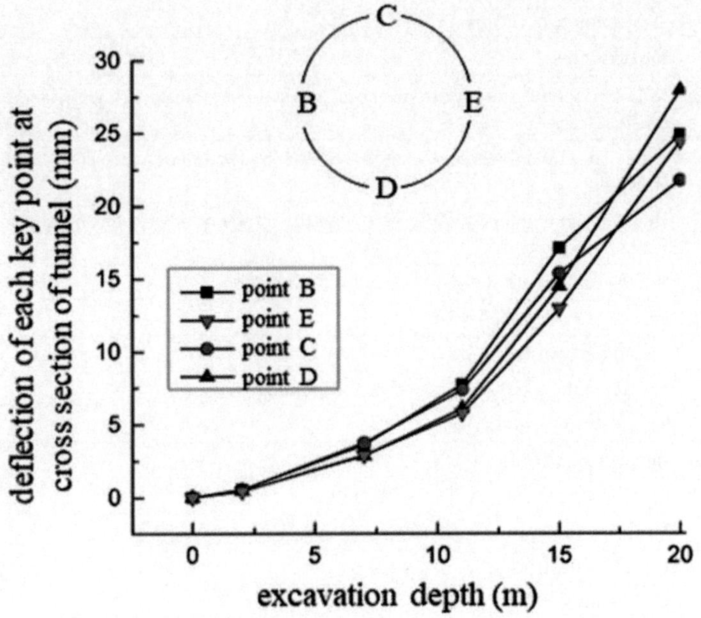

Fig. 9.48 Lateral displacement for key points of tunnels for different excavation depths

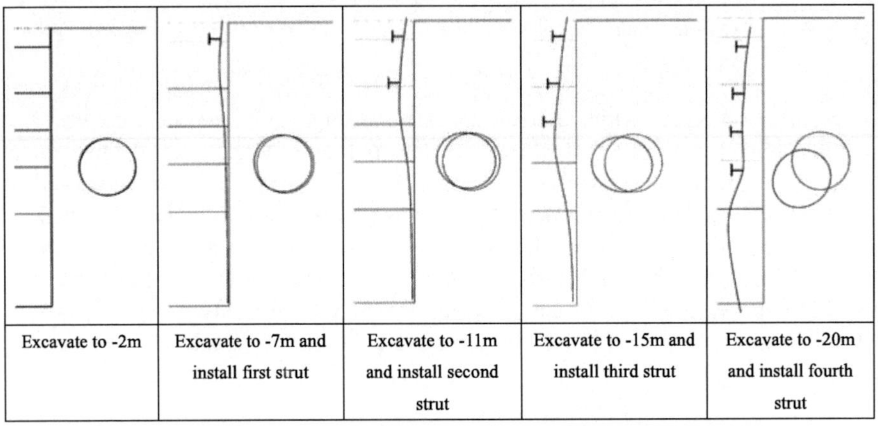

Fig. 9.49 Schematic diagram of structure deformation for different excavation depths

The deflection of retaining wall is restrained at the strut position. When the depth is less than 15 m (3/4 of H_e and location of the fourth strut), the existence of tunnel has little effect on the deflection of the wall. The horizontal displacement of the wall and tunnel is almost the same, similar to translation; when the depth is greater than 15 m, the deflection of the tunnel is obviously smaller than the diaphragm wall, which reflects the restraint effect of the soil below the base on the deflection of tunnel. When

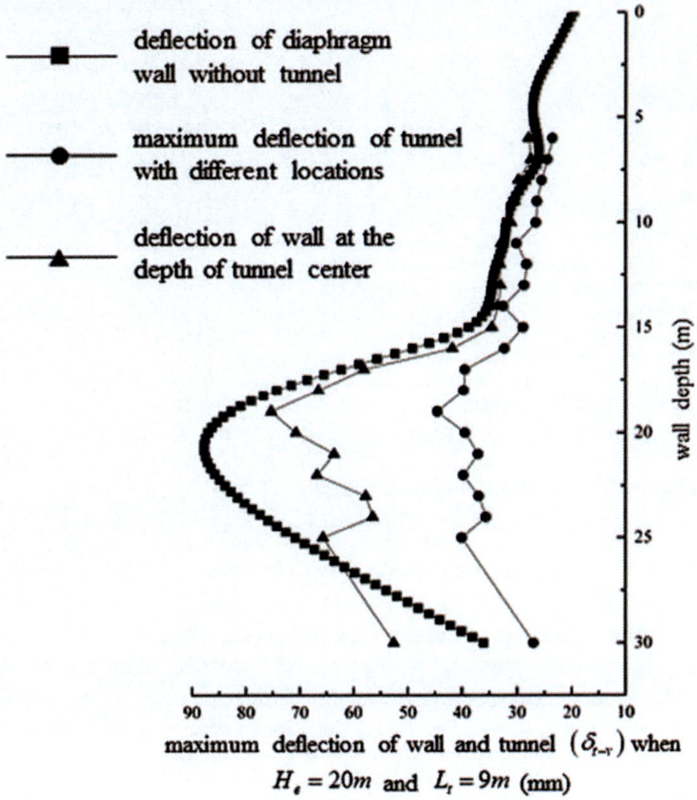

Fig. 9.50 Comparison of lateral wall and tunnel displacement for different tunnel depths

the tunnel is located at the depth of the toe of wall, the wall deflection is greater than the corresponding case without tunnel, which facilitates the deflection of the toe of wall. It can be seen that the interaction between tunnel, soil, and retaining wall is complex under excavation, which is significantly affected by the relative position of tunnel and diaphragm wall.

9.4.4.3 Influence of Excavation on the Tunnel Below the FEL

Figure 9.51 shows the influence of unloading of excavation on displacement of tunnel below the final elevation level (FEL) when H_e equals 20 m. H_b is defined as the vertical distance between the center of tunnel and the final excavation surface (the lower part is positive). In addition, Fig. 9.51 shows the maximum deflection of tunnel when L_t equals −5 m, −10 m, −15 m, −20 m, −25 m, and H_b equals 5 m, 7.5 m, 10 m, 15 m, 20 m, 25 m.

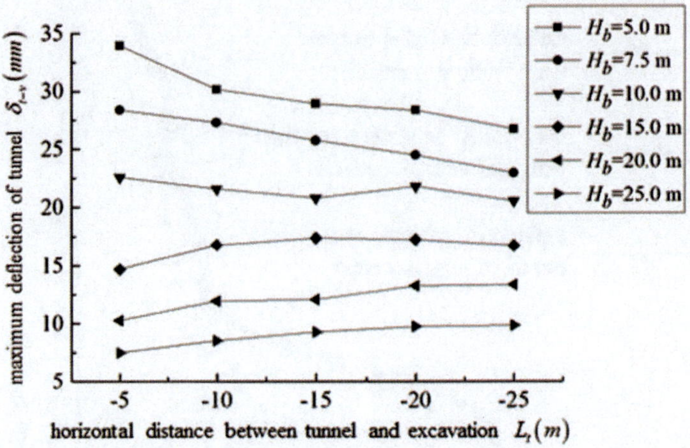

Fig. 9.51 Influence of different Lt and Hb on max. lateral tunnel displacement

From Fig. 9.51, it can be concluded that with increase of depth, the deflection of tunnel decreases. On the one hand, above the depth of diaphragm wall, the larger L_t is, the smaller the deflection of tunnel is. On the other hand, below the toe of the diaphragm wall, when H_b is greater than 15 m, the deflection of tunnel increases with the increase of L_t as the deflection of tunnel is almost unaffected by the diaphragm wall. At the same time, the larger L_t is, the closer tunnel is to the center of the base. Therefore, there is a significant basal heave, causing the increase of tunnel displacement.

9.4.4.4 Influence of Excavation on the Tunnel Beside the Excavation

In order to study the influence of excavation on the deflection of tunnel beside the excavation with different depths, the maximum deflection of tunnel with different depths is obtained for H_e of 20 m, L_t of 6 m and H_t ranging from 6 to 30 m. Figure 9.52 plots the relationship between the tunnel displacement and the wall deflection for different wall depths.

The deformation patterns of tunnel and wall are basically the same. When H_e equals 20 m, the deflection of wall and tunnel reaches the maximum simultaneously. When the wall depth is greater than 12 m and less than 16 m, the deflection of tunnel is greater than the wall at the same depth; when the tunnel depth is less than 12 m, the deflection of tunnel and wall are approximately equal, which is similar to translational displacement. When the tunnel depth is more than 16 m, the deflection of wall is obviously greater than the tunnel at the corresponding position. It is obvious that the interaction between tunnel, soil, and diaphragm wall is complex, significantly affected by the relative position of tunnel and diaphragm wall.

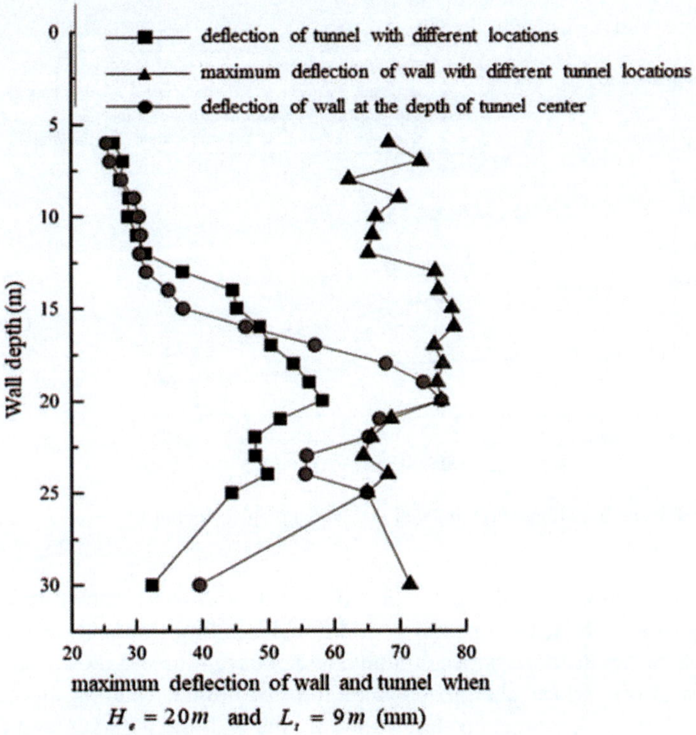

Fig. 9.52 Maximum lateral displacement of tunnel and wall along the wall length

9.4.4.5 Influence of Excavation on the Additional Bending Moment and Displacement of Tunnel

In order to systematically analyze the influence of excavation on additional bending moment and displacement of surrounding existing tunnels, the cases with different embedding depths and different horizontal distances (H_e = 15 m) are analyzed, respectively. The specific parameters of location are shown in Table 9.13.

Table 9.13 Ranges of position parameters L_t, H_b and H_e

Location	Symbol	Value
Beside wall	L_t(m)	6, 12, 18, 24, 30
	H_t (m)	7.5, 15, 22.5, 30
	Including 20 cases (5 × 4), as shown in Fig. 9.10	
At the base	L_t (m)	−10, −20
	H_b (m)	7.5, 15
	Including 4 cases (2 × 2), as shown in Fig. 9.10	

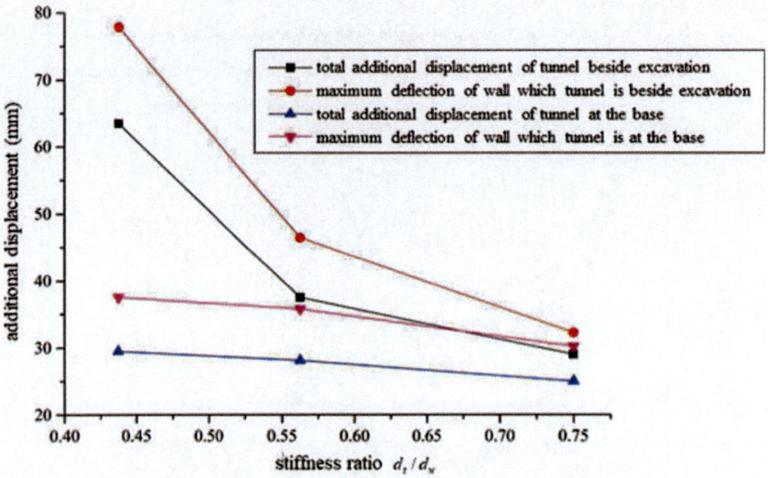

Fig. 9.53 Influence of stiffness ratio on wall and tunnel displacement (mm)

The bending moments of tunnels at different locations are shown in Fig. 9.53. According to Fig. 9.53, the maximum additional bending moment decreases with the increase of the distance between tunnel and excavation, and the closer tunnel is to the excavation surface, the greater maximum additional bending moment is. In addition, the bending moment of tunnel around the wall toe is larger, and the stress concentration occurs. The maximum negative bending moment of tunnel beside excavation is generally in horizontal direction, and the maximum positive bending moment is in vertical direction. On the contrary, the maximum negative bending moment of tunnel beside the wall is toward the direction of maximum deflection of diaphragm wall, while the maximum negative bending moment of tunnel around wall toe is toward the direction center of the base. The tunnel near the wall toe has almost no rotation, considering that the tunnel is balanced by the influence of two opposite directions of the basal heave and wall deflection. Figure 9.54 shows the deformation of tunnel. The maximum diameter of tunnel around the wall toe changes in the vertical and horizontal direction. The vertical direction is mainly of the extension deformation and the horizontal is of compression deformation, in contrast to the tunnel beside the wall. The rotation angle of tunnel beside wall is affected by the wall deflection. The smaller L_t is, the larger the rotation angle is. Additional bending moment plays an important role in assessing the safety of tunnel structure.

The distribution of additional displacement vectors of tunnels around the excavation is shown in Fig. 9.54. According to Fig. 9.54, the horizontal additional displacement of the tunnel beside the excavation is generally larger than the vertical additional displacement. The reason is that the deflection of the soil behind the wall caused by the deflection of diaphragm wall, which leads to the larger deflection of tunnel. The additional displacement of tunnel decreases with the increase of L_t. In addition, with increase of H_t, the horizontal additional displacement increases first and then

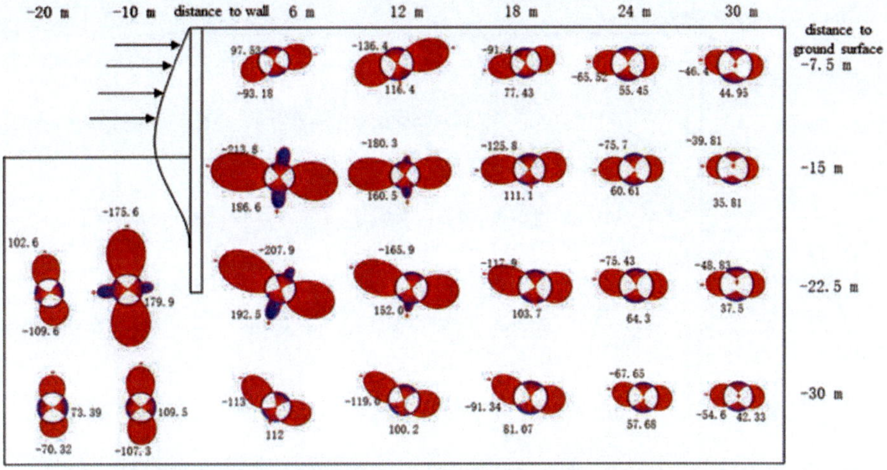

Fig. 9.54 Bending moment diagram of tunnels at different locations (kN·m/m)

decreases, and reaches the maximum at the final excavation face ($H_e = 15$ m), while the vertical additional displacement of tunnel decreases monotonously with increase of H_t. However, the direction of additional deformation of tunnel at the base is upper and toward to side excavation surface due to the basal heave of soil.

9.4.4.6 Influence of Relative Stiffness

To investigate the influence of relative stiffness, the wall stiffness is set a constant (the wall thickness $d_w = 0.8$ m) while the thickness of lining aries. The common tunnel lining thickness d_t equals 0.35, 0.45, and 0.6 m, respectively, when H_e equals 20 m. The stiffness ratio d_t/d_w is 0.4375, 0.5625, and 0.75, respectively. For the tunnel beside the wall, L_t equals 6 m and H_t equals 15 m. And for the tunnel at the base, L_t equals −6 m and H_b equals 7.5 m. The other parameters are identical except the stiffness ratio. As shown in Fig. 9.55, the increase of stiffness ratio can restrain the deflection of diaphragm wall and the additional displacement of tunnel. For the tunnel beside the wall, the increase of stiffness ratio can significantly reduce the deflection of diaphragm wall and the additional displacement of tunnel. Nevertheless, for the tunnel at the base, the stiffness ratio has minimal influence on the deflection of diaphragm wall and the additional displacement of tunnel.

9.4.5 Summary

This section examines the interaction among soil, tunnel, and diaphragm wall due to the excavation by numerical simulation. Based on the calculation results, the

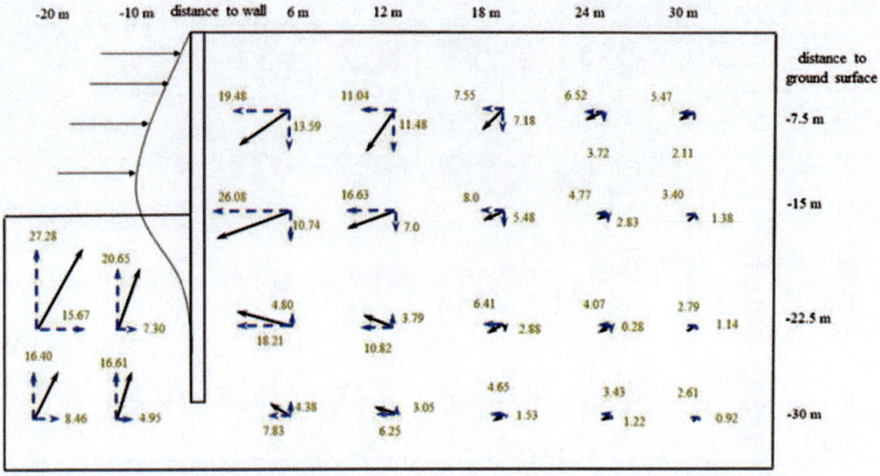

Fig. 9.55 Vector diagram of maximum Tunnel displacement at different locations (mm)

distributions of additional bending moment and displacement of tunnel with different locations were investigated, and the influence of stiffness ratio on the displacement of tunnel-excavation system was examined. The main conclusions are as follows:

(1) When the tunnel depth is greater than excavation depth while smaller than the wall depth, the greater L_t is, the smaller the deflection of tunnel is. When the tunnel depth is larger than excavation depth and wall depth, the deflection of tunnel increases with the L_t.
(2) The deflection of wall and tunnel reaches the maximum at the final excavation depth simultaneously. Meanwhile, the wall deflection at the depth of tunnel center is the maximum.
(3) The bending moment of tunnel around the wall toe is large, which leads to the stress concentration. The additional horizontal displacement of tunnel beside wall is generally larger than the vertical part.
(4) The total additional displacement of the tunnel decreases with increase of L_t. However, with increase of H_t, the additional horizontal displacement of tunnel increases first and then decreases, and finally reaches the maximum at the final excavation depth. On the contrary, the vertical additional displacement of tunnel decreases monotonously with the increase of H_t.
(5) Increasing stiffness ratio can be adopted to restrain deflection of diaphragm wall and additional displacement of tunnel.

9.5 Influence of Deep Braced Excavation on Adjacent Ground Structures

To evaluate the behavior of ground structures induced by adjacent braced excavation, a case study-based numerical analysis is presented. In the complex soil stress state encountered in the braced excavation, different stress states lead to diversion of principal stress directions. The stress-induced anisotropy is introduced to represent this soil characteristic, which is of great significance in assessing the excavation safety as well as its impact on the built environment [52]. This study reviewed a well-documented case history via Finite-Element analysis, adopting an advanced soil constitutive model NGI-ADP to simulate the stress-induced anisotropy of clay. The effects of various site soil conditions, especially the soil anisotropic degree, were explored to reveal the characteristics of adjacent infrastructures performances. Numerical modeling.

9.5.1 Numerical Modeling and Validation

9.5.1.1 Case Details

This study refers to a well-documented excavation project, the excavation of Sukhumvit Station. It is a Bangkok Mass Rapid Transit Underground Railway station, located in a building-intensive area. Surrounding infrastructures include 10 low-rise buildings (3–4 storeys) and 1 tall building with 45 stories as shown in Fig. 9.56. Geometries of the excavation is 200 m in length and 23 m in width, and the maximum excavation depth reaches GL-20.9 m, according to the plan view of the excavation site. The field monitoring system consists of 8 inclinometers and a series of surface settlement array, whose locations are marked in Fig. 9.56.

Geological conditions of the excavation site can be simplified as alternating horizontal layers of sand and clay, with an overlying made ground layer. The subsoil distribution and sectional view of the excavation project are shown in Fig. 9.57.

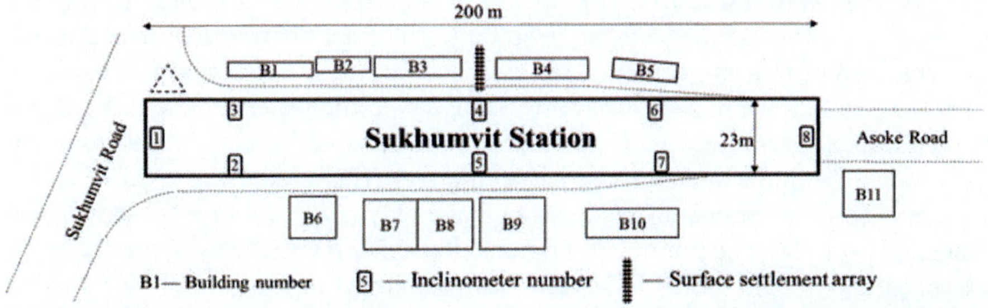

Fig. 9.56 Site conditions and instrumentations

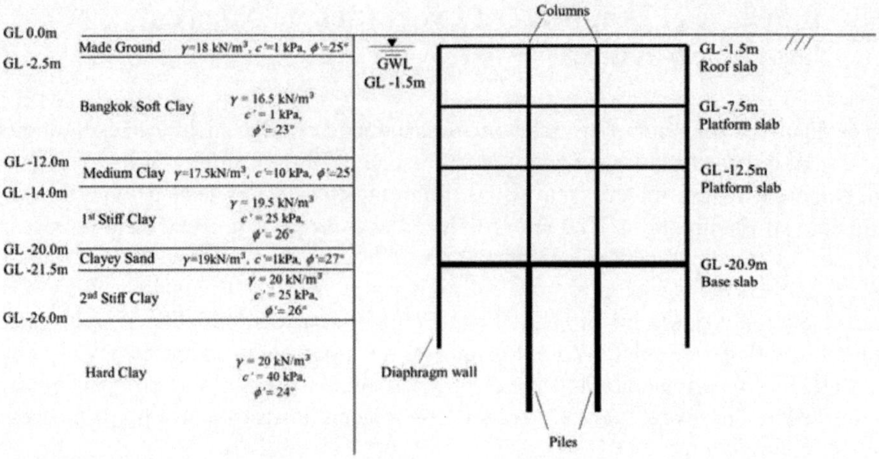

Fig. 9.57 Soil profile and station geometries

The clay layers cover dominant volumes behind the retaining wall. The retaining structures includes the diaphragm wall, temporary steel struts, platform slabs and base slab of the 3-storey underground structure, the pre-excavation bored piles and counterpart columns. The retaining wall reached the depth of GL-27.9 m, and the bored piles were penetrated into the depth of GL-34.0 m. The top-down method was employed to construct the underground station. The construction sequences included multiple staged excavation and structural member installation.

9.5.1.2 Numerical Model Establishment

Extensive analyses were conducted for the Sukhumvit excavation using finite-element software Plaxis. 3D model was established centered with the station pit and surrounding buildings. The model size was determined to be 300 m × 100 m × 50 m (x–y-z axes) was used for the analysis. The horizontal movement along the vertical boundaries was restrained while both the horizontal and vertical movements at the bottom were restrained. The vertical boundaries were extended far from the excavation to minimize the boundary restraints. The details of the groundwater draw-down measures of the project remain unclear according to related reports. Therefore, it is assumed that before each excavation step, the groundwater level was dewatered below the excavation surface, and the outside groundwater level remained constant.

For simulating the site soil, the hardening soil (HS) model was used to model the stress–strain responses of made ground and clayey sand. As for the clay layers (Bangkok soft clay, medium clay, first and second layer stiff clay, hard clay), the advanced anisotropic model NGI-ADP was employed to represent clay behavior. The NGI-ADP model was presented by the Norwegian Geotechnical Institute (NGI) and cored with the ADP concept. It is an anisotropic shear strength model where

Table 9.14 Input parameters of NGI-ADP model

Type	Parameter	Description	Unit
Stiffness	$G_u/{s_u}^A$	Ratio unloading/reloading shear modulus over (plane strain) active shear strength	–
	${\gamma_f}^C$	Shear strain at failure in triaxial compression	%
	${\gamma_f}^E$	Shear strain at failure in triaxial extension	%
	${\gamma_f}^{DSS}$	Shear strain at failure in direct simple shear	%
	γ_0	Poisson's ratio	–
Strength	${s_u}^A$	Reference active shear strength	kPa
	${s_u}^P/{s_u}^A$	Ratio of passive shear strength over active shear strength	–
	${s_u}^{DSS}/{s_u}^A$	Ratio of direct simple shear strength over active shear strength	–
Physical	γ	Unit weight	kN/m^3

a nonlinear stress path-dependent hardening relationship is adopted. The employed yield criterion is based on a translated approximated Tresca criterion. As for its formulation, the clay stress paths under active (A), direct simple shear (DSS), and passive (P) loadings are distinguished. The anisotropic stress–strain relation under three different loading modes are derived from the results of triaxial compression (TXC), direct simple shear (DSS), and triaxial extension (TXE) tests. Six independent parameters from the three different stress paths, including undrained shear strengths ${s_u}^{\mathrm{A}}$, ${s_u}^{\mathrm{DSS}}$, ${s_u}^{\mathrm{P}}$, and failure strains for triaxial compression ${\gamma_f}^{\mathrm{C}}$, direct simple shear ${\gamma_f}^{\mathrm{DSS}}$ and triaxial extension ${\gamma_f}^{\mathrm{E}}$, are defined as input to represent the strength and stiffness anisotropy. The model uses elliptical interpolation for plastic failure strains and shear strengths in arbitrary stress paths according to the above-mentioned six parameters. Dominant parameters defining soil anisotropy in the NGI-ADP model and their counterpart symbols are listed in Table 9.14. The soil properties and corresponding values according to two different constitutive models are listed in Table 9.15.

The most important input parameter in the NGI-ADP model to quantify the anisotropy degree of clay is the ${s_u}^P/{s_u}^A$ (i.e., the s_u ratio). It represents the undrained shear strength under passive loading state to that of active state, ranging from 0 to 1, value of 1 represents the ideal isotropic condition. The ${s_u}^{DSS}/{s_u}^A$ value is empirically calculated as follows:

$$\frac{s_u^{DSS}}{s_u^A} = \left(\frac{s_u^P}{s_u^A} + 1\right)/2 \tag{9.5}$$

According to previous studies, the s_u ratio falls within the normal range 0.5–0.7 for natural clay. The clay rigidity is characterized using the parameter $G_u/{s_u}^A$, and it stands for the ratio of unloading/reloading shear modulus over (plane strain) the active shear strength.

Table 9.15 Soil parameters used for numerical analyses

Depth (m)	Soil type	γ	NGI-ADP			HS model		
		(kN/m^3)	$s_u{}^A$(kPa)	$s_u{}^P/s_u{}^A$	$G_u/s_u{}^A$	c(kPa)	φ(°)	E_{50}^{ref} (MPa)
0–2.5	Made Ground	18	–	–	–	1	25	45.6
2.5–12	Bangkok soft clay	16.5	36.5	0.5	192	–	–	–
12.0–14	Medium clay	17.5	71.5	0.5	269	–	–	–
14–20	1st stiff clay	19.5	105.6	0.5	400	–	–	
20.0–21.5	Clayey sand	19	–	–	–	1	27	38
21.5–26	1st stiff clay	20	147.3	0.5	400	–	–	–
26–50	Hard clay	20	182.9	0.5	400	–	–	–

The earth retaining structural materials were idealized to be linear elastic. Their geometries and element types in simulation were listed in Table 9.16. According to Surarak et al. [53] the modulus E_c of concrete members was determined to be 28GPa, and E_s= 206GPa for steel members, with Poisson ratio of 0.15 for all structural members.

Additionally, the zero-thickness interfaces between the structures and soil were created to simulate the structure-soil interaction. The strength reduction factor of 0.9 for interface is assigned, characterizing the interface properties (including both strength and stiffness) according to a reduction rate of adjacent soil parameters in this study. Figure 9.58 shows the 3D FE model for Sukhumvit excavation for the Sukhumvit excavation. The simulation of buildings is composed of structural elements (plate, beam, and pile) and uniform loads. Only 10 ground structures were built in the numerical model since the Building-11 locates at the corner of the excavation, and the construction impact on B11 is negligible according to field data.

Table 9.16 Geometry and modeling element of retaining structures

Member	Size	Element type
Diaphragm wall	1 m thick	Plate
Base slab	1.8 m thick	Plate
Platform slab	1 m thick	Plate
Column	φ0.8 m	Beam
Pile	φ0.8 m	Embedded beam
Temporary strut	Area 218.69 cm^2	Beam

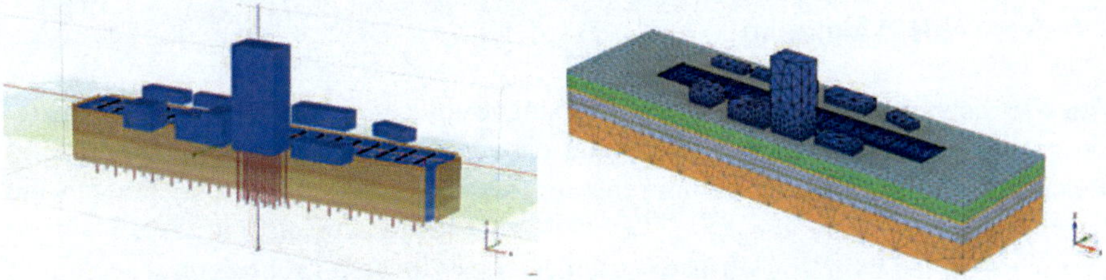

(a) retaining members (b) model mesh

Fig. 9.58 3D FE model for Sukhumvit excavation

According to the construction activities of Sukhumvit Station excavation [54], 11 simulation steps have been established grounded in the actual construction sequences as Table 9.17 lists. Firstly, the buildings in the site are considered to be erected a rather long time ago and in normal service. To focus on the environmental impact of braced excavation, the soil deformation caused by building servicing load and soil consolidation settlement before underground construction was eliminated.

The groundwater condition is shown in Fig. 9.57, and a rather high groundwater level of GL-1.5 m was observed in the excavation site. With the gradual excavation for the underground railway station, the groundwater drawdown was implemented simultaneously. The made ground and sand layers possess high permeability, while the clay layers share extreme low permeability coefficients. It should mention that the clay undrained type allows the possibility of conducting an undrained effective stress analysis, therefore, distinction was made between pore pressures and effective stresses during the FE analysis. The surrounding soil deformation was coupled with the pore pressure variation.

Table 9.17 Numerical simulation process with respect to construction sequences

Stage	Simulation process
1	K_0 consolidation process for a long time interval, supposing that the site buildings were built and in operation
2	Reset displacement of soil and structures; diaphragm wall wish-in-place
3	Bored piles penetration
4	Construction of upper column and fixation with bored pile at the junction point
5	First excavation to GL-1.0 m, activating structural elements of steel props
6	Excavation to GL-2.0 m, construction of roof slab. Removal of temporary prop
7	Excavation to level of GL-8.0 m, construction of second platform slab
8	Excavation to GL-13.0 m, construction of the third platform slab
9	Excavation to GL-20.9 m, construction of the base slab
10	Formation of composite column, replacement of column structural material from steel to concrete encased steel
11	Backfill overlying ground, activation of strip load

9.5.1.3 Model Validation

The significant effect of considering clay anisotropy can be reached compared with the calculated results from the isotropic case and field measurements, as Fig. 4 illustrates. The isotropic condition refers to that the s_u ratio for all clay layers were determined to be 1, and the $s_u^P/s_u^A = 0.5$ was set for the anisotropic case. The field data was derived from the surface settlement array located in the central section of the excavation as indicated in Fig. 9.56.

It can be seen from Fig. 9.59 that the computed trend of ground subsidence is consistent with the actual situation in the certain range (due to the limited monitoring area). The performance of anisotropic soil model can more satisfactorily predict the excavation response in contrast with the common isotropic hypothesis, which indicates that the NGI-ADP anisotropic model is conducive to capture the clay behavior encountered with the complex excavation situation.

As supplement of incomplete ground settlement profiles, the excavation induced maximum wall deflection values at two inclinometer locations were selected for anisotropic model validation. The comparison of computed values and measured values of Inc4 and Inc8 (central locations on both sides of excavation) at four excavation stages is shown in Fig. 9.60. The red dotted line represents the perfect match of anisotropic model predictions with the field data. While the area surrounded by the black dotted line is the allowable relative error range, indicating the computed results with less than 20% relative error are acceptable. Due to the corner effect, the deformation on the short side is substantially smaller than that on the long side, therefore, the values of Inc8 stay in a relative low level. All 8 points fall into the allowable range, showing the satisfactory performance of the numerical model. In general, according to the results shown two Figures, the feasibility and accuracy of

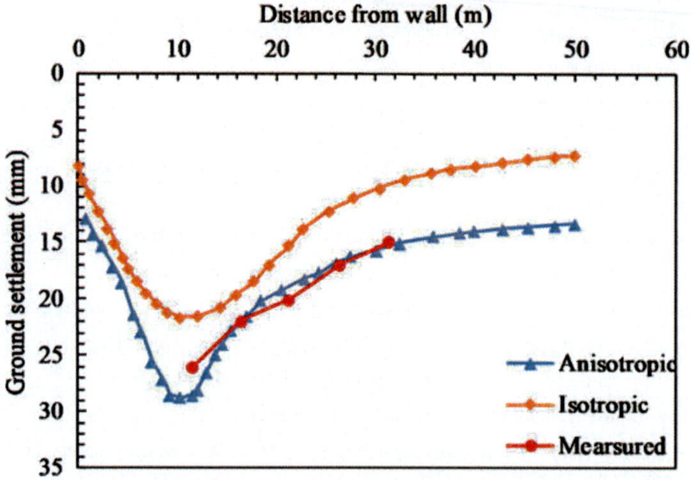

Fig. 9.59 Computed ground settlement profiles with reference to measured data

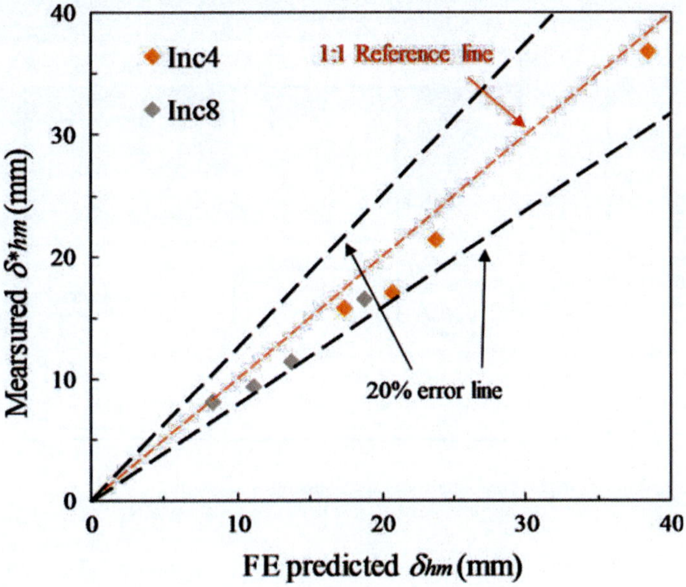

Fig. 9.60 Performances of anisotropic model at various excavation stages

using the advanced constitutive model NGI-ADP is verified. Subsequent parametric analysis adopting this model and relevant parameter can be conducted to solve the excavation induced deformation and adjacent infrastructures' stability problem.

9.5.2 Parametric Analyses

Based on the Sukhumvit Station excavation, parametric studies were carried employing the above-mentioned NGI-ADP model and relevant parameters. The investigated factors include the clay anisotropic degree, strength, and stiffness, aiming at qualitatively exploring the influence of various strength and stiffness properties of the anisotropic clay on the excavation responses.

9.5.2.1 Effects of Anisotropic Degree

Six s_u^P/s_u^A values from 1 to 0.01 were determined, representing that the clay varies from completely isotropic to extremely anisotropic.

Figure 9.61 shows the distribution of soil plastic points in the critical section (B9-B4) after completing the excavation. Three types of dots represent the failure elements under the condition of $s_u^P/s_u^A = 0.01$, 0.6, and 1. Under the multi-influence of building load and excavation unloading, numerous elements in the weak Bangkok

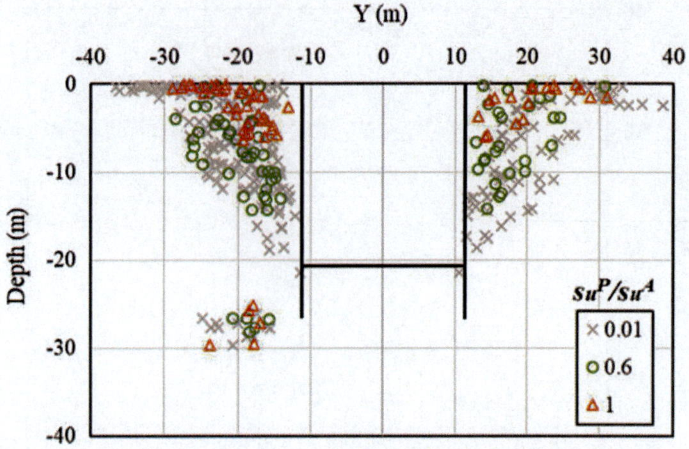

Fig. 9.61 Distribution of plastic points for various anisotropic degree

Soft Clay layer have reached its shear strength or even failed. There are limited numbers of plastic points at the end of pile foundation of the high-rise building (B9). As for the differences caused by clay anisotropic degree, the obvious broadening of the plastic zone can be observed. When all clay was assumed to be isotropic, the plastic zone was limited near the ground level. As the s_u ratio reduces, the plastic zone gradually grows deeper and approaches the excavation wall. It can be read that ignoring the clay anisotropy leads to underestimation of clay stress, clay in certain area which has reached plasticity or failure in the practical construction may become the critical cause of poor design.

Clay in plastic state implies that more deformation occurs under the same load. This leads to the similarity of soil displacement field and plastic zone distribution as shown in Fig. 9.62. With the soil anisotropy getting dominant, the displacement of the soil behind the retaining wall gradually become severe. When the clay was strongly anisotropic, which is not a common situation, the plastic zone can even develop into a continuous failure surface. The variation of the clay anisotropic degree directly results in the change of the clay strength under the non-active stress conditions, the soil elastic or plastic state consequently transforms, which further influences the corresponding deformation and displacement condition.

The excavation changes the stress displacement field of site soil, resulting in vertical and horizontal displacement. Influenced by the complex soil displacement situation, the ground structures, both lying on shallow foundation or supported by deep pile foundation, inevitably suffer varying degrees of distortion and displacement. The ground settlement and uneven settlement of adjacent buildings were chosen for illustrating the results.

It is indicated that the ground displacement gradually becomes severe as the soil turns to anisotropic. The ground settlement increases from 20 to 37 mm when the

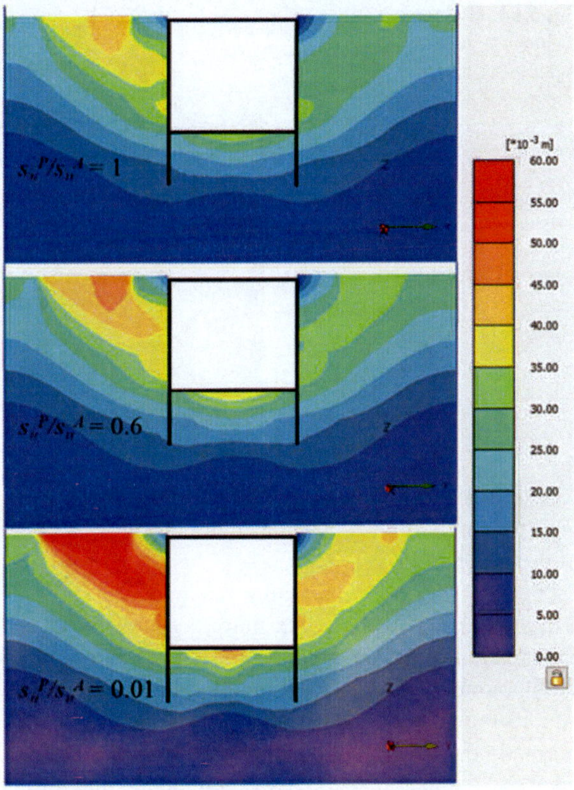

Fig. 9.62 Soil displacement field for various anisotropic degrees

clay varies from total isotropic to reasonable anisotropic. The building displacement shares the same increasing tendency. The reason why building settlement value exceeds ground settlement for all cases is that the two indices were derived from different locations. The ground settlement value is extracted from the central section of long side, as the settlement array lies in Fig. 9.56. While the building foundation, located at another cross-section, suffers vertical loads therefore induces greater displacement.

As for the building inclination, Fig. 9.63 shows the inclination patterns of buildings on both sides of excavation. The most critical section containing a pile-founded high-rise building and a shallow foundation building is sliced for analysis. It is worth mention that the buildings are inclined toward from the excavation. This phenomenon results from the distance between buildings and excavation edge. The maximum ground settlement occurs at the distance of 11 m according to Fig. 9.59, exactly close to the building side (near excavation side). The impact of braced excavation on adjacent buildings is evaluated through maximum settlement and uneven settlement degree. The uneven settlement degree is characterized by

$$i = \Delta / L \tag{9.6}$$

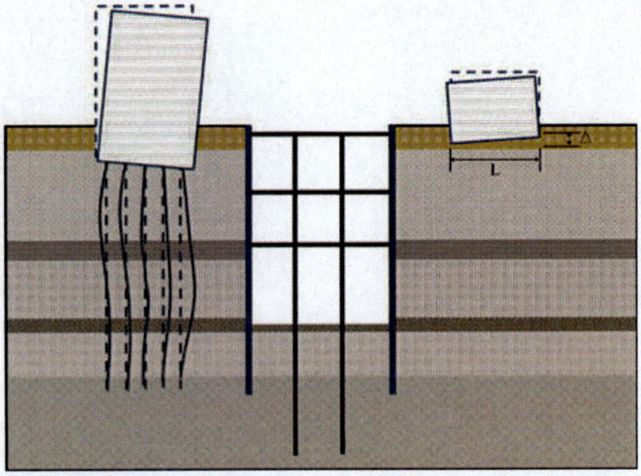

Fig. 9.63 Building inclination induced by excavation

i is the inclination of the building foundation; Δis the differential settlement between the two ends of the foundation, m; L is the distance between the two ends of the foundation, m

Table 9.18 lists the building responses as the clay anisotropic degree changes. Same as above results, maximum settlements of both shallow-founded and pile-founded building increase with the degree of anisotropy. The former one is substantially manipulated by the ground settlement, therefore, it shows higher increment. And its inclination reaches 1.41‰ as the $s_u{}^P/s_u{}^A = 0.01$ turns into 0.01, which is not normally encountered. For the higher building, the settlement stays in a controllable level due to the good behavior of embedded pile foundation. According to computed results in Table 9.18, when the soil presents anisotropic state, the deformation and inclination of both shallow-founded buildings and high-rise buildings are generally not critical to threaten its stability and normal serviceability.

Table 9.18 Adjacent buildings settlements for varying clay anisotropic degrees

$s_u{}^P/s_u{}^A$	B9		B4	
	Max settlement (mm)	Inclination degree(‰)	Max settlement (mm)	Inclination degree(‰)
1	20.89	0.92	25.89	0.71
0.8	24.94	0.97	35.94	0.83
0.6	26.21	1.24	40.21	0.97
0.4	27.64	1.34	45.64	0.98
0.2	28.24	1.25	57.24	1.39
0.01	30.47	1.2	59.47	1.41

9.5.2.2 Effects of Soil Strength

Considering the ground clay are normally anisotropic, i.e., $s_u^P/s_u^A = 0.6$, this section aims at discussing the development discrepancy of excavation consequences when the clays are characterized with various undrained shear strength. Three indices including ground movement, building displacement, as well as the deflection of retaining structures are the focus of excavation responses.

An index named "soil strength variation ratio" N_{su} is defined, representing the ratio of clay active undrained shear strength, i.e., s_u^A in the current analysis to that of the referencing Sukhumvit case. The N_{su} value greater than the unity indicates that the strength soft clay in this analysis is greater than the measured values in the practical project. Note that the clay undrained shear strengths under passive and direct simple shear state are also proportional varied according to the user-defined parameters, i.e., s_u^P/s_u^A and s_u^{DSS}/s_u^A. The clay shear modulus was set to be constant, as the parameter G_u/s_u^{An} changes in inverse proportion. The N_{su} of 0.5, 0.8, 1.2, and 1.5 are determined for this analysis.

In the case of constant stiffness, the soil strength directly determines its deformation. With the decrease of soil strength, a large number of soil elements reach the plastic stage in response of excavation unloading. Therefore, the displacement of surrounding soil generally increases according to Fig. 9.64. It is worth noticing that the displacement in the clay layer beneath the building grows sharply. One of the reasons is that clay in the location suffers part of the building load. On the other hand, influenced by the excavation effect, the passive loading state of the clay near the ground surface is more obvious. According to the above discussion on clay anisotropy, the significance of considering the strength reduction of passive loading soil is illustrated. Therefore, for excavations in soft clay with low shear strength, large deformation may occur in the clay near the ground surface, it should be alert incase of a continuous sliding surface.

The maximum ground settlement and counterpart building responses under the condition of various clay strengths are listed in Table 9.19. Without vertical load, the maximum value of ground subsidence nearly doubled as the N_{su} of all clay layers decreases from 1.5 to 0.5, while it keeps in the normal range. The weakening of clay strength shows little effect on the pile foundation embedded in deep hard clay layer. Excavation induced pile lateral displacement causes a small amount of building subsidence increment. While the building supported in the weak Bangkok Soft Clay near the ground surface shows large settlement and inclination when the $N_{su} < 1$. The inclination degree of 3.52‰ is encountered when the subsoil strength is relatively low. Under such circumstances, the bearing capacity of the foundation decreases, and crack can be induced on the building walls, the normal serviceability of the building cannot be guaranteed.

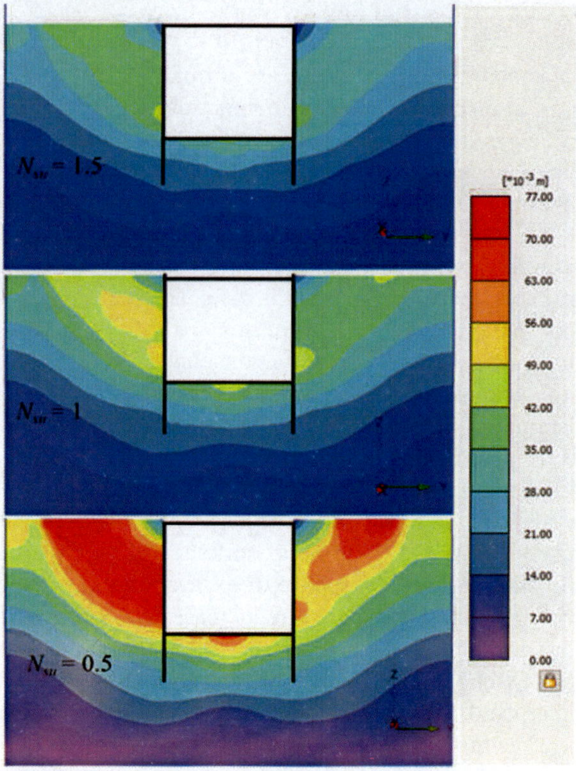

Fig. 9.64 Soil displacement field under various clay strength

Table 9.19 Ground and building settlements for varying clay strengths

N_{su}	Max ground settlement (mm)	B9		B4	
		Max settlement (mm)	Inclination degree (‰)	Max settlement (mm)	Inclination degree (‰)
0.5	45.92	30.33	0.97	72.38	3.52
0.8	33.3	27.62	0.82	64.72	2.47
1	29.88	26.495	0.73	43.49	1.13
1.2	26.94	23.35	0.87	40.35	1.26
1.5	24.3	20.48	0.76	33.59	1.10

9.5.2.3 Effects of Soil Stiffness

This section focuses on the role of clay stiffness in the development of excavation responses. The s_u ratio was assumed to be 0.6, a uniform value for all clay layers. Constant values for strength parameters were determined, and only value of the stiffness parameter G_u/s_u^A was altered for the discussion of clay rigidity. For better

Table 9.20 Excavation responses for varying clay stiffnesses

N_{Gu}	Max ground settlement (mm)	Max wall deflection (mm)	B9		B4	
			Max settlement (mm)	Inclination degree (‰)	Max settlement (mm)	Inclination degree (‰)
0.5	36.63	49.17	29.57	0.77	48.16	1.33
1	29.88	41.88	26.495	0.73	43.49	1.13
1.5	25.84	37.45	22.66	0.74	39.38	1.15
2	24.47	33.26	21.35	0.69	37.79	0.96

understanding, another index N_{Gu} is also proposed to represent the stiffness variation. According to computed results in Table 9.20, the excavation responses were not significantly affected by the variation of clay stiffness. As the N_{Gu} quadrupled from 0.5 to 2, the corresponding displacement values show little reduction. The settlement and inclination of important buildings (B9 and B4) were acceptable of all time, the ground surface subsidence in the no load area (central cross-section of excavation long side) remained uncritical, and small increment arose in the diaphragm wall deflection. It can be concluded that the clay stiffness plays insignificant role in controlling the subsoil behavior when its undrained shear strength remains constant.

9.5.3 *Optimization of Retaining System*

In light of parametric sensitivity analysis above, some effective ground improvement methods or retaining system optimization can be performed.

For the referencing Sukhumvit project, the excavation induced ground settlement and surrounding building displacements were well controlled due to the appropriate design and high-quality construction. However, when the poor design of excavation in soft clay is encountered, large deformation of retaining wall will occur during the construction process. Aimed at this problem, under the condition of determined excavation design, jet grouting technique can be employed before the excavation for strengthening the soft clay. With the known location of large deformation area in anisotropic clay, the mechanical properties of soft ground can be improved, therefore, leading to higher strength and less deformation during the braced excavation. Figure 9.65 shows the change of soil displacement field with the jet grout reinforcement, the effectiveness can be readily obtained. It should be mentioned that the focus lies on the improvement of mechanical and engineering properties of soil ameliorated by jet grouting, while the excess pore water pressure and slurry solidification process were not inspected in depth.

If the retaining scheme is enabled to be adjusted, it is suggested to install some temporary props in the middle depth of large excavation steps to prevent the diaphragm wall from local cracking or failure. Figure 9.66 reflects the significant

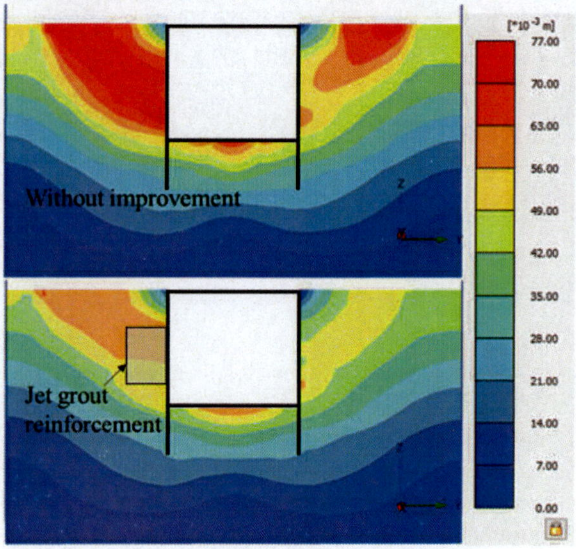

Fig. 9.65 Effect of Jet grouting reinforcement on control of the soil displacement

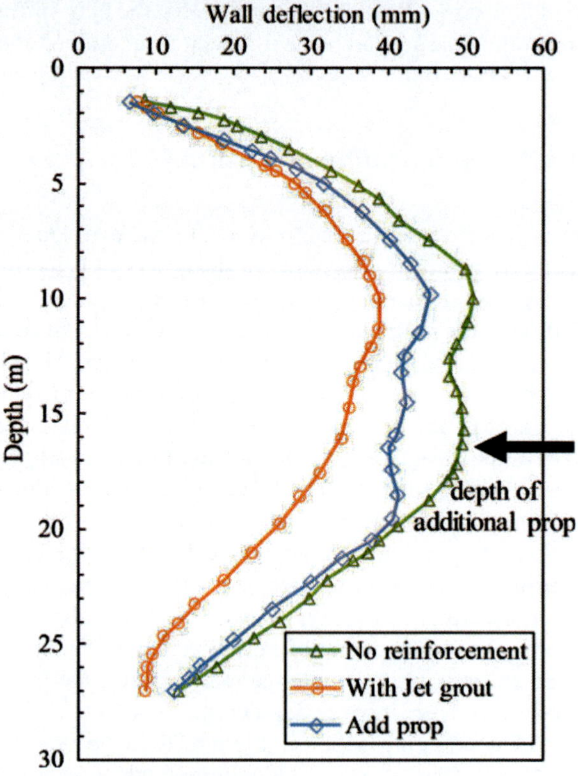

Fig. 9.66 Effect of two retaining reinforcement measures on controlling wall deflection

effect of the two retaining optimization measures on reducing the retaining wall deformation. Increasing a level of prop can directly push the diaphragm wall back at the specific position, while jet grouting can more effectively control the lateral displacement of the full depth. In addition, reducing the lateral displacement of retaining wall between the large-spacing platform slabs is essential to control the soil movements behind the wall, consequently, preventing the generation of serious continuous slip surface.

9.5.4 Summary

This study presents 3D FE analyses using the NGI-ADP soil constitutive model to assess the ground responses and adjacent building displacement for excavations in soft clays. The influence of soil strength and stiffness properties were also investigated. Main findings are as follows:

(1) The anisotropic degree influences significantly the soil displacements, which brings considerable differences in ground infrastructures. In comparison with the assumption that soil is isotropic, larger excavation responses are encountered in the anisotropic clay.

(2) For top-down method adopted in the case history, the critical location of diaphragm walls and surrounding soil lies at the center depth of the two platform slabs. In the case of very soft soil, continuous failure surface, which develops from the middle of retaining wall to the bottom of ground building, may occur. However, the clay stiffness has only marginal effects on the soil responses. Consequently, its impact on the buildings can be ignored.

(3) Jet grouting can be adopted before excavation to prevent larger deformation of soft clay. Alternatively, temporary prop level can be added to constrain the excessive wall deflection.

References

1. Zhang, R. H., Goh, A. T. C., & Zhang, W. G. (2019). System reliability assessment on deep braced excavation adjacent to an existing upper slope in mountainous terrain: A case study. *SN Applied Sciences, 1*, 876.
2. Zienkiewicz, O. C., Humpheson, C., & Lewis, R. W. (1975). Associated and non-associated visco-plasticity and plasticity in soil mechanics. *Geotechnique, 25*, 671–689.
3. Brinkgreve, R., & Bakker, H. (1991). Non-linear finite element analysis of safety factors. *Proceedings 7th International Conference Comp Methods and Advances in Geomech*, Cairns, Australia, pp. 117–122.
4. Lian, Z. Y., Han, G. C., & Kong, X. J. (2001). Stability analysis of excavation by strength reduction FEM. *Chinese Journal of Geotechnical Engineering, 26*, 5.
5. Cheng, Y. M., Lansivaara, T., & Wei, W. B. (2007). Two-dimensional slope stability analysis by limit equilibrium and strength reduction methods. *Computers & Geotechnics, 34*, 137–150.

6. Schneider-Muntau, B., Medicus, G., & Fellin, W. (2018). Strength reduction method in Barodesy. *Computers & Geotechnics, 95*, 57–67.
7. Zhang, W. G., & Goh, A. T. C. (2012). Reliability assessment on ultimate and serviceability limit states and determination of critical factor of safety for underground rock caverns. *Tunnelling and Underground Space Technology, 32*, 221–230.
8. Low, B. K., & Tang, W. H. (2007). Efficient spreadsheet algorithm for first-order reliability method. *Journal of Engineering Mechanics, 133*, 1378–1387.
9. Poulos, H. G., & Chen, L. T. (1996). Pile response due to unsupported excavation-induced lateral soil movement. *Canadian Geotechnical Journal, 33*, 670–677.
10. Goh, A. T. C., Teh, C. I., & Wong, K. S. (1997). Analysis of piles subjected to embankment induced lateral soil movements. *Journal of Geotechnical and Geoenvironmental Engineering, 123*, 792–801.
11. Yang, M. (2005). Numerical analysis of pile response due to unsupported excavation-induced lateral soil movement. *China Civil Engineering Journal, 38*, 91–96.
12. Miao, L. F., Goh, A. T. C., Wong, K. S., & Teh, C. I. (2006). Three-dimensional finite element analyses of passive pile behaviour. *International Journal for Numerical and Analytical Methods in Geomechanics, 30*, 599–613.
13. Zheng, G., Yan, Z. X., Lei, H. Y., & Lei, Y. (2007). Field observation and finite element numerical simulation analysis of effect on adjacent piles due to excavation. *Chinese Journal of Geotechnical Engineering, 29*, 638–643.
14. Chen, F. Q., Wang, J. W., & Liu, Y. C. (2008). Numerical analysis of pile response due to braced excavation-induced soil lateral movement. *Rock & Soil Mechanics, 29*, 1971–1976.
15. Li, L., Hu, X. X., Dong, G. H., & Liu, J. (2014). Three-dimensional numerical analyses of pile response due to braced excavation-induced lateral soil movement. *Applied Mechanics & Materials*, 580–583.
16. Liyanapathirana, D. S., & Nishanthan, R. (2016). Influence of deep excavation induced ground movements on adjacent piles. *Tunnelling and Underground Space Technology, 52*, 168–181.
17. Jiang, J., Chen, J., Ma, S., Shao, Y., & Tong, W. (2016). Three dimensional numerical analysis of pile response due to the excavation of foundation pit at dock. *Chinese Journal of Underground Space and Engineering, S1*, 246–253.
18. Li, L., Song, J., Zhang, J., Wei, X., & Shen, Y. (2017). Three-dimensional numerical analysis of pile responses due to multi-braced excavation induced lateral soil movement. *Journal of Disaster Prevention and Mitigation Engineering, 5*, 782–789.
19. Kenawi, M. A., & Ibrahim, M. G. (2017). Interaction analysis of supporting excavation piles and adjacent existing strap footing. *International Journal of Engineering Development and Research, 5*, 1045–1052.
20. Shakeel, M., & Ng, C. W. W. (2018). Settlement and load transfer mechanism of a pile group adjacent to a deep excavation in soft clay. *Computers and Geotechnics, 96*, 55–72.
21. Goh, A. T. C., Wong, K. S., Teh, C. I., & Wen, D. (2003). Pile response adjacent to braced excavation. *Journal of Geotechnical and Geoenvironmental Engineering, 129*, 383–386.
22. O'Rouke, T. D. (1981). Ground movements caused by braced excavation. J. Geotech. Geoenviron. *Journal of Geotechnical and Geoenvironmental Engineering, 107*, 1159–1178.
23. Zhang, R., Zheng, J., Pu, H., & Zhang, L. (2011). Analysis of excavation-induced responses of loaded pile foundations considering unloading effect. *Tunnelling and Underground Space Technology, 26*, 320–335.
24. Zheng, G., Liu, Q. C., & Deng, X. (2013). Numerical analysis of effect of excavation on underlying existing metro tunnel and deformation control. *Rock and Soil Mechanics, 34*(5), 10. (in Chinese).
25. Kuang, L. C. (2000). Influence of construction of deep foundation pit on tunnels of metro. *Chinese Journal of Geotechnical Engineering, 22*(3), 284–288. (in Chinese).
26. Chang, C. T., Sun, C. W., Duang, S. W., & Hwang, R. N. (2001). Response of a Taipei rapid transit system (TRTS) tunnel to adjacent excavation. *Tunnelling and Underground Space Technology, 16*(3), 151–158.

27. Sharma, J. S., Hefny, A. M., Zhao, J., & Chan, C. W. (2001). Effect of large excavation on deformation of adjacent MRT tunnels. *Tunnelling and Underground Space Technology, 16*(2), 93–98.
28. Jiang, H. S., & Hou, X. Y. (2002). The influence of deep excavation on adjacent metro tunnel in soft ground. *Industrial Construction, 32*(5), 53–56. (in Chinese).
29. Wang, W. D., Shen, J., Weng, Q. P., & Wu, J. B. (2006). Analysis and countermeasures of influence of excavation on adjacent tunnels. *Chinese Journal of Geotechnical Engineering, S1*, 1340–1345. (in Chinese).
30. Zheng, L. C., Wei, D. J., Zheng, J. F., & Lou, N. (2007). The practice of deformation monitoring of metro tunnel influenced by the construction of adjacent foundation pit. *Engineering of Surveying and Mapping, 16*(2), 47–50. (in Chinese).
31. Liu, T. J. (2008). Analysis on influence of deep excavation on tunnels of metro. *Modern Tunnelling Technology, S1*, 216–220. (in Chinese).
32. Gao, G. Y., Gao, M., Yang, C. B., & Yu, Z. S. (2010). Influence of deep excavation on deformation of operating metro tunnels and countermeasures. *Chinese Journal of Geotechnical Engineering, 32*(3), 453–459. (in Chinese).
33. Yan, J. Y. (2010). Design and construction of deep foundation pits near metro tunnels. *Chinese Journal of Geotechnical Engineering, S1*, 234–237. (in Chinese).
34. Hwang, R. N., Duann, S. W., Cheng, K. H., & Duann, S. W. (2011). Damages to metro tunnels due to adjacent Excavations. *International Symposium on Backwards Problem in Geotechnical Engineering and Monitoring of Geo-Construction, Proceeding of TC302 Symposium Osaka*, Osaka, Japan, 83–88.
35. Shao, H., & Wang, R. (2011). Monitoring data analysis on influence of operating metro tunnel by nearly excavation construction. *Chinese Journal of Underground Space and Engineering, S1*, 1403–1408. (in Chinese).
36. Xiao, T. G. (2011). Analysis on effect of deep-pit excavation and monitoring on neighboring metro tunnel in soft ground. *Chinese Journal of Underground Space and Engineering, 7*(5), 1013–1017. (in Chinese).
37. Wang, W. D., Xu, Z. H., & Wang, J. H. (2011). Statistical analysis of characteristics of ground surface settlement caused by deep excavations in Shanghai soft soils. *Chinese Journal of Geotechnical Engineering, 33*, 1659–1666.
38. Yuan, J., Liu, X. W., & Chen, W. L. (2012). Effect of construction of deep excavation in Hangzhou silty sand on adjacent metro tunnels and stations. *Chinese Journal of Geotechnical Engineering, 34*(S1), 398–403. (in Chinese).
39. Zou, W. B., Zhang, D. M., & Cai, X. W. (2012). Numerical simulation and real time monitoring of foundation pit excavation on deformation of adjacent MRT tunnel. *Chinese Journal of Underground Space and Engineering, 8*(S2), 1728–1731. (in Chinese).
40. Jiang, Z. H., & Zhang, Y. X. (2013). Calculation of influence on longitudinal deformation of adjacent tunnels due to excavation. *Journal of Civil, Architectural & Environmental Engineering, 35*(1), 7–11. (in Chinese).
41. Li, Y. S., & Yu, W. H. (2013). Analysis on influence of deep pit excavation on neighboring metro tunnel. *Chinese Journal of Underground Space and Engineering, 9*(2), 352–358. (in Chinese).
42. Tang, R., & Lin, B. H. (2014). Analysis on the impact of the foundation pit construction on neighboring metro tunnel. *Chinese Journal of Underground Space and Engineering, S1*, 1629–1634. (in Chinese).
43. Wen, Z. Y., Zhang, L. J., Chen, S., & Zhang, Y. N. (2014). The study on the influence of retaining wall of excavation deformation on adjacent metro tunnel. *Subgrade Engineering, 176*(5), 144–148. (in Chinese).
44. Xue, Y. S., & Huang, Y. L. (2014). Partition construction technology of deep and large-scale excavation in soft soil under complex condition. *Shanghai Construction Science & Technology, 1*, 36–39. (in Chinese).
45. Wang, L. F., Pang, J., Xu, Y. F., & Yang, K. F. (2016). Influence of foundation pit excavation on adjacent metro tunnels. *Rock and Soil Mechanics, 37*(7), 2004–2010. (in Chinese).

46. Yin, Y. Z., & Liu, B. (2016). Analysis of monitor of adjacent tunnel in processes of foundation excavation. *Architecture technology, 47*(9), 785–787. (in Chinese).
47. Chen, R., Meng, F., Li, Z., Ye, Y., & Ye, J. (2016). Investigation of response of metro tunnels due to adjacent large excavation and protective measures in soft soils. *Tunnelling and Underground Space Technology, 58*, 224–235.
48. Zhang, J., Wang, W. D., Li, J., & Xu, Z. H. (2017). Three-dimensional finite element analysis of the impact of zoned excavation on adjacent tunnels. *Building Structure, 2*, 90–95. (in Chinese).
49. Wang, D. J., Wang, W. T., Duan, L., Wang, J. W., & Xu, Q. W. (2017). Construction Influence of Pit Excavation on Underlying Subway Tunnel. *Chinese Journal of Underground Space and Engineering, 13*(s1), 223–232. (in Chinese).
50. Zheng, G., Yang, X., Zhou, H., Du, Y., Sun, J., & Yu, X. (2018). A simplified prediction method for evaluating tunnel displacement induced by laterally adjacent excavations. *Computers and Geotechnics, 95*, 119–128.
51. Zhang, R. H., Liu, H. L., & Zhang, W. G. (2018). Numerical investigation on tunnel responses induced by adjacent deep braced pit excavations. *Journal of Disaster Prevention and Mitigation Engineering, 38*(5), 857–866. (in Chinese).
52. Li, Y. Q., Zhang, W. G., Zhang, R. H. (2021). Numerical investigation on performance of braced excavation considering soil stress-induced anisotropy. *Acta Geotechnica*. https://doi.org/10.1007/s11440-021-01171-3.
53. Surarak, C., Likitlersuang, S., Wanatowski, D., Balasubramaniam, A., Oh, E., Guan, H. (2012). Stiffness and strength parameters for hardening soil model of soft and stiff Bangkok clays. *Soils and Foundations, 52*(4), 682–697.
54. Likitlersuang, S., Surarak, C., Wanatowski, D., Oh, E., & Balasubramaniam, A. (2013). Finite element analysis of a deep excavation: A case study from the Bangkok MRT. *Soils and Foundations, 53*, 756–773.

Chapter 10
Case Reports and Back Analysis

10.1 A Case Report of Excavations in BTG Residual Soils, Singapore

In a dense, built-up environment, damage to buildings and utilities may occur because of excessive ground movement arising from excavation activities nearby. Consequently, in Singapore, for deep excavation projects, stiff diaphragm walls and bored pile walls with multiple strut levels are commonly used to minimize lateral wall movement and ground settlement, even in fairly stiff residual soil deposits. However, if the residual soil permeability is high and the underlying rock is highly fissured or fractured, substantial groundwater drawdown and associated seepage-induced settlement may occur, as reported by Wen and Lin [1] and Goh and Zhang [2] for various deep excavation projects in Singapore.

Figure 10.1 illustrates the locations of the four cut-and-cover excavations for the construction of mass rapid transit stations, which form part of the Downtown Line 2 (DTL2) project that was recently completed. The stations are Cashew (CSH), Hillview (HLV), Bukit Panjang (BKP), and Beauty World (BTW). In this section, details of the subsurface conditions, excavation support system, field instrumentations, and observed excavation responses are presented, with focus on the large groundwater drawdown and associated ground settlements. The measured wall deflections, ground surface settlement values, settlement profiles, as well as strut forces, are also compared with empirical methods/charts from the literature. Specific issues encountered during excavation are presented, and the effectiveness of various groundwater control measures is discussed.

Figure 10.2 provides an overview of the site layout of the four stations. It is obvious that the four sites were all located in densely built-up urban environments. Therefore, the control of ground deformation is important in minimizing the impact on adjacent structures.

W. Zhang and H. Liu, *Design of Deep Braced Excavation and Earth Retaining Systems Under Complex Built Environment*,
https://doi.org/10.1007/978-981-16-5320-9_10

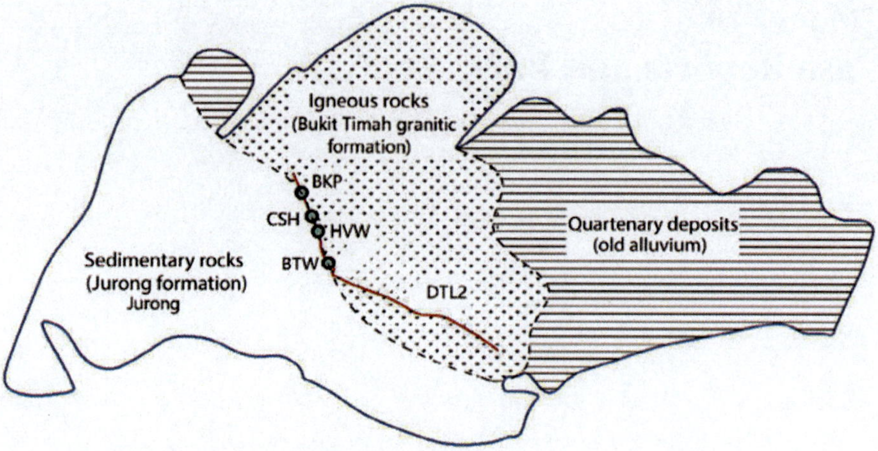

Fig. 10.1 Simplified geological map of Singapore and locations of the four sites

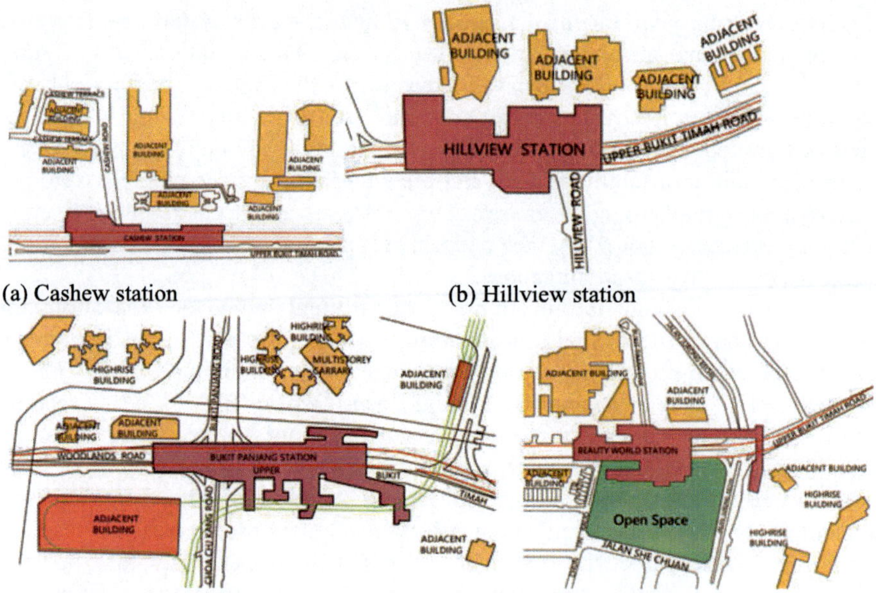

(a) Cashew station (b) Hillview station

(c) Bukit Panjang station (d) Beauty World Station

Fig. 10.2 Plan layout of the four stations

10.1.1 Subsurface Conditions

The ground conditions comprise Fill, Kallang Formation (Fluvial F1 and F2), Bukit Timah Granite (BTG) residual soil (denoted as GVI), completely weathered materials (GV), highly weathered materials (GIV), moderately weathered (GIII), slightly weathered (GII), and fresh (GI) Bukit Timah Granite rock. For the detailed properties of these geological units, Sharma et al. [3] are referred to. For brevity, this section only shows the geological soil profiles and cross-sectional view, as well as the SPT-N variation with depth for Cashew station. Figure 10.3a shows a typical cross-sectional view and the soil profile based on nearby boreholes which highlights the heterogeneous nature of the ground conditions that is commonly observed in the granitic residual soils of Singapore. This figure presents the typical geologic units, the generalized site stratigraphy, and the relative position between each strut level and geologic unit. Figure 10.3b summarizes the variation of SPT-N values with depth based on a number of borelogs. Generally, the ground subsurface profile is rather variable, as illustrated in Fig. 10.4. The permeability of the GVI residual soil was also very variable, in the range of 10^{-4}–10^{-8} m/s.

10.1.2 Excavation Support System Description

The excavation geometries and the ground conditions of the four sites are summarized in Table 10.1. In general, the GVI residual soil layer was thicker for the Cashew and Hillview sites compared with the other two sites. The retaining systems adopted for the four stations are shown in Table 10.2.

10.1.3 Construction Activities

The main excavation and construction activities of the four stations and the corresponding dates are summarized in Table 10.3. As the depths of excavation differed, for the Cashew and BTW sites, 5 levels of struts were installed. There were 6 and 7 levels of struts for the BKP and Hillview sites, respectively. The last row of Table 10.3 lists the groundwater control measures that were used to provide a hydraulic cut-off within the rock mass and to prevent significant groundwater drawdown. Fissure grouting was carried out at zones outside the excavation down to an adequate depth below the excavation base level. In addition, recharging wells were installed behind the excavation areas that are sensitive to settlements.

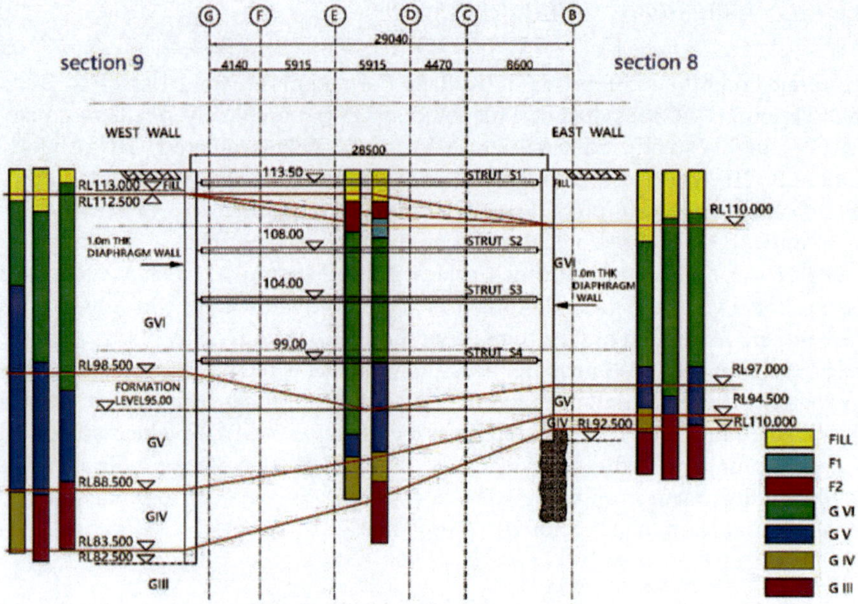

(a) Geological soil profiles and cross-sectional view

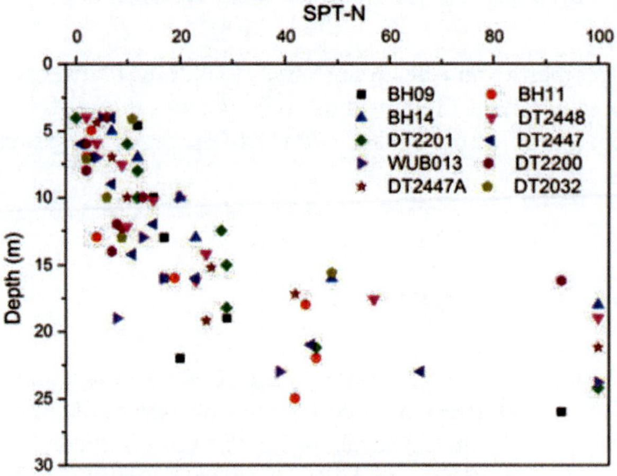

(b) SPT-N variation with depth

Fig. 10.3 Cashew station subsurface conditions

10.1.4 Field Instrumentation

Monitoring data was collected during the construction included diaphragm wall and secant bored pile wall deflections (using in-wall inclinometers), pore water pressures (using vibrating wire piezometers), ground settlements (using settlement markers),

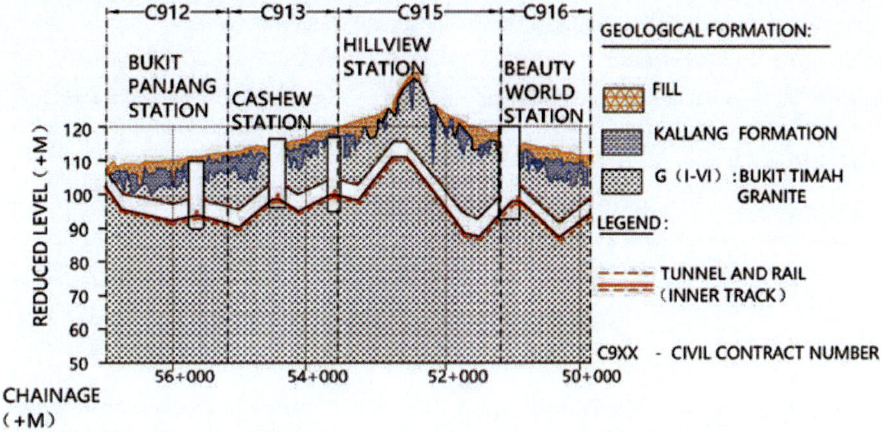

Fig. 10.4 Soil profile along DTL2 alignment for the four case studies (adapted from Goh and Zhang [2])

Table 10.1 A summary of excavation geometries and ground conditions for the four stations

Station	Max excav. depth (m)	Max excav. length (m)	Max excav. width (m)	Thickness of Fill/Kallang Formation (m) (average)	Thickness of GVI (m) (average)	Rockhead/mRL
Cashew	20	225	60	4.5–11 (4.8)	3–19 (11.7)	87.2–97
Hillview	24	190	48	0–9 (5.6)	3–25 (13.7)	87.1–105
Bukit Panjang	22	165	60	5–9 (7.4)	2.6–11 (7.1)	83.5–98.2
Beauty World	20	225	50	5.7–9 (7.4)	2.9–8.5 (5.0)	87.5–107.9

and strut loadings (using strain gauges and load cells). Data from the inclinometers and piezometers were obtained daily during excavation. For brevity, Fig. 10.5 only shows the plan view instrumentation layouts of two of the stations (Cashew and Hillview). Notations of the symbols are as follows: I—inclinometer; GWV—vibrating wire piezometer; LG—ground settlement marker; SG—strain gauge, and LC—load cell.

10.1.5 *Field Observations*

Field instrumentation responses including lateral wall deformations, ground settlements, piezometric level change, and strut loadings are presented in this section.

Table 10.2 A summary of retaining systems adopted for the four stations

Station	Wall system	Levels	Type of struts and preloading
Cashew	Diaphragm wall (thickness: 1 m, length: 25 m)	4	Strut HY 700, Double waler beam HY 700, Bracket HY 400; Strut: EA = $3.56\text{\textasciitilde}6.20\times10^6$ kN, preloadings in kN/m (1st: 100; 2nd: 300; 3rd and 4th: 350)
Hillview	Diaphragm wall (1 and 1.2 m thickness, length: 29 m)	6	Strut HY 700, Double wailer beam HY 650, Runner beam HR 350; Strut EA = $3.56\text{\textasciitilde}6.20\times10^6$ kN, preloadings in kN/m (1st level: 100; 2nd level: 200; 3rd level: 400; 4th, 5th, and 6th level: 500)
Bukit Panjang	Secant Bored Pill wall (1.2 m diameter with 1.5 m spacing, length: 21.5–35.3 m)	5	Both Reinforced Concrete (RC) and Steel Strut (SS) used with splays; Struts preloadings: RC1, SS, 2, 3, 4, and 5 = 400 kN/m, Strut EA, RC = 3.24×10^7 kN, SS = $1.02\text{\textasciitilde}1.26\times10^7$ kN
Beauty World	Secant Bored Pill wall (1.2 m diameter with 1.6 m spacing, length: 10.3–25.1 m)	4	Strut size: S1–2-UB 400×300× 94.3 kg/m; S2–2-UB 500×300× 128 kg/m; S3 and S4–2-UB 610× 324×155 kg/m, Preloading in kN (S1: 150, S2: 270, S3: 350, S4: 400)

10.1.5.1 Lateral Wall Movements

Figure 10.6 shows the typical lateral wall deflection plots for each of the cross-sections of the four stations. It is obvious that the maximum wall deflection increases as excavation proceeds. It should be noted that the shape of the deformation curves indicated a cantilever movement of the wall at the 1st or 2nd stages (S1 and S2).

Figure 10.7 shows the maximum lateral wall deflection for each excavation stage for all cross-sections of the four stations. The results show that from the excavation of the second layer soil (S2) onwards, the normalized maximum wall deflection is below 0.2% of the excavation depth H which indicates that the deflections of the diaphragm walls and the secant bored pile walls during subsequent excavation are generally small.

10.1.5.2 Ground Surface Settlements

Figure 10.8 shows the typical ground surface settlement profiles for different excavation stages for the Cashew and Hillview stations. It is obvious that the ground settlement increases with increasing excavation depth.

Table 10.3 A summary of excavation stages at the four stations

Activity	Stations and the corresponding dates			
	Cashew	BTW	BKP	Hillview
Stage S1: Excav. of 1st level of soil & installation of S1 struts	3/6/2011	18/3/2011	20/1/2012	17/6/2011
Stage S2: Excav. of 2nd level of soil & installation of S2 struts	2/9/2011	12/9/2011	3/3/2012	29/12/2011
Stage S3: Excav. of 3rd level of soil & installation of S3 struts	9/12/2011	9/1/2012	4/4/2012	12/5/2012
Stage S4: Excav. of 4th level of soil & installation of S4 struts	20/1/2012	29/3/2012	24/1/2013	22/6/2012
Stage S5: Excav. of 5th level of soil & installation of S5 struts	23/2/2012	5/5/2012	13/3/2013	10/8/2012
Stage S6: Excav. of 6th level of soil & installation of S6 struts	—	—	28/6/2013	23/8/2012
Stage S7: Excavation to final excavation level (FEL)	—	—	—	27/10/2012
Groundwater control measures	Fissure grouting, recharging	Recharging	Fissure grouting	Recharging

Figure 10.9 shows the ground surface settlement profiles for the final elevation stage of all the cross-sections of the four sites. It can be observed that the ground surface settlement profiles vary significantly, even for the same site, possibly because of the heterogeneous ground conditions of the sites. For the Cashew and Hillview sites, most of the measured maximum ground surface settlements occur within a distance of d = 1.0H where d is the distance from the retaining wall, but in some cases the maximum occurred further away at a distance of between d = 1.0H and d = 1.5H. Furthermore, the observed settlement profiles at some sections extended beyond d = 3.0H for the Cashew and Hillview sites. For the BKP and BTW sites, the maximum ground surface settlements occur within a distance of d = 1.0H and the settlement profiles did not extend beyond d = 3.0H.

10.1.5.3 Groundwater Level Changes

Figure 10.10 shows the piezometric level measurements for GWV 2008 & 2007 at different depths, at one of the sections at Cashew station. Figure 10.10a presents the

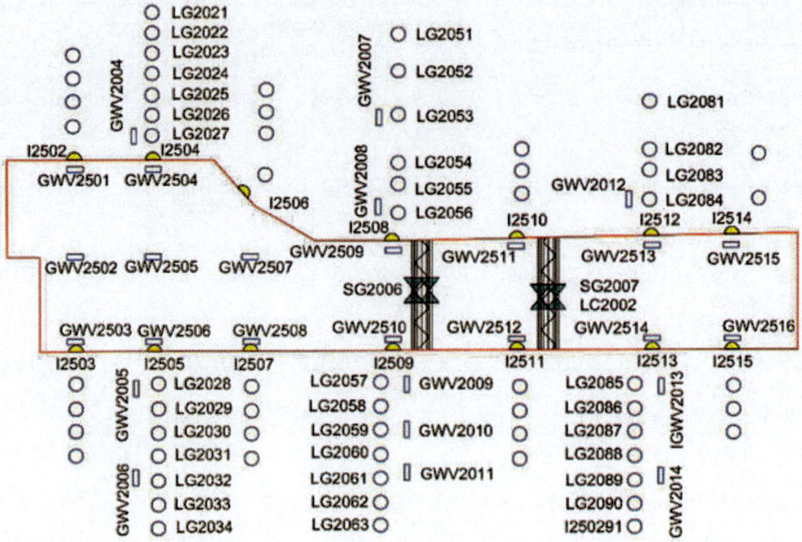

(a) Cashew

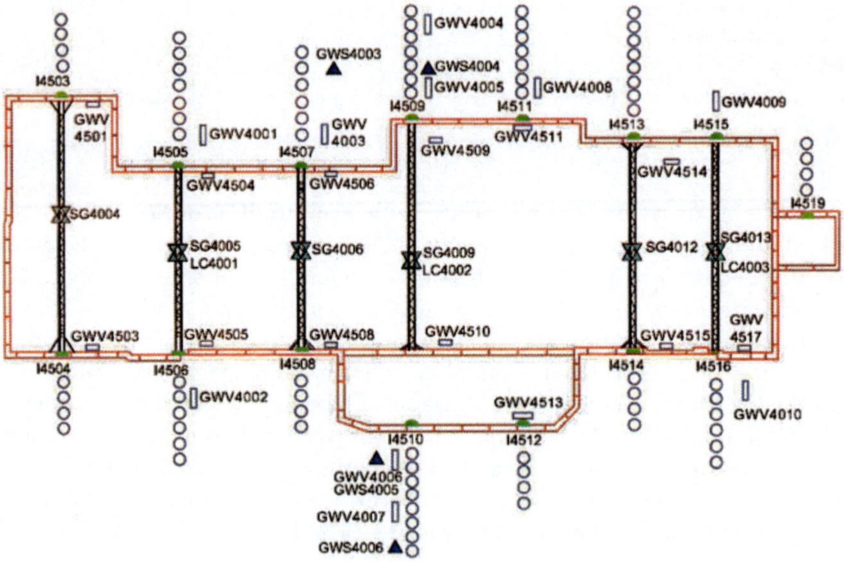

(b) Hillview

Fig. 10.5 Instrumentation layouts of Cashew and Hillview stations

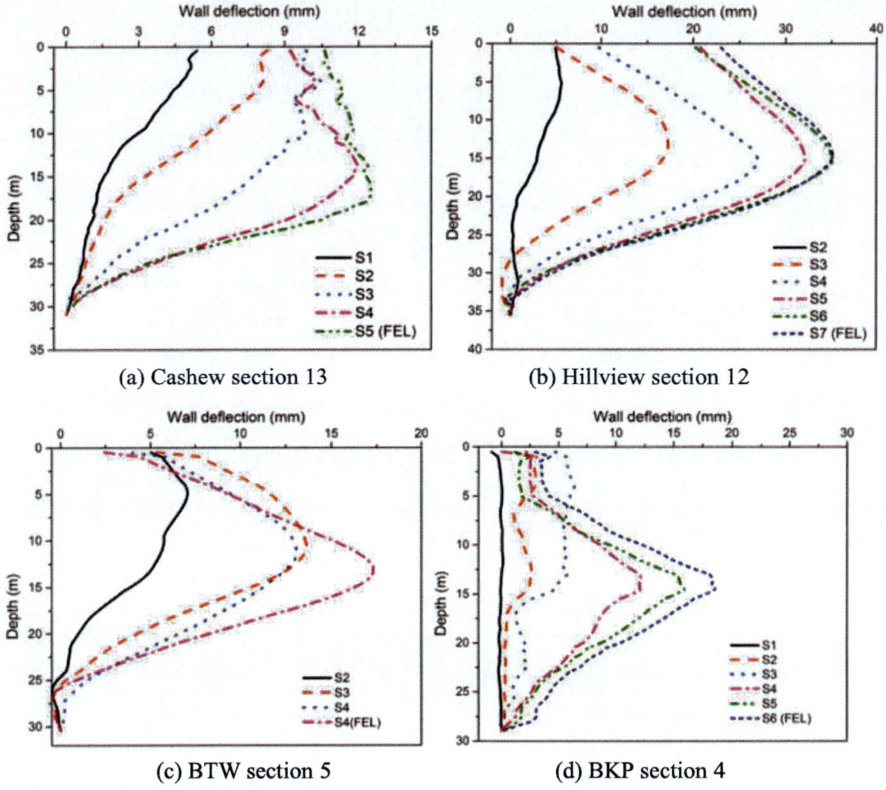

(a) Cashew section 13

(b) Hillview section 12

(c) BTW section 5

(d) BKP section 4

Fig. 10.6 Typical retaining wall deflection profiles

layout of the piezometers GWV 2008 & 2007 where GWV2008 is 15 m away from the wall and GWV is 28 m away from the wall. Figure 10.10b shows the recorded piezometric level changes with the construction day number for the top vibrating wire piezometers. It can be observed that for the two piezometers GWV2007-3 and GWV2008-3 located in the GVI layer, generally the two curves are quite similar up to day 280. Subsequently from day 280 to day 360, GWV2007-3 which is located further away from the wall showed significantly lower piezometric levels than GWV2008-3. A comparison of GWV2007-2 and GWV2008-2 in Fig. 10.10c shows that during the entire duration of the excavation activities especially after 200 days, the drawdown in GWV2007-2 which is further away from the wall is greater than that of GWV2008-2. For GW2007-1 and GWV2008-1, the differences in the groundwater drawdown were minimal as shown in Fig. 10.10d. The observed groundwater level measurements show the trend of smaller drawdown near the wall and larger drawdown further away from excavation. This may be because the fissure grouting/recharging which is carried out close to the wall was successful to prevent groundwater drawdown in this zone. However, since there was no recharging/fissure grouting away from the wall, this lead to the increased groundwater flow because of rock fissures in this zone. It

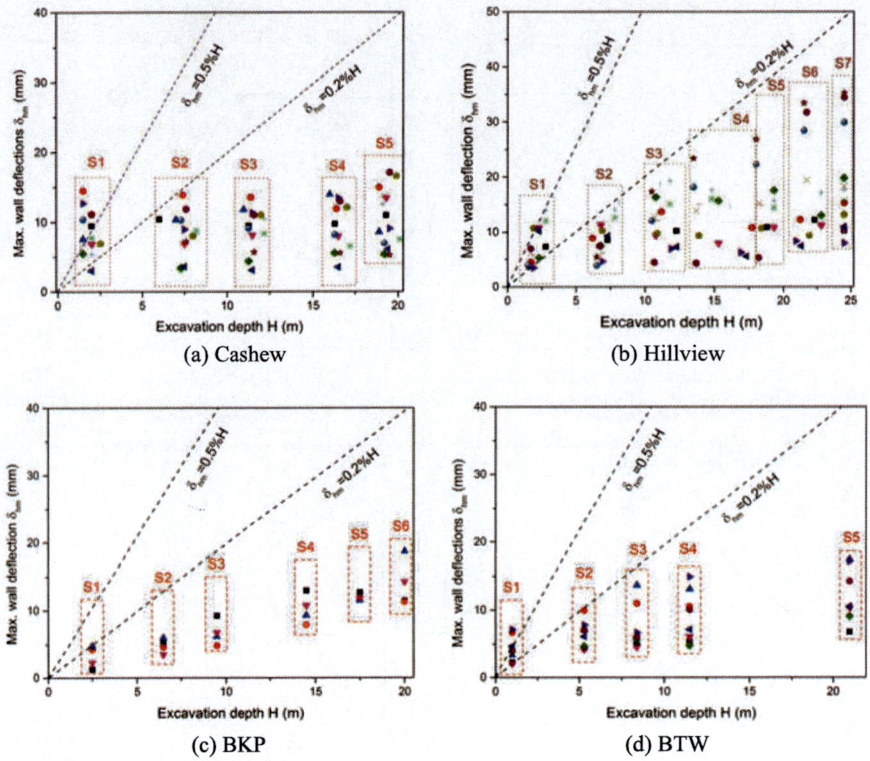

Fig. 10.7 Maximum wall deflection versus excavation depth

could also explain the unusual settlement profiles observed in Figs. 10.8 and 10.9b with smaller settlements near the wall and larger settlements further away from the wall.

Figure 10.11 plots the observed groundwater drawdown d_w behind the retaining wall for the four sites. Each data point corresponds to the average of the standpipe readings for the particular site. It is obvious that as excavation proceeds to the final excavation level (FEL), the groundwater table decreases. In addition, the groundwater drawdown for the Cashew and Hillview sites was more significant than that for the other two sites, which is consistent with the larger ground settlements observed at the Cashew and Hillview sites. Possible explanations with regard to this uncommon large drawdown include the interconnected fissures in the rock and the high permeability of the completely weathered residual soils.

Figure 10.12 shows the measured strut forces from the strain gauges for struts S2006 and S2007 of Cashew station at the different construction stages including the subsequent strut removal following the construction of the basement structure. After reaching a peak following excavation to the 1st stage soil level, loads for the 1st strut level (S2006-1 and S2007-1) decreased for subsequent excavation stages.

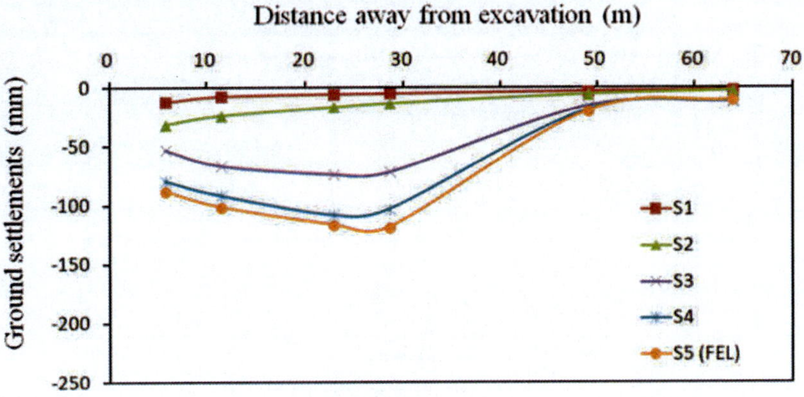

(a) Cashew section 9

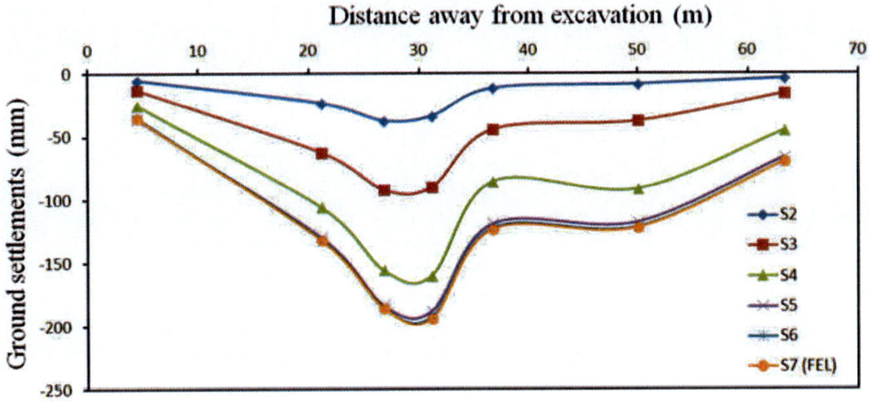

(b) Hillview section 5

Fig. 10.8 Typical ground surface settlement profiles for different stages

Strut loads for the 2nd level (S2006-2 and S2007-2) remained relatively constant after excavation to the 3rd soil level, while loads recorded at the 3rd (S2006-3 and S2007-3) and 4th strut levels (S2006-4 and S2007-4) continued increasing until excavation to the final formation level. The removal of strut 4 caused significant increase of the load in strut 3, but minimal changes to the load in struts 1 and 2. Similarly, the removal of strut 3 resulted in significant increase in the load in strut 2. Likewise, the removal of strut 2 caused significant increase in the load in strut 1.

Table 10.4 summarizes the maximum total loads of each strut level for S2006 and S2007 during excavation and after strut removal. Based on these maximum loads and the locations of each level strut, using the tributary area method proposed by Terzaghi and Peck [4], the normalized apparent earth pressure have been derived, as shown in Fig. 10.13. The maximum normalized pressure was approximately 0.38H where γ is the unit weight of the soil and H is the final excavation depth.

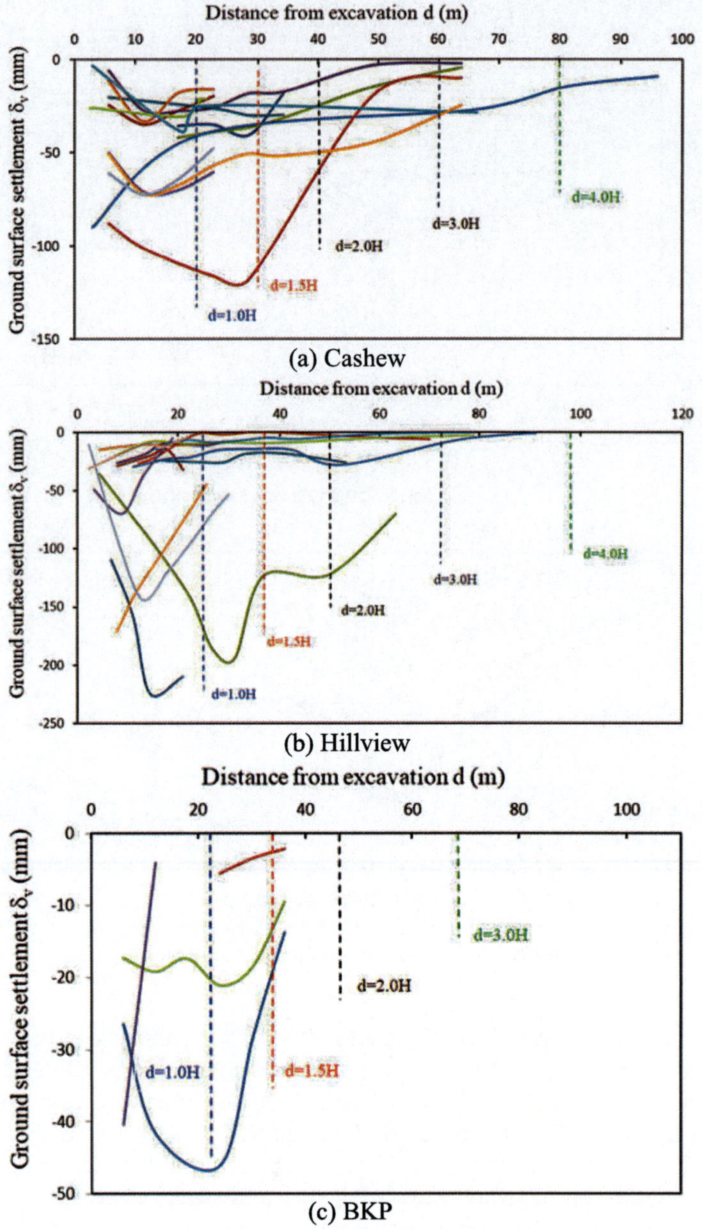

Fig. 10.9 Ground surface settlement profiles at final excavation stage of the four stations

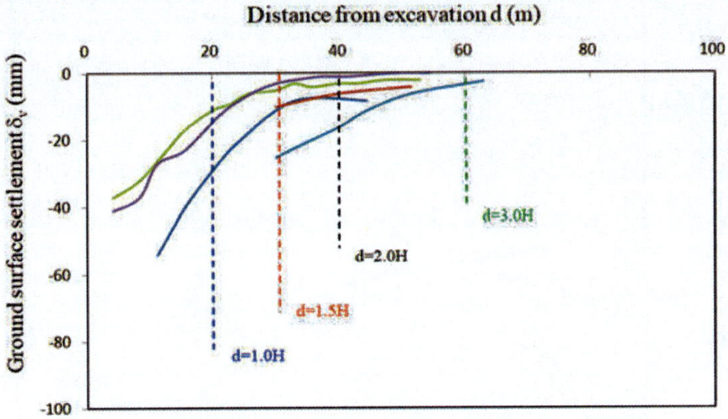

(d) BTW

Fig. 10.9 (continued)

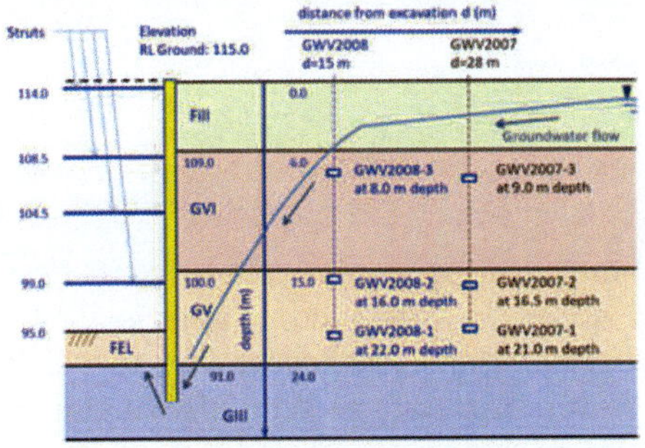

(a) Layout of GWV 2008 & 2007

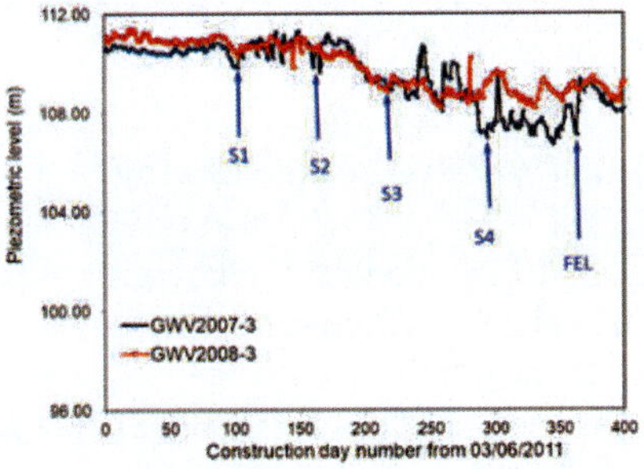

(b) Response of GWV2008-3 & 2007-3

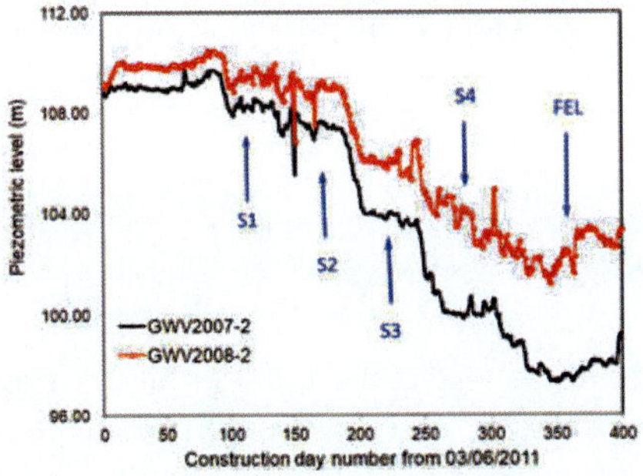

(c) Response of GWV2008-2 & 2007-2

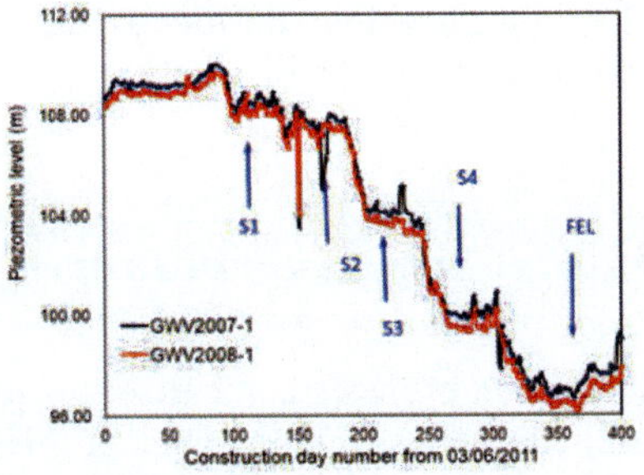

(d) Response of GWV2008-1 & 2007-1

Fig. 10.10 Groundwater level measurements for GWV 2008 & 2007 at different depths

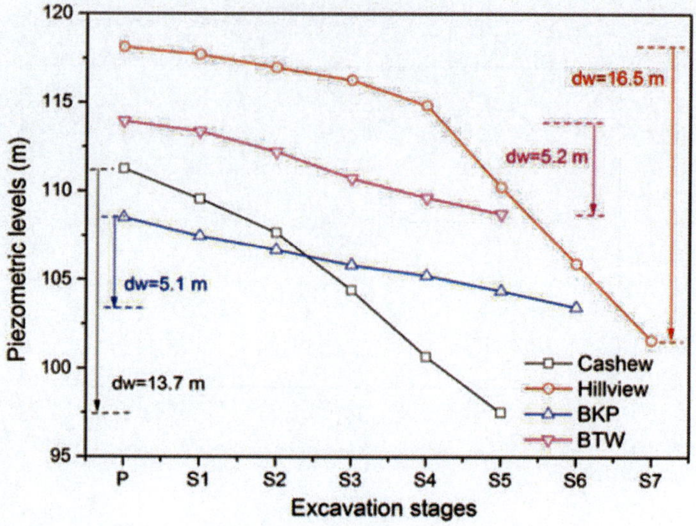

Fig. 10.11 The observed average groundwater drawdown

10.1.5.4 Effect of Recharging

Figure 10.14 shows the recharging well layouts for the Cashew and Hillview sites with well points denoted by the green solid circles, as well as the recharging effects. The number in red color denotes the maximum ground settlement corresponding to the cross-section. It is obvious that for most of the cross-sections, the maximum ground settlement is reduced as a result of recharging, compared with the value for the corresponding opposite section, particularly for Cashew station. Apart from Section 3 (denoted as S3) and 5 (S5) of Hillview station, both two plots indicate that recharging carried out during excavation is effective in reducing the ground surface settlement.

10.1.6 Comparison of Braced Excavation Responses

This section presents comparison of measured wall deflection, the ground surface settlement values and settlement profiles, and the strut forces with empirical methods (charts or figures) from the literature.

10.1.6.1 Characteristics of Lateral Wall Movements

The measured maximum wall deflections δ_{hm} from all available sections of the four sites have been plotted in Fig. 10.15 with respect to excavation depth H. There is

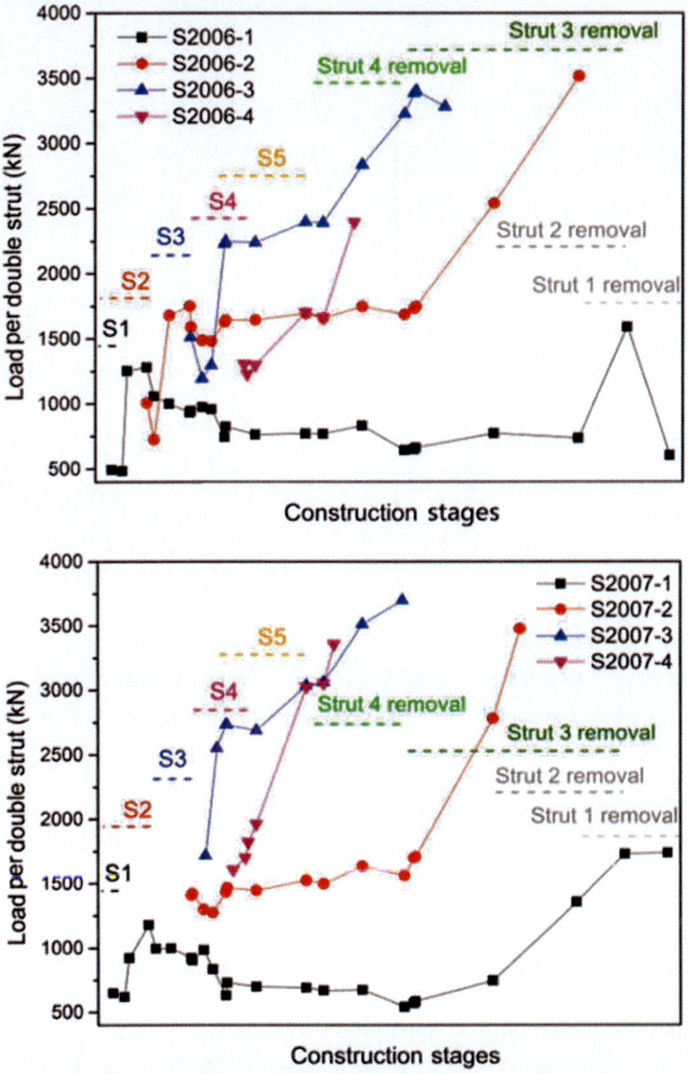

Fig. 10.12 Measured strut forces for Cashew cross-sections

Table 10.4 Maximum loadings for S2006 and S2007 struts (Cashew)

Strut No.	Maximum loads (kN)	Strut No.	Maximum loads (kN)
S2006-1	1593	S2007-1	1738
S2006-2	3517	S2007-2	3477
S2006-3	3403	S2007-3	3699
S2006-4	2396	S2007-4	3357

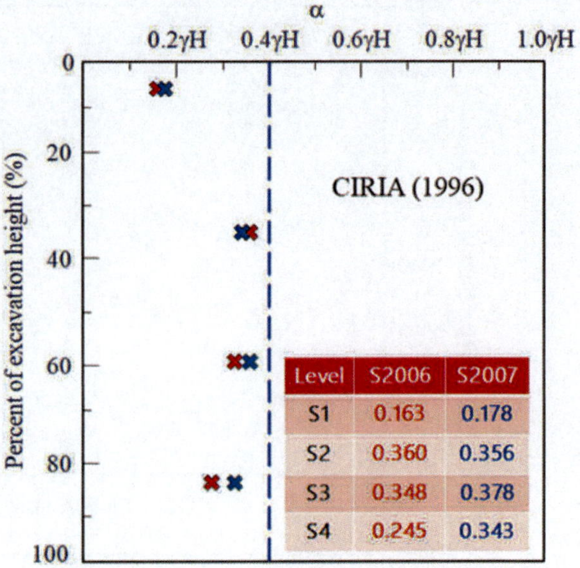

Fig. 10.13 Normalized apparent earth pressure versus normalized depth for Cashew sections

clear relationship between δ_{hm} and H. The maximum wall displacement is bounded by lines having ratio δ_{hm}/H between 0.025 and 0.15%, with an average value of 0.072%. Table 10.5 shows reported values of the normalized maximum wall deflection δ_{hm}/H for different soil types in different regions, including the BTG residual soil from this chapter. It is obvious that the normalized maximum wall deflections for excavations in BTG residual soil were significantly smaller than those for excavations in soft soil deposits and of the same order of magnitude as many of the reported excavations in the stiffer soil deposits.

10.1.6.2 Characteristics of Ground Surface Settlement Values

The measured maximum ground surface settlements δ_{vm} from all available sections of the four sites have been plotted in Fig. 10.16 with respect to excavation depth. The settlement values for Cashew and Hillview sites are much greater most probably due to the significant groundwater drawdown as described in Sect. 3.3. The maximum settlements are bounded by lines having normalized maximum settlement δ_{vm}/H ratios between 0.09 and 0.9%, with an average value of 0.27%. Table 10.6 shows comparison of reported values of δ_{vm}/H for different soil types in different regions. The ground surface settlements measured in BTG residual soils were larger than most of the previously reported excavations in both soft and stiff soils, possibly because of the large groundwater drawdown at the four sites.

Figure 10.17 plots the normalized maximum ground surface settlement versus the normalized maximum wall deflections based on the field measurements of the

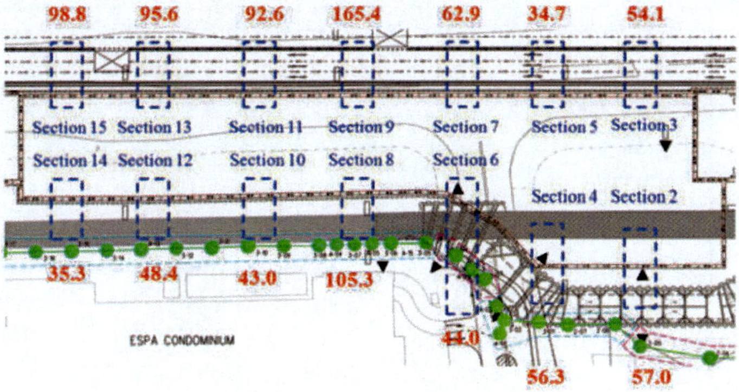

(a) Cashew station

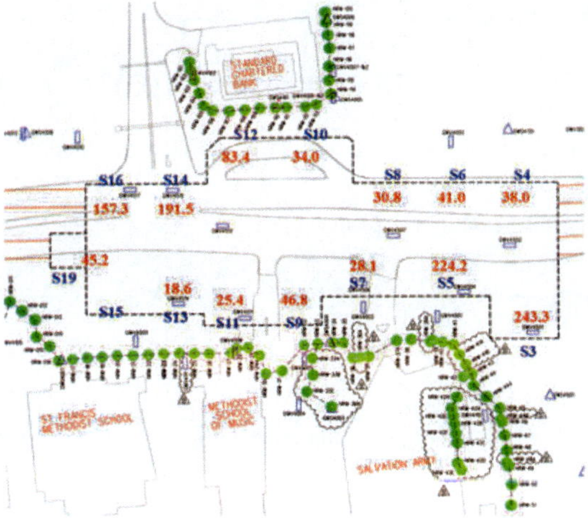

(b) Hillview station

Fig. 10.14 Recharging well layout and effects

four sites. For most of the cases, the ratio of δ_{vm}/δ_{hm} lies between 1.0 and 10.0, with an average value of 3.0, indicating that the maximum ground surface settlement is approximately 3 times the value of the maximum wall deflection. A comparison with the reported values of δ_{vm}/δ_{hm} for different soil types in different regions in Table 10.7 indicates that the δ_{vm}/δ_{hm} ratio for excavations in BTG residual soils were significantly larger than the previously reported cases. Again, this is likely to be because of the large groundwater drawdown at the four sites.

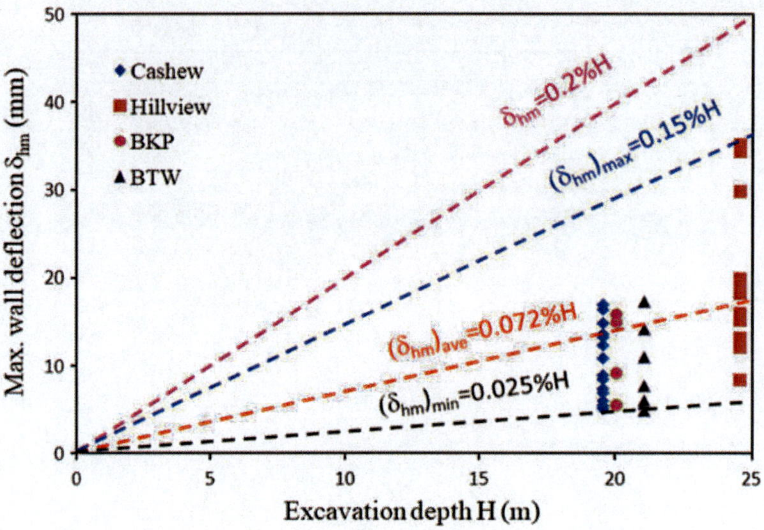

Fig. 10.15 Maximum wall deflections against excavation depth for the four sites

Table 10.5 Summary of reported δ_{hm}/H values

Data source	Soil layer, Country or region	δ_{hm}/H (%)
[5]	Soft clay, Taipei	0.20–0.50
[6]	Soft soil (c_u < 75 kPa)	0.5–1.0 (0.87)*
[7]	Soft soil, Shanghai	0.20–0.90
[8]	Soft soil in Old Alluvium, Singapore	0.15–0.60
[9]	Soft soil	<0.70
[10]	Soft soil, Shanghai	0.10–0.60
[11]	Soft clay, Istanbul	0.043–0.32
[8]	Medium soil in Old Alluvium, Singapore	0.20–0.70 (0.43)
[12]	Silty clay, Nanjing	0.08–0.32
[13]	Stiff soil, UK	0.15–0.20
[14]	Stiff soil, UK	0.13–0.40
[15]	Soft soils overlying residual soils, Singapore	0.04–0.35
[16]	Hard soil, Taipei	0.07–0.20
[17]	Stiff soils	0.05–0.25
[18]	Residual soils and soils overlying rocks	0.05–0.15
[8]	Stiff soil in Old Alluvium, Singapore	0.04–0.20
[19]	Stiff clay	0.05–0.25
[20]	Stiff clay	0.10–0.50
Present study	Bukit Timah Granite residual soil	0.025–0.15

* The number in the bracket indicates the average mean value

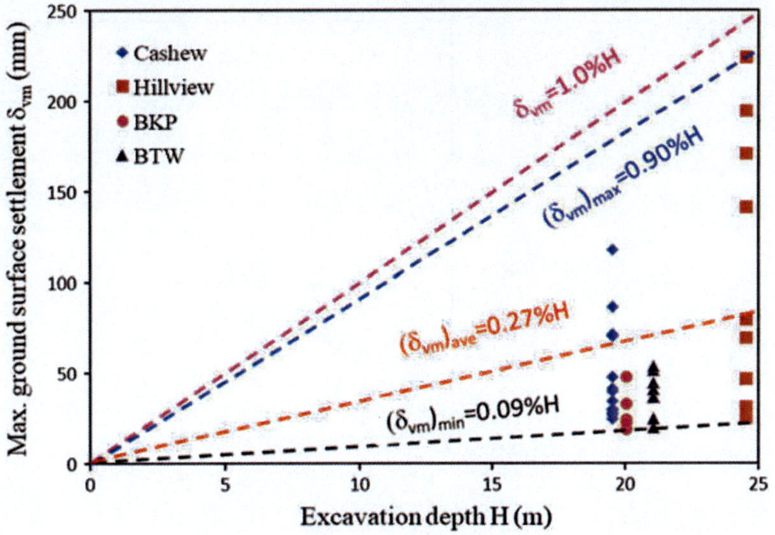

Fig. 10.16 Maximum ground surface settlement against excavation depth for the four sites

Table 10.6 Summary of reported δ_{vm}/H values

Data source	Soil layer, Country or region	δ_{vm}/H (%)
[5]	Soft clay, Taipei	0.13–0.36
[8]	Soft soil in Old Alluvium, Singapore	0.20–0.70 (0.47)
[6]	Soft soil (c_u < 75 kPa)	1.10
[21]	Soft soil, Shanghai	<0.42
[22]	Soft soil, Shanghai	0.10–0.60
[8]	Medium soil in Old Alluvium, Singapore	0.15–0.40 (0.26)
[8]	Silty clay, Beijing	0.08–0.32
[12]	Silty clay, Nanjing	0.04–0.19
[20]	Firm to stiff clays	0.10–0.40
[23]	Stiff clay, USA	<0.70
[19]	Stiff soil	<0.20
[14]	Stiff soil, UK	0.10–0.20
[13]	Stiff soil, UK	0.15
[15]	Soft soils overlying residual soils, Singapore	0.04–0.20
[17]	Stiff soil	<0.20
[8]	Stiff soil in Old Alluvium, Singapore	0.02–0.20 (0.06)
[24]	Silty sand, Hong Kong	0.20–0.50
Present study	Bukit Timah Granite residual soil	0.09–0.90

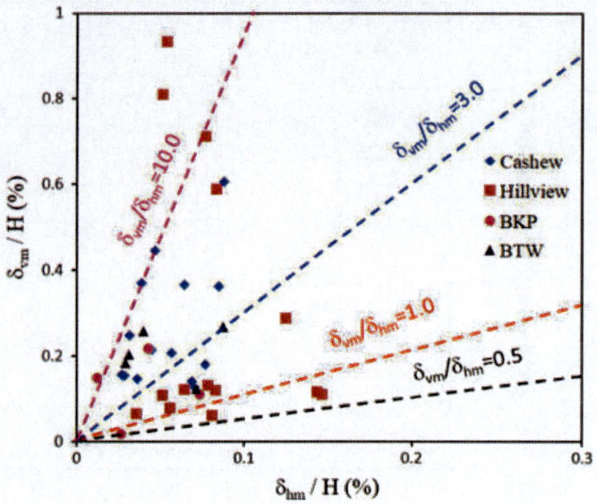

Fig. 10.17 Relationship between δ_{vm} and δ_{hm} for the four sites

Table 10.7 Summary of reported relationship between δ_{vm} and δ_{vm}

Data source	Soil layer	δ_{vm}/δ_{hm}
[25]	clays	0.5–1.0
[5]	Taipei soft clay	0.5–1.0
[21]	Shanghai soft soils	0.4–2.0
[26]	Soft to medium stiff clays	0.5–2.0
[20]	Firm to stiff clays	0.7–1.4
[8]	Old Alluvium of Singapore	0.6–1.1
Present study	Bukit Timah Granite residual soil	1.0–10.0

10.1.6.3 Characteristics of Ground Surface Settlement Profiles

Figure 10.18 shows the cross-sectional measurements (represented by different symbols) at the four sites together with the ground surface settlement envelopes proposed by Peck [23]. Also shown are the proposed ground surface settlement envelopes based on the measured data from this study. For cases with significant groundwater drawdown (> 13 m) and thickness of GVI exceeding 10 m, such as Cashew and Hillview, the upper bound of $\delta_{vm}/H = 10\%$, the knee point d/H = 1.5, with the influence zone extended to d/H = 3.0 was found to provide good estimates of the ground surface settlement. For BKP and BTW with thickness of GVI less than 10 m and significantly less groundwater drawdown, the upper bound is $\delta_{vm}/H = 0.5\%$ and the knee point is at d/H = 1.0, with the influence zone within d/H = 2.0.

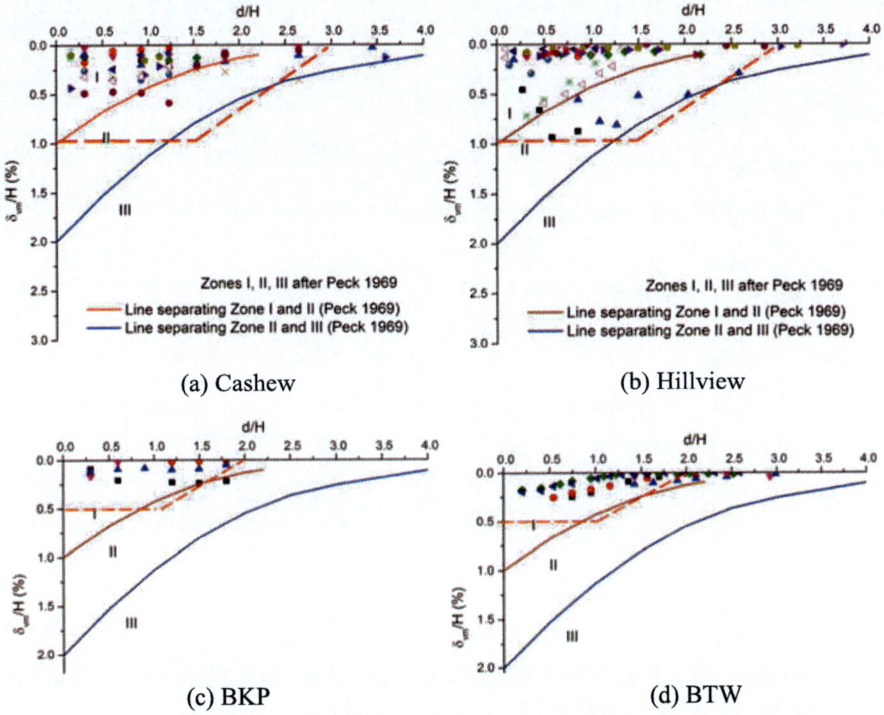

Fig. 10.18 Proposed settlement profiles for BTG residual soil sites

10.1.6.4 Characteristics of Strut Forces

Based on the maximum loadings of each level strut during excavation and after strut removal and its corresponding location for the four stations, using the tributary area method, similar apparent earth pressure diagrams (APD) have been developed for BTG residual soils, as shown in Fig. 10.19. Also plotted in Fig. 10.19 is the CIRIA recommended distributed prop load (DPL) of α=0.5 for stiff walls [27]. However, for the Bukit Timah Granite residual soils, based on the measured results, it is proposed to adopt α=0.4 instead of 0.5. Therefore, using this apparent earth pressure of $\sigma_h = 0.4\gamma H_e$ and Peck's tributary area method the force acting on each strut can be estimated.

10.2 Back Analysis

Finite-element method is a powerful tool for geotechnical problem analysis. The normal procedure of the analysis can be demonstrated as three steps: (1) build mechanical models, including geometry, constitutive model of soils and rocks and

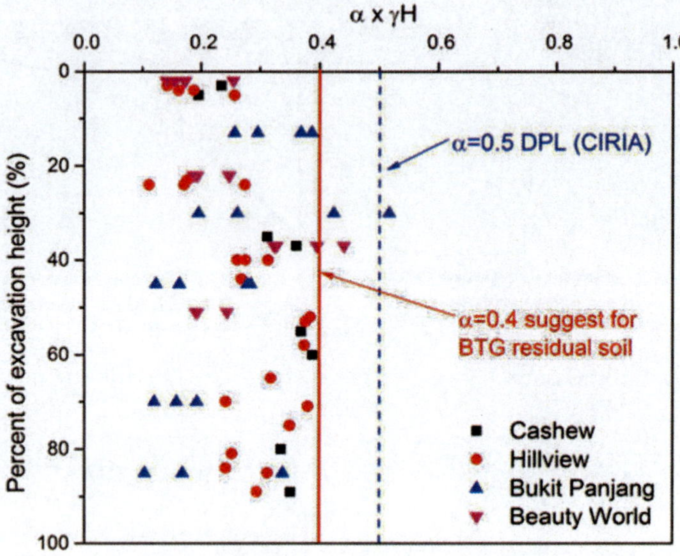

Fig. 10.19 Normalized apparent earth pressure for BTG residual soil

initial boundary conditions; (2) determine the parameters of soils and rocks; and (3) get the stress and deformation field. It can be called forward analysis.

However, the determination of parameters of soils and rock usually has uncertainty, making challenges for prediction of geotechnical performance. Because of the limitation of time and cost, the geological survey before the construction may be not comprehensive enough to reveal real condition of the whole construction field. On the other hand, the results of laboratory tests may fail to fully represent the in situ soils and the complicated soil stratum conditions because of disturbance caused by sampling. Also, some parameters like the small strain stiffness are difficult be determined via conventional tests. All of these factors lead FEM prediction fail to match the observed response.

Because braced excavations are generally carried out in stages, back analysis to update key soil parameters (such as the undrained shear strength and the initial tangent modulus in an excavation in clays) is generally realized in multiple stages; before the first excavation stage, the wall and ground responses are predicted with field tests and/or laboratory data. After the first stage excavation is completed and wall and/or ground responses are measured, the key soil parameters can be updated with the observed responses to refine the knowledge of the soil parameters. With the updated soil parameters, the wall and/or ground responses in the subsequent excavation stages may be predicted with improved fidelity. This process can be repeated stage by stage until the final excavation depth is reached.

10.2.1 MARS Back Analysis Models

The data used to build MARS model is also extracted from the numerical analysis described in Sect. 4.4.1. While in this section, the targets are relative soil stiffness ratio E_{50}/c_u logarithm of the system stiffness ln (S), instead of retaining wall deflection δ_h.

10.2.1.1 MARS Model for E_{50}/c_u

Based on the forward analysis results described in the previous section, a MARS model has been developed for inverse analysis to estimate the soil relative stiffness ratio E_{50}/c_u as a function of seven input parameters: γ, B, H_e, T, ln(S), δ_h, and c_u/σ_v. Herein the measured wall deflection δ_h is used as an input parameter to back-calculate E_{50}/c_u.

The dataset was separated randomly into a training set of 775 patterns and a testing set of 257 patterns. The MARS model with the highest R^2 value and less BFs for the testing data set is considered to be the optimal. The optimal MARS model adopted 28 BFs of linear spline functions.

The observed versus predicted 1:1 plots of E_{50}/c_u are shown in Fig. 10.20. The developed model predicts slightly higher estimations for low E_{50}/c_u values and slightly lower estimations for higher values of E_{50}/c_u. Figure 10.21 presents the histogram plots of the relative errors e (defined as the ratio of the difference between the MARS predicted and the target E_{50}/c_u to the target value, in percentage). It is

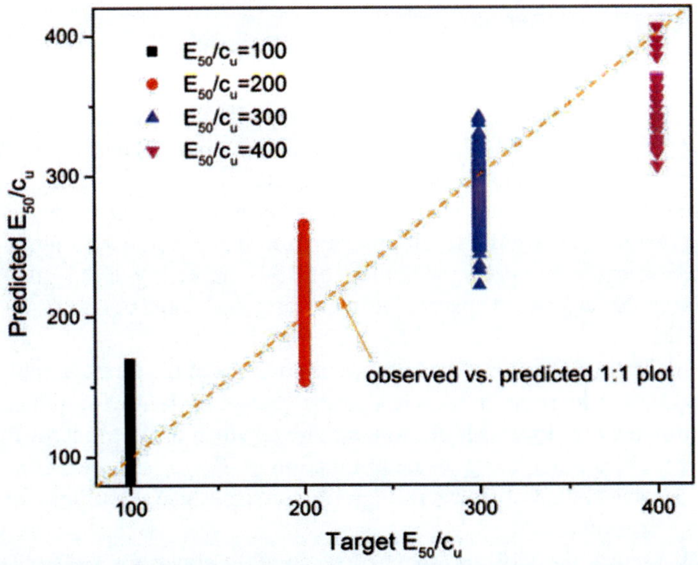

Fig. 10.20 Observed vs. predicted 1:1 plots of E_{50}/c_u

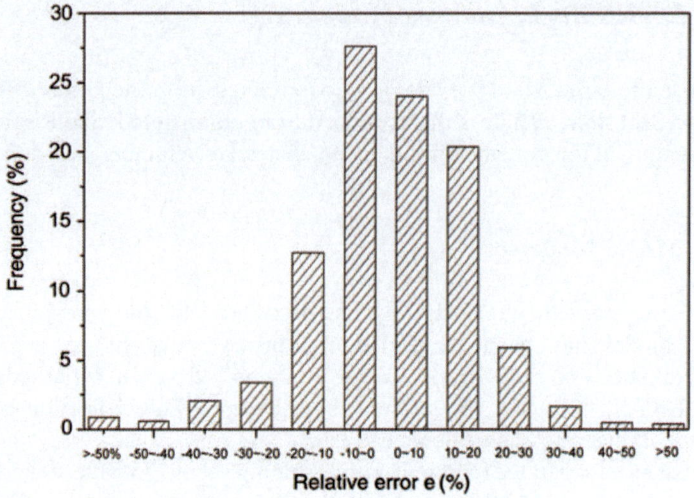

Fig. 10.21 Histogram of relative errors for the MARS E_{50}/c_u model

obvious that most of the MARS estimations of the data patterns fell within ±20% of the target values. In addition, it should be noted that MARS predictions still have a large range of variations with respect to the same group of targets. On the other hand, it can be acceptable since the target outputs are of category nature (types) while the estimations based on the developed model are numerical values. Some typical training and testing data sets together with the MARS predictions are listed in Table 10.8 and 10.9, respectively.

Table 10.10 displays the ANOVA decomposition of the developed MARS model. The first column lists the ANOVA function number. The second column gives an indication of the importance of the corresponding ANOVA function, by listing the GCV score for a model with all BFs corresponding to that particular ANOVA function removed. The third column provides the standard deviation of this function. It also gives an indication of its relative importance to the overall model and can be interpreted in a manner similar to the standardized regression coefficient in a linear model. The fourth column gives the number of BFs comprising of the ANOVA function. The last column gives the particular input variables associated with the ANOVA function.

Figure 10.22 gives the plot of the relative importance of the input variables, which is evaluated by the increase in the GCV value caused by removing the considered variables from the developed MARS model. The results indicates that the three most important variables influencing the determination of E_{50}/c_u are the calculated wall deflection δ_h, the system stiffness in logarithmic scale $\ln(S)$, and the soil relative shear strength ratio c_u/σ'v.

Table 10.11 lists the BFs and their corresponding equations for the developed MARS model. It is observed from Table 4 that interactions have occurred between BFs since exactly half of the basis functions are products of two spline functions (from

Table 10.8 Some typical training data for MARS E_{50}/c_u identification model

B (m)	T (m)	H_e (m)	$c_u/\sigma'v$	(mm)δ_h	ln(S)	γ (kN/m^3)	Target E_{50}/c_u	MARS predicted E_{50}/c_u
30	30	11	0.34	26	8.846	19.0	400	390
30	30	11	0.34	35	7.313	19.0	400	405
40	25	14	0.34	48	8.176	19.0	300	281
30	25	20	0.34	70	7.313	17.0	300	280
40	25	20	0.25	76	8.176	19.0	300	274
30	30	20	0.29	77	8.176	19.0	300	286
30	30	20	0.34	77	7.313	19.0	300	300
40	25	20	0.25	98	7.313	19.0	300	281
40	30	20	0.29	98	8.176	17.0	300	283
30	30	11	0.29	54	7.313	19.0	300	303
40	25	14	0.29	69	8.176	19.0	200	211
40	35	20	0.34	100	8.176	19.0	200	201
50	25	20	0.34	101	7.313	17.0	200	232
30	30	20	0.34	101	8.176	15.0	200	229
40	30	17	0.34	92	8.176	15.0	300	287
40	30	20	0.21	115	8.846	19.0	200	215
50	25	14	0.29	81	8.176	17.0	200	210
40	35	14	0.34	81	6.097	19.0	300	313
40	30	17	0.34	99	8.176	19.0	100	102
40	30	17	0.21	107	8.846	19.0	200	225
40	30	17	0.29	108	8.176	15.0	300	291
30	30	20	0.25	134	8.176	15.0	200	224
40	30	11	0.25	74	8.176	17.0	300	290
40	30	20	0.34	135	7.313	15.0	300	260
30	30	20	0.29	135	7.313	17.0	200	208
30	30	20	0.34	298	6.097	15.0	100	101
40	30	20	0.25	293	6.097	17.0	200	200
50	35	14	0.29	204	6.097	17.0	200	219
40	35	11	0.25	159	7.313	19.0	100	87
50	25	11	0.25	150	7.313	17.0	100	95

BF15 to BF28). The presence of interactions suggests that these two models are not simply additive and that interactions play an important role in building accurate model for soil parameter identification. The equation of this optimal MARS model is given by

Table 10.9 Some typical testing data for MARS E_{50}/c_u identification model

B (m)	T (m)	H_e (m)	c_u/σ'v	(mm)δ_h	ln(S)	γ (kN/m^3)	Target E_{50}/c_u	MARS predicted E_{50}/c_u
30	30	11	0.29	37	8.176	19.0	400	395
40	30	11	0.34	42	8.176	19.0	300	326
40	25	17	0.34	65	8.176	19.0	200	207
30	30	14	0.34	55	7.313	19.0	300	290
60	30	14	0.29	76	8.846	17.0	200	246
40	30	20	0.25	115	8.176	17.0	300	269
30	30	17	0.25	98	8.176	19.0	200	230
40	25	20	0.29	119	6.097	19.0	300	276
30	30	17	0.34	112	7.313	15.0	300	263
40	30	20	0.25	132	8.176	15.0	300	275
40	25	17	0.34	114	7.313	19.0	100	114
30	30	17	0.29	125	8.176	17.0	100	102
30	30	17	0.34	167	6.097	19.0	100	87
40	30	14	0.29	138	6.097	19.0	200	211
30	25	11	0.29	110	6.097	17.0	200	207
50	25	11	0.29	74	7.33	17.0	300	293

Table 10.10 ANOVA decomposition for MARS E_{50}/c_u model

Function	GCV	STD	#basis	Variable(s)
1	1711.6	23.61	2	B
2	4915.2	58.34	2	T
3	5724.9	54.83	2	H_e
4	9107.5	54.66	2	c_u/σ'v
5	25,614.0	134.44	2	δ_h^*
6	12,670.7	98.30	2	ln(S)
7	2584.5	37.73	2	γ
8	1290.3	19.34	2	T & H_e
9	1027.5	12.38	2	T & c_u/σ'v
10	944.8	8.98	2	T & γ
11	955.4	9.62	2	H_e &ln(S)
12	1029.3	12.46	2	c_u/σ'v &ln(S)
13	1289.3	17.04	1	δ_h &ln(S)
14	983.2	10.42	1	δ_h& γ
15	1355.8	20.12	2	ln(S)& γ

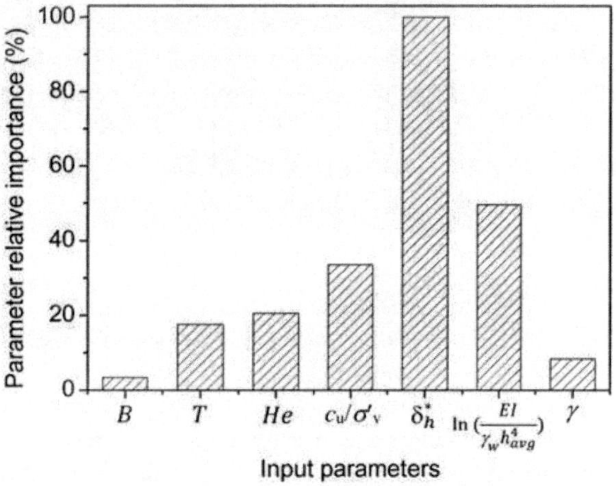

Fig. 10.22 Relative importance of the input variables selected in the MARS E_{50}/c_u model

Table 10.11 Basis functions and corresponding equations of MARS model for E_{50}/c_u

BF	Equation	BF	Equation
BF1	max(0, δ_h–132)	BF15	BF2 × max(0,ln(S)-7.313)
BF2	max(0, 132–δ_h)	BF16	BF6 × max(0, T–30)
BF3	max(0, ln(S)–7.313)	BF17	BF6 × max(0, 30–T)
BF4	max(0, 7.313–ln(S))	BF18	BF9 × max(0, 54–δ_h)
BF5	max(0, H_e–17)	BF19	BF4 × max(0, γ–17)
BF6	max(0, 17–H_e)	BF20	BF4 × max(0, 17–γ)
BF7	max(0, $c_u/\sigma'v$–0.25)	BF21	BF4 × max(0, $c_u/\sigma'v$–0.29)
BF8	max(0, 0.25–$c_u/\sigma'v$)	BF22	BF4 × max(0, 0.29–$c_u/\sigma'v$)
BF9	max(0, γ–17)	BF23	BF7 × max(0, T–30)
BF10	max(0, 17–γ)	BF24	BF7 × max(0, 30–T)
BF11	max(0, T–30)	BF25	BF11 × max(0, γ–17)
BF12	max(0, 30–T)	BF26	BF11 × max(0, 17–γ)
BF13	max(0, B–40)	BF27	BF4 × max(0, H_e–14)
BF14	max(0, 40–B)	BF28	BF4 × max(0, 14–H_e)

$$\begin{aligned} E_{50}/c_u = {} & 290 - 1.97 \times \text{BF1} + 3.45 \times \text{BF2} - 133.26 \times \text{BF3} + 106.75 \times \text{BF4} + 8.61 \times \text{BF5} \\ & - 19.86 \times \text{BF6} - 1377 \times \text{BF7} + 1575 \times \text{BF8} - 17.64 \times \text{BF9} + 37.12 \times \text{BF10} \\ & + 15.80 \times \text{BF11} - 23.47 \times \text{BF12} + 1.2 \times \text{BF13} - 4.67 \times \text{BF14} + 0.67 \times \text{BF15} \\ & - 1.69 \times \text{BF16} + 2.16 \times \text{BF17} + 2.75 \times \text{BF18} - 19.37 \times \text{BF19} + 31.61 \times \text{BF20} \\ & - 372 \times \text{BF21} + 967 \times \text{BF22} - 53.95 \times \text{BF23} + 85 \times \text{BF24} - 2.42 \times \text{BF25} \\ & + 2.8 \times \text{BF26} + 3.23 \times \text{BF27} - 8.9 \times \text{BF28} \end{aligned} \quad (10.1)$$

10.2.1.2 Excavation History Validations on E_{50}/c_u Model

To validate this proposed MARS model for soil parameter identification, a total of 12 well-documented excavation case histories from various countries as listed in Table 10.12 were analyzed. To visualize the quality of prediction, predicted stiffness ratio E_{50}/c_u values by MARS model are compared with the target values for the 12 cases listed in Table 5 and shown in Fig. 10.23. Figure 10.23 also plots the range of E_{50}/c_u by varying δ_h by + 15 and −15%. Table 10.12 and Fig. 10.23 indicate that the developed MARS model is able to predict reasonably well the soil stiffness ratios for the case histories considered, even with considerable variability in the wall deflection measurements.

Figure 10.24 gives the plot of the probability density of E_{50}/c_u for a typical case (Bugis MRT station) derived from Monte Carlo Simulation with 1 000 000 iterations assuming that the coefficient of variation COV of both δ_h and $c_u/\sigma'v$ are 0.15. Both δ_h and $c_u/\sigma'v$ follow the normal distribution, the mean values of which are 150 and 0.25, respectively. The other five input variables are deterministic and the values can be found in Table 5. As can be seen from Fig. 10.24, the variation of E_{50}/c_u follows the normal distribution. The most probable E_{50}/c_u value is 147, very close to the target value of 150.

10.2.1.3 MARS Model for ln($mathbf S$)

A second MARS model was also developed for inverse parameter estimation of the wall system stiffness. This model will assist engineers to determine the appropriate wall size during the preliminary design phase. The same patterns as used for E_{50}/c_u model are adopted for training and testing of the MARS ln(S) model, respectively. The optimal MARS model adopted 28 BFs of linear spline functions. For comparison, Fig. 10.25 plots the R^2 values of the testing data sets for the MARS ln(S) model with different BFs (from 14 to 48). The observed versus predicted 1:1 plots of ln(S) are shown in Fig. 10.26. Figure 10.27 presents the histogram plots of the relative errors. It is obvious that most of the MARS estimations of the data patterns fell within ±10% of the target values. Some typical training and testing data sets together with the MARS predictions are listed in Table 10.13 and 10.14, respectively.

Table 10.12 Summary of excavation case histories validating MARS E_{50}/c_u model

Case number and name	B (m)	T (m)	H_e (m)	$\frac{c_u}{\sigma'v}$	δ_{hM}[a](mm)	ln(S)	γ (kN/m^3)	μ_w	Target $\frac{E_{50}}{C_u}$[b]	MARS predictions	References
1. Form case	35	27	18.5	0.34	62	7.30	19.0	0.8	200	264	[5]
2. Taiwan Power Company	60	13.5	14.7	0.30	63	6.65	19.0	0.9	150	212	[28]
3. Shandao Temple	21.5	26.5	18.5	0.30	36.7	7.82	18.7	0.8	250	391	[29]
4. Xinyi planning zone	41	27	14.45	0.34	78	7.02	18.7	0.8	150	192	[30]
5. Bugis MRT	21	35	18	0.25	135	8.18	16.5	0.9	150	164	[8]
6. Lavender	24	16	15.7	0.25	32	7.96	17.0	0.8	150	172	[31]
7. MRT line in Singapore	20	20	16	0.25	38.6	8.11	17.6	0.8	150	191	[32]
8. Muni Metro Turnback	16	20.5	13.1	0.22	48	7.31	16.5	0.8	250	273	[33]
9. Lurie	64	7.4	11.8	0.25	66	5.85	18.9	0.8	165	205	[34]
10. Yanchang Road	18.1	15.5	15.2	0.30	65	5.77	18.0	0.9	190	183	[10]
11. Pudian Road	20.4	15.5	16.5	0.30	71	6.12	18.0	0.9	190	132	[10]
12. Shanghai Bank building	43	19.3	15.2	0.30	67.4	6.57	18.6	0.9	190	250	[35]

[a] As shown in Eq. (5), $\delta_{h,M} = \delta_h^* \times \mu_w$, $\delta_{h,M}$ is the measured wall deflection, mainly used as inputs for validation purposes based on case histories

[b] Herein target means that the values are from either lab/field tests and reported by the various authors or back analysis

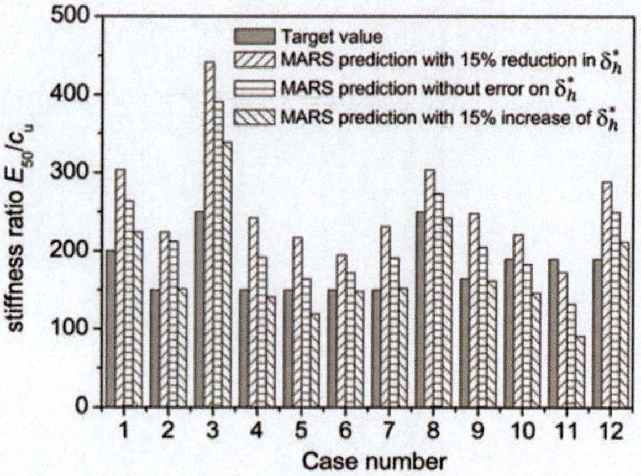

Fig. 10.23 Target and predicted E_{50}/c_u values by developed MARS model

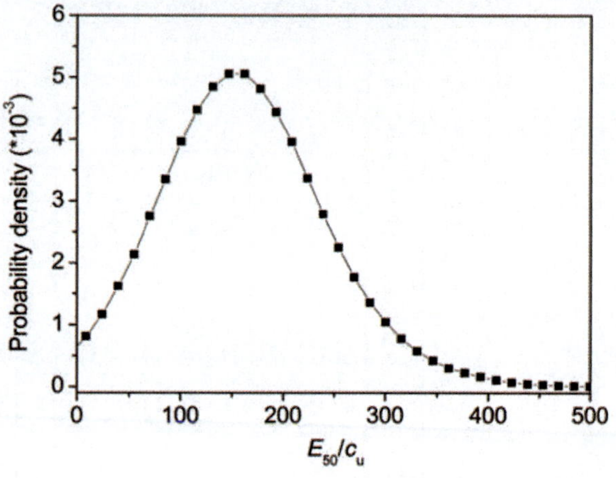

Fig. 10.24 Distribution of the predicted E_{50}/c_u values based on Eq. 10.1 for Bugis MRT

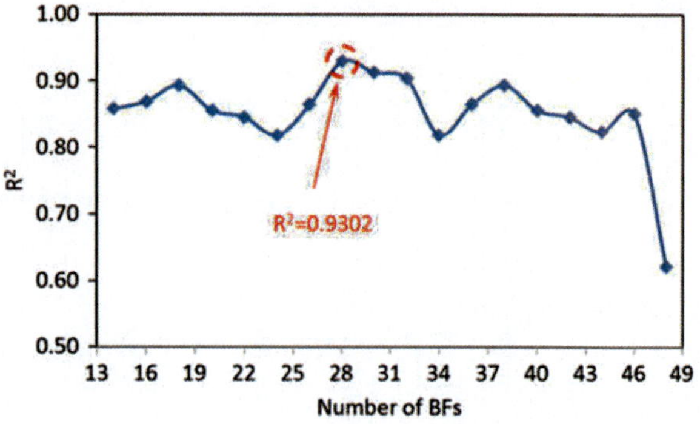

Fig. 10.25 R^2 for different number of BFs for ln(S) model

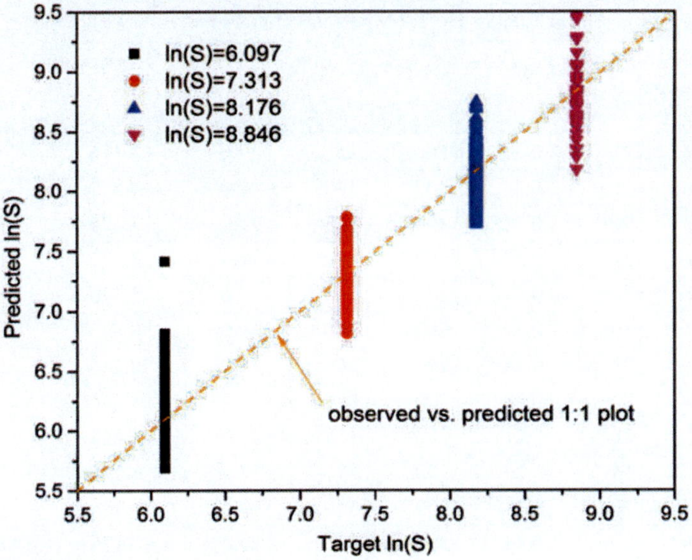

Fig. 10.26 Observed versus predicted 1:1 plots of ln(S)

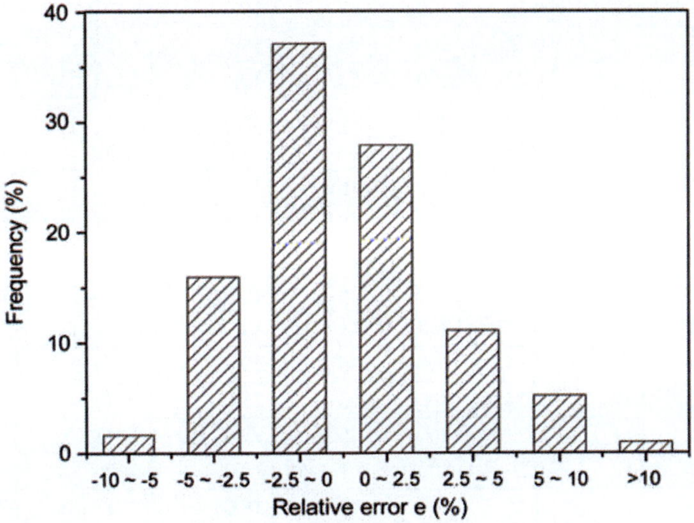

Fig. 10.27 Histogram of relative errors for MARS ln(S) model

Table 10.13 Some typical training data for MARS ln(S) model

B (m)	T (m)	H_e (m)	$c_u/\sigma'v$	E_{50}/c_u	(mm)δ_h	γ (kN/m^3)	Target ln(S)	MARS predicted ln(S)
40	25	20	0.34	100	95	19.0	8.176	8.319
30	25	20	0.34	100	110	17.0	7.313	7.559
30	30	20	0.34	100	115	17.0	8.176	8.120
30	30	14	0.34	100	98	17.0	8.176	8.169
40	25	20	0.34	200	68	19.0	8.176	8.109
30	30	17	0.34	200	58	19.0	8.846	8.721
30	30	20	0.25	200	86	19.0	8.846	9.153
40	30	20	0.29	200	87	19.0	8.846	8.849
40	30	20	0.21	200	115	19.0	8.846	9.029
50	25	14	0.29	200	81	17.0	8.176	8.251
40	25	20	0.34	300	57	19.0	8.176	8.092
30	30	20	0.34	300	63	19.0	8.176	8.001
30	25	20	0.34	300	70	17.0	7.313	7.499
30	30	20	0.29	300	77	19.0	8.176	8.048
40	25	17	0.25	300	72	19.0	8.176	8.235
30	30	17	0.25	300	83	19.0	8.176	8.145
30	30	11	0.29	300	54	19.0	7.313	7.315
40	25	17	0.25	300	142	19.0	6.097	5.856
40	30	11	0.21	300	93	17.0	8.176	8.030
40	30	17	0.25	300	181	19	6.097	6.142
30	30	14	0.34	300	151	15.0	6.097	5.951
40	30	20	0.25	300	202	19.0	6.097	6.044
30	30	11	0.34	400	26	19.0	8.846	8.699
30	30	17	0.34	400	41	19.0	8.846	8.810
40	30	20	0.34	400	75	19.0	7.313	7.332
40	30	17	0.34	400	68	19.0	7.313	7.351
30	30	11	0.29	400	45	19.0	7.313	7.260
30	30	17	0.29	400	61	19.0	8.176	8.070
40	35	17	0.25	400	123	19.0	7.313	7.399
40	30	14	0.25	400	90	19.0	7.313	7.292

Table 10.15 lists the BFs and their corresponding equations for the developed MARS ln(S) model. It is observed from Table 10.15 that interactions have occurred between BFs since exactly 12 out of the 28 BFs are products of two spline functions. The presence of interactions suggests that these two models are not simply additive

Table 10.14 Some typical testing data for MARS ln(S) model

B (m)	T (m)	H_e (m)	$c_u/\sigma'v$	E_{50}/c_u	(mm)δ_h	γ (kN/m^3)	Target ln(S)	MARS predicted ln(S)
30	30	20	0.34	100	106	19.0	8.176	8.151
30	30	20	0.29	100	122	19.0	8.176	8.310
40	25	17	0.34	100	114	19.0	7.313	7.012
50	25	20	0.34	100	136	17.0	7.313	7.357
30	30	20	0.29	200	75	19.0	8.846	8.904
30	25	17	0.34	200	78	17.0	7.313	7.281
50	25	17	0.29	100	151	17.0	7.313	7.381
50	25	20	0.29	200	116	17.0	7.313	7.442
40	30	11	0.34	200	64	19.0	7.313	7.441
30	25	17	0.21	200	126	17.0	7.313	7.442
30	25	20	0.29	200	150	17.0	6.097	5.955
40	35	17	0.34	200	128	17.0	7.313	7.337
40	25	17	0.29	300	63	19.0	8.176	8.004
30	30	14	0.34	300	54	19.0	8.176	8.219
40	25	20	0.34	300	94	19.0	6.097	6.075
40	35	20	0.25	400	147	19.0	7.313	7.297

Table 10.15 Basis functions and corresponding equations of MARS ln(S) model

BF	Equation	BF	Equation
BF1	max(0, 46–δ_h)	BF15	max(0, 118–δ_h)
BF2	max(0, $c_u/{}'v$–0.25)	BF16	BF13 × max(0, E_{50}/c_u–200)
BF3	max(0, 0.25–$c_u/{}'v$)	BF17	BF13 × max(0, 200–E_{50}/c_u)
BF4	max(0, E_{50}/c_u–300)	BF18	max(0, δ_h–46) × max(0, E_{50}/c_u–200)
BF5	max(0, 300–E_{50}/c_u)	BF19	max(0, T–30)
BF6	max(0, δ_h–46) × max(0, H_e–14)	BF20	max(0, 30–T)
BF7	max(0, γ–17)	BF21	BF13 × max(0, T–30)
BF8	max(0, 17–γ)	BF22	BF13 × max(0, 30–T)
BF9	max(0, δ_h–46) × max(0, 30–T)	BF23	max(0, 190–δ_h)
BF10	max(0, B–40)	BF24	BF2 × max(0, δ_h–74)
BF11	max(0, 40–B)	BF25	BF2 × max(0, 74–δ_h)
BF12	max(0, H_e–17)	BF26	BF23 × max(0, 200–E_{50}/c_u)
BF13	max(0, 17–H_e)	BF27	BF5 × max(0, γ–17)
BF14	max(0, δ_h–118)	BF28	BF5 × max(0, 17–γ)

and that interactions play an important role in building MARS $\ln(S)$ model. The equation of this optimal MARS model is given by

$$\begin{aligned}\ln(S) = {} & 7.055 + 0.038 \times BF1 - 16.977 \times BF2 + 17.522 \times BF3 - 0.004 \times BF4 + 0.007 \times BF5 \\ & - 0.0007 \times BF6 - 0.318 \times BF7 + 0.439 \times BF8 - 0.0004 \times BF9 + 0.024 \times BF10 \\ & - 0.046 \times BF11 + 0.149 \times BF12 - 0.225 \times BF13 - 0.007 \times BF14 + 0.03 \times BF15 \\ & - 0.0002 \times BF16 + 0.0003 \times BF17 + 210^{-5} \times BF18 + 0.1 \times BF19 - 0.152 \times BF20 \\ & - 0.013 \times BF21 + 0.015 \times BF22 + 0.010 \times BF23 + 0.040 \times BF24 + 0.259 \times BF25 \\ & + 710^{-5} \times BF26 + 0.0007 \times BF27 - 0.0008 \times BF28 \end{aligned} \quad (10.2)$$

10.2.1.4 Validations on ln(*S*) Model

The same excavation case histories as used for validating the E_{50}/c_u model are adopted for verifying the developed MARS $\ln(S)$ model. The MARS predicted $\ln(S)$ values are compared with the target ones for cases listed in Table 10.16 and shown in Fig. 10.28. Table 10.16 and Fig. 10.28 indicate that the developed MARS model is able to provide reasonable estimates of the wall stiffness for the case histories considered.

10.2.2 *FORM EXCEL Spreadsheet Method*

The back analysis framework is based on a Microsoft Excel spreadsheet first-order reliability method (FORM) proposed by Low [36] and the polynomial regression model in Sect. 4.4.1.

Low [36] has shown that a Microsoft Excel spreadsheet can be used to perform the minimization and determine β. As an alternative to the FORM method, MCS is adopted for comparison. In this study, the Microsoft Excel add-in program @RISK (http://www.palisade.com) was used.

The differences, defined as $g(x_i)$, between the PR predicted maximum wall deflection $\delta_{h,PR}$ and the field measurement $\delta_{h,M}$ for each excavation stage are minimized though $g(x_i) = \delta_{h,PR} - \delta_{h,M}$ under condition of

$$d = \sqrt{x_1'^2 + x_2'^2 + +x_n'^2} = \left(\mathbf{x}'^T \mathrm{x}\right)^{\prime 1/2} \quad (10.3)$$

where a primed dimensionless variable $x_i'^2$ is defined as $x_i'^2 = \left(x_i - \mu_{x_i}\right)/\sigma_{x_i}$, in which each soil variable x_i is defined in terms of its mean μ_{x_i} and its standard deviation σ_{x_i}. $\mathbf{x}'$ is the vector of the $x_1^{'}$s and the superscript T indicates the transpose of a matrix or vector, x_i^* is the design point which minimizes d and satisfies $g(x_i) = 0$.

Table 10.16 Summary of excavation case histories validating MARS ln(S) model

Case name	B (m)	T (m)	He (m)	$\frac{c_u}{\sigma'v}$	$\frac{E_{50}}{c_u}$	(mm)δ_h	γ (kN/m^3)	μ_w	Target ln(S)	MARS predicted ln(S)	References
1. F-foundation pit	35	27	18.5	0.34	200	62	19.0	0.8	7.30	7.142	[5]
2. Taiwan Power Company	60	13.5	14.7	0.30	150	63	19.0	0.9	6.65	7.786	[28]
3. Shandao Temple	21.5	26.5	18.5	0.30	250	36.7	18.7	0.8	7.82	8.761	[29]
4. Xinyi planning zone	41	27	14.45	0.34	150	78	18.7	0.8	7.02	7.201	[30]
5. Bugis MRT	21	35	18	0.25	150	135	16.5	0.9	8.18	8.054	[8]
6. Lavender	24	16	15.7	0.25	150	32	17.0	0.8	7.96	9.729	[31]
7. MRT line in Singapore	20	20	16	0.25	150	38.6	17.6	0.8	8.11	9.343	[32]
8. Muni Metro Turnback	16	20.5	13.1	0.22	250	48	16.5	0.8	7.31	8.253	[33]
9. Lurie	64	7.4	11.8	0.25	165	66	18.9	0.8	5.85	7.440	[34]
10. Yanchang Road	18.1	15.5	15.2	0.30	190	65	18.0	0.9	5.77	5.891	[10]
11. Pudian Road	20.4	15.5	16.5	0.30	190	71	18.0	0.9	6.12	5.575	[10]
12. Shanghai Bank building	43	19.3	15.2	0.30	190	67.4	18.6	0.9	6.57	7.218	[35]

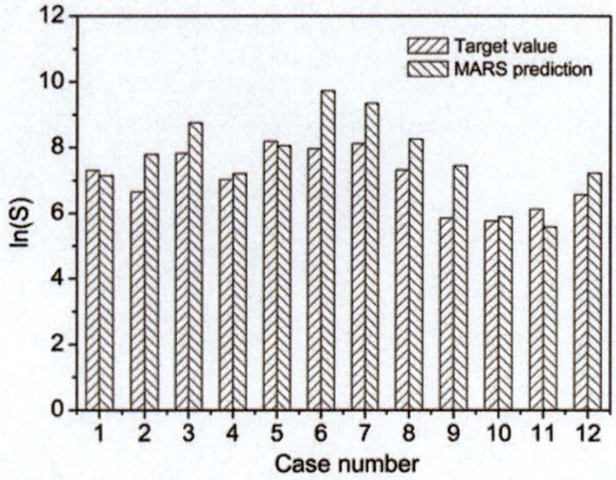

Fig. 10.28 Target and predicted ln(S) values by developed MARS model

The problem is then a constrained minimization that can be summarized as the requirement to minimize d subject to the constraint that $g(x_i) = 0$ is satisfied.

10.2.2.1 Spreadsheet Back Analysis Procedure

Generally speaking, the constrained minimization can be performed through a spreadsheet or mathematical software such as MATLAB. In this study, the design point x_i^* and thus the minimal d value were obtained by using the EXCEL spreadsheet's built-in optimization routine SOLVER.

Figure 10.29 shows the example spreadsheet setup for deriving the updated soil parameters through minimizing the difference between the actual measured wall deflection $\delta_{h,M}$ and the estimated value $\delta_{h,PR}$ obtained using this PR model.

The spreadsheet-based procedure for estimating the posterior distribution can be summarized as follows:

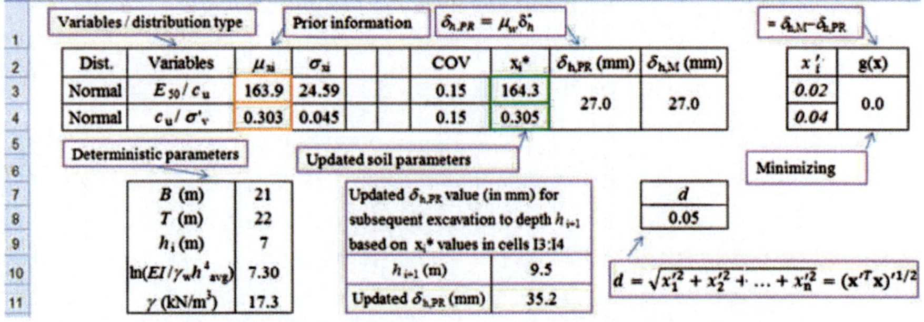

Fig. 10.29 Example spreadsheet for updating of soil parameters

Step 1: The semi-empirical PR model is firstly implemented in the spreadsheet, as shown in cells J3:K4 in Fig. 10.29. For the excavation case histories presented later, γ, B, H for each excavation stage, ln $EI/\gamma_w h_{avg}^4$, and T are simply treated as constants since they are either deterministic geometrical properties or can be determined fairly precisely. These five constant inputs are entered in cells D7:D11. For simplicity, the distribution types of the random parameters c_u/σ'_v and E_{50}/c_u are assumed as normal. The prior mean and standard deviation (SD) of input parameters c_u/σ'_v and E_{50}/c_u are entered into cells D3:D4 and E3:E4, respectively; the COVs for the two soil parameters are entered into cells H3:H4. Cells L3:L4 contain the measured wall deflection corresponding to excavation to depth H_i. The x_i' vector in Cells O3:O4 contains equations for $(x_i - \mu_{x_i})/\sigma_{x_i}$ as defined for Eq. 10.3.
Step 2: In this step, the posterior mean of the two input parameters are obtained by minimizing the difference between $\delta_{h,M}$ and $\delta_{h,PR}$, which is entered into Cells P3:P4. The design point x_i^* value in Cells I3:I4 was obtained by using the EXCEL spreadsheet's built-in optimization routine SOLVER to minimize the cell, by changing the x_i^* values, under the constraint that the performance function $g(x_i) = 0$. Prior to invoking the SOLVER search algorithm, the x_i^* values were set equal to the original mean values (150, 0.25) of the two variables. Iterative numerical derivatives and directional search for the design point x_i^* are automatically carried out in the spreadsheet environment with "Automatic scaling", "Quadratic for Estimate", "Central for Derivatives", and "Newton for search" set as default options. It can be observed from Fig. 10.29 that, based on the measured $\delta_{h,M}$ or 27.0 mm at an excavation depth H_i of 7.0 m and the prior mean values of (163.9, 0.303) of the two soil parameters, the updated soil parameter are (164.3, 0.305) and the corresponding forward prediction of the wall deflection at the subsequent stage (where $H_{i+1} = 9.5m$) is 35.2 mm (Cell I11).
Step 3: These procedures are repeated stage by stage till the end of excavation (where H equals the final excavation depth) and final pair of c_u/σ'_v and E_{50}/c_u values and be determined.

10.2.2.2 Excavation Case Histories

It should be noted that in the early stages (the first or second stage) of excavation using a diaphragm wall, the deformation of the wall generally exhibits a cantilever shape and then changes into a concave shape at later stages (e.g., after the first or second stage) [37]. Thus, as pointed out by Juang et al. [38], back analysis for the purpose of updating soil parameters based on field observations in the early stages of excavation may not be meaningful because of the inevitable transition of deformation patterns. Furthermore, the effectiveness of back analysis in the early stages is also compromised by the fact that the wall deflections in these early stages are generally small and are more prone to measurement errors. Thus, back analysis for the purpose of updating soil parameters beginning from stage 3 excavation is more reliable.

A total of seven well-documented staged excavation case histories from various countries as listed in Table 10.17 were analyzed. Table 10.18 lists the depth of excavation at various stages for each case history.

Utilizing the spreadsheet outlined in Fig. 10.29, the predicted wall deflections $\delta_{h_{PR}}$ versus the field observations for the seven case histories are shown in Figs. 10.30, 10.31, 10.32, 10.33, 10.34, 10.35, 10.36, respectively. Take TNEC and Far East Enterprise cases, for example. The mean values of these predictions versus the field observations were plotted in Figs. 10.30a and 10.31a, in which the predictions made prior to excavation, referred to herein as the as-design predictions, are denoted with the square symbol. These are the predictions made without any field measurements and involve no updating of the soil parameters. As can be seen from Figs. 10.30a and 10.31a, the as-design predictions using the initially assumed prior information deviated from the 1:1 line (the predicted deflections versus the observed values). Here,

Table 10.17 Summary of excavation case histories

Case name	B (m)	T (m)	H_e (m)	c_u/σ'_v	E_{50}/c_u	ln(S)	γ (kN/m^3)	δ_{hm} (mm)	References
TNEC	35	27	18.5	0.34	200	7.30	19.0	108	[37]
Far east enterprise	70	24	20	0.31	325	6.44	19.0	124	[39]
Bugis MRT	21	35	18	0.25	150	8.18	16.5	53	[40]
Farrer park	21	22	17.5	0.25	150	7.3	17.3	135	[41]
HDR-4 subway	12.2	22	12.2	0.25	150	6.3	19.1	165	[42]
Muni metro	16	20.5	13.1	0.2	250	7.31	16.5	48	[33]
Pudian road	20.4	15.5	16.5	0.3	190	6.12	18.0	71	[9]

Table 10.18 Depths of excavation at various stages for excavation case histories examined

Case name	Depth of excavation h (m)							
	Stage 1	Stage	Stage	Stage	Stage	Stage	Stage	Stage
TNEC	2.8	4.9	8.6	11.8	15.2	17.3	19.7	–
Far east enterprise	5.0	8.6	12.4	15.4	16.9	20.0	–	–
Bugis MRT	2.0	4.5	7.0	9.5	12.5	15.0	17.0	–
Farrer park	2.2	5.0	8.7	11.0	12.8	14.5	16.0	18.0
HDR-4 subway	4.7	7.9	11.3	12.2	–	–	–	–
Muni metro	0.5	2.1	5.3	8.5	13.1	–	–	–
Pudian road	7.2	11.1	14.2	16.5	–	–	–	–

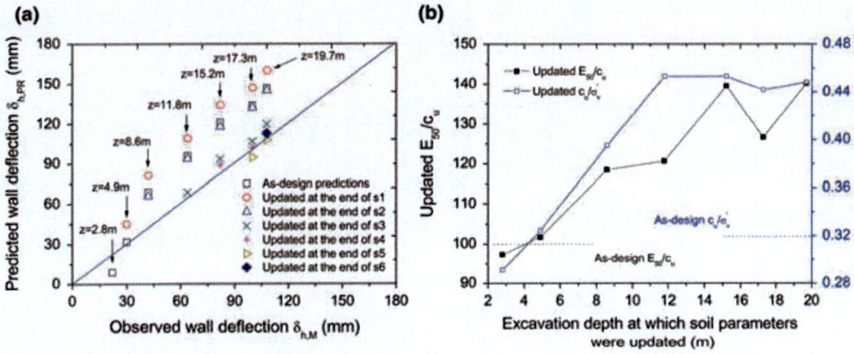

Fig. 10.30 Updating at various excavation depths for TNEC case: **a** updated maximum wall deflection; **b** updated input parameters

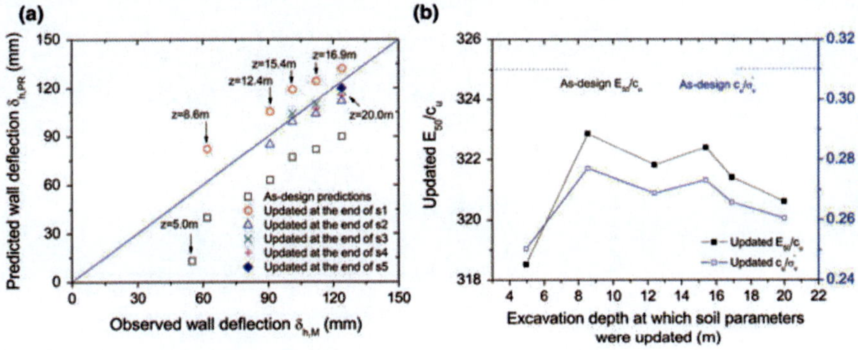

Fig. 10.31 Updating at various excavation depths for Far East Enterprise case: **a** updated maximum wall deflection; **b** updated input parameters

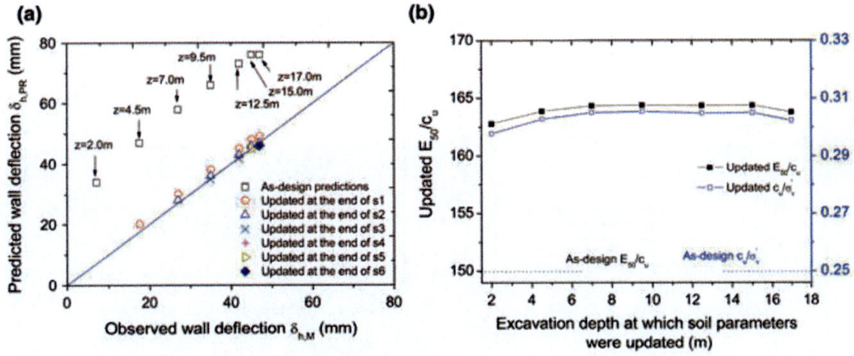

Fig. 10.32 Updating at various excavation depths for Farrer Park case: **a** updated maximum wall deflection; **b** updated input parameters

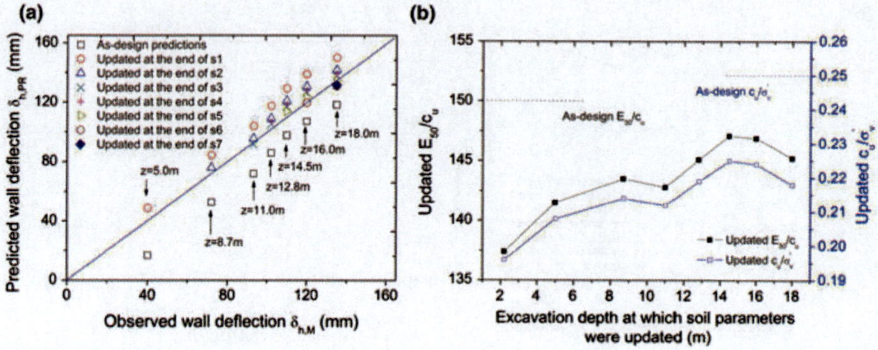

Fig. 10.33 Updating at various excavation depths for Bugis station case: **a** updated maximum wall deflection; **b** updated input parameters

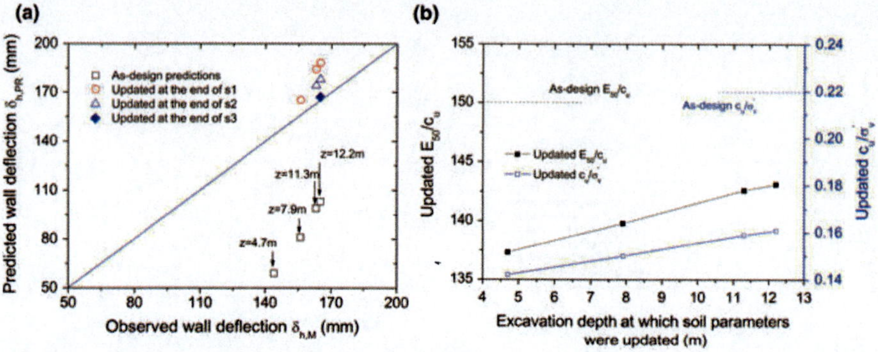

Fig. 10.34 Updating at various excavation depths for HDR-4 subway case: **a** updated maximum wall deflection; **b** updated input parameters

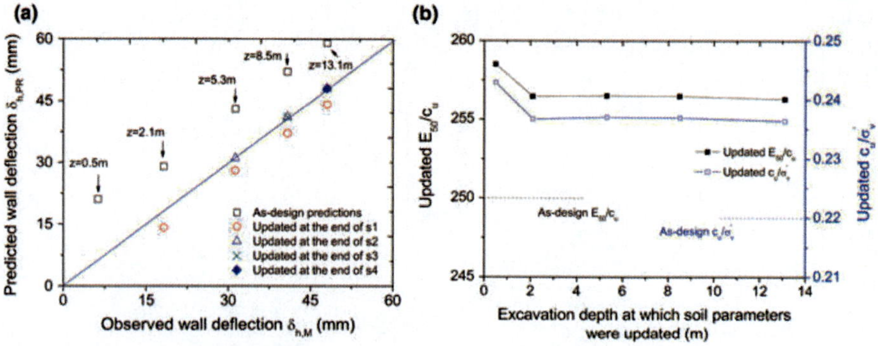

Fig. 10.35 Updating at various excavation depths for Muni Metro case: **a** updated maximum wall deflection; **b** updated input parameters

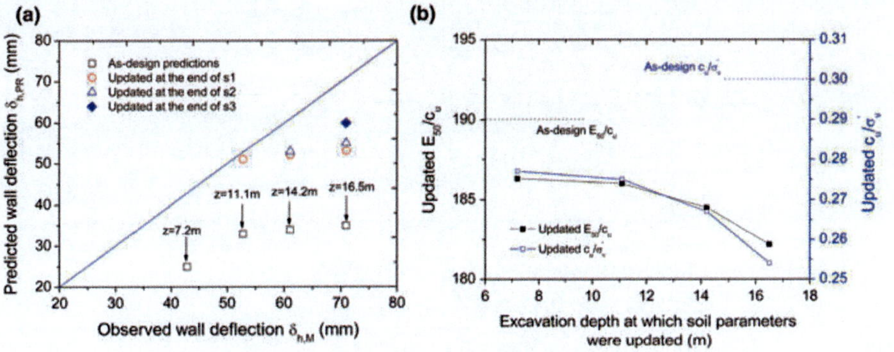

Fig. 10.36 Updating at various excavation depths for Pudian case: **a** updated maximum wall deflection; **b** updated input parameters

the observed wall deflections are those measured at the end of stages 1–7 (6 for Far East Enterprise case). The predicted maximum wall deflections are those predicted prior to stages 1–6, respectively. It should be noted that prior to each excavation stage, the wall deflection predictions are made for various excavation depths that are anticipated at all subsequent stages. That is, prior to stage 3, predictions are made at depths corresponding to end of stages 3–7, respectively. It is obvious that predictions updated at the end of stages 1 and 2 (s1 and s2), labeled with circle in red and triangle in blue, respectively, also deviated from the 1:1 line. As mentioned previously, back analysis for the purpose of updating soil parameters is more meaningful for stage 3 (s3) onwards as a result of the inevitable change of deformation patterns for s1 and s2. From stage 3 onwards, the updated predictions were very close to the 1:1 line. For the TNEC case, the updated wall deflection at the end of stage 6 is 113 mm, much closer to the observed 108 mm, compared with the 147 mm based on the as-design estimation. For the Far East Enterprise case, the benefit from updated soil parameters is more obvious, as the updated wall deflection at the end of stage 5 is 120 mm, much closer to the measured 124 mm. Similar trends can be observed for the Bugis station case, in which the continuously updated wall deflection at the end of stage 7 is 132 mm, much closer to the measured 135 mm than the as-design prediction of 118 mm.

The updated mean values of E_{50}/c_u and c_u/σ_v' are shown in Figs. 10.30b, 10.31b, 10.32b, 10.33b, 10.34b, 10.35b, 10.36b, respectively. It is obvious that both E_{50}/c_u and c_u/σ_v' change as the excavation proceeds. It should be noted that the soil parameters updated with field observations through back analysis are not necessarily the true values of these parameters, because the wall response in braced excavations is also influenced by workmanship and other construction issues. Nevertheless, updated soil parameters allow for more accurate estimations of the wall deflection in the subsequent stages of excavation. In addition, forward predictions of excessive deflections in subsequent stages will allow the designer time to develop remedial measures to prevent damage to adjacent buildings and utilities.

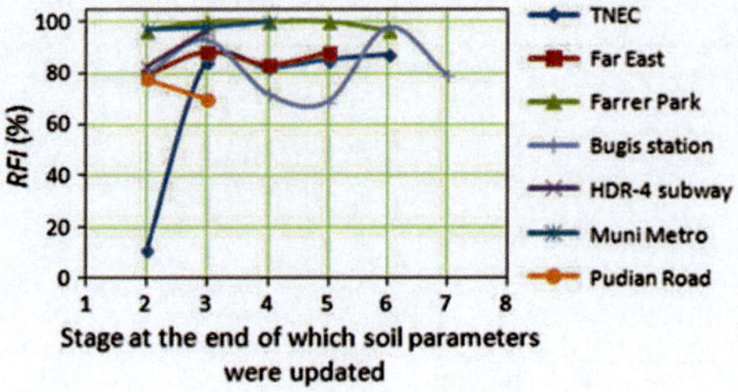

Fig. 10.37 *RFI* for various excavation stages of the seven case histories

To quantify the effectiveness of the updating procedure and assess the reliability of the predictions, the model statistic Relative Fit Improvement RFI [43] is adopted. The RFI indicates in percentage the improvement in the forward prediction from back analysis in the previous excavation stage compared with the prediction made during the as-design phase, and is defined as

$$\mathrm{RFI}_{i+1}(\%) = \frac{abs\left(\delta_{h,PR_{i+1\_asdesign}} - \delta_{h,M_{i+1}}\right) - abs\left(\delta_{h,PR_{i+1\_updating}} - \delta_{h,M_{i+1}}\right)}{abs\left(\delta_{h,PR_{i+1\_asdesign}} - \delta_{h,M_{i+1}}\right)} \times 100 \tag{10.4}$$

in which $\delta_{h,PR_{i+1\_asdesign}} - \delta_{h,M_{i+1}}$ is the anticipated wall deflection at stage $i + 1$ based on the as-design estimation; $\delta_{h,PR_{i+1\_updating}} - \delta_{h,M_{i+1}}$ is the updated predictions at stage $i + 1$ based on the recalibrated soil parameters in stage i; $\delta_{h,M_{i+1}}$ is the field observations at stage $i + 1$. *abs*() is the absolute value function. It is obvious that higher RFI values indicate better estimations of wall deflections obtained from back analysis.

Figure 10.37 plots the RFI values for the various excavation stages of the seven case histories. It is obvious that most of the cases benefited from updating soil parameters that starts at the end of stage 2. For cases Far East Enterprise, Farrer Park, HDR-4 Subway, Bugis station, and Muni Metro, the RFI values are reasonably high for every stage considered. The improvement of wall deflection prediction is marginal at the end of stage 2 for the TNEC case. However, from the end of stage 3, the effect of back analysis in improving prediction accuracy is more obvious. For Pudian Road case, the RFI value is 78 at the end of stage 2. It slightly decreases to 70 in the subsequent stage.

References

1. Wen, D. Z., & Lin, K. Q. (2002). The effect of deep excavation on pore water pressure changes in the Old Alluvium and under-drainage of marine clay in Singapore. In *Geotechnical Aspects of Underground Construction in Soft Ground*.
2. Goh, A. T. C., & Zhang, W. G. (2014). An improvement to MLR model for predicting liquefaction-induced lateral spread using multivariate adaptive regression splines. *Engineering Geology, 170*, 1–10.
3. Sharma, J. S., Hefny, A. M., Zhao, J., & Chan, C. W. (2001). Effect of large excavation on deformation of adjacent MRT tunnels. *Tunnelling and Underground Space Technology, 16*(2), 93–98.
4. Terzaghi, K., & Peck, R. B. (1967). *Soil mechanics in engineering practice*. John Wiley & Sons.
5. Ou, C. Y., Hsien, P. G., & Chiou, D. C. (1993). Characteristics of ground surface settlement during excavation. *Canadian Geotechnical Journal, 30*, 758–767.
6. Moormann, C. (2004). Analysis of wall and ground movements due to deep excavations in soft soil based on a new worldwide database. *Soils and Foundations, 44*, 87–98.
7. Liu, X. W., Shi, Z. Y., Yi, D. Q., & Wu, S. M. (1999). Deformation characteristics analysis of braced excavation on soft clay. *Chinese Journal of Geotechnical Engineering, 21*, 456–460.
8. Li, W. (2002). *Braced excavation in old alluvium in Singapore*. Singapore: Nanyang Technological University.
9. Wang, Z. W., Ng, C. W. W., & Liu, G. B. (2005). Characteristics of wall deflections and ground surface settlements in Shanghai. *Canadian Geotechnical Journal, 42*, 1243–1254.
10. Wang, J. H., Xu, Z. H., Chen, J. J., & Wang, W. D. (2005). Deformation properties of diaphragm wall due to deep excavation in Shanghai soft soil. *Chinese Journal of Underground Space and Engineering, 1*, 485–489.
11. Cetin, D. (2016). Performance of soil-nailed and anchored walls based on field-monitoring data in different soil conditions in Istanbul. *Acta Geotechnica Slovenica, 13*, 48–63.
12. Li, D., Li, Z., & Tang, D. (2015). Three-dimensional effects on deformation of deep excavations. *Geotechnical Engineering, Proceedings of the Institution of Civil Engineers, 168*, 551–562.
13. Fernie, R., & Suckling, T. (1996). Simplified approach for estimating lateral movement of embedded walls in UK ground. In *Proceedings of the International Symposium on Geotech nical Aspects of Underground Construction in Soft Ground* (pp. 131–136). Rotterdam, The Netherlands: A.A. Balkema.
14. Carder, D. R. (1995). Ground movements caused by different embedded retaining wall construction techniques. Wokingham, UK: Transport Research Laboratory.
15. Wong, I. H., Poh, T. Y., & Chuah, H. L. (1997). Performance of excavations for depressed expressway in Singapore. *Journal of Geotechnical and Geoenvironmental Engineering, 123*, 617–625.
16. Wu, P. Z., Wang, M. J., & Peng, Y. R. (1997). Discussion on deformation of diaphragm wall. In *Proceedings of the 7th Conference on Current Researches in Geotechnical Engineering* (pp. 601–608).
17. Long, M. (2001). Database for retaining wall and ground movements due to deep excavations. *Journal of Geotechnical and Geoenvironmental Engineering, 127*, 203–224.
18. Yoo, C. (2001). Behavior of braced and anchored walls in soils overlying rock. *Journal of Geotechnical and Geoenvironmental Engineering, 127*, 225–233.
19. Clough, G. W., & O'Rourke, T. D. (1990). Construction induced movements of in situ wall. Geotechnical Special Publication (pp. 439–470).
20. Ali, J., & Khan, A.Q. (2018). Behaviour of anchored pile wall excavations in clays. In *Proceedings of the Institution of Civil Engineers-Geotechnical Engineering*.
21. Wang, J. H., Xu, Z. H., & Wang, W. D. (2010). Wall and ground movements due to deep excavations in Shanghai soft soils. *Journal of Geotechnical and Geoenvironmental Engineering, 136*, 985–999.

22. Wang, W. D., Xu, Z. H., & Wang, J. H. (2011). Statistical analysis of characteristics of ground surface settlement caused by deep excavations in Shanghai soft soils. *Chinese Journal of Geotechnical Engineering, 33*, 1659–1666.
23. Peck, R.B. (1969). Deep excavation and tunneling in soft ground. In *Proceedings of the 7th International Conference on soil Mechanics and Foundation Engineering*, Mexico City, pp. 225–290.
24. Leung, E. H. Y., & Ng, C. W. W. (2007). Wall and ground movements associated with deep excavations supported by cast in situ wall in mixed ground conditions. *Journal of Geotechnical and Geoenvironmental Engineering, 133*, 129–143.
25. Mana, A. I., & Clough, G. W. (1981). Prediction of movements for braced cuts in clay. *Journal of the Geotechnical Engineering Division, ASCE, 107*, 759–777.
26. Goldberg, D. T., Jaworski, W. E., & Gordon, D. (1976). *Lateral support systems and underground, vol 2.* Design Fundamentals Construction.
27. Twine, D., & Roscoe, H. (1999). *Temporary propping of deep excavations-guidance on design, CIRIA C517*. London: CIRIA.
28. Moh, Z. C., & Song, T. F. (2013). Performance of diaphragm walls in deep foundation excavations. In *First International conferences on case histories in geotechnical engineering* (pp. 1335–1343). Missouri University of Science and Technology.
29. Fang, M. L. (1987). A deep excavation in Taipei Basin. In *Ninth Southeast Asian Geotechnical Conference* (pp 35–42). Bangkok.
30. Fang, T. C., Tsai, Y. Y., Su, T. C., Tsung, P., & Seeley, P. (2004). A case study on time-dependent displacement of diaphragm wall induced by creep of soft clay. In *Proceedings of 5th Cross-strait Geotechnics Seminars* (pp. 283–290). Taipei.
31. Lim, K.W., Wong, K.S., Orihara, K., & Ng, P.B., 2003. Comparison of results of excavation analysis using WALLUP, SAGE CRISP, and EXCAV9. In *Proceedings Singapore Underground* (pp. 83–94).
32. Goh, A. T. C., Wong, K. S., Teh, C. I., & Wen, D. (2003). Pile response adjacent to braced excavation. *Journal of Geotechnical and Geoenvironmental Engineering, 129*, 383–386.
33. Koutsoftas, D. C., Frobenius, P., Wu, C. L., Meyersohn, D., & Kulesza, R. (2000). Deformations during cut-and cover construction of MUNI Metro Turnback project. *Journal of Geotechnical and Geoenvironmental Engineering, 126*, 344–359.
34. Kung, G. T. C., Hsiao, E. C. L., & Juang, C. H. (2007). Evaluation of a simplified small-strain soil model for analysis of excavation-induced movements. *Canadian Geotechnical Journal, 44*, 726–736.
35. Xu, Z. H., Wang, W. D., Wang, J. H., & Shen, S. L. (2005). Performance of deep excavation retaining wall in Shanghai soft deposit. *Lowland Technol Int, 7*, 31–43.
36. Low, B. K. (1996). Practical probabilistic approach using spreadsheet. In *Uncertainty in the geologic environment—from theory to practice* (pp. 1284–1302). Madison, WI.
37. Ou, C. Y., Liao, J. T., & Lin, H. D. (1998). Performance of diaphragm wall constructed using top-down method. *Journal of Geotechnical and Geoenvironmental Engineering, 124*, 798–808.
38. Juang, C. H., Luo, Z., Atamturktur, S., & Huang, H. (2013). Bayesian updating of soil parameters for braced excavations using field observations. *Journal of Geotechnical and Geoenvironmental Engineering, 139*(3), 395–406.
39. Hsieh, P. G., & Ou, C. Y. (1998). Shape of ground surface settlement profiles caused by excavation. *Canadian Geotechnical Journal, 35*, 1004–1017.
40. Shirlaw, J. N., & Wen, D. (1999). Measurements of pore pressure changes and settlements due to a deep excavation in Singapore marine clay. *Proc Field Measurements in Geomechanics* (pp. 241–246). Balkema.
41. Li, W. (2001). *Braced excavation in old alluvium in Singapore.* Nanyang Technological University.
42. Finno, R. J., & Harahap, I. S. (1991). Finite-element analyses of HDR-4 excavation. *Journal of Geotechnical Engineering, 117*, 1590–1609.
43. Finno, R. J., Bryson, S., & Calvello, M. (2002). Performance of a stiff support system in soft clay. *Journal of Geotechnical and Geoenvironmental Engineering, 128*, 660–671.

Chapter 11
Excavation Failure Cases and Analysis

11.1 Introduction

In recent years, excavation accidents have occurred continuously and many accidents have occurred in cities in China such as Shanghai, Guangzhou, Hangzhou, and Taipei. These accidents include the collapse of foundation pits, water seepage, ground surface settlement, tilting of adjacent buildings, leaking of nearby water or gas pipes, and even explosions. The accidents warn us that the construction of the underground excavation has huge risks and potential danger must be taken care of. The cost of the accident is high and painful. Engineers should learn lessons from the accident cases. The following excavation failures selected are known for their impact on the geotechnical community.

11.2 Nicoll Highway Collapse, Singapore, 2004

A deep excavation for railway construction collapsed along with a section of 6 lane highway in Singapore. Four workers died. This type of construction had been used for more than 30 years in Singapore with few accidents. Such a large collapse was a big shock. There was a detailed inquiry which identified that the initiation of the failure was a poorly detailed connection in the strutting system and the total collapse was due to the inability of the system to withstand the failure of one row of strutting. Moreover, there were many other contributory factors that were identified, such as incorrect use of a computer program used to design the earth retaining system, erroneous back analyses, deficient monitoring, failure to observe stop work limits, and many more including an over-riding culture of complacency. This chapter briefly summarizes the events, the findings of the Committee of Inquiry, and the remedial actions taken, and sets out some of the lessons learned.

W. Zhang and H. Liu, *Design of Deep Braced Excavation and Earth Retaining Systems Under Complex Built Environment*,
https://doi.org/10.1007/978-981-16-5320-9_11

11.2.1 Case Description

Around 3:30 pm on April 20, 2004, the 33.5 m deep cut and cover excavation required for the construction of tunnels between Nicoll Highway and Boulevard stations (as shown in Fig. 11.1) collapsed. The collapse was catastrophic, resulting in four fatalities and a lot of damage (as shown in Fig. 11.2), which delayed the completion of the CCL1 stage by about four years.

The Singapore Government appointed a Committee of Inquiry (COI) to ascertain the causes and circumstances of the incident. The COI Report, submitted to the Government in May 2005 and made public on the Internet, is the main source of information for this chapter.

When the incident occurred, the surrounding ground collapsed into the excavation area. The area of the collapsed zone was approximately 100 by 130 and 30 m deep. Its edge was only about 10 m away from the closest building-Golden Mile Complex. The Nicoll Highway and the approach slab before the abutment of the Merdeka Bridge over the Kallang River were also damaged. A storm drain located south of the cut and cover tunnel was the main drainage outlet, conveying water to the Kallang River. When the collapse occurred, the river water rushed into the collapse area. Several key utilities, including power electric mains, gas mains, and water mains were broken by the collapse, causing a fire. All this represented a challenge for the following remedial works.

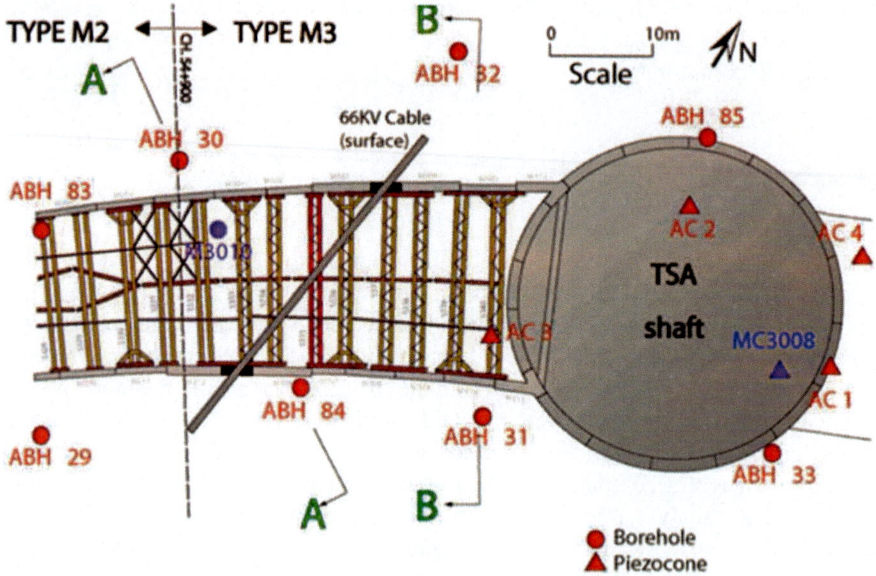

Fig. 11.1 Plan showing location of diaphragm wall panels, 9th level strutting system and site investigation [1]

Fig. 11.2 The surrounding ground deformation of the collapsed excavation

The COI report states: "Warnings of the approaching collapse were present from an early stage but these were not taken seriously". Indeed, as early as August 2003, it became clear that something was wrong with the design. First, in a launch shaft for the tunnel boring machines, across the river east from the collapsed area, measured wall deflection exceeded 500 mm, while the calculated design level was only 190 mm. This caused damage to the retaining walls, ground settlements, and cracks at a cricket field nearby.

Next, in January 2004, the design deflection level was exceeded and cracks appeared in the retaining structures in the cut and cover excavation west of the collapsed area. The design deflection level had to be revised upwards from 222 to 522 mm (from 125 to 313 mm closer to the collapse area). This was accompanied by numerous complaints received from owners of neighboring properties, including ground settlements of a driveway, cracks in buildings, and other building damages. Manpower resources from the C824 project team were deployed to handle the repair works. Finally, in the collapsed area itself, the inclinometer I-104 indicated that the original design level of 145 mm in the southern diaphragm wall was exceeded as early as February 2004 (Fig. 11.3). At that moment, installation of the sixth strut level was in progress.

The COI concluded, which in our computer-dominated age sounds like a paradox: a wrong use of numerical modeling in geotechnical design, together with some structural errors, were the main causes of the Nicoll Highway collapse [1].

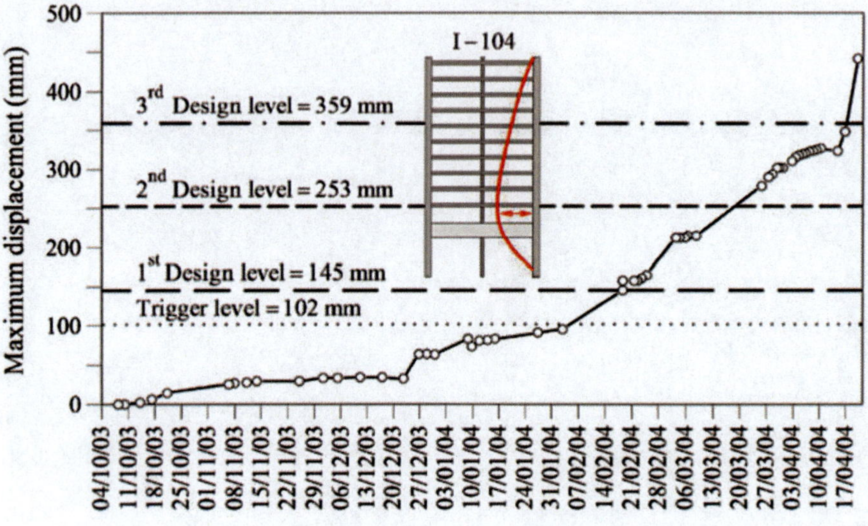

Fig. 11.3 Displacement of the southern wall in time: measurements for inclinometer I-104

11.2.2 Lessons Learned

The final report of the Committee of Inquiry into the collapse of the Nicoll Highway was submitted on May 11, 2005. Faced with the monumental task of piecing together information from 193 witnesses and volumes of documents, the COI elucidated the crucial errors and persons responsible for the collapse of Nicoll Highway. In the COI's words: "Warnings of the approaching collapse were present from an early stage but these were not taken seriously". It concludes that "the death of four persons was the direct result of the collapse" and that "the Nicoll Highway collapse could have been prevented". The following recommendations were made by the Committee of Inquiry to avoid similar incidents in the future.

(1) Effective risk management
The potential for major accidents, whether due to the construction process or deficiencies in design, must be recognized and expeditiously controlled. It is inappropriate to leave the control of risk wholly to contractors. Owner's and builder's management must seek a balance between production pressures and quality and safety goals.

(2) Robustness of design
A robust design is essential. This robustness is provided by identifying the hazards and checking that the proposed design can adequately withstand them. The design should have sufficient redundancy to prevent a catastrophic collapse in the event of a failure of any particular element. Temporary works for deep excavation should be given the same respect as permanent works [1].

(3) Numerical modeling in geotechnical design
Numerical analysis should supplement and not supplant sound engineering judgment and practice. Those who perform geotechnical numerical analysis must have a fundamental knowledge of soil mechanics and a clear understanding of numerical modeling.

(4) Back analysis
A proper back analysis should be done with an understanding of why the design is not performing as originally predicted, and not just to increase the design levels. The input parameters (e.g., soil properties) should be adjusted to allow for the displacements and forces to fit the measurements. If these adjusted input parameters fall outside the meaningful range, the analysis model is probably wrong, and should not be used for further predictions.

11.3 Xianghu Metro Station Collapse, Hangzhou, China, 2008

Xianghu Station is the starting point of Hangzhou Metro Line 1. This station was constructed using the bottom-up construction method in eight excavations from north to south. The collapsed position is the second excavation on the northern side (N2 excavation), which is near the junction of Fengqing Road and Xiangxi Road as shown in Fig. 11.4. The excavation occupies a planning area of about 2313 m^2, with approximately 21.5 m in width and 107.6 m in length. The maximum digging depth is up to 16.2 m. The accident happened at the No. 2 foundation pit. The foundation pit is 107.8 long, 21.05 m wide and the depth is 15.7–16.3 m.

At around 15:15 on November 15, 2008, a large-scale collapse accident occurred on the west side of the North 2 Foundation Pit as shown in Fig. 11.5. The collapsed area is 75 m long, about 20 m wide, and 15 m deep. More than 11 vehicles in progress fall into the collapsed area. At least 50 people are buried in the collapse, causing 21 deaths, 4 injuries, and direct economic losses is more than 49.62 million yuan. This is the most serious accident that happened in the history of China's subway construction.

On November 17, 2008, there was a crack on the ground at the southern end of the collapsed foundation pit. In order to facilitate the rescue work, the houses near the accident have been completely knocked down.

11.3.1 *Case Analysis*

Through field investigation and preliminary analysis, the reasons for the excavation collapse were the misuse of the soil parameters, over-excavation, incorrect installation of steel struts, invalid monitoring data, and inadequate ground improvement [2].

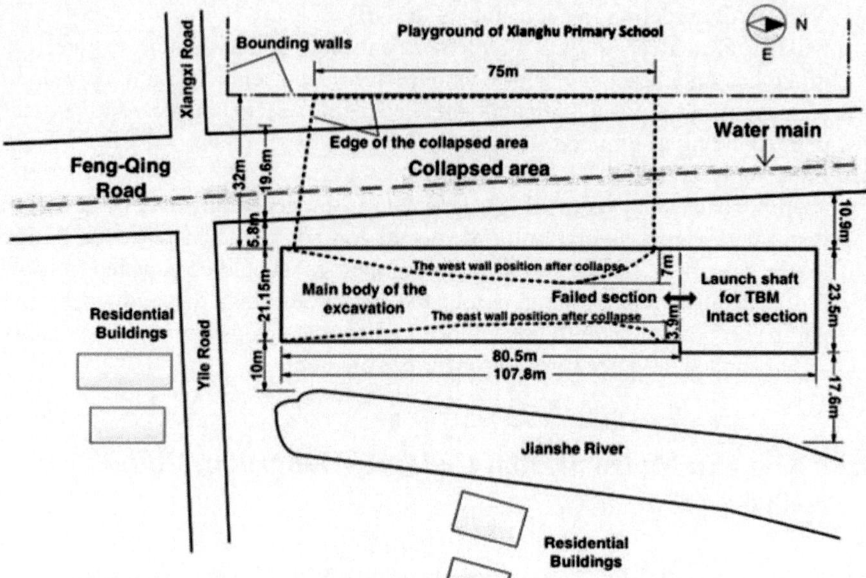

Fig. 11.4 Plan of the N2 excavation and collapse-influenced area

The soil condition of Hangzhou is abnormal, according to the field investigation, the accident was in silty clay. As the main road, Fengqing Avenue aside the foundation pit has a large traffic and a large load, which has a significant influence on the load-bearing wall on the west side of the foundation pit, causing the entire foundation pit to collapse, furthermore, a rare continuous rainfall in Hangzhou in October increased the instability of sand [3].

The whole design did not consider the influence of the surrounding environment. The demanded demolition and transfer of nearby house was not carried out. The construction method should be closely integrated with the stratum and the surrounding environment, however, no good construction method has been developed in this project, since a cheaper one has been chosen, and caused a serious damage to the environment, and many problems occurred, it ends up as the most expensive method.

The construction was in a rush, in order to save money and time, the construction period and cost was unreasonable as shown in Table 11.1, these are some of the main reasons that affected the safety and quality. During the project, there is a collapsed sewage pipe leaking for a long time, which causes the formation of the slipping surface, consequently, it may cause the collapse with the help of continuous rain.

There are some main causes for the collapse:

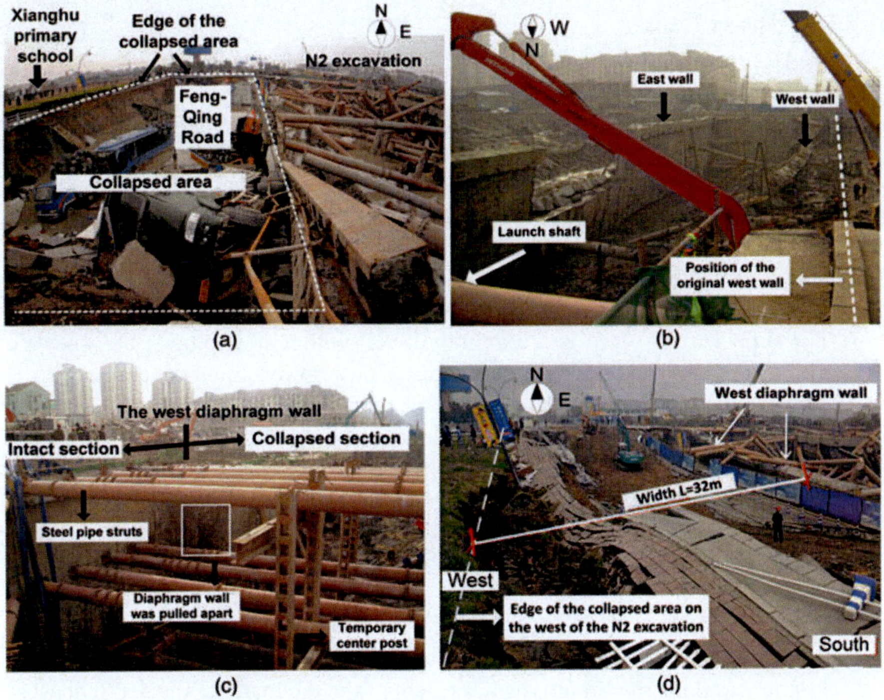

Fig. 11.5 Site scenes of the accident

Table 11.1 Construction details of the N2 excavation

Construction details	Date
Diaphragm wall construction	April 6 to August 15, 2008
Dewatering began	September 5, 2008
Soil excavation began	October 2, 2008
Section 11.1: foundation floor and four level struts completed	October 2 to November 5, 2008
Section 11.2: paving the bottom with concrete cushion four level struts completed	November 13, 2008
Section 11.3: clearing the bottom and four level struts completed	November 13, 2008
Section 11.4: excavated to the bottom, three level struts complete, installing the fourth-level struts	November 15, 2008
Section 11.5: excavated to the bottom, three level struts complete, the fourth-level struts not installed	November 15, 2008
Section 11.6: excavating the soil from GL − 12.7 to GL − 15.7 m, the third-level struts completed	November 15, 2008
N2 excavation collapsed	November 15, 2008 at 3:00p.m

11.3.1.1 Direct Causes

(1) Over-excavation:
The design document stipulates that when the foundation pit is excavated to 0.5 m below the support design elevation, the excavation must be stopped, and the support should be set in time, and no over-excavation is allowed.
However, during the excavation, when the last strut was supposed to install, some sections have been dug to the bottom of the foundation pit, and the strut installation and the construction of the concrete floor are not in time.

(2) Weak links in the support system:
The supporting system is a diaphragm wall with a steel pipe strut, however, the design report did not include a detailed description about the connection of the supporting steel pipe strut and the diaphragm wall, as well as an example of the joint of the strut. There are no corresponding technical requirements put forward regarding the joint, nor were welding requirements for steel strut and diaphragm wall embedded parts. No welding was performed.
The excessive lateral displacement of sectional wall together with large axial forces and severely eccentric of the struts, resulting in instability of the support system.

(3) Monitoring failed:
Before November 15th, the maximum settlement of the ground has reached 316 mm, and the maximum displacement of the inclinator at the depth of 18 m is 43.7 mm. On November 13th, the maximum lateral displacement of the inclinator has reached 65 mm, both of which have already exceeded the alarm value, but none of them have alarmed.

11.3.1.2 Indirect Causes:

(1) The original plan of the designer is reinforcing the foundation pit by strip-shaped mixing pile group at the bottom; however, the contractor suggested to consolidate the base by dewatering instead of using reinforcement measures. The design report proposed:

   (a) The groundwater in the foundation pit is reduced to 1 m below the lowest point of the structural internal members and 3 m below the bottom of the foundation pit;
   (b) The dewatering test should be carried out before excavation, and on-site dewatering should be done 4 weeks in advance;
   (c) When the ground subsidence exceeds the alarm value, the dewatering should be stopped and reported in time.

   But none of these were fully implemented in the actual construction.

(2) The supervisor does not stop the serious problems that disobey the design codes and specifications.

(3) The results of physical and mechanical properties test are reliable. However, the design of the supporting system has not taken into consideration of the comprehensive design and reasonable recommendation according to the characteristics of the local soft soil, which caused the selected parameters too large and reduced the factor of safety of the foundation pit system.

11.3.2 Lessons Learned

After field investigation and preliminary analysis, five reasons for excavation collapse were concluded: (1) using the average value of CQ and not the standard value of CU as the strength parameter; (2) canceling the fourth-level struts; (3) poor connection of pipe struts with coupling beams and the diaphragm wall; (4) insufficient monitoring instruments and ignoring the unusual performance of the excavation; and (5) using artesian wells for consolidation instead of jet grouting [2].

(1) This collapse accident once again proves that excavation of soils should be based on the principles of layered and zoned construction, symmetrical and balanced excavation, early cast of basal slabs, and over-excavation is strictly prohibited, especially in soft soil areas. This project was not implemented in accordance with these principles, which is one of the direct causes of the accident.
(2) When using a retaining wall with inner support structure in thick soft soil area, the wall must consider stability problems in the deep area, including wall settlement, uplift, and "toe kick" damage, it should be calculated and verified by a variety of methods. If the requirements cannot be met, the insert depth of walls should be deepened or the passive area should be reinforced. The design of this project was not firm enough, and passive area reinforcement was not carried out.
(3) If the inner support is made of steel truss or steel pipe strut, a requirement of joint structure and welding should be proposed properly to avoid eccentricity and instability; and the first inner support should be supported by reinforced concrete.
(4) The selection of the shear strength parameters should be matched with the calculation purpose and the safety factor. For the calculation of earth pressure, the related calculation of "toe kick" and "overturning" should choose the results of consolidation quick shear test or triaxial CU test; the calculation of basal heave, sliding, and overall stability should choose the results of quick direct shear test, triaxial UU test or cross plate shear strength test; when soft clay is playing an important role, the design should be based on comprehensive consideration, not just the average, but also the maximum and minimum data.
(5) The monitoring should be carried out strictly, independently, and without interference. When the alarm value is reached, it must be timely reported and must not be concealed. The supervision department should be familiar with geotechnical engineering and should strictly stop some major violations.

Fig. 11.6 Collapse on the south side of the foundation pit

11.4 Guangzhou Haizhu City Square Foundation Pit Collapse

The foundation pit of Zhuhai City Plaza collapsed at around 12:20 on July 21, 2005. The pit has been put into construction since October 2002, but it has been suspended several times in the middle until it collapsed on July 21, 2005, which lasted for 2 years and 9 months.

After the collapse on the south side of the foundation pit, the strut at southeast corner failed, causing the east 20 m deep retaining wall unsupported, and posing a threat to the safety of the surrounding environment (as shown in Fig. 11.6). The foundation piles of the Seafarer Hotel, which at the south of collapsed pit were broken and displaced, and the cap was suspended in the air, resulting in the building collapsed near the foundation pit (as shown in Fig. 11.7); the foundation pile of the residential building No. 1 near the pit was exposed and failed, and soil under the caps were hollowed out (as shown in Fig. 11.8).

Fig. 11.7 The collapse of the building near the pit

Fig. 11.8 The ground surface settlement and cap failure caused by pit collapse

11.4.1 Case Analysis

According to Chen et al. [4], the main technical reasons for this collapse are:

(1) It can be concluded from the geological exploration data that the site consists of rock layer with argillaceous siltstone mingled with interlayers of strongly weathered soil within the depth of excavation; during the excavation, it is found that the rock mass on the south side was inclined to the pit, and there is water seepage and mud flow in the strongly weathered interlayer; according to the deformation monitoring data, it has shown obvious signs of the collapse on the south side of the foundation pit. However, none of these problems have drawn due attention, and no effective measures have been taken, which is the main cause of the accident.

(2) The initial design depth of the foundation pit is 17.0 m, but the actual excavation depth is 20.3 m. The 3.3 m deep over-excavation not only caused the insufficient of the anchoring force, but also makes the 20 m deep retaining wall a suspended wall and cannot be adjusted, resulting in a potential danger.

(3) The time of intermittent construction of the foundation pit is 2 years and 9 months. It is far longer than the period of one year according to the code concerning temporary construction, which leads to the softening of the soil layers because of seepage, and the reduction of anchoring force of anchor (cable) because of the corrosion of steel components.

(4) When the foundation pit collapses, there is still one backhoe excavator, one crane, and one 16 m^2 dump truck on the south side of the foundation pit near the Seafaring hotel. When the foundation pit has been in danger, these construction loads are also one of the causes of the collapse of the foundation pit.

11.4.2 Lessons Learned

The following situation should be strictly avoided:

1. Over-excavation: the original design depth is 17 m, but the actual pit has been excavated to 20.3 m, caused the retaining wall resulting in hanging wall;
2. Overtime: the service period of the foundation pit support structure is one year, and the actual excavation have been nearly three years;
3. Overloading: overloading at the top of pit such as dump truck, crane, and backhoe excavator;
4. Ignoring geological response: the rock layer is shallow and inclined into the pit. The designers use the software to design and check the foundation pit, and ignored the fact that the rock layer was inclined 25°. It is found that the displacement of the wall on the south side was too large, so this part of the wall was reinforced with anchor cables but the range was too small that only covers 20–30 m along the west side of the south. Constructors believe that the insufficient reinforcement is caused by unreasonable design [4].

11.5 Shanghai Metro Line 4 Seepage Failure

In the early morning of July 1, 2003, a great amount of water and quicksand burst into the section of cross-river tunnel of Shanghai Rail Transit Line 4 [5], which was a connection passage used to connect the upper and lower lines (as shown in Fig. 11.9). It caused about 270 m tunnel damaged and the ground settlement reached a maximum 7 m in the surrounding area, which caused three buildings heavily inclined. Even worse, the adjacent flood wall of Huangpu River collapsed and caused piping (as shown in Fig. 11.10). Fortunately, due to the timely alarm, all the personnel in the tunnel and nearby building were safely evacuated without causing casualties. However, the accident caused direct economic losses of around 150 million.

Due to the influence of ground settlement, the 30-m-long cut-off wall at Dongjiadu road section was sunk and cracked (as shown in Fig. 11.11), which caused threat on flood control. The city rescue headquarters mobilized the armed police force to rush to the scene to take measures such as using sandbags and plugging, and controlled the danger in a timely manner. In addition, in order to protect the buildings in the surrounding areas, the technical measures were adopted such as grouting to minimize the surging of underground quicksand. In order to balance the water and earth pressure inside and outside the tunnel, a method of closing the tunnel opening and injecting water is adopted. A large number of grouting filling and reinforcement were carried out on roads and important buildings, and large settlements on the ground were back-filled, and all the damaged buildings were demolished (as shown in Fig. 11.12).

Fig. 11.9 Aerial view of accident site

Fig. 11.10 The collapsed flood wall of Huangpu River

Fig. 11.11 Ground settlement

11.5.1 Cause Analysis

The excavation sequence of the shaft and the bypass channel is wrong. The failure of the refrigeration equipment causes the rising of temperature and the underground pressurized water causes sandblasting. These three disadvantages eventually lead to the accident. For projects with high construction risks like this, it is an important cause of the accident that there is lacking of emergency response plan, the on-site management and supervision were out of control [5].

Fig. 11.12 Heavily inclined buildings

(1) Excavation sequence error
There was a large shaft above the tunnel, and other two small shafts 8–9 m below the large shaft are excavated to connect the already formed tunnel. According to the construction practice, the bypass channel should be dug first and then the shaft. However, the construction sequence has been changed, which is very likely to cause collapse. When the accident occurred, one small shaft had been dug and the other shaft was excavated about 2 m.

(2) The incorrect application of freezing method
Before the accident, the contractor, the project department of Shanghai Branch of China Coal Mine Engineering Co., Ltd., arbitrarily adjusted the original construction design while the expert group determined that the adjustment did not strictly follow the relevant provisions of the freezing construction process, resulting in weak layers of the frozen soil in the adjacent channel construction. The adjusted scheme reduced the average temperature requirement for frozen soil, from −10 °C in the original scheme to −8 °C; the number of vertical freezing tubes at the bypass channel decreased from 24 in the original scheme to 22, while for the original 7 vertical freezing pipes of 25 m deep, 4 of them were shortened to 14.25 m, and 3 were shortened to 16 m, resulting in weak frozen soil in the intersection below the waistline of the bypass channel and the under tunnel. A row of 6 frozen inclined holes with a hole spacing of 1 m was adopted, although the length of the frozen hole is increased, the number is too small and the spacing is too large, so that the freezing effect is insufficient to resist the water and soil pressure of the corresponding part.

(3) On-site management dereliction
On the morning of June 28th, the small refrigerator of the down line of the tunnel broke down, and the cooling was stopped for 7.5 h. At about 2 pm, the construction crew installed a hydrological observation hole in the down-line tunnel and found that there was always pressure water leaking. Although some measures such as blocking the tunneling surface with wood were adopted, the effect was not good. At 3 o'clock on the morning of the 29th, the water pressure measured at the water valve was close to the confined water pressure of external

seventh layer. The danger was first revealed, but no one at the scene reported the situation to the contractor and supervisor, resulting in a gradual increase in the risk.

It is in such a dangerous situation that at 0:00 on July 1st, although the deputy manager was aware of the serious potential dangers in the frozen soil layer of the connection passage, and the project has been suspended, he still determined to arrange the construction personnel to partially remove the wood sealing at the excavation face besides the frozen soil, and use the wind chisel to cut the hole with a diameter of 0.2 m to prepare the concrete conveying pipe. It is from this hole that the water emerges, and the water sand continuously rushes in from the lower right corner of the tunneling face and the side wall, so that the sealing is invalid and eventually causes an accident.

11.5.2 Repair Plan

In August 2004, the repair work of Dongjiadu section of Line 4 was restarted. After comparing the method of the in-situ repair and rerouting scheme, the in-situ repair scheme was finally adopted. After overcoming a series of technical problems such as the 65 m ultra-deep underground continuous wall, 38 m ultra-deep underground obstacle clearing, ultra-deep foundation reinforcement in complex strata, ultra-deep confined water dewatering, 41 m ultra-deep foundation pit excavation, tunnel pumping cleanup, large section freezing dug, the Dongjiadu section of Shanghai Line 4 was re-opened on July 9, 2007.

Both east and west bulkhead were built by using freezing method as cut-off wall, after the debris in the tunnel were removed, R.C. bulkhead was built to protect the unaffected tunnels from the cut and cover recovery works (as shown in Fig. 11.13).

11.6 Excavation Collapse at 14th and H Streets, N.W. Washington, D.C.

On November 19, 1990, at approximately 8:30 p.m., the structural steel support system of a 47-foot-deep open excavation site at 14th and H Streets, N.W., Washington, D.C., collapsed causing a cave-in of several thousand cubic yards of soil [6]. The collapse caused a loss of the soldier beam and lagging wall in the immediate area of the collapse. The collapse caused the internal support system to slide and fall into the open excavation site. The edge of the internal support system slide 35 feet westward from the eastern perimeter. The sidewalks along 14th and H Streets were lost, as well as an alleyway between the United Press International (UPI) building and the excavation site. Due to the time of day, there were no fatalities or injuries in this collapse, although there was a potential for a significant number. Figures 11.14 and 11.15 are photographs taken of the site after the collapse.

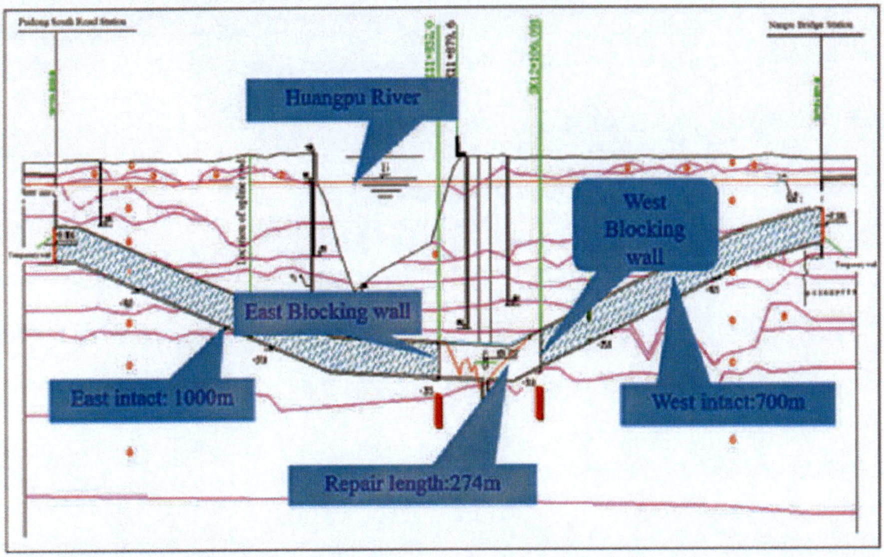

Fig. 11.13 In-situ repair scheme of Dongjiadu section of Line 4

Two collapses were happened in different days in this project. The initial collapse happened on November 19, 1990 and included the entire 20 feet width of the sidewalk for almost the entire 150 feet length of the excavation. The sliding of the 14th Street side of the internal support system then pulled away the internal support system that was holding up the alley on the north side and H Street on the south side. As a result, the entire 20 feet width of the alleyway for almost the entire 114 foot length of the internal support system caved in (as shown in Fig. 11.16). Similarly, the entire 20 foot width of the sidewalk for a length of about 114 feet along H Street also caved in (as shown in Fig. 11.17).

After the initial collapse, the general contractor started to refill the excavation along the 14th Street side, to stabilize the nearly vertical failure surface and to prevent further cave-in. The refilling operation was successful in preventing further cave-in along 14th Street. This operation was not sufficient along H Street. On November 20, at about 1:40 p.m., 17 h after the initial collapse, a 10 foot wide and 100 foot long section of H Street collapsed into the excavation site, along with remaining wooden sidewalk protection (as shown in Fig. 11.18).

11.6.1 Case Analysis

Through field investigation and geotechnical analysis, the reasons for the excavation collapse were as follows:

Fig. 11.14 Overall view of the collapsed site (looking toward west)

(1) The collapse of the excavation occurred due to the failure of certain structural members of the internal support system on the north and south walls.
(2) The north and south wall outlookers and their welded connections at all three tiers were not properly proportioned to resist the expected load.
(3) The sloping wale on the south wall at the top tier between soldier beams 63 and 64 was not properly proportioned to resist the expected load.
(4) The wales at all three tiers on the east wall between soldier beams 37 to 40 were not properly proportioned for the expected load.
(5) The failure load, the soil pressure immediately preceding the collapse, was determined to be lower than the expected load for which the structure was designed by the shoring subcontractor for the project [6].

Fig. 11.15 Overall view of the collapsed site (looking toward east)

11.6.2 Lesson Learned

The following situations should be taken into consideration during the period of construction:

1. Insufficient bearing capacity of the support system: the internal support system erected at the construction site, was not capable of resisting the soil load which could be reasonably expected. Certain members and their connections were not adequately designed to suit the field conditions for the expected loads.
2. Strengthening the detection of high stress areas: the spacer beams (outlooker beams) between the wale and the soldier beams and their connections and the sloping wale were determined to be the most highly stressed members and connections of the support system and contributed to the collapse.

Fig. 11.16 Failure soil surface along the north wall after the initial collapse

Fig. 11.17 Failure soil surface along the south wall after the initial collapse

Fig. 11.18 Failure soil surface along the south wall after the second collapse

11.7 Temporary Excavation System Collapse at Klang Valley in Peninsular Malaysia

This case history involved construction of a two-storey basement in an urban area. Figure 11.19 shows the location of the project site and the adjacent land lot which had been affected due to ground distresses.

Temporary shoring structure consisting of CBP wall propped by raking struts against lower basement slab was proposed by the contractor to provide peripheral support for a 10.5 m deep excavation at western boundary of the project site. The designed CBP wall was 16 m long 750 mm diameter bored pile with top cut off

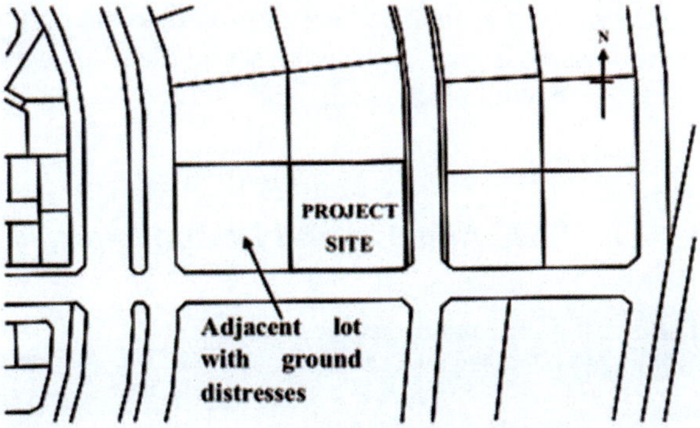

Fig. 11.19 Site location [7]

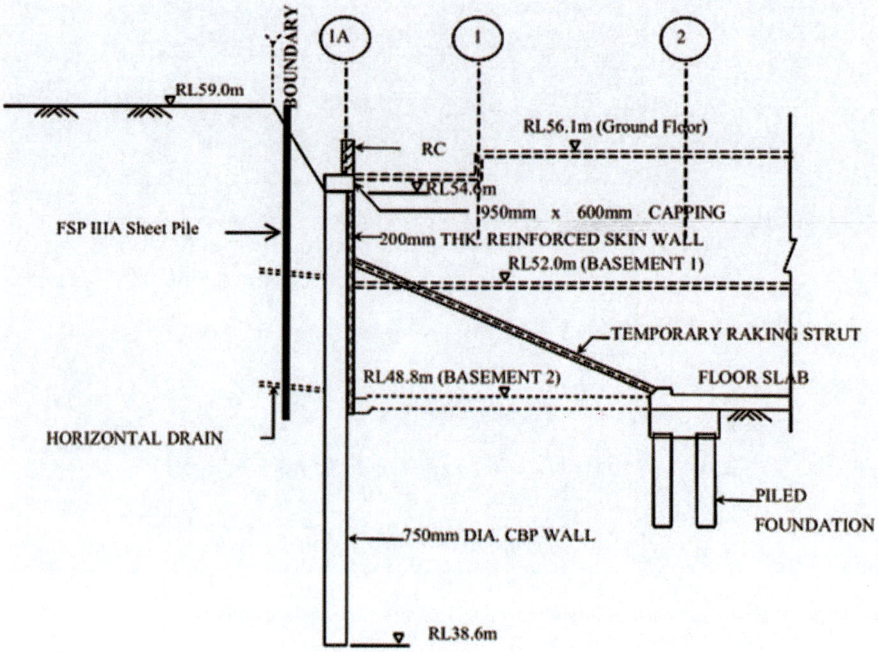

Fig. 11.20 Cross-section of the proposed alternative retaining walls

level at RL 54.6 m. In order to facilitate the CBP installation, 12 m long temporary steel sheet piles (type FSP IIIA) were installed at retained ground at RL59.0 m with about 0.7 m offset from project site boundary to provide sufficient working platform and as the temporary shoring support for the exposed 4.4 m temporary excavation (from RL59.0 to RL54.6 m). Figure 11.20 shows the details and cross-section of the proposed alternative temporary shoring system and permanent retaining wall.

It was observed that continuous tension cracks appeared at varying distance away along the sheet pile wall of approximately 73 m long. The tension cracks were more distant away from the sheet pile wall at the southern end and became closer to the sheet pile walls toward northern end. At the time of inspection, more extensive excavation was carried out at the southern portion than the northern portion. The platform for the backyard car park of the adjacent lot had shown ground subsidence with tension cracks as shown in Figs. 11.21 and 11.22.

11.7.1 Case Analysis and Lesson Learned

The investigation results and lessons learnt from this case study revealed the following findings [7]:

Fig. 11.21 Site conditions of adjacent lot after the incidence of wall movement and ground distresses (Southward view)

(1) The distressed ground was located over a natural valley where thicker fill was placed over the previous soft deposits without proper engineering treatment to form building platform. Soft deposits at the lower part of the valley and potential concentrated underground seepage are common in hilly terrain and should not be overlooked. Desk study of pre-development ground contours to identify potential geotechnical problems is highly recommended.

(2) Filling over valleys without proper site clearing, removal of unsuitable soft deposits and compaction could result in highly unstable backfill for any open excavation. It is important to thoroughly investigate the subsoil condition beneath the fill. Otherwise, proper treatment to the low-lying ground before the development earthworks shall not always be assumed.

(3) Occurrence of tension cracks during initial open excavation and installation of sheet piles suggested that the underlying subsoil and at the valley area are inherently vulnerable to ground disturbance and hence are prompted to distressing.

(4) The existence of soft compressible material at the valley area was further confirmed during additional subsurface investigation and localized deep pile cap excavation when reaching the final excavation level.

(5) Original topographical features are the important design consideration for excavation stability and remedial strategy. In these case studies, natural valley with

Fig. 11.22 Site conditions of adjacent lot after the incidence of wall and ground distresses (Northward view)

soft deposits was not detected during design stage causing cost escalation as a result of costly remedial work.

(6) Perched groundwater regime can occur in backfilling over natural valley leading to unfavorable behavior of backfill.

(7) Excessive removal of passive berm with localized pile cap excavation without installing the planned strutting resulted in loss of wall support and led to excessive ground movement.

11.8 Shangcheng Road Station of Metro Line 9 Collapse, Shanghai

Shangcheng Road Station of Metro Line 9 is located in Pudong New Area, which lies along Shangcheng Road, with South Pudong Road in the west. As shown in Fig. 11.23, the metro station in soft soils collapsed under construction. Extremely large wall deflections up to 320 mm took place within few days. This led to the wracking of the supporting system and consequent failure of the excavation. The field measurements show that as excavation went down to a deep level, the diaphragm walls developed deep-seated movements (bulging profiles) toward the excavation

Fig. 11.23 Failure of overexcavated metro station in soft clay in China

side [8]. The excavation-induced wall deflections almost increased linearly with time until the concrete floor slabs were cast. Thereafter, the deflections decreased gradually over time and almost stopped completely after completion of roof floors. And following excavation to a deeper level, the measured ground settlements almost increased linearly with time. During the construction of floor slabs between days 55 and 80, the observed ground settlements still continuously increased linearly at similar rates and did not show a tendency of decreasing.

Figure 11.24 presents the site plan of the project along with the instrumentation layouts for field monitoring. And the site conditions can be approximately known via Fig. 11.24.

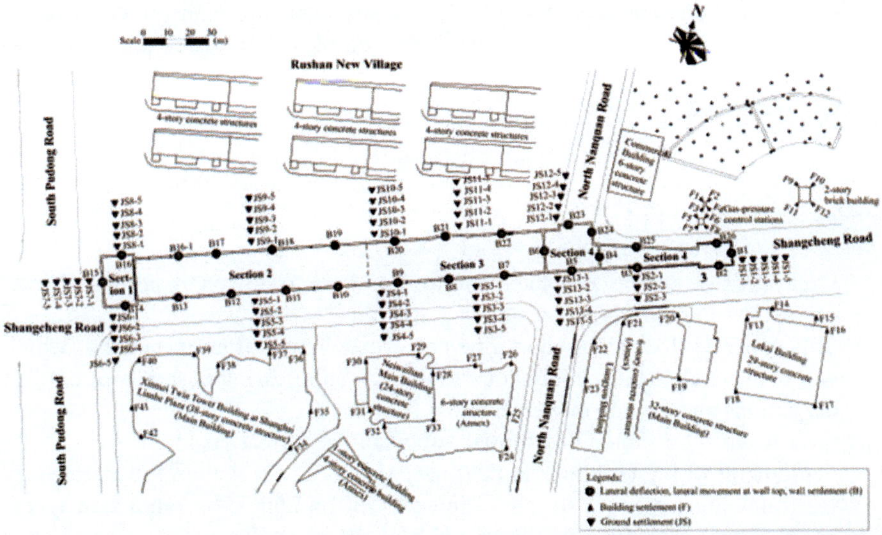

Fig. 11.24 Site plan of the project site

11.8.1 Case Analysis

Failure investigations including ground survey, visual inspection, and numerical analyses were implemented to figure out the cause, mechanism of the failure after the failure of overexcavated metro station. And the main failure reasons are as follows:

1. The structures failed probably because of over-excavation and improperly popped steel pipe struts [9].
2. To accelerate the speed of excavation, the contractor removed the soils at deep layers without propping struts timely as required by design. As a result, extremely large wall deflections up to 320 mm took place within few days. This led to the wracking of the supporting system and consequent failure of the excavation.
3. There may exist some imperfections of the construction quality of the supporting system.
4. The actual construction does not meet the design requirements. At Sects. 11.2 and 11.3, the excavation was designed to be propped by five levels of steel pipes. To reduce project cost and shorten construction duration time, the contractor did not follow the design but adopted a new excavation scheme in which the pit was propped by only four levels of struts. This move led to over-excavation at each level of Sects. 11.2 and 11.3.
5. The connection of steel support is not considered properly, which contributes to the damage of the connection and the failure of the supporting system.
6. In most excavation cases in the Shanghai area, it was observed that the reinforcement layers below excavation bases just functioned as curtains for cutting off seepage of waterflow owing to the usually difficult quality of compaction grouting because of the techniques and complex soil conditions. This did not constrain lateral wall movements. And over time this can destabilize the retaining wall.

11.8.2 Lesson Learned

1. Excavation should be completed within short time, followed by propping struts promptly and properly;
2. Stiff concrete floor slabs should be constructed right after excavation, which will help stabilize the soil-structure system by suppressing postexcavation wall deflections and ground movements;
3. Over-excavation should be avoided, especially at a great depth.
4. Settlements of the buildings in the proximity of an excavation are affected by the following factors: a structural integrity of buildings; b. foundation types; and c. distances between buildings and excavation pits. Therefore, the buildings with better structural integrity and/or resting on deep foundations and/or further away from excavation pits would experience much smaller and more uniform settlements.

11.9 Retaining Structure Failure, Nanchang, 2010

The project location is in the city of Nanchang, Jiangxi, China. The excavation is for the construction of metro station (Zhujiang Road station), whose location is indicated in Fig. 11.25. The station is 467 m long and is located 800 m from the Ganjiang River. The minimum distance from residential area is 47 m. The station is two-story underground station and so needed a large excavation to be built. The excavation is divided into different sections: launch shafts, intermediate segments, and standard segments. The final excavation depth is approximately 15.5–17.8 m. The ground surface is 20 m high, the approximately average of groundwater level is 15 m. The site profile is characterized by thin silty clay, thick muddy clay, and very thick sand layers (10–12 m) including medium, coarse and gravelly sand, strongly weathered siltstone, moderately weathered siltstone. The field investigation showed that the site did not contain any unfavorable geological phenomena. Because the site is near a river (Ganjiang River), near residential area, the groundwater table is very high and there is also a very thick sand layer that is not only highly permeable, but also has a high water content, making the excavation work quite difficult.

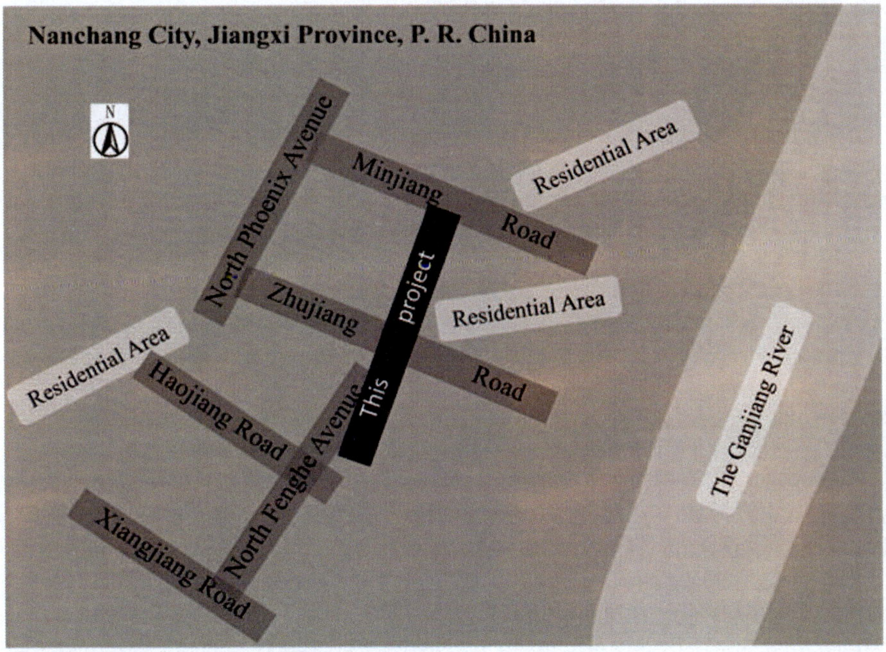

Fig. 11.25 Location of the proposed project

Fig. 11.26 Failure at the east of the north launch shaft: **(a)** ground collapse; **(b)** sand boil

11.9.1 Failure of the Retaining Structure

The geologic conditions of the site (high groundwater level, thick sand), finally lead to poorly performance of waterproof wall of the excavation pit as shown in Fig. 11.26. A few hours after the excavation reached the maximum depth, in the north launch shaft, large seepage and sand boils suddenly developed at the western cast-in-place piles in the north launch shaft at depth of 16 m, and with the rapid leaking of water and sand, the ground surface collapsed also. The failure keeps on until few hours. The overall amount of displaced sand was 140 cube meters. Few days later, the same phenomenon has been observed at the east side of the north launch shaft with displaced sand of approximately 70 cube meters. Although the accident was dangerous, the struts have not excessively deformed [10].

11.9.2 Reasons for the Failure

Prior to the project and during the construction, the site has been monitored and all the results have not indicated any failure in advance. The values obtained from groundwater level, maximum lateral deflection of cast-in-place piles, and ground settlement are stable or much lower than the critical values. To know more about the failure reasons, field and laboratory tests have been conducted on the waterproof effectiveness of the three-axis mixing piles and the bislurry grouting. The results of the tests show that around JN15 mixing pile, there was a high coefficient of permeability due to the big amount of silt and sands around this area. The water was then easily flowing and the sand between mixing piles displaced by the water pressure. The cracks due to earth and water pressure gradually increased and developed rapidly sand boils. Ground collapsed then after the sand was flooded. This shows that the three-axis mixing pile was not the appropriate choice to address the conditions of this project site.

Bislurry grouting was an optimal method to overcome the highly permeable thick sand layers, however, it was necessary to think about the low permeable clay and silt layers. The reinforcement was not well done in those layers, which cause the sand boils likely to develop.

Because piping failure in silt and sands is common, the engineers of this project could have paid more attention to overcome this. Luckily, there have not been any life loss or people injured with the accident [10].

11.9.3 Remedial Measures and Lessons Learned

Because the accident was sudden, they first took emergency solutions to stop sand boils. They filled the space between the piles with cotton and fasten piles with steel mesh, dumped large amount of quick-drying cement into the space between piles to keep the cotton from being washed away, dumped viscous slurry onto the cotton to stop the quicksand in the leakage area, backfilled with soil around the leakage area to reduce soil and water pressure difference in the excavation pit, filled the ground collapse area with C20 to avoid movement or buckling failure of the retaining structure.

After the leakage stopped, the high-pressure rotating-jet grouting piles were constructed between the leakage points and collapsed area to strengthen the soil (Fig. 11.27). Also, some pipes were installed to grout the bislurry to strengthen the soil and form waterproof curtain. After cleaning the sandbags and backfilling in the excavation pit, pipes for bislurry are installed layer by layer. The grouting pressure was 3–5 Mpa.

There is no doubt that the accident could have been avoided if the high-pressure rotating-jet grouting method had been used at the beginning of the project instead of the bislurry grouting method. The bislurry grouting method is appropriate for preventing seepage due to high permeability of the coarse sand. However, in soils with low permeability, the bislurry grouting is inappropriate for creating a waterproof wall, and high-pressure rotating-jet grouting should be considered based on the failure and remedial measures in this documented case. The upper clay and silty sand layers did not receive serious evaluation during the initial design in this case, whereas the very thick sand layers were taken into consideration and were the focus of a large amount of work. Consequently, neglecting the upper clay and silty sand led to the unexpected failure. However, experience regarding similar excavation work involving thick sand layers and high groundwater levels was acquired during this project [10].

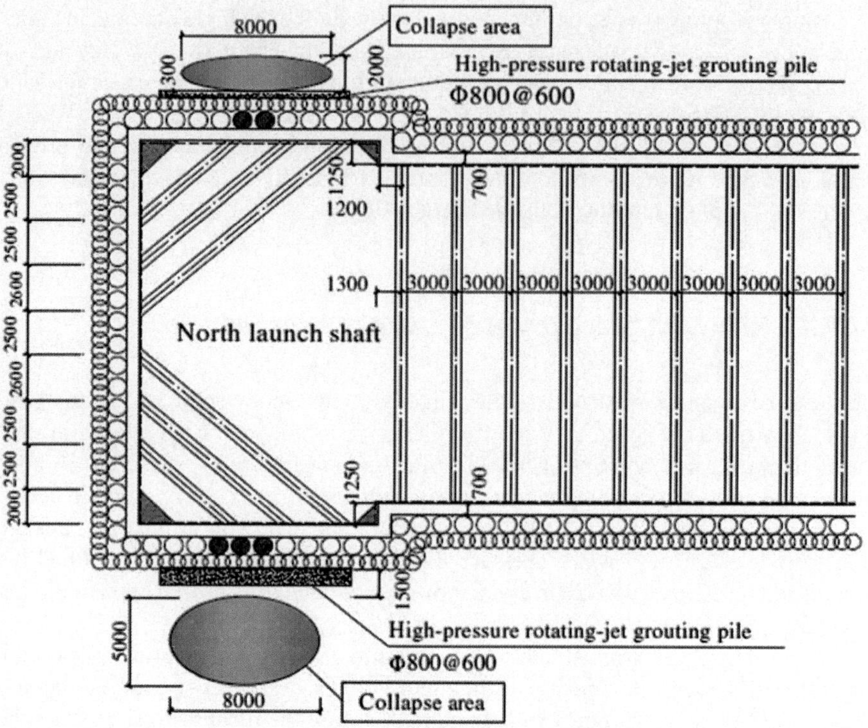

Fig. 11.27 Collapse area and high-pressure rotating-jet grouting pile

11.10 Nanning Greenland Central Square Foundation Pit Collapse, 2019

11.10.1 Case Condition

Nanning Greenland central square is located in the center of Qingxiu district, surrounded by high speed railway business district, CBD business district, and several business circles. The project covers an area over 190 thousand square meters, with a total construction area of about 1.1 million square meters. In the southeast area of the project, it is known as Lot D, a super high-rise structure with a height of more than 200 m will be constructed there. In mid-2018, the foundation pit of 200 m high-rise began to be excavated downward, with a depth of 21 m. After reaching the depth of 21 m, the basement structure will be constructed upward. The site soil conditions comprised of: uplying 5 m thick plain fill; a 6.2 m thick strongly weathered mudstone; followed by a layer of strongly weathered argillaceous siltstone, whose thickness is 8.5 m; and an underpart moderately weathered mudstone, which is more than 12 m thick. The slope of foundation pit adopts layered slope and retaining. Pile and anchoring with different strengths and arrangements were employed as excavation retaining measure.

11.10.2 Failure Discription

At about 18:00 p.m. on June 6, according to the monitoring data, the horizontal displacement of the retaining wall was found to be constantly increasing and the cracks widened; by about noon on June 7, the cumulative displacement had reached about 50 cm and the crack width was 15 cm. At a similar time in the afternoon of June 8, the foundation pit collapsed suddenly, and about 4000 square meters of soil cracked and fell as Fig. 11.28 shows. As can be seen in Fig. 11.29, the ground that originally cracked and subsided has become a big pit. The width of the crack is about 40 cm, the influence distance is about 20 m, and the wall of the construction site has cracked.

Fig. 11.28 Situation of foundation pit collapse in Nanning

Fig. 11.29 Pavement cracking caused by foundation pit collapse

11.10.3 Cause Analysis

According to the preliminary analysis of official experts, due to the deformation of the foundation pit support at the construction site, coupled with the long-term leakage of water pipes, the soil around the foundation pit was hollowed out. The local soil of the collapse area was soft, which led to the burst of water pipes as Fig. 11.30 shows, consequent failure of the foundation pit anchor cable structure occurred, and finally caused the collapse accident.

With the development of accident investigation, some experts thought that the foundation pit may have the following problems: (1) Unreasonable design, for deep foundation pit excavation, the retaining system with higher stiffness, i.e., retaining wall with internal props should be choosed; (2) Inadequate geological investigation, there could be unfavorable differences between the investigated soil and the actual site condition; (3) Poor construction quality, there may exists problems like pile anchor not in place [11].

For this type of deep and large space braced excavation, it is very important to control the deformation of foundation pit wall, therefore, it was a major design error to adopt pile anchor retaining system instead of wall with internal prop. Moreover, when the increase of deformation lead to water pipe rupture, on the one hand, it

Fig. 11.30 Water pipe rupture caused by soil deformation

increased the water and soil pressure, on the other hand, it also greatly reduced the anchoring force. If the measures were not taken on time, it was very likely to collapse, and this foundation pit accident is worth learning.

11.11 Other Cases

International Conferences on Case Histories in Geotechnical Engineering aim to share knowledge gained from geotechnical projects and Missouri University of Science and Technology provides many volumes of the conference proceedings. We can find some case histories involving failure in deep excavation, such as:

Boones and Westland [12], (https://scholarsmine.mst.edu/icchge/5icchge/session05/30);

Rosenvinge and Teodore [13] (https://scholarsmine.mst.edu/icchge/7icchge/session17/4);

Rahimi et al. [14] (https://scholarsmine.mst.edu/icchge/3icchge/3icchge-session05/16).

References

1. Whittle, A. J., & Davies, R. V. (2006). June. Nicoll Highway collapse: Evaluation of geotechnical factors affecting design of excavation support system. *International Conference on Deep Excavations, 28*, 30–36.
2. Ma, Y., de Jong, M., Koppenjan, J., Xi, B., & Mu, R. (2012). Explaining the organizational and contractual context of subway construction disasters in China: The case of Hangzhou. *Policy and Society, 31*(1), 87–103.
3. Gong, X. N., & Zhang, X. C. (2012). Excavation collapse of Hangzhou subway station in soft clay and numerical investigation based on orthogonal experiment method. *Journal of Zhejiang University Science, 13*(10), 760–767.
4. Chen, W., Wu, Y. J., & Peng, Z. Z. (2006). Monitoring of rescue of a foundation pit and technical analysis of its collapse in GuangZhou. *Chinese Journal of Underground Space and Engineering, 2*(6), 1034–1039. (in Chinese).
5. Qu, C. S. (2010). Dewatering of the Dongjiadu tunnel restoration on Shanghai metro line 4. *Chinese Journal of Geotechnical Engineering, 32*(2), 339–342. (in Chinese).
6. Jin, J. S., Ayub, M., & Liu, F. (1991). Investigationof November 19, 1990, Excavation Collapse at 14th and H Streets, N.W. Washington, D.C. Washington, DC: U.S. Department of Labor, Occupational Safety and Health Administration. (p. 109).
7. Liew, S. S., & Loh, Y. E. (2011). Two case studies of collapsed temporary excavation using contiguous bored pile wall. In *Proceeding of CIEHKIEY IEM Tripartied Seminar (Deep Excavation)* (p. 21)
8. Tan, Y., Wei, B., Zhou, X., & Diao, Y. (2014). Lessons learned from construction of Shanghai metro stations: Importance of quick excavation, prompt propping, timely casting, and segmented construction. *Journal of Performance of Constructed Facilities, 29*(4), 04014096.
9. Tan, Y., & Wei, B. (2011). Performance of an overexcavated metro station and facilities nearby. *Journal of Performance of Constructed Facilities, 26*(3), 241–254.
10. Feng, S. J., & Lu, S. F. (2015). Failure of a retaining structure in a metro station excavation in Nanchang City, China. *Journal of Performance of Constructed Facilities, 30*(4), 04015097.
11. Gong, X. N. (2019). Investigation and reflection on the collapse of foundation pit in Nanning. *Chinese Ground Improvement, 1*(1), 95–96. (in Chinese).
12. Boone, S. J., & Westland, J. (2004). Design, construction, and performance of a deep braced excavation. In *International conference on Case Histories in Geotechnical Engineering*. (p. 29).
13. Theodore Von-Rosenvinge, I. V. (2013). Tied-Back Wall Failure. Boston, MA: International Conference on Case Histories in Geotechnical Engineering. (p. 4).
14. Rahimi, M. M., Karwaj, C., & Deb, P. K. (1993). Failure of sewerage mains constructed in soft estuarine deposit. In *International conference on Case Histories in Geotechnical Engineering*. (p. 16).

Appendix
Database of Propped and Anchored Deep Excavation

W. Zhang and H. Liu, *Design of Deep Braced Excavation and Earth Retaining Systems Under Complex Built Environment*,
https://doi.org/10.1007/978-981-16-5320-9

No	Location	Soil at dredge level	*St* (kPa)	*H* (m)	*h* (m)	*h*/*H*	*s*	*FoS*	*EI* (N m^2)	Wall type	Construction method	Support type	δ_h (mm)	δ_v (mm)	References
1	Oslo Enerhaughen, Norway	Soft clay	20	8	17	1	2.5	1.34	45,000	Sheet	No data (NA)	Multi-prop	40	106	[1]
2	Oslo Telecom, Norway	Soft clay	20	8.5	10	1	2.25	0.9	35,850	Sheet	NA	Multi-prop	80	93	[2]
3	Oslo Gronland 2, Norway	Soft clay	25	11.5	26	1	3.75	1.3	73,800	Sheet	NA	Multi-prop	100	178	[3]
4	Oslo Vaterland 3, Norway	Soft clay	34	12	26	1	2	1.26	73,800	Sheet	NA	Multi-prop	125	114	[4]
5	Oslo Gunnerus, Norway	Soft clay	35	10.5	18	1	2	1.21	82,350	Sheet	NA	Multi-prop	320	600	[5]
6	Vasteras, Sweden	Soft clay	30	6.3	12.5	1	2.1	1.5	17,000	Sheet	NA	Multi-anchor	100	175	[6]
7	Gothenburg, Sweden	Soft clay	17	4	10	1	2	1.6	17,000	Sheet	NA	Single-anchor	400	410	[6]
8	Chicago Subway, USA	Soft clay	35	19	11	1	3	0.96	50,000	Sheet	NA	Multi-prop	60	NA	[7]
9	Chicago A, USA	Soft clay	35	9.4	NA	1	3.05	1.67	55,250	Sheet	NA	Multi-prop	64	NA	[8]
10	Chicago C, USA	Soft clay	35	8.8	NA	1	1.98	1.18	55,250	Sheet	NA	Multi-prop	56	NA	[8]
11	Chicago D, USA	Firm clay	70	5.7	11	1	2.44	1.55	55,250	Sheet	NA	Multi-prop	13	NA	[8]
12	Chicago E, USA	Firm clay	70	10.67	11	1	3.04	1.55	1,106,000	Diaphragm	NA	Multi-prop	38	NA	[8]
13	Chicago F, USA	Firm clay	70	10.67	11	1	3.81	1.25	55,250	Sheet	NA	Multi-prop	89	NA	[8]
14	Chicago G, USA	Firm clay	70	12.34	11	0.89	3.96	1.41	69,000	Sheet	NA	Multi-prop	178	NA	[8]
15	Chicago H, USA	Firm clay	70	10.67	11	1	3.05	1.4	566,340	Diaphragm	NA	Multi-prop	69	NA	[8]
16	HDR-4 Chicago, USA	Firm clay	30	12.2	16	1	2.4	1.1	161,000	Sheet	NA	Multi-prop	190	250	[9]
17	Washington, USA	Soft clay	30	9.1	18	1	2.2	0.81	50,160	Soldier	NA	Multi-prop	254	NA	[10]
18	Bowlin Point, N.Y., USA	Stiff clay	40	9.8	9.8	1	1.96	2.4	50,000	Sheet	NA	Multi-prop	80	NA	[11]

(continued)

(continued)

No	Location	Soil at dredge level	St (kPa)	H (m)	h (m)	h/H	s	FoS	EI (N m^2)	Wall type	Construction method	Support type	δ_h (mm)	δ_v (mm)	References
19	Davidson 1, San Francisco, USA	Soft clay	10	9.1	21	1	3	0.83	72,500	Sheet	NA	Multi-prop	254	NA	[12]
20	Islais 2, San Francisco, USA	Soft clay	13	9.1	21	1	3	1.22	55,250	Sheet	NA	Multi-prop	38	NA	[13]
21	Embarcadero III	Soft clay	30	13.7	27	1	3.4	0.99	80,000	Sheet	NA	Multi-prop	150	NA	[5]
22	Levi Strauss San Francisco, USA	Soft clay	35	14	19.5	1	2.74	1.3	80,000	Sheet	NA	Multi-prop	190	NA	[14]
23	SNBB San Francisco, USA	Soft clay	35	14	19.5	1	2.74	1.3	4,528,466	Diaphragm	NA	Multi-prop	22	NA	[14]
24	3rd Har Tun Boston, USA	Soft clay	35	15.8	30	1	2.63	1.2	72,500	Sheet	NA	Multi-prop	150	NA	[15]
25	H'Fok B Singapore	Soft clay	15	7.3	30	1	1.83	0.87	75,700	Sheet	NA	Multi-prop	235	250	[16]
26	Tokyo Airport, Japan	Soft clay	35	11	15	1	2.75	1.64	172,000	Sheet	NA	Multi-prop	300	NA	[17]
27	Mexico City, Mexico	Soft clay	25	9	20	1	1.8	0.95	50,640	Sheet	NA	Multi-prop	155	NA	[18]
28	Shanghai, China	Soft Clay	8	33.6	NA	1	2.5	2	567,000	Diaphragm	NA	Multi-prop	40	15	[19]
29	Shanghai, China	Soft Clay	8	33.6	NA	1	2.5	2	567,000	Diaphragm	NA	Multi-prop	42	21	[19]
30	Boston, USA	Soft Clay	35	13.4	NA	1	3.35	1.5	1,200,000	Diaphragm	NA	Multi-prop	97	NA	[20]
31	UOB Singapore	Soft clay	30	13	30	1	2.6	>3	4,320,000	Diaphragm	NA	Multi-prop	56	130	[21]
32	H'Fok A Singapore	Soft clay	15	7.3	19	1	1.83	>3	75,700	Sheet	NA	Multi-prop	60	NA	[16]
33	CTC Singapore	Soft clay	20	12	37	1	2	>3	57,440	Sheet	NA	Multi-prop	188	150	[22]
34	Somerset Singapore	Peats/silts	15	15.2	10	1	3.8	>3	540,000	Diaphragm	NA	Multi-prop	20	NA	[23]
35	MOE I2 Singapore	Soft clay	18	6.8	24	1	1.7	>3	45,436	Sheet	NA	Multi-prop	330	NA	[24]
36	MOE I9 Singapore	Soft clay	18	6.4	12	1	1.6	>3	45,436	Sheet	NA	Multi-prop	100	NA	[24]
37	Singapore Bugis, Singapore	Soft clay	40	18.3	30	1	2.29	>3	4,320,000	Diaphragm	NA	Multi-prop	160	NA	[25]

(continued)

(continued)

No	Location	Soil at dredge level	*St* (kPa)	*H* (m)	*h* (m)	*h*/*H*	*s*	*FoS*	*EI* (N m^2)	Wall type	Construction method	Support type	δ_h (mm)	δ_v (mm)	References
38	Singapore CBD, Singapore	Soft clay	13	15	17	1	2.5	>3	70,000	Sheet	NA	Multi-prop	145	100	[26]
39	Singapore Parking, Singapore	Soft clay	35	9.5	12		5.9	>3	540,000	Diaphragm	NA	Single-prop	70	NA	[27]
40	Taiwan Airline, China	Firm clay	45	9.6	9	0.94	2.4	>3	18,850	Contiguous	NA	Multi-prop	22	22	[28]
41	Taiwan Power, China	Firm clay	120	16.2	15	0.93	3.24	>3	540,000	Diaphragm	NA	Multi-prop	80	56	[28]
42	Taiwan Quen M, China	Firm clay	47	10.7	8	0.75	2.68	>3	540,000	Diaphragm	NA	Multi-prop	70	35	[28]
43	Taiwan Tax, China	Soft clay	38	7.65	8	1	1.91	>3	40,000	Sheet	NA	Multi-prop	69	41	[28]
44	Taiwan F-foundation pit, China	Soft clay	35	18.45	20	1	2.64	>3	1,280,000	Diaphragm	NA	Multi-prop	60	42	[28]
45	Taiwan Cathay, China	Firm clay	90	21	12	0.57	2.63	>3	857,500	Diaphragm	NA	Multi-prop	62	31	[28]
46	Bangkok A, Thailand	Soft clay	35	9.8	15	1	3.1	>3	1,378,420	Diaphragm	NA	Multi-prop	50	NA	[29]
47	Bangkok C, Thailand	Soft clay	35	18.5	12	1	4.6	>3	1,378,420	Diaphragm	NA	Multi-prop	30	NA	[29]
48	Bangkok E, Thailand	Soft clay	35	7.2	12	1	1.8	>3	50,000	Sheet	NA	Multi-prop	220	NA	[29]
49	Oslo Vaterland 1, Norway	Soft clay	25	11	16	1	2	>3	73,800	Sheet	NA	Multi-prop	220	270	[3]
50	Oslo Vaterland 2, Norway	Soft clay	20	11	16	1	2	>3	73,800	Sheet	NA	Multi-prop	140	260	[30]
51	Oslo Studenterlu, Norway	Soft clay	40	16	37	1	5.3	>3	2,500,000	Diaphragm	NA	Multi-prop	42	65	[31]

(continued)

(continued)

No	Location	Soil at dredge level	St (kPa)	H (m)	h (m)	h/H	s	FoS	EI (N m^2)	Wall type	Construction method	Support type	δ_h (mm)	δ_v (mm)	References
52	Oslo Jerbanetorget, Norway	Soft clay	20	10	35	1	5	>3	2,500,000	Diaphragm	NA	Multi-prop	20	NA	[31]
53	Oslo Bank of Norway, Norway	Soft clay	20	16	18	1	3.2	>3	2,500,000	Diaphragm	NA	Multi-prop	16	62	[32]
54	Eastbourne 1, UK	Soft clay	35	11	15	1.00	10.00	>3	2,500,000	Diaphragm	NA	Single-prop	61	NA	[33]
55	Eastbourne 2, UK	Soft clay	35	14	15	1.00	13.00	>3	2,500,000	Diaphragm	NA	Single-prop	15	NA	[33]
56	Pietrafitta, Italy	Soft–hard clay	40	5.5	20	1.00	7.80	>3	42,000	Sheet	NA	Multi-prop	71	NA	[34]
57	Chicago, USA	Soft clay	35	13.4	15	1.00	4.46	>3	1,055,000	Sheet	NA	Multi-prop	150	NA	[33]
58	Inland Steel Chicago, USA	Soft clay	35	11	19	1.00	2.00	>3	50,000	Sheet	NA	Multi-prop	55	NA	[34]
59	Osaka A, Japan	Soft clay	35	20.6	25	1.00	3.00	>3	2,500,000	Diaphragm	NA	Multi-prop	78	NA	[35]
60	Lake zone, Mexico	Soft clay	25	15.7	20	1.00	2.62	>3	2,500,000	Diaphragm	NA	Multi-prop	135	NA	[5]
61	Shanghi-Jin Mao, China	Soft–hard clay	10	19.65	36	1.00	3.93	>3	2,500,000	Diaphragm	NA	Multi-prop	81	NA	[36]
62	Shanghi-Heng Long, China	Soft–hard clay	10	18.2	29	1.00	3.64	>3	2,500,000	Diaphragm	NA	Multi-prop	99	NA	[36]
63	Shanghi, China	Soft–hard clay	10	17.85	24	1.00	3.57	>3	2,500,000	Diaphragm	NA	Multi-prop	129	NA	[37]
64	River Wall M'Boro, UK	Soft clay	35	9.5	12	1.00	4.25	>3	177,660	Sheet	NA	Single-anchor	125	NA	[38]
65	Detroit, USA	Soft clay	35	7	10	1.00	11.20	>3	83,400	Sheet	NA	Single-anchor	45	NA	[33]
66	TP, Bogota, Colombia	Soft clay/silt	15	16	34	1.00	3.75	>3	540,000	Diaphragm	NA	Multi-anchor	125	1000	[39]
67	Newton Singapore	Soft clay	18	14.5	12	1.00	3.63	>3	2,500,000	Diaphragm	Top-down	Undifferentiated multi-support	110	220	[40]

(continued)

(continued)

No	Location	Soil at dredge level	*St* (kPa)	*H* (m)	*h* (m)	*h/H*	*s*	*FoS*	*EI* (N m^2)	Wall type	Construction method	Support type	δ_h (mm)	δ_v (mm)	References
68	Taiwan Chi Ching, China	Soft clays	30	13.9	15	1.00	2.78	>3	857,500	Diaphragm	Top-down	Undifferentiated multi-support	65	65	[28]
69	Taiwan, China	Firm clay	60	20	24	0.60	3.33	>3	857,500	Diaphragm	Top-down	Undifferentiated multi-support	135	67	[28]
70	Oslo Christiana, Norway	Soft clay	35	9.6	23	1.00	3.00	>3	483,600	Sheet	Top-down	Undifferentiated multi-support	48	100	[41]
71	Shanghai, China	Soft clay	30	14.5	50	1.00	3.00	>3	2,500,000	Diaphragm	Top-down	Multi-prop	35	45	[28]
72	Illinois, USA	Soft clay	35	7.5	10.1	1.00	5.00	>3	625,000	Sheet	Top-down	Multi-prop	40	190	[42]
73	NTUH, Taipei, China	Soft Soil	36	15.7	18	1.00	2.40	>3	680,000	Diaphragm	Top-down	Multi-anchor	80	NA	[43]
74	TCC, Taipei, China	Soft Soil	35	12.5	22	1.00	2.00	>3	340,000	Diaphragm	Top-down	Multi-anchor	32	NA	[43]
75	TCAC, Taipei, China	Soft Soil	35	20	24	1.00	2.12	>3	4,200,000	Diaphragm	Top-down	Multi-anchor	50	NA	[43]
76	Shanghai, China	Soft clay	40	33.6	NA	1.00	2.50	>3	1,215,000	Diaphragm	Bottom-up	Multi-prop	27	6	[19]
77	Shanghai, China	Soft clay	40	33.6	NA	1.00	2.50	>3	1,215,000	Diaphragm	Bottom-up	Multi-prop	31	NA	[19]
78	Shanghai, China	Very soft clay	20	9	NA	1.00	4.50	>3	60,000	Deep soil mix	Top-down	Multi-prop	10	16	[44]
79	Chicago, USA	Soft–medium stiff clay	36	12.8	12.8	1	4.5	>3	131,220	Sheet	Top-down	Multi-anchor	90	74	[45]
80	Chicago, USA	Soft–medium stiff clay	46	10.8	10.5	0.97	4.5	>3	131,220	Sheet	Top-down	Multi-anchor	45	NA	[45]
81	Case 2	Soft clay	35	10	42	1	2.5	>3	447,000	Diaphragm	Bottom-up	Multi-prop	22	NA	[46]
82	Hsinkuang, Pakistan	Soft clay	35	16	55	1	2.9	>3	709,000	Diaphragm	Bottom-up	Multi-prop	83	NA	[46]
83	Taipei Gas, China	Soft clay	35	18.1	46	1	2.5	>3	2,067,000	Diaphragm	Bottom-up	Multi-prop	76	NA	[46]
84	Tzuchyang	Soft clay	35	13.6	50	1	3.2	>3	709,000	Diaphragm	Top-down	Multi-prop	53	NA	[46]
85	Case 10	Soft clay	35	12.3	40.5	1	2.7	>3	447,000	Diaphragm	Bottom-up	Multi-prop	59	NA	[46]

(continued)

(continued)

No	Location	Soil at dredge level	*St* (kPa)	*H* (m)	*h* (m)	*h/H*	*s*	*FoS*	*EI* (N m^2)	Wall type	Construction method	Support type	δ_h (mm)	δ_v (mm)	References
86	Baisern	Soft clay	35	12.3	37	1	3.2	>3	447,000	Diaphragm	Bottom-up	Multi-prop	39	NA	[46]
87	MRT-1	Soft clay	35	16.8	45	1	3.2	>3	1,059,000	Diaphragm	Bottom-up	Multi-prop	30	NA	[46]
88	MRT-2	Soft clay	35	16.4	52	1	3.1	>3	1,059,000	Diaphragm	Bottom-up	Multi-prop	41	NA	[46]
89	MRT-3	Soft clay	35	12.4	47.9	1	3.4	>3	2,067,000	Diaphragm	Top-down	Multi-prop	22	NA	[46]
90	MRT-4	Soft clay	35	16.2	45	1	2.4	>3	2,067,000	Diaphragm	Top-down	Multi-prop	49	NA	[46]
91	Subway 8	Soft clay	35	28.8	50	1	2.6	>3	3,572,000	Diaphragm	Bottom-up	Multi-prop	69	NA	[46]
92	Subway 9	Soft clay	35	26.4	50	1	3.1	>3	3,572,000	Diaphragm	Bottom-up	Multi-prop	55	NA	[46]
93	Subway 10	Soft clay	35	21.7	50	1	2.6	>3	3,572,000	Diaphragm	Bottom-up	Multi-prop	41	NA	[46]
94	Sinyi, China	Soft clay	35	12.3	46	1	2.6	>3	447,000	Diaphragm	Bottom-up	Multi-prop	46	NA	[46]
95	Taiwan Sugar, China	Soft clay	35	13.2	45	1	3.2	>3	1,059,000	Diaphragm	Bottom-up	Multi-prop	58	NA	[46]
96	Tai Kai	Soft clay	35	12.6	48	1	2.7	>3	447,000	Diaphragm	Bottom-up	Multi-prop	60	NA	[46]
97	Case 12	Soft clay	35	13.7	36	1	3.1	>3	709,000	Diaphragm	Bottom-up	Multi-prop	48	NA	[46]
98	Subway 1	Soft clay	35	14.5	46	1	2.5	>3	1,059,000	Diaphragm	Bottom-up	Multi-prop	35	NA	[46]
99	Subway 2	Soft clay	35	19.4	45.1	1	2.8	>3	2,751,000	Diaphragm	Bottom-up	Multi-prop	60	NA	[46]
100	Subway 3	Soft clay	35	19.4	50	1	2.4	>3	2,751,000	Diaphragm	Bottom-up	Multi-prop	62	NA	[46]
101	Subway 4	Soft clay	35	16.2	50	1	3.4	>3	2,067,000	Diaphragm	Top-down	Multi-prop	47	NA	[46]
102	Subway 5	Soft clay	35	15.5	45.8	1	2.4	>3	3,572,000	Diaphragm	Bottom-up	Multi-prop	36	NA	[46]
103	Subway 6	Soft clay	35	12.7	45.8	1	3.1	>3	3,572,000	Diaphragm	Bottom-up	Multi-prop	29	NA	[46]
104	Subway 7	Soft clay	35	19.9	45.8	1	3.1	>3	3,572,000	Diaphragm	Bottom-up	Multi-prop	52	NA	[46]
105	Syed Alwi, Singapore	Soft clay	35	7.8	16	1	3.5	>3	447,000	Diaphragm	Bottom-up	Multi-prop	48	NA	[46]
106	Lavender, Singapore	Soft clay	35	15.7	20.5	1	2.6	>3	2,067,000	Diaphragm	Bottom-up	Multi-prop	31	NA	[46]
107	BIF, Boston, USA	Marine clay	50	19.4	15	0.77	3.2	>3	60,000	Deep soil mix	NA	Multi-prop	41	NA	[47]

(continued)

(continued)

No	Location	Soil at dredge level	St (kPa)	H (m)	h (m)	h/H	s	FoS	EI (N m^2)	Wall type	Construction method	Support type	δ_h (mm)	δ_v (mm)	References
108	Gu Bei Station, Shanghai, China	Soft clay	26	14.5	50	1	3.5	>3	1,280,000	Diaphragm	Top-down	Multi-prop	20	39	[48]
109	Block 37, Chicago, USA	Soft clay	35	15	NA	1	3.75	>3	180,000	Diaphragm	Top-down	Multi-prop	45	30	[49]
110	Shanghai, China	Soft clay	35	22.8	NA	1	3.8	>3	1,200,000	Diaphragm	Bottom-up	Multi-prop	80	8	[50]
111	Shanghai, China	Soft clay	35	15.5	NA	1	2.7	>3	600,000	Diaphragm	Top-down	Multi-prop	40	20	[51]
112	Salerno, Italy	Sand and gravel	150	8	1	0.13	4	>3	540,000	Diaphragm	Top-down	Multi-prop	28	11	[52]
113	Longwood, Boston, USA	Stiff clay	60	23	1	0.04	3.1	>3	1,200,000	Diaphragm	Top-down	Multi-anchor	20	4	[53]
114	Longwood, Boston, USA	Stiff clay	60	23	1	0.04	3.1	>3	1,200,000	Diaphragm	Top-down	Multi-anchor	65	12	[53]
115	Longwood, Boston, USA	Stiff clay	60	23	1	0.04	3.1	>3	1,200,000	Diaphragm	Top-down	Multi-anchor	65	21	[53]
116	Marmaris, Turkey	Silty sand	65	6	3.5	0.58	2	>3	289,000	Secant	Bottom-up	Multi-anchor	29	NA	[54]
117	Marmaris, Turkey	Silty sand	65	6	3.5	0.58	2	>3	289,000	Secant	Bottom-up	Multi-anchor	49	NA	[54]
118	Shanghai, China	Soft clay	35	12.2	NA	1	3	>3	1,200,000	Diaphragm	Top-down	Undifferentiated multi-support	40	25	[55]
119	Constance, Germany	Soft clay	20	9.9	25	1	5	>3	91,140	Sheet	Bottom-up	Multi-prop	113	160	[56]
120	Nangang, Taipei, China	Soft clay	20	11.6	21	1	2.9	>3	5,570,000	Diaphragm	Bottom-up	Multi-prop	114	83	[57]
121	Shihuduan, Taipei, China	Soft clay	35	12.1	NA	1	3	>3	460,844	Diaphragm	Bottom-up	Undifferentiated multi-support	94	68	[57]
122	Hepingshang, Taipei, China	Soft clay	35	12.1	NA	1	3	>3	460,844	Diaphragm	Bottom-up	Undifferentiated multi-support	71	NA	[57]
123	Jyru Road, China	Soft clay	35	9.4	NA	1	2.4	>3	460,844	Diaphragm	Bottom-up	Undifferentiated multi-support	54	43	[57]

(continued)

(continued)

No	Location	Soil at dredge level	*St* (kPa)	*H* (m)	*h* (m)	*h*/*H*	*s*	*FoS*	*EI* (N m^2)	Wall type	Construction method	Support type	δ_h (mm)	δ_v (mm)	References
124	Neihu Road, Taipei, China	Soft clay	35	12.3	NA	1	3.1	>3	687,908	Diaphragm	Bottom-up	Undifferentiated multi-support	38	NA	[57]
125	Jianbei, Taipei, China	Soft clay	35	16	NA	1	3.2	>3	687,908	Diaphragm	Bottom-up	Undifferentiated multi-support	52	NA	[57]
126	Duenbei, Taipei, China	Soft clay	35	12	NA	1	4	>3	460,844	Diaphragm	Top-down	Undifferentiated multi-support	28	15	[57]
127	CPC, Taipei, China	Soft clay	35	31.7	NA	1	4	>3	4,534,547	Diaphragm	Top-down	Undifferentiated multi-support	192	NA	[57]
128	CPC, Taipei, China	Soft clay	35	31.7	NA	1	4	>3	1,454,766	Diaphragm	Top-down	Undifferentiated multi-support	123	NA	[57]
129	Taipei 101, China	Soft clay	35	21.7	NA	1	4.3	>3	2,321,688	Diaphragm	Top-down	Undifferentiated multi-support	106	NA	[57]
130	Taipei 101, China	Soft clay	35	21.7	NA	1	3.6	>3	2,321,688	Diaphragm	Top-down	Undifferentiated multi-support	54	NA	[57]
131	UPIB, Taipei, China	Soft clay	35	32.6	NA	1	4.1	>3	4,534,547	Diaphragm	Top-down	Undifferentiated multi-support	74	30	[57]
132	Jungshan, Taipei, China	Soft clay	35	26.6	NA	1	3.8	>3	2,321,688	Diaphragm	Top-down	Undifferentiated multi-support	122	NA	[57]
133	Kuntsevo Plaza, Russia	Sand and clay	35	25	NA	1	4.2	>3	1,200,000	Diaphragm	Bottom-up	Multi-anchor	8	13	[58]
134	Shanghai, China	Soft soil	35	15	40	1	3.7	>3	1,200,000	Diaphragm	Top-down	Multi-prop	70	136	[59]
135	Shanghai, China	Soft soil	35	14.5	40	1	3.7	>3	1,200,000	Diaphragm	Top-down	Multi-prop	21	16	[59]
136	Shanghai, China	Soft soil	35	15.2	40	1	3.7	>3	1,200,000	Diaphragm	Top-down	Multi-prop	21	16	[59]
137	Brescia, Italy	Clay and sand	75	18.5	1	0.05	2.6	>3	1,200,000	Diaphragm	NA	Multi-anchor	40	38	[60]
138	YBCC, Shanghai, China	Soft clay	35	15.9	NA	1	4.4	>3	680,000	Diaphragm	Top-down	Multi-prop	30.7	NA	[61]
139	Bangkok, Thailand	Soft clay	18	5.6	NA	1	2.9	>3	1,200,000	Diaphragm	Top-down	Single-prop	38	NA	[62]

(continued)

(continued)

No	Location	Soil at dredge level	*St* (kPa)	*H* (m)	*h* (m)	*h*/*H*	*s*	*FoS*	*EI* (N m^2)	Wall type	Construction method	Support type	δ_h (mm)	δ_v (mm)	References
140	Patras, Greece	Silty sand and clay	125	7.5	2	0.27	3.8	>3	1,200,000	Diaphragm	Bottom-up	Single-anchor	20	NA	[63]
141	Nangang, Taipei, China	Soft clay	20	11.6	7	1	2.9	>3	5,570,000	Diaphragm	Bottom-up	Multi-prop	217	178	[57]
142	CFSC, Norwich, UK	Sand and gravel	50	12	10	0.83	6	>3	340,000	Contiguous	Top-down	Berm and support	20	NA	[64]
143	CFSC, Norwich, UK	Sand and gravel	150	12	2	0.17	6	>3	340,000	Contiguous	Top-down	Berm and support	20	NA	[64]
144	CFSC, Norwich, UK	Sand and gravel	50	12	10	0.83	3	>3	340,000	Contiguous	Top-down	Multi-prop	8	NA	[64]
145	CFSC, Norwich, UK	Sand and gravel	150	12	2	0.17	3	>3	340,000	Contiguous	Top-down	Multi-prop	8	NA	[64]
146	CFSC, Norwich, UK	Sand and gravel	50	12	10	0.83	4	>3	340,000	Contiguous	Top-down	Multi-prop	6	NA	[64]
147	CFSC, Norwich, UK	Sand and gravel	150	12	2	0.17	4	>3	340,000	Contiguous	Top-down	Multi-prop	6	NA	[64]
148	John Rog. Quay, Dublin, Ireland	Soft silt	35	7	12.5	1	7	>3	690,135	Secant	NA	Single-anchor	24	NA	[65]
149	Portmarnoch, Dublin, Ireland	Soft sandy silt	18	4	6	1	4	>3	191,000	Secant	NA	Single-anchor	3	NA	[65]
150	Singapore River, Singapore	Stiff clay	200	29	20	0.69	3.3	>3	1,280,000	Diaphragm	NA	Multi-prop	79	NA	[5]
151	Singapore Havelock, Singapore	Stiff clay	200	16.5	16.4	0.99	4	>3	1,280,000	Diaphragm	NA	Multi-prop	40	NA	[5]
152	Singapore CBD, Singapore	Stiff clay	70	14.7	14	0.95	2.1	>3	60,000	Sheet	NA	Multi-prop	280	100	[5]
153	Singapore CE II, Singapore	Stiff clay	125	10	NA	0.9	4	>3	1,280,000	Diaphragm	NA	Multi-prop	22	15	[66]

(continued)

(continued)

No	Location	Soil at dredge level	*St* (kPa)	*H* (m)	*h* (m)	*h*/*H*	*s*	*FoS*	*EI* (N m^2)	Wall type	Construction method	Support type	δ_h (mm)	δ_v (mm)	References
154	Singapore CE II, Singapore	Stiff clay	125	10	NA	0.9	4	>3	73,500	Sheet	NA	Multi-prop	32	30	[66]
155	Singapore CE II, Singapore	Stiff clay	125	11	NA	0.9	4	>3	73,500	Sheet	NA	Multi-prop	48	25	[66]
156	Singapore CE II, Singapore	Stiff clay	125	14	NA	0.9	4	>3	100,000	Contiguous	NA	Multi-prop	37	18	[66]
157	Singapore CE II, Singapore	Stiff clay	125	15	NA	0.9	4	>3	685,000	Sheet	NA	Multi-prop	88	75	[66]
158	Singapore CE II, Singapore	Stiff clay	125	17	NA	0.9	4	>3	100,000	Contiguous	NA	Multi-prop	73	58	[66]
159	Singapore CE II, Singapore	Stiff clay	125	17	NA	0.9	4	>3	73,500	Sheet	NA	Multi-prop	75	25	[66]
160	Singapore CE II, Singapore	Stiff clay	125	17.5	NA	0.9	4	>3	685,000	Sheet	NA	Multi-prop	43	NA	[66]
161	Singapore CE II, Singapore	Stiff clay	125	18.5	NA	0.9	4	>3	100,000	Contiguous	NA	Multi-prop	73	NA	[66]
162	Singapore CE II, Singapore	Stiff clay	125	19	NA	0.9	4	>3	685,000	Sheet	NA	Multi-prop	50	NA	[66]
163	Singapore Mstory, Singapore	Clayey sand	500	17.3	17.2	0.99	2.88	>3	2,500,000	Diaphragm	NA	Multi-prop	50	NA	[22]
164	Singapore interchange, Singapore	Stiff clay	500	20	15	0.75	4	>3	1,338,750	Contiguous	NA	Multi-prop	40	NA	[5]
165	Singapore canal, Singapore	Stiff clay	140	6.5	6	0.92	5	>3	50,000	Sheet	NA	Single-prop	85	120	[5]
166	Bangkok B, Thailand	Stiff clay	140	15.5	15	0.97	5.1	>3	1,378,420	Diaphragm	NA	Multi-prop	45	NA	[29]

(continued)

(continued)

No	Location	Soil at dredge level	*St* (kPa)	*H* (m)	*h* (m)	*h/H*	*s*	*FoS*	*EI* (N m^2)	Wall type	Construction method	Support type	δ_h (mm)	δ_v (mm)	References
167	Bangkok D, Thailand	Stiff clay	140	16	14	0.88	3.2	>3	2,500,000	Diaphragm	NA	Multi-prop	25	NA	[29]
168	Japan 1, Japan	Stiff clay	140	13.75	12	0.87	6.88	>3	63,450	Sheet	NA	Multi-prop	120	NA	[33]
169	Sheung Wan HK CDG, China		350	30	19	0.63	2.73	>3	4,320,000	Diaphragm	NA	Multi-prop	20	23	[5]
170	Singapore CE II, Singapore	Stiff clay	125	12	NA	0.9	4	>3	1,280,000	Diaphragm	NA	Multi-anchor	15	17	[66]
171	Quai Gloria France	Sands	150	12	9	0.75	10	>3	1,378,420	Diaphragm	NA	Single-anchor	39	NA	[5]
172	Hartford, Conn., USA	Dense gravels	150	7	13.5	0.75	5	>3	26,250	Soldier	NA	Single-anchor	20	NA	[5]
173	C11A1 Boston, USA	Glacial till	95	20	8	0.4	2.9	>3	70,100	Soldier	Top-down	Multi-prop	40	33	[67]
174	C15A1 Boston, USA	Medium stiff clay	95	21.3	NA	0.7	5.2	>3	730,000	Soldier	Top-down	Multi-prop	23	34	[67]
175	C17A2 Boston, USA	Medium stiff clay	95	17	NA	0.7	5.7	>3	1,000,000	Soldier	Top-down	Multi-prop	39	22	[67]
176	FXWS, Chicago, USA	Medium stiff clay	95	12.2	NA	0.7	4	>3	1,020,000	Secant	Top-down	Multi-prop	38	41	[68]
177	Kings Inn St., Dublin, Ireland	Dense gravel	188	5.7	3.9	0.68	5.7	>3	381,700	Secant	NA	Single-prop	6	NA	[69]
178	Croydon, UK	London clay	85	11.4	2	0.18	5	>3	500,000	Secant	NA	Multi-prop	20	NA	[5]
179	Holborn, UK	London clay	120	11	3.5	0.32	8.65	>3	1,169,950	Secant	NA	Single-prop	12	NA	[5]
180	Minster Court, UK	London clay	120	9	5	0.56	7.3	>3	1,280,000	Diaphragm	NA	Berm and support	17	NA	[5]
181	Britannic Hse, UK	London clay	120	14	1	0.07	4.67	>3	1,280,000	Diaphragm	NA	Berm and support	60	34	[5]
182	Chelsea, UK	London clay	100	13	4	0.31	4.33	>3	312,500	Diaphragm	NA	Berm and support	27	NA	[5]

(continued)

(continued)

No	Location	Soil at dredge level	St (kPa)	H (m)	h (m)	h/H	s	FoS	EI (N m^2)	Wall type	Construction method	Support type	δ_h (mm)	δ_v (mm)	References
183	Walthamstow, UK	London clay	140	7.9	1.4	0.18	7.4	>3	8,437,500	Secant	NA	Multi-prop	18	20	[5]
184	Barbican, UK	London clay	180	16	2	0.13	4	>3	8,437,500	Diaphragm	NA	Multi-prop	10	NA	[5]
185	Charing Cross, UK	London clay	180	11	5	0.45	2.75	>3	1,280,000	Diaphragm	NA	Multi-prop	35	20	[5]
186	John Lewis KUT, UK	London clay	140	12	2.5	0.21	3.8	>3	1,280,000	Diaphragm	NA	Berm and support	20	NA	[5]
187	Victoria Emb, UK	London clay	140	18	6	0.33	6	>3	2,717,000	Secant	NA	Berm and support	34	28	[5]
188	London, UK	London clay	140	8	2	0.25	5.2	>3	312,500	Diaphragm	NA	Single-prop	3	NA	[33]
189	Guildhall, UK	London clay	140	6.5	2	0.31	3	>3	312,500	Diaphragm	NA	Single-prop	9	NA	[33]
190	Vauxhall, UK	London clay	140	14.5	2	0.14	3.63	>3	540,000	Diaphragm	NA	Multi-prop	22	NA	[33]
191	Chingford, UK	London clay	140	8	2	0.25	9.2	>3	2,291,750	Secant	NA	Single-prop	21	NA	[33]
192	Mark Lane, UK	London clay	140	7	2	0.29	3.5	>3	250,000	Contiguous	NA	Multi-prop	8	NA	[33]
193	Mark Lane, UK	London clay	140	5.5	2	0.36	6.3	>3	250,000	Contiguous	NA	Single-prop	14	NA	[33]
194	JLE III, UK	London clay	140	8	2	0.25	8.2	>3	540,000	Diaphragm	NA	Single-prop	10	NA	[33]
195	Malden Way, UK	London clay	80	7.5	2	0.27	5.5	>3	2,544,700	Contiguous	NA	Single-prop	36	NA	[33]
196	Bermondsey, UK	W1R beds	300	19.5	0	0	6	>3	2,500,000	Diaphragm	NA	Multi-prop	13	NA	[5]
197	Canada Water, UK	W1R beds	300	17	7	0.41	5.7	>3	805,158	Secant	NA	Multi-prop	15	NA	[5]
198	Humber Bridge, UK	Kimmer clay	250	24.5	NA	0	4.92	>3	1,280,000	Diaphragm	NA	Multi-prop	21	NA	[5]
199	Cambridge, UK	Gault clay	120	10	2	0.2	3.3	>3	540,000	Diaphragm	NA	Multi-prop	13	10	[5]
200	Channel Tunnel, UK	Gault clay	140	6.5	0	0	4	>3	73,500	Sheet	NA	Multi-prop	45	NA	[5]
201	Lyon, France	Stiff clay	80	8	0	0	3.2	>3	540,000	Diaphragm	NA	Multi-prop	14	NA	[5]
202	Dublin-Jervis, Ireland	Glacial till	501	9.7	3	0.31	8.5	>3	1,254,800	Secant	NA	Single-prop	3	0	[5]
203	Dublin-Clarend, Ireland	Glacial till	501	6.2	1	0.16	5	>3	3895	Soldier	NA	Single-prop	7	0	[5]

(continued)

(continued)

No	Location	Soil at dredge level	St (kPa)	H (m)	h (m)	h/H	s	FoS	EI (N m^2)	Wall type	Construction method	Support type	δ_h (mm)	δ_v (mm)	References
204	Dublin-M&S, Irealnd	Glacial till	501	7.2	3	0.42	6	>3	58,500	Sheet	NA	Single-prop	5	2	[5]
205	MBTA, Boston, USA	Stiff clay	140	15.2	0	0	3.36	>3	1,908,880	Diaphragm	NA	Multi-prop	25	13	[5]
206	Oakland, USA	Stiff clay	140	19	NA	0	3.36	>3	63,500	Sheet	NA	Multi-prop	25	13	[70]
207	Houston, USA	Stiff clay	140	18.3	3	0.16	6.1	>3	704,900	NA	NA	Multi-prop	18	NA	[70]
208	Seattle, USA	Stiff clay	300	23.8	0	0	2.64	>3	63,500	Sheet	NA	Multi-prop	114	NA	[70]
209	West. Station, Seattle, USA	Stiff clay	140	15.2	0	0	3.8	>3	1,780,000	Contiguous	NA	Multi-prop	5	5	[5]
210	Pion. Square, Seattle, USA	Stiff clay	140	21.9	0	0	5.5	>3	1,126,000	Secant	NA	Multi-prop	14	5	[5]
211	Washington, USA	Hard clay	300	17	4	0.24	11	>3	1,535,850	Secant	NA	Multi-prop	26	NA	[5]
212	Washington, USA	Hard clay	300	25	4	0.16	6.25	>3	63,500	Sheet	NA	Multi-prop	28	NA	[5]
213	Washington, USA	Stiff clay	140	15	4.5	0.3	7.5	>3	20,000	Contiguous	NA	Multi-prop	14	7	[5]
214	Houston-Exxon, USA	Stiff clay	130	16.2	0	0	6.5	>3	7425	Soldier	NA	Multi-prop	9	NA	[71]
215	Houston-1Shell, USA	Stiff clay	120	18	3	0.17	6	>3	841,550	Contiguous	NA	Multi-prop	25	NA	[72]
216	Houston-Coker, USA	Stiff clay	375	17.1	0	0	3.5	>3	103,600	Sheet	NA	Multi-prop	28	NA	[72]
217	Tiong Bahru, Singapore	Stiff clay	300	15.3	1.5	0.1	4.6	>3	24,900	Soldier	NA	Multi-prop	25	NA	[23]
218	Singapore CE II, Singapore	Stiff clay	175	9	1	0.11	4	>3	26,250	Soldier	NA	Multi-prop	30	10	[66]
219	Singapore CE II, Singapore	Stiff clay	175	9	1	0.11	4	>3	26,250	Soldier	NA	Multi-prop	30	NA	[66]

(continued)

(continued)

No	Location	Soil at dredge level	*St* (kPa)	*H* (m)	*h* (m)	*h*/*H*	*s*	*FoS*	*EI* (N m^2)	Wall type	Construction method	Support type	δ_h (mm)	δ_v (mm)	References
220	Singapore CE II, Singapore	Stiff clay	175	9	1	0.11	4	>3	26,250	Soldier	NA	Multi-prop	35	NA	[66]
221	Singapore CE II, Singapore	Stiff clay	175	10.2	1	0.1	4	>3	26,250	Soldier	NA	Multi-prop	10	5	[66]
222	Singapore CE II, Singapore	Stiff clay	175	10.5	1	0.1	4	>3	26,250	Soldier	NA	Multi-prop	20	9	[66]
223	Singapore CE II, Singapore	Stiff clay	175	11.5	1	0.09	4	>3	26,250	Soldier	NA	Multi-prop	17	10	[66]
224	Singapore CE II, Singapore	Stiff clay	175	11.5	1	0.09	4	>3	26,250	Soldier	NA	Multi-prop	10	15	[66]
225	Singapore CE II, Singapore	Stiff clay	175	10	1	0.1	4	>3	73,500	Sheet	NA	Multi-prop	8	7	[66]
226	Singapore CE II, Singapore	Stiff clay	175	10	1	0.1	4	>3	73,500	Sheet	NA	Multi-prop	9	12	[66]
227	Singapore CE II, Singapore	Stiff clay	175	10	1	0.1	4	>3	73,500	Sheet	NA	Multi-prop	12	15	[66]
228	Singapore CE II, Singapore	Stiff clay	175	10	1	0.1	4	>3	73,500	Sheet	NA	Multi-prop	37	20	[66]
229	Singapore CE II, Singapore	Stiff clay	175	11	1	0.09	4	>3	73,500	Sheet	NA	Multi-prop	20	15	[66]
230	Singapore CE II, Singapore	Stiff clay	175	11.5	1	0.09	4	>3	73,500	Sheet	NA	Multi-prop	23	10	[66]
231	Singapore CE II, Singapore	Stiff clay	175	11.5	1	0.09	4	>3	73,500	Sheet	NA	Multi-prop	42	12	[66]
232	Singapore CE II, Singapore	Stiff clay	175	12	1	0.08	4	>3	73,500	Sheet	NA	Multi-prop	15	25	[66]
233	Singapore CE II, Singapore	Stiff clay	175	17	1	0.06	4	>3	73,500	Sheet	NA	Multi-prop	17	15	[66]

(continued)

(continued)

No	Location	Soil at dredge level	*St* (kPa)	*H* (m)	*h* (m)	*h*/*H*	*s*	*FoS*	*EI* (N m^2)	Wall type	Construction method	Support type	δ_h (mm)	δ_v (mm)	References
234	Singapore CE II, Singapore	Stiff clay	175	17	1	0.06	4	>3	73,500	Sheet	NA	Multi-prop	22	27	[66]
235	Singapore CE II, Singapore	Stiff clay	175	11	1	0.09	4	>3	100,000	Contiguous	NA	Multi-prop	7	12	[66]
236	Singapore CE II, Singapore	Stiff clay	175	11	1	0.09	4	>3	100,000	Contiguous	NA	Multi-prop	7	10	[66]
237	Singapore CE II, Singapore	Stiff clay	175	11.5	1	0.09	4	>3	100,000	Contiguous	NA	Multi-prop	7	10	[66]
238	Singapore CE II, Singapore	Stiff clay	175	12	1	0.08	4	>3	100,000	Contiguous	NA	Multi-prop	20	NA	[66]
239	Singapore CE II, Singapore	Stiff clay	175	12	1	0.08	4	>3	100,000	Contiguous	NA	Multi-prop	20	NA	[66]
240	Singapore CE II, Singapore	Stiff clay	175	12.5	1	0.08	4	>3	100,000	Contiguous	NA	Multi-prop	15	18	[66]
241	Singapore CE II, Singapore	Stiff clay	175	13.5	1	0.07	4	>3	100,000	Contiguous	NA	Multi-prop	7	15	[66]
242	Singapore CE II, Singapore	Stiff clay	175	13.5	1	0.07	4	>3	100,000	Contiguous	NA	Multi-prop	8	20	[66]
243	Singapore CE II, Singapore	Stiff clay	175	13.5	1	0.07	4	>3	100,000	Contiguous	NA	Multi-prop	17	NA	[66]
244	Singapore CE II, Singapore	Stiff clay	175	20	1	0.05	4	>3	100,000	Contiguous	NA	Multi-prop	15	NA	[66]
245	Singapore CE II, Singapore	Stiff clay	175	21.5	1	0.05	4	>3	100,000	Contiguous	NA	Multi-prop	25	NA	[66]
246	Singapore CE II, Singapore	Stiff clay	175	13.5	1	0.07	4	>3	1,280,000	Diaphragm	NA	Multi-prop	15	15	[66]
247	Singapore CE II, Singapore	Stiff clay	175	14.5	1	0.07	4	>3	1,280,000	Diaphragm	NA	Multi-prop	27	23	[66]

(continued)

(continued)

No	Location	Soil at dredge level	*St* (kPa)	*H* (m)	*h* (m)	*h*/*H*	*s*	*FoS*	*EI* (N m^2)	Wall type	Construction method	Support type	δ_h (mm)	δ_v (mm)	References
248	Waterloo, UK	Med gravel	75	5.78	2	0.35	5.8	>3	540,000	Diaphragm	NA	Berm and support	20	NA	[5]
249	Eastbourne, UK	Gravel/sand	163	11	0	0	7	>3	2,500,000	Diaphragm	NA	Multi-prop	60	NA	[33]
250	Buffalo, Canada	Dense sand/gravel	150	11	NA	0	3.67	>3	63,585	Sheet	NA	Multi-prop	63	NA	[70]
251	Ontario, Canada	VD sand	250	15.2	NA	0	3.04	>3	42,000	Sheet	NA	Multi-prop	230	NA	[5]
252	Lyon-P Kleb, France	Sandy gravel	85	6.75	3.5	0.52	6.75	>3	57,700	Sheet	NA	Single-prop	7	NA	[5]
253	Lyon-R Ney, France	Sandy gravel	125	9.95	6.5	0.65	3.1	>3	107,000	Diaphragm	NA	Multi-prop	14	NA	[5]
254	Lyon-S. Gam., France	Sandy gravel	135	10.7	3	0.28	5.2	>3	185,000	Diaphragm	NA	Single-prop	60	NA	[5]
255	Karlshrue, Germany	Sands	75	5	0	0	3.75	>3	2033	Sheet	NA	Single-prop	5	NA	[5]
256	Maas, Rotterdam, Netherlands	Silts/sands	260	21	0	1	17	>3	1,717,900	Sheet	NA	Single-prop	32	NA	[5]
257	Lisbon-Carlos, Portugal	Clay/sands	60	13.8	5.5	0.4	2.76	>3	540,000	Diaphragm	NA	Multi-prop	43	NA	[5]
258	Sao Paulo, ES1, Brazil	Residual soil	40	9	0	1	3.6	>3	71,700	Soldier	NA	Multi-prop	15	NA	[5]
259	Sao Paulo, ES2. Brazil	Residual soil	40	19	0	1	3.8	>3	71,700	Soldier	NA	Multi-prop	18	NA	[5]
260	Argyle Station, China	Residual soil	250	18.7	7	1	6.2	>3	2,500,000	Diaphragm	NA	Multi-prop	29	58	[5]
261	Han River, Seoul, Japan	Wth. rock	175	25	13.5	0.54	2.78	>3	60,640	Secant	NA	Multi-prop	10	NA	[5]
262	YMCA, London, UK	London clay	140	16	1	0.06	10	>3	540,000	Diaphragm	NA	Multi-anchor	20	NA	[5]
263	Neasden, UK	London clay	150	8	1	0.13	2	>3	540,000	Diaphragm	NA	Multi-anchor	52	54	[5]

(continued)

(continued)

No	Location	Soil at dredge level	*St* (kPa)	*H* (m)	*h* (m)	*h*/*H*	*s*	*FoS*	*EI* (N m^2)	Wall type	Construction method	Support type	δ_h (mm)	δ_v (mm)	References
264	Oresund-Sydh, Denmark	Boulder clay	300	10	5	0.5	7	>3	48,125	Soldier	NA	Single-anchor	55	NA	[5]
265	Copenhagen, Denmark	Stiff clays	175	11	1	0.09	5.5	>3	1,000,000	Contiguous	NA	Single-anchor	2	NA	[5]
266	Lisbon-DD Ave., Portugal	Stiff clay	175	17	2.6	0.15	3.25	>3	19,250	Soldier	NA	Multi-anchor	24	NA	[73]
267	Lisbon-Colom, Portugal	Stiff clay	200	14	4	0.29	3.25	>3	14,435	Soldier	NA	Multi-anchor	90	NA	[73]
268	Lisbon-Ivens, Portugal	Stiff clay	140	9	0	0	3.25	>3	19,250	Soldier	NA	Multi-anchor	4	NA	[73]
269	Colomb., Seattle, USA	Stiff clay	250	37	0	0	1.75	>3	763,400	Secant	NA	Multi-anchor	15	10	[5]
270	University St. Sta., Seattle, USA	Stiff clay	140	18.3	0	0	2.6	>3	1,126,000	Contiguous	NA	Multi-anchor	13	2	[5]
271	Seattle, USA	Stiff clay	140	23	0	0	1.5	>3	718,690	Soldier	NA	Multi-anchor	23	5	[5]
272	Houston-Herm, USA	Stiff clay	75	9	0	0	3	>3	203,900	Contiguous	NA	Multi-anchor	20	5	[71]
273	Houston-Bank, USA	Stiff clay	130	16.8	3	0.18	3.35	>3	203,900	Contiguous	NA	Multi-anchor	29	NA	[71]
274	Houston-FCB, USA	Stiff clay	75	9.1	0	0	8	>3	209,650	Contiguous	NA	Multi-anchor	20	NA	[71]
275	Houston-Smith, USA	Stiff clay	150	15.5	0	0	3	>3	140,135	Contiguous	NA	Multi-anchor	20	NA	[71]
276	Houston-Texas, USA	Stiff clay	100	16	3	0.19	3.2	>3	13,860	Contiguous	NA	Multi-anchor	30	NA	[71]
277	Houston-Cullen, USA	Stiff clay	75	8.2	0	0	7	>3	210,470	Contiguous	NA	Multi-anchor	20	NA	[71]
278	Houston-321, USA	Stiff clay	75	9.1	0	0	3	>3	315,050	Contiguous	NA	Multi-anchor	20	NA	[71]
279	Washington, USA	Stiff clay	140	15.2	3	0.2	3	>3	19,000	Soldier	NA	Multi-anchor	51	NA	[5]

(continued)

(continued)

No	Location	Soil at dredge level	*St* (kPa)	*H* (m)	*h* (m)	*h*/*H*	*s*	*FoS*	*EI* (N m^2)	Wall type	Construction method	Support type	δ_h (mm)	δ_v (mm)	References
280	State Trans, Boston, USA	Stiff clay	140	9.1	0	0	3.36	>3	567,450	Diaphragm	NA	Multi-anchor	20	31	[74]
281	60 State St., Boston, USA	Stiff clay	140	9.1	0	0	3.36	>3	567,450	Diaphragm	NA	Multi-anchor	19	31	[74]
282	Davis Square, Boston, USA	Stiff clay	140	17.1	0	0	3.36	>3	567,450	Diaphragm	NA	Multi-anchor	28	43	[74]
283	1 Memorial, Boston, USA	Stiff clay	140	8.2	0	0	3.36	>3	567,450	Diaphragm	NA	Multi-anchor	25	31	[74]
284	Harvard Square, Boston, USA	Till dense	300	15.7	0	0	5.2	>3	1,908,000	Diaphragm	NA	Multi-anchor	10	NA	[5]
285	Harvard Square, Boston, USA	Till	300	15.7	0	0	6.2	>3	1,908,001	Diaphragm	NA	Multi-anchor	11	NA	[5]
286	Boston, USA	Stiff clay	140	18.9	2	0.11	2.7	>3	347,430	Soldier	NA	Multi-anchor	89	NA	[5]
287	Salt Lake City, USA	Stiff clay	140	12.5	3	0.24	3.2	>3	127,436	Soldier	NA	Multi-anchor	25	NA	[5]
288	Singapore CE II, Singapore	Stiff clay	175	12.2	1	0.08	4	>3	26,250	Soldier	NA	Multi-anchor	10	8	[66]
289	Singapore CE II, Singapore	Stiff clay	175	12.5	1	0.08	4	>3	26,250	Soldier	NA	Multi-anchor	18	6	[66]
290	Singapore CE II, Singapore	Stiff clay	175	12.5	1	0.08	4	>3	26,250	Soldier	NA	Multi-anchor	14	12	[66]
291	Singapore CE II, Singapore	Stiff clay	175	16	1	0.06	4	>3	73,500	Soldier	NA	Multi-anchor	15	NA	[66]
292	Singapore CE II, Singapore	Stiff clay	175	15	1	0.07	4	>3	100,000	Soldier	NA	Multi-anchor	8	NA	[66]
293	Singapore CE II, Singapore	Stiff clay	175	17	1	0.06	4	>3	100,000	Soldier	NA	Multi-anchor	35	NA	[66]
294	Singapore CE II, Singapore	Stiff clay	175	18.5	1	0.05	4	>3	100,000	Soldier	NA	Multi-anchor	27	NA	[66]

(continued)

(continued)

No	Location	Soil at dredge level	*St* (kPa)	*H* (m)	*h* (m)	*h*/*H*	*s*	*FoS*	*EI* (N m^2)	Wall type	Construction method	Support type	δ_h (mm)	δ_v (mm)	References
295	Singapore CE II, Singapore	Stiff clay	175	19	1	0.05	4	>3	100,000	Soldier	NA	Multi-anchor	23	NA	[66]
296	A1(M), UK	Sand	120	9.3	NA	0	10.6	>3	104,785	Sheet	NA	Single-anchor	20	NA	[5]
297	Hatfield, UK	Gravels	100	9.3	3	0.32	6.8	>3	96,440	Sheet	NA	Single-anchor	25	27	[5]
298	Paris-13e, France	Sands	220	17.4	0	0	4.35	>3	540,000	Diaphragm	NA	Multi-anchor	20	NA	[5]
299	Calais, France	Sands	360	24	4.5	0.19	8	>3	5,881,600	Diaphragm	NA	Multi-anchor	58	NA	[5]
300	Le Havre, France	Sand/gravel	500	16.5	9	0.55	10.5	>3	4,320,000	Diaphragm	NA	Single-anchor	12	NA	[5]
301	Geneva, Le Mail, Switzerland	Sand/gravel	190	14.8	4	0.27	2.5	>3	2,500,000	Diaphragm	NA	Multi-anchor	13	NA	[5]
302	Berlin P Platz DB, Germany	Sands	225	18	3	0.17	15	>3	5,184,000	Diaphragm	NA	Single-anchor	42	10	[5]
303	Berlin-Hofgarten, Germany	Sands	214	17	3	0.18	2.83	>3	1,280,000	Diaphragm	NA	Multi-anchor	36	NA	[5]
304	Berlin, Germany	Sands	230	18.5	3	0.16	0.75	>3	4,320,000	Soldier	NA	Multi-anchor	15	5	[5]
305	Berlin, Germany	Sands	155	12.3	3	0.24	3.6	>3	631,000	Secant	NA	Multi-anchor	27	NA	[5]
306	SONY, Berlin, Germany	Sand/gravel	180	14.3	5	0.35	3.58	>3	1,280,000	Diaphragm	NA	Multi-anchor	50	NA	[5]
307	Grauholz, Switzerland	Sands/silts	150	17	0	0	2.4	>3	483,095	Contiguous	NA	Multi-anchor	20	NA	[5]
308	Johannsburg, South Africa	Firm silt	70	18.3	0	0	3.05	>3	2581	Soldier	NA	Multi-anchor	18	22	[5]
309	Milwaukee, USA	Sands	65	5	0	0	1.67	>3	60,000	Deep soil mix	NA	Multi-anchor	15	NA	[5]
310	Norwich, UK	Chalk	75	18	1.2	0.07	2.57	>3	919,700	Contiguous	NA	Multi-anchor	9	NA	[5]
311	Dartford, UK	Chalk	150	9	0	0	2	>3	1,280,000	Diaphragm	NA	Multi-anchor	15	NA	[5]
312	Bell Common, UK	London clay	130	9	4	0.44	8.9	>3	2,330,250	Secant	Top-down	Undifferentiated single-support	25	25	[5]

(continued)

(continued)

No	Location	Soil at dredge level	*St* (kPa)	*H* (m)	*h* (m)	*h*/*H*	*s*	*FoS*	*EI* (N m^2)	Wall type	Construction method	Support type	δ_h (mm)	δ_v (mm)	References
313	New Palace Yard, UK	London clay	150	18.5	2	0.11	3.08	>3	2,500,000	Diaphragm	Top-down	Undifferentiated multi-support	30	20	[5]
314	British Library, UK	London clay	200	24.4	3	0.12	5	>3	2,571,750	Secant	Top-down	Undifferentiated multi-support	30	30	[5]
315	Nat Gal Ext, UK	London clay	140	10	4.2	0.42	7	>3	618,000	Secant	Top-down	Undifferentiated single-support	10	2	[5]
316	Aldersgate, UK	London clay	250	23	8	0.35	3.3	>3	2,500,000	Diaphragm	Top-down	Undifferentiated multi-support	33	18	[5]
317	Limehouse, UK	W&R beds	300	16	4	0.25	4	>3	4,320,000	Diaphragm	Top-down	Undifferentiated multi-support	5	NA	[5]
318	PO Square Boston, UK	Till	140	23.4	4	0.17	3.3	>3	1,822,500	Diaphragm	Top-down	Undifferentiated multi-support	52	45	[5]
319	HK&S Bank, China	Decom. granite	200	16	5	0.31	4	>3	2,500,000	Diaphragm	Top-down	Undifferentiated multi-support	48	25	[5]
320	Rowes Whr, Boston, USA	Stiff clay	140	16.8	5	0.3	3.36	>3	1,106,125	Diaphragm	Top-down	Undifferentiated multi-support	19	NA	[75]
321	75 State St., Boston, USA	Stiff clay	140	19.8	3	0.15	3.36	>3	1,106,125	Diaphragm	Top-down	Undifferentiated multi-support	51	102	[75]
322	125 Sum. St., Boston, USA	Stiff clay	140	18.3	0	0	3.36	>3	1,106,125	Diaphragm	Top-down	Undifferentiated multi-support	15	10	[75]
323	Ashford International Station, UK	Atherfield clay	250	20	2.8	0.14	10	>3	800,000	Contiguous	Top-down	Multi-prop	25	NA	[75]
324	Chicago, USA	Glacial Clays	100	12.2	4.9	0.4	4	>3	118,600	Secant	Top-down	Multi-anchor	30	NA	[63]
325	Chicago, USA	Glacial clays	100	12.2	4.9	0.4	4	>3	118,600	Secant	Top-down	Multi-anchor	25	NA	[68]
326	Seoul Metro Line 7, Japan	Weathered soil	250	24.8	13.1	0.53	1.7	>3	340,000	Soldier	Bottom-up	Multi-anchor	16	20	[76]

(continued)

(continued)

No	Location	Soil at dredge level	*St* (kPa)	*H* (m)	*h* (m)	*h*/*H*	*s*	*FoS*	*EI* (N m^2)	Wall type	Construction method	Support type	δ_h (mm)	δ_v (mm)	References
327	Athens Metro, Greece	Hard sandy clay	400	26	6	0.23	3	>3	285,000	Contiguous	Top-down	Multi-anchor	31	13	[77]
328	Cincon Station, Istanbul, Turkey	Hard-dense silty clay	213	32.5	0	0	5.4	>3	4,320,000	Diaphragm	Top-down	Multi-prop	11	18	[78]
329	Harrods, London, UK	London clay	130	23	0	0	3	>3	680,000	Diaphragm	Top-down	Undifferentiated multi-support	27	30	[79]
330	Heuston, Dublin, Ireland	Dublin boulder clay	100	14	0	0	7	>3	340,000	Secant	NA	Single-anchor	80	NA	[80]
331	TOOBA, Tehran, Iran		250	16.5	8	0.48	4.1	>3	340,000	Contiguous	Top-down	Multi-prop	20	NA	[81]
332	Los Angeles Metro, USA	Sand with silt	324	18	0	0	4	>3	340,000	Soldier	Top-down	Multi-prop	51	NA	[82]
333	London, UK	London clay	130	20	1	0.05	5	>3	340,000	Secant	Top-down	Multi-prop	10	NA	[83]
334	Swainswick Bypass, UK	Lias clay	150	7.9	4	0.51	4	>3	680,000	Diaphragm	Bottom-up	Single-prop	16	NA	[84]
335	St. Johns Road, Dublin, Ireland	Medium dense gravel	250	8	1	0.13	7.5	>3	245,450	Contiguous	NA	Single-prop	23	NA	[85]
336	South Lotts Road, Dublin, Irel'd	Dense gravel	250	6.5	3.5	0.54	6	>3	381,700	Secant	NA	Single-prop	19	NA	[65]
337	DPT Southern C&C, Dublin, Irel'd	Dense sand and gravel	263	22	4.5	0.2	9	>3	4,320,000	Diaphragm	NA	Multi-prop	4	NA	[65]
338	DPT Southern C&C, Dublin, Irel'd	Loose sand and gravel	253	18	8.2	0.46	7.5	>3	4,320,000	Diaphragm	NA	Multi-prop	9	NA	[65]
339	Spencer Dock, Dublin, Ireland	Gravel	120	9.6	7.5	0.78	9.6	>3	644,126	Secant	NA	Single-anchor	3	NA	[65]
340	Smithfield, Dublin, Ireland	Dense gravel	260	11	3	0.27	7	>3	1,130,300	Diaphragm	NA	Single-anchor	4	NA	[69]

(continued)

(continued)

No	Location	Soil at dredge level	*St* (kPa)	*H* (m)	*h* (m)	*h*/*H*	*s*	*FoS*	*EI* (N m^2)	Wall type	Construction method	Support type	δ_h (mm)	δ_v (mm)	References
341	Railway St., Dublin, Ireland	Dens gravel	220	7.2	4	0.56	8	>3	381,700	Secant	NA	Single-prop	11	NA	[69]
342	Clancy Barracks, Ireland	Dense gravel	210	7.2	4	0.56	7.2	>3	644,126	Secant	NA	Single-anchor	12	NA	[69]
343	Leinster House, Dublin, Ireland	Dublin boulder clay	300	6	0	0	6	>3	381,700	Secant	NA	Single-prop	3	NA	[65]
344	Hilton House, Dublin, Ireland	Gravel/Dublin boulder clay	300	6.3	4	0.63	6.3	>3	1,254,800	Secant	NA	Single-prop	1	NA	[65]
345	Grand Canal, Dublin, Ireland	Dublin boulder clay	300	4	2	0.5	3.5	>3	347,000	Secant	NA	Single-anchor	5	NA	[65]
346	TCD Library-Lecky, Ireland	Dublin boulder clay	300	7.2	0	0	2.5	>3	347,000	secant	NA	Multi-anchor	3	NA	[65]
347	Pearce St., Dublin, Ireland	Upper Brown boulder clay	170	12	3	0.25	8	>3	372,750	Secant	NA	Multi-prop	9	NA	[85]
348	Cherrywood, Dublin, Ireland	Dublin boulder clay	300	8.5	1	0.12	8.5	>3	878,355	Contiguous	NA	Single-anchor	8	NA	[85]
349	Mater Hospial, Dublin, Ireland	Dublin boulder clay	300	8	0.6	0.08	7.75	>3	690,135	Secant	NA	Single-anchor	6	NA	[85]
350	Rathmines, Dublin, Ireland	Dublin boulder clay	300	7	1	0.14	6	>3	265,000	Contiguous	NA	Single-anchor	3	NA	[85]
351	Burlington Rd., Dublin, Ireland	Dublin boulder clay	300	10	2	0.2	2	>3	690,135	Secant	NA	Single-anchor	2	NA	[85]
352	Terrenure, Dublin, Ireland	Dublin boulder clay	300	3.2	0.3	0.09	3.2	>3	25,700	Sheet	NA	Single-prop	2	NA	[85]
353	TCD Library-Nassau, Ireland	Dublin boulder clay	300	8.6	0	0	3.5	>3	347,000	Secant	NA	Multi-anchor	7	NA	[85]

(continued)

(continued)

No	Location	Soil at dredge level	*St* (kPa)	*H* (m)	*h* (m)	*h*/*H*	*s*	*FoS*	*EI* (N m^2)	Wall type	Construction method	Support type	δ_h (mm)	δ_v (mm)	References
354	Ely Place, Dublin, Ireland	Dublin boulder clay	300	3.5	0	0	3.5	>3	201,300	Secant	NA	Single-prop	3	NA	[85]
355	Harcourt St., Dublin, Ireland	Dublin boulder clay	300	4.5	0	0	3	>3	381,700	Secant	NA	Single-prop	5	NA	[85]
356	Balbriggan, Dublin, Ireland	Gravel and clay	140	5.7	2.5	0.44	5.7	>3	644,126	Secant	NA	Single-anchor	9	NA	[85]
357	Westgate, Dublin, Ireland	Dublin boulder clay	300	14	2	0.14	14	>3	690,135	Secant	NA	Single-anchor	7	NA	[85]
358	Tallaght Centre, Dublin, Ireland	Dublin boulder clay	300	11	1.5	0.14	6	>3	293,620	Contiguous	NA	Single-anchor	18	NA	[85]
359	TCD-Biosciences, Ireland	Dublin boulder clay	300	10.5	4	0.38	6.7	>3	644,125	Secant	NA	Single-anchor	11	NA	[85]
360	TCD-Biosciences, Ireland	Dublin boulder clay	300	10.5	4	0.38	3	>3	644,125	Secant	NA	Multi-anchor	9	NA	[85]
361	TCD-Biosciences, Ireland	Dublin boulder clay	300	10.5	4	0.38	3.6	>3	644,125	Secant	NA	Multi-anchor	7	7	[85]
362	Monte Vetro, Ireland	Dublin boulder clay	300	13.7	2	0.15	12	>3	568,400	Secant	NA	Single-prop	3	NA	[85]
363	Monte Vetro, Ireland	Dublin boulder clay	300	17.4	8	0.46	2.9	>3	568,400	Secant	NA	Multi-anchor	4	NA	[85]
364	Barrow St., Dublin, Ireland	Dublin boulder clay	300	11.4	4	0.35	2.9	>3	568,400	Secant	NA	Multi-anchor	1	NA	[85]
365	TCD Sports Centre, Ireland	Dublin boulder clay	300	6.9	1	0.14	5.6	>3	644,126	Secant	NA	Single-anchor	11	NA	[85]
366	TCD-Crann Building, Ireland	Dublin boulder clay	300	6.9	1	0.14	2.3	>3	644,126	Secant	NA	Multi-anchor	7	NA	[85]
367	DPT Northern C&C, Ireland	Dublin boulder clay	300	12	1	0.08	10.2	>3	4,320,000	Diaphragm	NA	Single-prop	5	NA	[85]

(continued)

(continued)

No	Location	Soil at dredge level	*St* (kPa)	*H* (m)	*h* (m)	*h*/*H*	*s*	*FoS*	*EI* (N m^2)	Wall type	Construction method	Support type	δ_h (mm)	δ_v (mm)	References
368	DPT Northern C&C, Ireland	Dublin boulder clay	300	17	1	0.06	5.5	>3	4,320,000	Diaphragm	NA	Multi-prop	6	NA	[85]
369	DPT-Shaft WA2, Ireland	Dublin boulder clay	300	25	1	0.04	12	>3	8,437,500	Diaphragm	NA	Single-prop	12	NA	[85]
370	Mespil Road, Dublin, Ireland	Dublin boulder clay	300	11	2	0.18	8	>3	1,288,250	Secant	NA	Single-prop	14	6	[85]
371	CTRL-Chart Road, UK	Atherfield clay	80	8	2.5	0.31	6	>3	340,000	Contiguous	Bottom-up	Undifferentiated single-support	43	NA	[86]
372	CTRL-Advance, UK	Atherfield clay	95	9.8	0	0	6	>3	340,000	Contiguous	Top-down	Undifferentiated single-support	18	NA	[86]
373	CTRL-Chart Road, UK	Atherfield clay	135	14	1.4	0.1	6	>3	340,000	Contiguous	Bottom-up	Multi-prop	40	NA	[86]
374	CTRL-Maidstone Railway, UK	Atherfield clay	135	14	0	0	6	>3	340,000	Contiguous	Top-down	Single-prop	31	NA	[86]
375	CTRL-Greensands Way, UK	Atherfield clay	100	10	0	0	6	>3	340,000	Contiguous	Bottom-up	Undifferentiated single-support	20	NA	[86]
376	CTRL-Gasworks Lane, UK	Atherfield clay	100	10.5	5.8	0.55	4.5	>3	340,000	Contiguous	Top-down	Multi-prop	17	NA	[86]
377	CTRL-Cattlemarket, UK	Atherfield clay	95	9.6	4	0.42	4.5	>3	340,000	Contiguous	Bottom-up	Multi-prop	19	NA	[86]
378	Shanghai, China	Silty clay	25	4.5	5	1	3.5	>3	129,000	Secant	NA	Single-prop	10	NA	[87]
379	Shanghai, China	Silty clay	35	7	5	1	3	>3	129,000	Secant	NA	Multi-prop	16	NA	[87]
380	Savoy, Limerick, Ireland	Boulder clay	300	6.4	2	0.31	6.4	>3	191,000	Secant	NA	Single-anchor	7	NA	[80]
381	Savoy, Limerick, Ireland	Boulder clay	300	5.8	2	0.34	6.4	>3	191,000	Secant	NA	Single-anchor	2	NA	[80]
382	Main St., Cavan, Ireland	Boulder clay	300	9	2	0.22	4.5	>3	8000	Contiguous	NA	Multi-anchor	45	NA	[80]

(continued)

(continued)

No	Location	Soil at dredge level	*St* (kPa)	*H* (m)	*h* (m)	*h*/*H*	*s*	*FoS*	*EI* (N m^2)	Wall type	Construction method	Support type	δ_h (mm)	δ_v (mm)	References
383	Dundalk Cellar, Ireland	Boulder clay	300	8.7	2	0.23	8.7	>3	176,000	Secant	NA	Single-anchor	6	NA	[80]
384	Dundalk Cellar, Ireland	Boulder clay	300	8.1	2	0.25	8.1	>3	176,000	Secant	NA	Single-anchor	7	NA	[80]
385	Midleton, Ireland	Boulder clay	300	6	2	0.33	6	>3	206,000	Secant	NA	Single-anchor	6	NA	[80]
386	Kilkenny, Ireland	Boulder clay	300	7.7	2	0.26	7.7	>3	254,000	Contiguous	NA	Single-prop	5	NA	[80]
387	Kilkenny, Ireland	Boulder clay	300	6.9	2	0.29	6.9	>3	354,000	Contiguous	NA	Single-prop	2	NA	[80]
388	Portlaoise SC, Ireland	Boulder clay	300	12	2	0.17	6	>3	247,000	Secant	NA	Multi-anchor	17	NA	[80]
389	Portlaoise SC, Ireland	Boulder clay	300	7	2	0.29	7	>3	247,000	Secant	NA	Single-anchor	4	NA	[80]

References

1. Finno, R. J., & Roboski, J. F. (2005). Three-dimensional responses of a tied-back excavation through clay. *Journal of Geotechnical and Geoenvironmental Engineering, 131*, 273–282.
2. NGI (1962). Measurements at a strutted excavation, Oslo Subway, Vaterland 3, km 1450, Technical Report No. 5. NGI, Oslo, Norway.
3. NGI (1962). Measurements at a strutted excavation, Oslo Subway, Enerhaugen South, km 1982, Technical Report No. 3. NGI, Oslo, Norway.
4. NGI (1962). Measurements at a strutted excavation, Oslo Subway, Vaterland 2, km 1408., Technical Report No. 6. NGI, Oslo, Norway.
5. Bryson L. S. & Zapata-Medina, D. G. (2012). Method for estimating system Stiffness for excavation support walls. *Journal of Geotechnical and Geoenvironmental Engineering, 138*, 1104–1115.
6. Rahimi, M. M., Karwaj, C., & Deb, P. K. (1993). Failure of sewerage mains constructed in soft estuarine deposit. *International conference on Case Histories in Geotechnical Engineering, 16*.
7. Broms, B. B., & Stille, H. (1976). Failure of anchored sheet pile walls. *Journal of the Geotechnical Engineering Division, ASCE, 102*, 235–251.
8. Hu, Z. F., Yue, Z. Q., Zhou, J., & Tham, L. G. (2003). Design and construction of a deep excavation in soft soils adjacent to the Shanghai Metro tunnels. *Canadian Geotechnical Journal, 40*(5), 933–948.
9. Miyoshi, M. (1977). Mechanical behavior of temporary braced wall, In: Proc 9th Int conf on soil mechanics and foundation engineering, Tokyo, pp. 655–658.
10. Flaate, K. S. (1966). *Stresses and movement in connection with braced cuts in sand and clay*. Urbana, USA: University of Illinois.
11. Swanson, P. G., & Larson, T. W. (1990). Shoring failure in soft clay. In: J. Lambe (ed.) *Design and Performance of Earth Retaining Structures*. ASCE, 551–561.
12. Murphy, D. J., Woolworth, R. S., & Clough, G. (1975). Temporary excavation in varved clay. *Journal of the Geotechnical Engineering Division, ASCE, 101*, 279–295.
13. Zhang, W. G., Wang, W., Zhou, D., Goh, A. T. C., & Zhang, R. H. (2018). Influence of groundwater drawdown on excavation responses—A case history in Bukit Timah granitic residual soils. *Journal of Rock Mechanics and Geotechnical Engineering, 10*, 856–864.
14. Clough, G. W., & Reed, M. W. (1984). Measured behaviour of braced wall in very soft clay. *Journal of Geotechnical Engineering, 110*.
15. Tait, R. G., & Taylor, H. T. (1975). Rigid and flexible bracing systems on adjacent sites. *Journal of the Construction Division, ASCE, 101*, 365–376.
16. Cacoilo, D., Tamaro, G., & Edinger, P. (1998). Design and performance of a tied back sheet pile wall in soft clay. Design and construction of earth retaining systems: Proceedings of a conference. ASCE, pp. 14–25.
17. Davies, R. V., & Walsh, N. M. (1983). Excavations in Singapore marine clays. In: *International Seminar on Construction Problems in Soft Soils*. Singapore: Nanyang Technological Institute.
18. Tanaka, H. (1996). Undrained shear strength for passive earth pressure in an excavation in soft clay. In: *Proceedings of the International Symposium Geotechnical Aspects of Underground Construction in Soft Ground*. Rotterdam, the Netherlands: A.A. Balkema, pp. 213–218.
19. Rodriguez, J. M., & Flamand, C. L. (1969). Strut loads recorded in a deep excavation in clay. In: *Proceedings of the 7th International Conference on Soil Mechanics and Foundation Engineering*, Vol. 2. Mexico City, 459–468.
20. Tan, Y., & Wei, B. (2012). Observed behaviours of a long and deep excavation constructed by cut-and-cover technique in Shanghai soft clay. *Journal of Geotechnical and Geoenvironmental Engineering, 138*, 69–88.
21. Sen, K., Alostaz, Y., Pellegrino, G., & Hagh, A. (2004). Support of deep excavation in soft clay: A case history study. In: *Proceedings of the 5th International Conference on Case Histories in Geotechnical Engineering*. Missouri University of Science and Technology.

22. Wallace, J., Ho, C., & Long, M. (1992). Retaining wall behaviour for a deep basement in Singapore marine clay. In: *Proceedings of the International Conference on Retaining Structures.* London: Thomas Telford, pp. 195–204.
23. Lee, F., Yong, K., Quan, K., & Chee, K. (1998). Effect of corners in strutted excavations: Field monitoring and case histories. *Journal of Geotechnical and Geoenvironmental Engineering, 124*, 339–349.
24. Leonard, M. S., W. H., & DeLabrusse, P. (1987). The design and performance of the temporary works for Somerset and Tiong Bahru stations. In: *Mass Rapid Transit System: Proceedings of the Singapore Mass Rapid Transit Conference.* Singapore: Mass Rapid Transit Corp, pp. 141–145.
25. Tan, S., Tan, S., & Chin, Y. (1985). A braced sheet pile excavation in Singapore marine clay. In: *Proceedings of the 11th International Conference on Soil Mechanics and Foundation Engineering*, Vol. 5. Rotterdam, the Netherlands: A.A. Balkema, pp. 1671–1674.
26. Hulme, T., Potter, J., & Shirlaw, N. (1989). Singapore mass rapid transit system: Construction. *Proceedings of the Institution of Civil Engineers, 86*(4) 709–70.
27. Broms, B. B., Wong, I. H., & Wong, K. S. (1986). Experience with finite-element analysis of braced excavations in Singapore. In: *Proceedings of the 2nd International Symposium on Numerical Methods in Geomechanics.* Rotterdam, the Netherlands: A.A. Balkema, pp. 309–324.
28. Powrie, W., & Kantartzi, C. (1996). Ground response during diaphragm wall installation in clay: centrifuge model tests. *Geotechnique, 46*, 725–739.
29. Vuillemin, R. J., & Wong, H. (1991). Deep excavation in urban environment: 3 examples. In: *Proceedings of the 10th European Conference on Soil Mechanics and Foundation Engineering*, Vol. 2. Rotterdam, the Netherlands: A.A. Balkema, pp. 843–847.
30. NGI, N. G. I. (1962). Measurements at a strutted excavation, Oslo Subway, Vaterland 1, km 1373 Technical Report No. 5. NGI, Oslo, Norway.
31. Balasubramaniam, A., Bergado, D., Chai, J. S., & Sutabur, T. (1994). Deformation analysis of deep excavations in Bangkok. In: *Proceedings of the 13th International Conference on Soil Mechanics and Foundation Engineering*, Vol. 2. Rotterdam, the Netherlands: A.A. Balkema, pp. 909–915.
32. Karlsrud, K. (1981). Performance and design of slurry walls in soft clay. In: *Proceedings of ASCE Spring Convention.* New York, pp. 81–147.
33. Li, D., Li, Z., & Tang, D. (2015). Three-dimensional effects on deformation of deep excavations. *Geotechnical Engineering, Proceedings of the Institution of Civil Engineers, 168*, 551–562.
34. Roti, J., & Friis, J. (1985). Diaphragm wall performance in soft clay excavation. In: *Proceedings of the 11th International Conference on Soil Mechanics and Foundation Engineering*, Vol. 4. Rotterdam, the Netherlands: A.A. Balkema, pp. 2073–2078.
35. Rampello, S., Tamagini, C., & Calabresi, G. (1992). Observed and predicted response of a braced excavation in soft to medium clay. In: *Predictive Soil Mechanics, Proceedings of the Wroth Memorial Symposium.* London: Thomas Telford, pp. 544–561.
36. Tamano, T. F. S., Mizutani, S., Tsuboi, H., & Hisatake, M. (1996). Earth and water pressures acting on a braced excavation in soft ground. In: *Proceedings of the International Symposium Geotechnical Aspects of Underground Construction in Soft Ground.* Rotterdam, the Netherlands: A.A. Balkema, pp. 201–212.
37. Zhao, X., Gong, J., Chen, Z., & Bao, Y. (1999). Design and practice on special deep and large excavation engineering in Shanghai. In: *Proceedings of the 12th European Conference on Soil Mechanics and Geotechnical Engineering.* Rotterdam, the Netherlands: A.A. Balkema, pp. 239–244.
38. Onishi, K., & Sugawara, T. (1999). Behaviour of an earth retaining wall during deep excavation in Shanghi soft ground. *Soils and Foundations, 39*, 89–97.
39. Baggett, J. K., & Buttling, S. (1977). Design and in-situ performance of a sheet pile wall. In: *Proceedings of the 9th International Conference on Soil Mechanics and Foundation Engineering*, Vol. 2. Tokyo, Japan, pp. 3–8.

40. Maldonado, R. (1998). Big digs in the lacustrine soil of Bogotá, Colombia. In: J. R. Lambrechts, R. Hwang, A. Urzua (eds.). *Big Digs Around the World.* ASCE, pp. 252–272.
41. Nicholson, D. (1987). The design and performance of the retaining walls at Newton Station. In: *Mass Rapid Transit System: Proceedings of the Singapore Mass Rapid Transit Conference.* Singapore: Mass Rapid Transit Corp, 147–154.
42. Finstad, J. (1991). Royal Christianina Hotel—Basement with permanent sheet pile wall, up and down method. In: Proceedings of the 4th International Conference on Piling and Deep Foundations, Vol. 1. A.A. Balkema, pp. 387–392.
43. Ou, C. Y., Teng, F. C., & Wang, I. W. (2008). Analysis and design of partial ground improvement in deep excavations. *Computers and Geotechnics, 35*(4), 576–584.
44. Li, J. J., & Wang, W. D. (2011). Design and construction of deep excavation engineering adjacent to the subway tunnel. *Journal of Railway Engineering Society, 158*(11), 104–111. (in Chinese).
45. Ulrich, E. J., Jr. (1989). Tieback supported cuts in over-consolidated soils. *Journal of Geotechnical Engineering, 115*, 521–545.
46. Kung, G. T. C. (2003). *Surface settlement induced by excavation with consideration of small-strain behavior of Taipei silty clay.* Chinese Taiwan: University of Science and Technology.
47. Blackburn, J., & Finno, R. (2007). Three-dimensional responses observed in an internally braced excavation in soft clay. *Journal of Geotechnical and Geoenvironmental Engineering, 133*, 1364–1373.
48. O'Rourke, T., & McGinn, A. (2006). Lessons learned for ground movements and soil stabilization from the Boston Central Artery. *Journal of Geotechnical and Geoenvironmental Engineering, 132*, 966–989.
49. Hong, Y., Ng, C., Liu, G., & Liu, T. (2015). Three-dimensional deformation behaviour of a multi-propped excavation at a '"greenfield"' site at Shanghai soft clay. *Tunnelling and Underground Space Technology, 45*, 249–259.
50. Mu, L., & Huang, M. (2016). Small strain based method for predicting three-dimensional soil displacements induced by braced excavation. *Tunnelling and Underground Space Technology, 52*, 12–22.
51. J, S., G, L., P, H., CWW., N., . (2015). Interaction between a large-scale triangular excavation and adjacent structures in Shanghai soft clay. *Tunnelling and Underground Space Technology, 50*, 282–295.
52. Liu, G., Ng, C., & Wang, Z. (2005). Observed performance of a deep multistrutted excavation in Shanghai soft clays. *Journal of Geotechnical and Geoenvironmental Engineering, 131*, 1004–1013.
53. Bilotta, E., Ramondini, M., & Viggiani, C. I. (2004). Monitoring an excavation in an urban area. In: *Proceedings of the 5th International Conference on Case Histories in Geotechnical Engineering.* Missouri University of Science and Technology.
54. Konstantakos, D., Whittle, A., Regalado, C., & Scharner, B. (2004). Control of ground movements for a multi-level-anchored, diaphragm wall during excavation. In: *Proceedings of the 5th International Conference on Case Histories in Geotechnical Engineering.* New York.
55. Nalcakan, M., Tekin, M., Tönuk, G., & Ergun, U. (2004). Behaviour of a watertight anchored retaining wall in soft soil conditions. In: *Proceedings of the 5th International Conference on Case Histories in Geotechnical Engineering.* Missouri University of Science and Technology.
56. Dong, Y., Harve, H., Houlsby, G., & Xu, Z. (2013). 3D FEM modelling of a deep excavation case considering small-strain stiffness of soil and thermal shrinkage of concrete. In: *Proceedings of the 7th International Conference on Case Histories in Geotechnical Engineering.* Missouri University of Science and Technology.
57. Becker, P., Gebreselassie, B., & Kempfert, H. (2008). Back analysis of a deep excavation in soft lacustrine clays. In: *Proceedings of the 6th International Conference on Case Histories in Geotechnical Engineering.* Missouri University of Science and Technology.
58. Chen, S., Ho, C., & Gui, M. (2014). Diaphragm wall displacement due to creep of soft clay. *Proceedings of the Institution of Civil Engineers – Geotechnical Engineering*, vol. 167, pp. 297–310.

59. Mothersille, D., Duzceer, R., Gokalp, A., & Okumusoglu, B. (2015). Support of 25 m deep excavation using ground anchors in Russia. *Proceedings of the Institution of Civil Engineers – Geotechnical Engineering*, vol. 168, pp. 281–295.
60. Xu, Z. H., Zhang, J., & Chen, C. (2013). A case history of a deep foundation pit constructed by zoned excavation method in Shanghai soft deposit. In: *Proceedings of the 7th International Conference on Case Histories in Geotechnical Engineering*. Missouri University of Science and Technology.
61. Sanzeni, A., Colleselli, F., Mino, M., & Merline, A. (2013). Behavior prediction and monitoring of a deep excavation in the historic center of Brescia. In: *Proceedings of the 7th International Conference on Case Histories in Geotechnical Engineering*. Missouri University of Science and Technology.
62. Jia, J., Wang, J. H., Liu, C. P., Zhang, L. L., & Xie, X. L. (2008). Behaviour of an excavation adjacent to a historical building and metro tunnels in Shanghai soft clays. In: *Proceedings of the 6th International Conference on Case Histories in Geotechnical Engineering*. Missouri University of Science and Technology.
63. Thasnanipan, N., Aye, Z. Z., & Submaneewong, C. (2004). Construction of diaphragm wall support underground car park in historical area of Bangkok. In: *Proceedings of the 5th International Conference on Case Histories in Geotechnical Engineering*. Missouri University of Science and Technology.
64. Zekkos, D. P., Athanasopoulos, A. G., & Athanasopoulos, G. A. (2004). Deep supported excavation in difficult ground conditions in the city of Patras, Greece – Measured vs. predicted behavior. In: *Proceedings of the 5th International Conference on Case Histories in Geotechnical Engineering*. Missouri University of Science and Technology.
65. Bowden, A., & Lees, A. (2010). Design and performance of a basement in Norwich, UK. *Proceedings of the Institution of Civil Engineers – Geotechnical Engineering*, vol. 163, pp. 55–64.
66. Wong, I. H., Poh, T. Y., & Chuah, H. L. (1997). Performance of excavations for depressed expressway in Singapore. *Journal of Geotechnical and Geoenvironmental Engineering, 123*, 617–625.
67. Hashash, Y. M. A., & Whittle, A. J. (2002). Mechanism of load transfer and arching for braced excavation in clay. *Journal of Geotechnical and Geoenvironmental Engineering, 128*, 198–197.
68. Hsieh, P. G., & Ou, C. Y. (1998). Shape of ground surface settlement profiles caused by excavation. *Canadian Geotechnical Journal, 35*, 1004–1017.
69. Long, M., Menkiti, C., Skipper, J., Brangan, C., & Looby, M. (2010c). Retaining wall behaviour in Dublin's estuarine deposits, Ireland. *Proceedings of the Institution of Civil – Engineers Geotechnical Engineering*, vol. 165, pp. 351–365.
70. Peck, R. B. (1969). Deep excavation and tunneling in soft ground. *Proceedings of the 7th International Conference on soil Mechanics and Foundation Engineering*, Mexico City, pp. 225–290.
71. Becker, J. M., & Haley, M. X. (2011). Up/Down construction-decision making and performance. *Design & Performance of Earth Retaining Structures*, 170–189.
72. Long, M., Daynes, P., Donohue, S., & Looby, M. (2012). Retaining wall behaviour in Dublin's fluvio-glacial gravel, Ireland. *Proceedings of the Institution of Civil Engineers – Geotechnical Engineering*, vol. 165, pp. 289–307.
73. Ulrich, E. (1989). Internally braced cuts in overconsolidated soils. *Journal of Geotechnical Engineering, 115*, 504–520.
74. Whittle, A. J., Hashash, Y. M. A., & Whitman, R. V. (1993). Analysis of deep excavation in Boston. *Journal of Geotechnical Engineering, 119*, 69–90.
75. Correia, A., N. M., d. C. G. (1997). Performance of three Berlin type retaining walls. In: *Proceedings of the 14th International Conference on Soil Mechanics and Foundation Engineering*, Vol. 2. A.A. Balkema, pp. 1297–1300.
76. Becker, J. M., & Haley, M. X. (1990). Up/Down construction-decision making and performance. *Design & Performance of Earth Retaining Structures*, pp. 170–189.

77. Lee, J. (2009). An application of three-dimensional analysis around a tunnel portal under construction. *Tunnelling and Underground Space Technology, 24,* 731–738.
78. Kulesza, R., Boussoulas, N., & Marr, W. (2008). Deep excavations in hard sandy clays for stations and shafts of the Athens Metro Stavros Extension. *Proceedings of the 6th International Conference on Case Histories in Geotechnical Engineering.* Missouri University of Science and Technology.
79. Sevencan, O., Ozaydin, K., & Kilic, H. (2013). Numerical analysis of soil deformations around deep excavations. In: *Proceedings of the 7th International Conference on Case Histories in Geotechnical Engineering.* Missouri University of Science and Technology.
80. Fong, F., Standing, J., & Bourne-Webb, P. (2014). Building response to adjacent deep basement construction. *Proceedings of the Institution of Civil Engineers –Geotechnical Engineering,* vol. 167, pp. 130–143.
81. Long, M., O'Leary, F., Ryan, M., & Looby, M. (2013). Deep excavation in Irish glacial deposits. *Proceedings of the 18th International Conference on Soil Mechanics and Geotechnical Engineering.* International Society for Soil Mechanics and Geotechnical Engineering, pp. 2039–2042.
82. Haeri, M., Sasar, M., & Afshari, K. (2013). Deep excavation on 3 sides of a 21 story building: Accounts of a successful deep excavation project. In: *Proceedings of the 7th International Conference on Case Histories in Geotechnical Engineering.* Missouri University of Science and Technology.
83. Roth, W., Su, B., Vanbaarsel, J., & Lindquist, E. (2008). Effect of high in-situ stress on braced excavations. In: *Proceedings of the 6th International Conference on Case Histories in Geotechnical Engineering.* Missouri University of Science and Technology.
84. Chapman, T., & Green, G. (2004). Observational Method looks set to cut city building costs. *Proceedings of the Institution of Civil Engineers – Civil Engineering,* vol. 157, pp. 125–133.
85. Gourvenec, S., Powrie, W., E. K. D. M. (2002). Three-dimensional effects in the construction of a long retaining wall. *Proceedings of the Institution of Civil Engineers – Geotechnical Engineering,* vol. 155, 163–173.
86. Long, M., Brangan, C., Menkiti, C., Looby, M., & Casey, P. (2012). Retaining walls in Dublin boulder clay, Ireland. *Proceedings of the Institution of Civil Engineers – Geotechnical Engineering,* vol. 165, pp. 247–266.
87. Roscoe, H., & Twine, D. (2010). Design and performance of retaining walls. *Proceedings of the Institution of Civil Engineers – Geotechnical Engineering,* 163, pp. 279–290.
88. Faheem, H., Cai, F., Ugai, K., & Hagiwara, T. (2003). Two-dimensional base stability of excavations in soft soils using FEM. *Computers and Geotechnics, 30,* 141–163.
89. Luo, Z., Atamturktur, S., Cai, Y., & Juang, C. H. (2012). Simplified approach for reliability-based design against Basal-Heave failure in braced excavations considering spatial effect. *Journal of Geotechnical and Geoenvironmental Engineering, 138,* 441–450.
90. Wu, S., Ou, C., Ching, J., & Juang, C. (2010). Reliability-base design for basal heave in an excavation considering spatial variability. GeoFlorida 2010. *Advances in Analysis, Modelling and Design, 199,* 1914–1922.
91. Wong, K., & Goh, A. (2002). Basal heave stability for wide excavations. *Proceedings 3rd International Symp Geotech Aspects of Underground Construction in Soft Ground,* Toulouse, pp. 699–704.
92. Han, C. Y., Chen, J. J., Xia, X. H., & Wang, J. H. (2014). Three-dimensional stability analysis of anisotropic and non-homogeneous slopes using limit analysis. *Journal of Central South University, 21,* 1142–1147.
93. Huang, M. S., & Qin, H. L. (2009). Upper-bound multi-rigid-block solutions for bearing capacity of two-layered soils. *Computers and Geotechnics, 36,* 525–529.
94. Huang, M. S., Tang, Z., & Yuan, J. Y. (2018). Basal stability analysis of braced excavations with embedded walls in undrained clay using the upper bound theorem. *Tunnelling and Underground Space Technology, 79,* 231–241.
95. Qin, H. L., Chen, Z. Y., & Liu, L. P. (2012). Basal stability analysis for excavations in soft clay based on upper bound method. *Chinese Journal of Geotechnical Engineering, 34,* 1611–1619. (in Chinese).

96. Qin, H. L., Huang, M. S., & Wang, Y. J. (2010). Application of Monte Carlo search technique to bearing capacity calculations by upper bound method. *Rock and Soil Mechanics, 31*, 3145–3150. (in Chinese).
97. Hong, L., & Zhang, W. G. (2020). Application of progressive search algorithm in upper bound basal stability for braced excavations in soft clay. *Journal of Civil and Environmental Engineering, 42*(6), 46–53.
98. Rackwitz, R. (2000). Reviewing probabilistic soils modelling. *Computers and Geotechnics, 26*, 199–223.
99. Hwang, R. N., Lee, T. Y., Chou, C. R., & Su, T. C. (2012). Evaluation of performance of diaphragm walls by wall deflection paths. *Journal of Geotechnical Engineering, 7*, 1–12.
100. Konstantakos, D. C. (1998). *Measured performance of slurry walls*. USA: Massachusetts Institute of Technology.
101. Zhang, R. H., Zhang, W. G., Goh, A. T. C., Hou, Z. J., & Wang, W. (2018). A simple model for ground surface settlement induced by braced excavation subjected to a significant groundwater drawdown. *Geomechanics and Engineering, 16*(6), 635–642.
102. Goh, A. T. C., Zhang, R. H., Wang, W., Wang, L., Liu, H. L., & Zhang, W. G. (2020). Numerical study of the effects of groundwater drawdown on ground settlement for excavation in residual soils. *Acta Geotechnica, 15*, 1259–1272.
103. Zhang, R. H., Goh, A. T. C., Zhou, T. Q., & Zhang, W. G. (2021). ReLiability assessment of excavation-induced ground surface settlement with ground water drawdown considering spatial variability. *Journal of Civil and Environmental Engineering*. https://doi.org/10.11835/J.issn.2096-6717.2020.170I
104. Phoon, K. K., & Kulhawy, F. H. (1999). Characterization of geotechnical variability. *Canadian Geotechnical Journal, 36*, 612–624.
105. Chowdhury, S. S., Deb, K., & Sengupta, A. (2015). Behavior of underground strutted retaining structure under seismic condition. *Earthquakes and Structures, 8*, 1147–1170.
106. Goh, A. T. C., Zhang, F., Liu, H. L., Zhang, W. G., & Zhou, D. (2018). Numerical analysis on strut responses due to one-strut failure for braced excavation in clays. *Proceedings of the 2nd International Symposium on Asia Urban GeoEngineering*.
107. Zhang, W. G., Hong, L., Li, Y. Q., Zhang, R. H., & Goh, A. T. C. (2020). Assessment of effect of jet grouting slabs on responses for deep braced excavations. *Underground Space*. https://doi.org/10.1016/j.undsp.2020.02.002
108. Zhang, W. G., Li, Y. Q., Goh, A. T. C., & Zhang, R. H. (2020). Numerical study of the performance of jet grout piles for braced excavations in soft clay. *Computers and Geotechnics*. https://doi.org/10.1016/j.compgeo.2020.103631
109. Ignat, R., Baker, S., Larsson, S., & Liedberg, S. (2015). Two- and three- dimensional analyses of excavation support with rows of dry deep mixing columns. *Computers and Geotechnics, 66*, 16–30.
110. Surarak, C., Likitlersuang, S., Wanatowski, D., Balasubramaniam, A., Oh, E., & Guan, H. (2012). Stiffness and strength parameters for hardening soil model of soft and stiff Bangkok clays. *Soils and Foundations, 52*(4), 682–697.
111. Xiao, H., Tang, J., Li, Q., & Luo, Q. (2003). Analysis of multi-braced earth retaining structures. *Proceedings of the Institution of Civil Engineers – Structures and Buildings*, vol. 156, pp. 307–318.